P. Vidhyasekaran, PhD

Concise Encyclopedia of Plant Pathology

Pre-publication
REVIEWS,
COMMENTARIES,
EVALUATIONS . . .

D1232632

"This book is a massive production that presents a vast amount of background and current information in an accessible format. It is a tremendous compilation of information that, because of its broad scope, will be useful to both specialists and generalists. It will be a particularly valuable reference for scientists and policymakers without a specialist training in plant pathology who want access to information on most topics in plant pathology."

David I. Guest, PhD
Associate Professor and Reader,
School of Botany,
The University of Melbourne,
Australia

"The *Concise Encyclopedia of Plant Pathology*, by P. Vidhyasekaran, is a much-needed companion to university teaching of plant pathology. It

has an excellent coverage of all topics taught under the banner of plant pathology, from the history of the field to current control possibilities such as plant defense activators. I often teach by reference to case studies; this usually retains the interest of students but also requires further reading on aspects not detailed in the case studies. The *Concise Encyclopedia of Plant Pathology* will provide structured, focused, and up-to-date background information for students, spurring them to additional reading. It will also find a place on the shelf of any research scientist in the field of plant pathology as a rich source of information."

Caroline Mohammed, PhD
Senior Research Scientist,
CSIRO Forestry
and Forest Products;
Senior Lecturer,
University of Tasmania

Concise Encyclopedia
of Plant Pathology

FOOD PRODUCTS PRESS®
Crop Science
Amarjit S. Basra, PhD
Senior Editor

Heterosis and Hybrid Seed Production in Agronomic Crops edited by Amarjit S. Basra

Intensive Cropping: Efficient Use of Water, Nutrients, and Tillage by S. S. Prihar, P. R. Gajri, D. K. Benbi, and V. K. Arora

Physiological Bases for Maize Improvement edited by María E. Otegui and Gustavo A. Slafer

Plant Growth Regulators in Agriculture and Horticulture: Their Role and Commercial Uses edited by Amarjit S. Basra

Crop Responses and Adaptations to Temperature Stress edited by Amarjit S. Basra

Plant Viruses As Molecular Pathogens by Jawaid A. Khan and Jeanne Dijkstra

In Vitro Plant Breeding by Acram Taji, Prakash P. Kumar, and Prakash Lakshmanan

Crop Improvement: Challenges in the Twenty-First Century edited by Manjit S. Kang

Barley Science: Recent Advances from Molecular Biology to Agronomy of Yield and Quality edited by Gustavo A. Slafer, José Luis Molina-Cano, Roxana Savin, José Luis Araus, and Ignacio Romagosa

Tillage for Sustainable Cropping by P. R. Gajri, V. K. Arora, and S. S. Prihar

Bacterial Disease Resistance in Plants: Molecular Biology and Biotechnological Applications by P. Vidhyasekaran

Handbook of Formulas and Software for Plant Geneticists and Breeders edited by Manjit S. Kang

Postharvest Oxidative Stress in Horticultural Crops edited by D. M. Hodges

Encyclopedic Dictionary of Plant Breeding and Related Subjects by Rolf H. G. Schlegel

Handbook of Processes and Modeling in the Soil-Plant System edited by D. K. Benbi and R. Nieder

The Lowland Maya Area: Three Millennia at the Human-Wildland Interface edited by A. Gómez-Pompa, M. F. Allen, S. Fedick, and J. J. Jiménez-Osornio

Biodiversity and Pest Management in Agroecosystems, Second Edition by Miguel A. Altieri and Clara I. Nicholls

Plant-Derived Antimycotics: Current Trends and Future Prospects edited by Mahendra Rai and Donatella Mares

Concise Encyclopedia of Temperate Tree Fruit edited by Tara Auxt Baugher and Suman Singha

Landscape Agroecology by Paul A Wojkowski

Molecular Genetics and Breeding of Forest Trees edited by Sandeep Kumar and Matthias Fladung

Concise Encylcopedia of Plant Pathology by P. Vidhyasekaran

Testing of Genetically Modified Organisms in Foods edited by Farid E. Ahmed

Concise Encyclopedia of Bioresource Technology edited by Ashok Pandey

Agrometeorology: Principles and Applications of Climate Studies in Agriculture by Aarpal S. Mavi and Graeme J. Tupper

Concise Encyclopedia
of Plant Pathology

P. Vidhyasekaran, PhD, FNA

Food Products Press®
The Haworth Reference Press
Imprints of The Haworth Press, Inc.
New York • London • Oxford

Published by

Food Products Press® and The Haworth Reference Press, imprints of The Haworth Press, Inc., 10 Alice Street, Binghamton, NY 13904-1580.

Cover design by Marylouise E. Doyle.

Library of Congress Cataloging-in-Publication Data
Vidhyasekaran, P.
    Concise encyclopedia of plant pathology / P. Vidhyasekaran.
        p. cm.
Includes bibliographical references and index.
    ISBN 1-56022-942-X (hard cover : alk. paper)—ISBN 1-56022-943-8 (pbk. : alk. paper)
    1. Plant diseases—Encylopedias. I. Title.
    SB728.V525 2003
    632'.3'03—dc21
                                                                                                    2003002492

CONTENTS

COMMON DISEASE SYMPTOMS, ASSESSMENT, AND DIAGNOSIS

CROP DISEASE EPIDEMIOLOGY

CROP DISEASES

ABOUT THE AUTHOR

P. Vidhyasekaran, PhD, FNA, is the former Director of the Center for Plant Protection Studies, Tamil Nadu Agricultural University, Coimbatore, India. He is the author of eight books from leading publishers in the United States and India. He has published more than 400 articles in 32 international journals as well as several book chapters. Dr. Vidhyasekaran has served on the editorial boards of several journals and has been elected President of the Indian Society of Plant Pathologists. He is a Fellow of the National Academy of Agricultural Sciences and several other scientific societies, and a member of the New York Academy of Sciences. Dr. Vidhyasekaran has won several national awards as well as the Tamil Nadu Best Scientist Award. He has served as a visiting scientist in 14 countries, including the United States, Philippines, and Denmark. He is the author of *Bacterial Disease Resistance in Plants: Molecular Biology and Biotechnological Applications* (Haworth).

Foreword

Plant diseases cause billions of dollars in crop losses each year throughout the world. Historically, uncontrolled plant disease epidemics have resulted in great human suffering through famine and resultant social upheaval. Perhaps the best-known example is the Irish famine of 1846, in which two million people died and many more emigrated from their homeland. The famine was caused by the destruction of the potato crop, the staple food in Ireland at that time, by *Phytophthora infestans,* the causal agent of late blight disease. The efforts of early scientists such as De Bary, Speerschneider, and Kühn to find the causes of late blight and other plant diseases established plant pathology as a critical discipline in the struggle to prevent catastrophic crop failure.

The science of plant pathology seeks to understand and work to manage such devastating outbreaks of plant diseases. It describes (1) what causes disease, (2) how pathogens interact with plants to cause disease, (3) how diseases spread and develop into epidemics, (4) the scope of losses caused by disease, and (5) disease management. It is a very broad field, incorporating plant anatomy, physiology, breeding and genetics, virology, bacteriology, mycology, nematology, molecular biology, biochemistry, mathematics, computer modeling, and statistics. It is a daunting task indeed to pull all of this together in one book.

Professor Vidhyasekaran has written an encyclopedia that describes the breadth of subjects in the field of plant pathology. It will serve as a beginning reference for students and professionals in plant pathology as well as allied fields. Each section provides an overview of a given area, which will serve as both an introduction and a starting point for further studies. It also comprehensively describes modern plant pathology. It includes classification of pathogens based on DNA analyses and protein patterns, molecular diagnostic techniques, computer-aided decision support systems, remote sensing, digital image analysis, microbial pesticides, plant activators, molecular marker-assisted breeding, gene pyramiding, genetic engineering, in vitro breeding, and molecular plant pathology. Professor Vidhyasekaran and The Haworth Press are to be commended for providing this excellent resource.

Sally A. Miller
Professor, Department of Plant Pathology
The Ohio State University
Ohio Agricultural Research and Development Center

Preface

Globally, diseases cause heavy crop losses amounting to several billion dollars annually. Some diseases have shattered the economies of nations. Late blight of potato is known to be responsible for the Irish famine of 1845-1847 when about two million people died because of starvation in Ireland. Canker is the most serious disease in *Citrus* spp. and is widely prevalent in Florida, Georgia, Alabama, Louisiana, South Carolina, Texas, and Mississippi. Eradication programs have been initiated by several states since 1915. By 1991, more than 20 million citrus trees were destroyed at a cost of about $94 million. Still the Asiatic citrus canker is prevalent in several parts of the United States. Disease management is still a challenging task, and plant pathology is the most important subject in crop science. This encyclopedia covers the entire field of plant pathology. It defines the numerous subjects covered as well as describes them. It is designed to stand as a compendium of current knowledge about the topics in plant pathology. This encyclopedia will serve as a single, easy-to-use reference work including a full range of subject areas associated with plant pathology. Each section in this encyclopedia serves as a comprehensive overview of a given area, providing both breadth of coverage for students and depth of coverage for research professionals.

This book provides the latest nomenclature and classification of bacterial, fungal, viral, and phytoplasma pathogens and the systematic position of each crop pathogen (more than 1,500 pathogens). It describes disease and crop loss assessment models, remote sensing, digital image analysis, disease forecasting models, computer-aided decision support systems, and plant clinics. The book consists of detailed sections on microbial pesticides, induced systemic resistance, mycorrhiza, seed health testing, indexing plant-propagation materials, a complete list of fungicides, ready-formulated mixtures of fungicides, fungicide resistance, modern fungicide application equipments, bactericides, and viricides. This comprehensive book on modern plant pathology describes classification of pathogens based on DNA analyses and protein profiles, molecular diagnostic techniques, commercial development of microbial pesticides, plant products and plant activators, and novel breeding techniques using molecular marker-assisted selection and gene pyramiding. This book also describes molecular biology of plant-pathogen interactions and host defense mechanisms and in vitro selection and genetic engineering technologies. It also provides a list of plant pathology terminology. It is unique in providing a complete list of diseases of all major crops cultivated in tropics, semitropics, and temperate regions of the world and the accepted names, synonyms, and anamorphic and teleo-morphic names of pathogens (over 910 fungal, 315 viral, and 200 bacterial pathogens) with authority. It is intended for a broad international audience of graduate and undergraduate students, university faculty, public and private sector research scientists, extension specialists, and development workers engaged in plant protection. This book will be an indispensable resource for plant pathologists, mycologists, bacteriologists, virologists, botanists, and graduate-level students in these disciplines.

Birth and Development of Plant Pathology

Plant pathology is the science or study of plant diseases. The word "disease" means any impairment of normal physiological function of plants, producing characteristic symptoms. A symptom is a phenomenon accompanying something and is regarded as evidence of its existence. Disease is caused by pathogen. A pathogen is any agent that can cause disease. Plant pathology describes: (1) what causes disease, (2) how the disease is caused, (3) how the disease spreads, causing epidemic, (4) how much loss the disease can cause, and (5) how to manage the disease.

HISTORY

Ancient History

Occurrence of diseases in plants has been recognized from ancient times. Around 2000 B.C., from the Babylonian kingdoms, a disease called samana in barley was mentioned in a farm almanac. Books of the Old Testament that date to the eighth and fifth centuries B.C. contain references to the blasting and mildew of crops as great scourges of humankind. In the third century B.C., several Greek writings began to contain references to plant diseases. A Greek named Cleidemus is often identified as the first plant pathologist and grandfather of plant pathology according to McNew (1963). He made observations on diseases of grapes, figs, and olives. The Greek philosopher Theophrastus (c. 372-c. 287 B.C.) was the first to study and write about diseases of trees, cereals, and legumes. He is known as the father of botany.

A Roman author, Marcus Terentius Varro (116-27 B.C.) in his *Rerum Rusticarum* mentions the god Robigus among the deities to be propitiated. According to Varro, Robigus is the "rust god," who had to be depended upon to protect cereal crops from rust attack. Propitiatory ceremonies, the Robigalia, were performed in April or May each year when rust often first became noticeable. Other gods such as Flora, Ceres (who protected grain crops), Bacchus (grapes), and Minerva (olives) also had to be propitiated. In his writing *Historia Naturalis,* Pliny the Elder (23-79 A.D.) described many control measures against diseases. He mentioned treating cereal seeds with wine or a concoction of cypress leaf extract to control mildew. Palladius (around fourth to fifth centuries) wrote about plant diseases in his *De Re Rustica.* He described differences in the susceptibility of different cultivars to diseases and suggested sanitation through removal of diseased plants.

A German, Gottfried von Franken (circa 1200 A.D.), in his *Pelzbuch,* suggested control measures against cankers of cherry and other fruit diseases. A chapter on grapevine diseases is found in the book titled *Ruralium Commodorum Libri XII* written by Petrus Crescentius (1230-1320) during the thirteenth century. *Kreuterbuch* by German writer Jerome Bock (1498-1554) contains an illustration of wheat plants affected by loose smut. Robert Hooke (1635-1703) in England developed a microscope and described teliospores of *Phragmidium mucronatum* on rose in his book *Micrographia,* which was written in 1665. In 1679, Marcello Malpighi (1628-1694) in his *Anatome Plantarum,* described aecia of *Gymnosporangium* on hawthorn. In 1675, Dutch naturalist Antoni van Leeuwenhoek (1632-1723) discovered bacteria while working with a microscope he had built.

Eighteenth Century

In 1705 Joseph Pitton de Tournefort (1656-1708), a French botanist, reported in a journal article that fungi could reproduce by bodies similar to seeds and could incite disease in plants under humid conditions in the greenhouse. In 1727, Stephen Hales (1677-1761) from England published on powdery mildew of hop. Henri Louis Duhamel du Monceau (1700-1782), in France described *Helicobasidium purpureum* in 1728. The Italian Micheli (1679-1737) published his book *Nova Plantarum Genera* in 1729 and introduced the generic name *Puccinia* in his book. In 1753, Swedish botanist Carolus Linnaeus (1707-1778) developed the binomial system of nomenclature in botany and published *Species Plantarum.* He included ten genera of fungi in his book.

In 1755, French scientist Mathieu Tillet (1714-1791) added black dust from bunted wheat to seed from healthy wheat and observed that bunt was more prevalent in plants produced from such seed than from nondusted seed. He showed the efficacy of seed treatment in the control of wheat bunt, and he recommended washing seeds in lye, drying, sprinkling with lye, and dusting with lime. This is the first seed treatment recommended to control diseases. Fontana (1730-1805) from Italy described teliospores and uredospores of wheat rust in 1767. In the same year, Tozzetti (1712-1783) from Italy described rust diseases. In 1774, Danish entomologist Johann Christian

Fabricius (1745-1808) published a system of classification of fungal pathogens in which he arranged pathogens by classes, genera, and species.

Nineteenth Century

In 1801, Christian Hendrik Persoon (1761-1836) from France published his book *Synopsis Methodica Fungorum;* this was the first reliable systematic account of fungi. Benedict Prevost (1755-1819) in 1807 gave the first experimental proof that a fungus could cause disease in a plant. He recommended copper sulfate as seed treatment to control bunt disease in wheat. In 1817, Thomas Andrew Knight (1759-1838) reported that sulfur controlled scab *(Venturia pirina)* in pear trees. In 1821 Elias Magnus Fries (1794-1878) published *Systema Mycologicum* providing classification of fungi. John Robertson (1824) from Ireland showed that mildew (*Sphaerotheca pannosa* var. *persicae*) is controlled by a mixture of sulfur in soapsuds. Control of peach leaf curl *(Taphrina deformans)* with lime and sulfur was demonstrated by Knight in 1842. Miles Joseph Berkeley (1803-1889) in 1845 published papers on diseases of cereals and vegetables. In 1853, Heinrich Anton De Bary (1831-1888) worked on smut and rust fungi, downy mildews, and late blight of potato. He discovered two alternate hosts for the rusts. Speerschneider (1857) proved that *Phytopthora infestans* was the cause of potato late blight. Julius Gotthelf Kühn (1825-1910) published the first textbook of plant pathology in 1858.

Thomas Jonathan Burrill (1839-1916) was the first person to show that bacteria can cause plant diseases, and in 1878 he described fire blight of pear and other fruits. In 1884, Robert Koch (1843-1910) developed his postulates, which stipulate conditions for describing an organism as the cause of a disease. In 1885, Alexis Millardet (1838-1902) developed Bordeaux mixture as a fungicide. In 1886, Adolph Mayer (1843-1942) worked on mosaic of tobacco and called it mosaic disease of tobacco. He showed that the causal agent was transmissible to healthy plants in juice extract. Smith (1891) was working on peach yellows disease and reported that the causal agent was contagious and was bud transmitted. In 1891, Joseph Charles Arthur (1850-1942) identified wheat varieties resistant to scab disease. In 1892, Russian scientist Dmitrii Iosifovich Ivanowski (1864-1924) showed that tobacco mosaic causal agent is filterable through bacteriological filters. In 1898, Dutch scientist Martinus Willem Beijerinck (1851-1931) called the causal agent "contagium vivum fluidum." He used the word "virus" to describe the contagium. He demonstrated that it was graft transmissible. Jakob Eriksson (1848-1931) discovered the physiologic races in rust pathogens in 1896.

MODERN PLANT PATHOLOGY

During the twentieth century, plant pathology developed as a full-fledged science. Biffin (1905) bred the first resistant host cultivar using Mendelian principles. Allard (1914) partially purified tobacco mosaic virus (TMV). Antigenic nature of plant viruses was discovered by Dvorak in 1928. In 1928, Purdy developed antiserum against viruses, according to Corbett (1964). Holmes (1929) developed the TMV local lesion assay. Wendell Meredith Stanley (1904-1971) obtained crystals of TMV in 1935 and received a Nobel Prize for crystallizing TMV in 1946. Ernest Athearn Bessey (1877-1957) authored the first American textbook of mycology, which was published in 1936.

During the first 50 years of the century, most of the contributions in the field of mycology were by traditional and taxonomically inclined mycologists. By midcentury, the approach had shifted from a traditional taxonomic focus to one that stressed the biology of fungi (Sequeira, 2000). Fungal physiology became a subject of great interest and host-parasite interactions were studied in detail (Keen, 2000). In 1910, Lewis Ralph Jones (1864-1945) suggested the role of pectic enzymes in plant disease development. Subsequently, the role of enzymes and toxins in disease development has been demonstrated. Several toxins were isolated and purified in the 1950s. Meehan and Murphy (1947) identified a host-specific toxin from *Helminthosporium victoriae*. Since then, several host-specific toxins have been identified from various fungal pathogens.

In 1942, Harold Flor developed the gene-for-gene concept. Flor hypothesized that the genes which condition the reaction of the host could be identified only by their interaction with specific strains of the parasite, while those which condition pathogenicity in the parasite could be identified only by their interaction with specific varieties of the host (Flor, 1942). Several disease-resistant genes have been identified. Newer techniques to develop durable resistance have been initiated. The major breakthrough in this field is successful cloning of resistance genes. Johal and Briggs (1992) cloned the first disease-resistant gene in 1992. Molecular breeding for disease resistance is the modern method of disease management. In 1998, Williamson and his associates demonstrated that a single cloned disease-resistant gene, *Mi* from tomato, confers resistance against a nematode and an insect (Rossi et al., 1998).

Van der Plank (1963) published a book titled *Plant Diseases—Epidemics and Control;* his work facilitated the rapid development of the field of epidemiology. Several disease-forecasting modules have been developed (Jones,

1998). Rapid advances have been made in the field of virology. The presence of satellite viruses was first reported by Kassanis in 1962. In 1968, Shepherd and his associates discovered that cauliflower mosaic virus is a DNA virus (Shepherd et al., 1968). Satellite virus is dependent on its virus for its replication. Satellite RNAs were first reported by Schneider in 1969. In 1971, Diener described viroids. Several molecular diagnostic techniques have been developed to detect plant diseases. Plant viral genomes have been cloned (Boyer and Haenni, 1994). Transgenic plants expressing capsid protein of a virus were first developed in 1986 by Beachey and his co-workers (Abel et al., 1986). Transgenic plants expressing satellite RNAs, viral replicase genes, and defective movement protein genes have all been developed (Zaitlin and Palukaitis, 2000).

Molecular plant pathology has emerged as an attractive field in the latter part of the century. In 1940, Muller and Borger proposed the phytoalexin theory. Several signal molecules have been identified from pathogens and host. Albersheim and his associates identified a highly active elicitor from a fungal cell wall in 1984 (Sharp et al., 1984). Elicitor molecules, which are the products of avirulence genes, have been characterized. The first avirulence gene was cloned from *Pseudomonas syringae* pv. *glycinea* by Staskawicz et al. in 1984. Several kinds of pathogenesis-related proteins have been identified. Genes encoding them have been cloned, and transgenic plants expressing them have been developed. In 1991, Broglie and colleagues developed transgenic plants expressing the *PR-3* gene. More than 100 fungicides have been developed during the twentieth century. The development of systemic fungicides with activity against broad spectrum of fungal pathogens was a major breakthrough in the history of plant pathology. Plant (defense) activators with simultaneous action against fungal, bacterial, and viral diseases have been identified. Biocontrol agents, which are capable of inducing systemic resistance, have been developed as commercial products and their efficacy in control of a wide range of diseases in commercial farms has been demonstrated (Vidhyasekaran, 2001). All these facets of modern plant pathology are described in detail in this book.

Plant pathology is now developing very rapidly. Plant pathologists are using many modern tools such as gene cloning, genetic engineering, molecular diagnostics, molecular-assisted breeding, remote sensing, digital imaging, and molecular manipulation of signaling systems to develop both plants with built-in resistance to pathogens and sustainable integrated disease management systems in various cropping systems. This book describes all of these novel approaches in the field of modern plant pathology.

REFERENCES

Abel, P. P., Nelson, R. S., De, B., Hoffman, N., Rogers, S. G., Fraley, R. T., and Beachy, R. N. (1986). Delay of disease development in transgenic plants that express the tobacco mosaic virus coat protein gene. *Science*, 232:738-743.

Allard, H. A. (1914). The mosaic disease of tobacco. *US Dep Agric Bull*, 40:1-33.

Biffin, R. H. (1905). Mendel's laws of inheritance and wheat breeding. *J Agric Sci*, 1:4-48.

Boyer, J. C. and Haenni, A. L. (1994). Infectious transcripts and cDNA clones of RNA viruses. *Virology*, 198:415-426.

Broglie, K., Chet, I., Holliday, M., Cressman, R., Biddle, P., Knowlton, S., Mauvais, C. J., and Broglie, R. (1991). Transgenic plants with enhanced resistance to the fungal pathogen *Rhizoctonia solani*. *Science*, 254:1194-1197.

Corbett, M. K. (1964). Introduction. In M. K. Corbett and H. D. Sisler (Eds.), *Plant Virology*. University of Florida Press, Gainsville.

Diener, T. O. (1971). Potato spindle tuber "virus." IV. A replicating, low molecular weight RNA. *Virology*, 45:411-428.

Flor, H. H. (1942). Inheritance of pathogenicity in *Melampsora lini*. *Phytopathology*, 32:557-564.

Holmes, F. O. (1929). Local lesions in tobacco mosaic. *Bot Gaz*, 87:39-55.

Johal, G. S. and Briggs, S. P. (1992). Reductase activity encoded by the HM1 disease resistance gene in maize. *Science*, 258:985-987.

Jones, D. G. (1998). *The Epidemiology of Plant Diseases*. Kluwer Academic Publishers, Dordrecht, pp. 460.

Kassanis, B. (1962). Properties and behaviour of virus depending for its multiplication on another. *J Gen Virol*, 27:477-488.

Keen, N. T. (2000). A century of plant pathology: A retrospective view on understanding host-parasite interactions. *Annu Rev Phytopathol*, 38:31-48.

McNew, G. L. (1963). The ever-expanding concepts behind 75 years of plant pathology. In S. Rich (Ed.), *Perspectives of Biochemical Plant Pathology. Conn. Agric Expt Bulletin*, 633:163-183.

Meehan, F. and Murphy, H. C. (1947). Differential phytotoxicity of metabolic byproducts of *Helminthosporium victoriae*. *Science*, 106:270-271.

Muller, K. O. and Borger, H. (1940). Experimentelle Untersuchungen uber die Phytophthora-Resistenz der Kartoffel. Zugleich ein Beitrag zum Problem der "erworbenen Resistenz" in Pflanzenreich. *Arb Biol Reichsanst Land Forstwirtsch*, 23:189-231.

Robertson, J. (1824). On the mildew and some other diseases incident to fruit trees. *Trans Hort Soc London*, 5:175-185.

Rossi, M., Goggin, F. L., Milligan, S. B., Kolashian, I., Ullman, D. E., and Williamson, V. M. (1998). The nematode resistance gene *Mi* of tomato confers resistance against the potato aphid. *Proc Natl Acad Sci USA*, 95:9750-9754.

Schneider, I. R. (1969). Satellite-like particle of tobacco ringspot virus that resembles tobacco ringspot virus. *Science*, 166:1627-1629.

Sequeira, L. (2000). Legacy for the millennium: A century of progress in plant pathology. *Annu Rev Phytopathol,* 38:1-7.

Sharp, J. K., McNeil, M., and Albersheim, P. (1984). The primary structure of one elicitor-active and seven elicitor-inactive hexa (β-D-glucopyranosyl)–D-glucitols isolated from the mycelial walls of *Phytophthora megasperma* f. sp. *glycinea.* *J Biol Chem,* 259:11321-11336.

Shepherd, R. J., Wakeman, R. J., and Romanko, R. R. (1968). DNA in cauliflower mosaic virus. *Virology,* 36:150-152.

Smith, E. F. (1891). Additional evidence on the communicability of peach yellows and peach rosette. *US Dep Agric Div Veg Pathol Bull,* 1:1-65.

Speerschneider, J. (1857). Die Ursache der Erkrankung der Kartoffelknolle durch ein Reihe Experimente beweisen. *Bot Zeit,* 14:121-125.

Staskawicz, B. J., Dahlbeck, D., and Keen, N. T. (1984). Cloned avirulence gene of *Pseudomonas syringae* pv. *glycinea* determines race-specific incompatibility on *Glycine max* (L.). Merr. *Proc Natl Acad Sci USA,* 81:6024-6028.

Van der Plank, P. E. (1963). *Plant Diseases: Epidemics and Control.* Academic Press, New York.

Vidhyasekaran, P. (2001). Induced systemic resistance for the management of rice fungal diseases. In S. Sreenivasaprasad and R. Johnson (Eds.), *Major Fungal Diseases of Rice: Recent Advances.* Kluwer Academic Publishers, the Netherlands, pp. 347-358.

Zaitlin, M. and Palukaitis, P. (2000). Advances in understanding plant viruses and virus diseases. *Annu Rev Phytopathol,* 38:117-143.

CAUSAL ORGANISMS

2

Bacteria

STRUCTURE OF BACTERIAL CELLS

Bacteria are primitive organisms classified as *prokaryotes,* with a primitive type of nucleus lacking a clearly defined membrane. Most of the genetic information in a bacterial cell is carried on a single chromosome with double-stranded deoxyribonucleic acid (DNA) in a closed circular form. A *chromosome* is the carrier of genes (the factors that control inherited traits). *Genes* are deoxyribonucleic acids that are linked and aligned on the chromosome. Chromosomes are located in the chromatin body of the bacteria. In addition, some bacterial cells contain extrachromosomal DNA as plasmids. *Plasmids* are circular DNA molecules that are generally dispensable and not essential for cell growth and division. However, they confer traits such as antibiotic resistance or pathogenicity on the host organism. Plasmids replicate independently of the chromosomes and can pass from one bacterial cell to another with ease. Some bacteria contain *episomes,* which are also autonomous and dispensable genetic elements similar to plasmids. Unlike plasmids, however, episomes can exist even as integrated with the chromosome. Generally, the bacteria containing plasmids do not have episomes and vice versa. *Transposons,* mobile DNA segments that can insert into a few or several sites in a genome, are found in some bacteria. Transposons are transposable genetic elements (the word transpose means "alter the positions of" or "interchange") that are capable of moving between prokaryotes and eukaryotes.

Bacterial cells have either no organelles or poorly developed organelles. All bacteria except *Streptomyces* are unicellular. Bacteria are rod shaped, round (spherical, ovoid, or ellipsoidal), or spiral (helices). All plant-pathogenic bacteria are rod shaped. They measure 0.5-3.5 µm in length and 0.3-1.0 µm in diameter. Each bacterial cell consists of a cell wall and a compound membrane called the cytoplasmic membrane, which encloses the cell's protoplasm. Outside the cell wall is a *slime layer.* The slimy material may remain firmly adhered as a discrete covering layer of each cell, or it may part freely from the cell as it is formed. The former, a thick, well-developed slime layer, is called a capsule, while the latter is called free slime. Most plant-pathogenic bacteria have one or more flagella. Flagella may occur at one or both ends (polar flagella) or all over the surface of the bacterium (peritrichous flagella). Flagella may allow some movement of the bacteria.

The bacterial cell wall is composed of a *peptidoglycan*. The peptidoglycan is also called a mucopeptide or murein. *Murein* is composed of six different components such as N-acetylglycosamine, N-acetylmuramic acid, L-alanine, D-alanine, D-glutamic acid, and either L-lysine or meso-diaminopimelic acid. The rigid peptidoglycan layer is located between the cytoplasmic membrane and a multiple-tract layer. The multiple-tract layer is composed of lipoprotein and lipopolysaccharide complexes. The external multiple-tract layer, the rigid peptidoglycan layer, and the cytoplasmic membrane constitute the bacterial envelope.

Immediately below the cell wall is the cytoplasmic membrane. The membrane contains about 75 percent protein, 20 to 30 percent lipid, and about 2 percent carbohydrate. The membrane is semipermeable and controls the passage of nutrients and metabolites into and out of the cell. Besides the cytoplasmic membrane, many bacteria possess other intracellular membrane systems such as mesosomes or chondrioids. The *mesosome* structure is formed by an invagination of the cytoplasmic membrane. Mesosomes serve for compartmentalization and integration of biochemical systems.

Cytoplasm is the cell material within the cytoplasmic membrane. The cytoplasm can be divided into the area rich in ribonucleic acid (RNA), the chromatinic area (nuclear area) rich in DNA, and the fluid portion with dissolved nutrients. Bacteria do not have a characteristic nucleus. They contain bodies within the cytoplasm that are regarded as a nuclear structure and DNA is confined to this area. Because it is not a discrete nucleus, it is called a *chromatin body*.

Ribosomes are globular structures found in the cytoplasm. They are composed of about one-third protein and two-thirds RNA. Ribosomes are designated 30S, 50S, 70S, etc., depending upon their size. The size is determined by the rate, measured in Svedberg units, at which a particle sediments when it is centrifuged at high speed in an ultracentrifuge. Ribosomes that act in clusters are called polyribosomes or polysomes. *Polysomes* may contain three to 70 ribosomes depending upon the bacteria. Ribosomes are the sites of protein synthesis.

Vacuoles are cavities in the cytoplasm that contain a fluid called cell sap. As the cell approaches maturity, some of the water-soluble reserve food materials manufactured by the cell are dissolved in the sap. Insoluble constituents precipitate out as cytoplasmic inclusion bodies. Volutin, glycogen, and fat globules are common cytoplasmic inclusions. Volutins are metachromatic granules. They are found to localize in the vacuoles of mature forms. They contain inorganic polyphosphate, lipoprotein, RNA, and magnesium. They may serve as phosphate storage structures. Glycogen accumulates in the cytoplasm at the ends of the cell in the form of granules. Lipids are found in the cytoplasm of bacteria in the form of fat globules.

Fimbriae and *pili* are hairlike structures that are attached to the bacterial cells as appendages. Fimbriae are common in plant-pathogenic bacteria. Pili are considered to be sex organs. They mediate conjugation (mating) of bacteria. They may also serve as adsorption organs for bacteriophages.

Actinomycetes are now classified as a group of bacteria. They produce individual, small, bacteria-like spores and funguslike vegetative mycelium.

REPRODUCTION OF BACTERIA

The predominant mode of reproduction in bacteria is *binary fission*. The bacterial cell divides into two daughter cells. This is an asexual process. A transverse wall develops across the middle of the bacterial cell. When new cell wall material has developed, the cells separate. During this process, the DNA condenses into an amorphous mass, which elongates and becomes dumbbell shaped before it divides into two equal pieces. These pieces serve as the nuclei of the daughter cells. This process is repeated every 20 minutes and the bacteria multiply in logarithmic proportion. However, the multiplication of the bacteria is limited by the exhaustion of available nutrients and/or the accumulation of toxic metabolic products.

Sometimes, bacteria may reproduce sexually by *conjugation*. Genetic material of one cell is transferred to another cell during conjugation. The two cells are genetically different. The donor cell transfers part of its genome (set of genes) to the recipient cell. Donor strains are designated F+ or Hfr (males, high frequency cells). The recipient strains are designated F- (females). The F factor is called the sex factor. It is a type of episome. Episomes are genetic elements that can exist in two alternative states: autonomous in the cytoplasm or incorporated in the chromosome. In the former condition, the episome is free to multiply independently of cell division. In the latter, it is replicated only when the chromosome is replicated. F factor controls the ability of the cells to act as gene donors. In the process of conjugation, the donor cell injects a chromosomal thread into the recipient cells, and the genes along this thread enter in a definite order, one after another, as they occur in a particular donor strain. Conjugation is one type of genetic recombination. Genetic recombination refers to any process leading to the formation of a new individual that derives some of its genes from one parent and some from another, genetically different, parent. The recipient cell, which receives genes from a donor, is called a "recombinant."

NOMENCLATURE OF BACTERIAL PATHOGENS

The nomenclature of plant bacterial pathogens has been revised in recent years according to the *International Code of Nomenclature of Bacteria* (ICSB, 1992). A list of all valid names of plant-pathogenic bacteria from 1864 to 1995 was published by Young et al. (1996). More than 330 bacterial pathogens were listed. According to Schaad et al. (2000), the names of some of these bacterial pathogens still need to be modified. Accepted names (Young et al., 1996) of some important bacterial pathogens are provided here. Names suggested by Schaad et al. (2000) are given within parentheses and marked with an asterisk (*).

Apple fire blight: *Erwinia amylovora* (Burrill 1882) Winslow et al. 1920
Banana moko wilt: *Ralstonia solanacearum* (Smith 1896) Yabauuchi, Kosako, Yano, Hotta, and Nishiuchi 1995
Bean bacterial brown spot: *Pseudomonas syringae* pv. *syringae* van Hall 1902
Bean common blight: *Xanthomonas axonopodis* pv. *phaseoli* (Smith 1897) Vauterin, Hoste, Kersters, and Swings 1995 (*X. campestris* pv. *phaseoli* [Smith 1897] Dye 1978)*
Bean halo blight: *Pseudomonas savastanoi* pv. *phaseolicola* (Burkholder 1926) Gardan, Bollet, Abu Ghorrah, Grimont, and Grimont 1992 (*P. syringae* pv. *phaseolicola* [Burkholder 1926] Young, Dye, and Wilkie 1978)*
Cabbage and cauliflower black rot: *Xanthomonas campestris* pv. *campestris* (Pammel 1985) Dowson 1939
Citrus canker: *Xanthomonas axonopodis* pv. *citri* (Hasse 1915) Vauterin, Hoste, Kersters, and Swings 1995 (*X. campestris* pv. *citri* [Hasse 1915] Dye 1978)*
Citrus greening: Citrus greening organism (Nonculturable, phloem-restricted, Gram-negative bacteria) *Candidatus* Liberobacter africanum Jagoueix et al. 1994 and *Candidatus* Liberobacter asiaticum Jagoueix et al. 1994
Citrus stubborn: *Spiroplasma citri* Saglio, L'hospital, Lafleche, Dupont, Bove, Mouches, Rose, Coan, and Clark 1986
Citrus variegated chlorosis: *Xylella fastidiosa* Wells, Raju, Weisburg, Mandelo-Paul, and Brenner 1987

Corn Stewarts disease (bacterial wilt): *Pantoea stewartii* (Smith 1898) Mergaert et al.,1993

Cotton bacterial blight (black arm): *Xanthomonas axonopodis* pv. *malvacearum* (Smith 1901) Vauterin, Hoste, Kersters, and Swings 1995 (*X. campestris* pv. *malvacearum* [Smith 1901] Dye 1978)*

Peach bacterial canker: *Pseudomonas syringae* pv. *syringae* van Hall 1902

Peach bacterial spot: *Xanthomonas arboricola* pv. *pruni* (Smith 1903) Vauterin, Hoste, Kersters, and Swings 1995

Pear fire blight: *Erwinia amylovora* (Burrill 1882) Winslow, Broadhurst, Buchanan, Krumwiede, Rogers, and Smith 1920

Plum bacterial canker: *Pseudomonas syringae* pv. *syringae* van Hall 1902

Plum bacterial spot: *Xanthomonas arboricola* pv. *pruni* (Smith 1903) Vauterin, Hoste, Kersters, and Swings 1995

Potato bacterial wilt (brown rot): *Ralstonia solanacearum* (Smith 1896) Yabauuchi, Kosako, Yano, Hotta, and Nishiuchi 1995

Potato blackleg and bacterial soft rot: *Erwinia carotovora* ssp. *atroseptica* (van Hall 1902) Dye 1969, *E. carotovora* ssp. *carotovora* (Jones 1901) Bergey, Harrison, Breed, Hammer, and Huntoon 1923, and *E. chrysanthemi* Burkholder, McFadden, and Dimock 1953

Potato common scab: *Streptomyces scabies* (ex Thaxter 1892) Lambert and Loria 1989

Potato ring rot: *Clavibacter michiganensis* ssp. *sepedonicus* (Spieckermann and Kotthoff 1914) Davis, Gillaspie, Vidaver, and Harris 1984

Rice bacterial blight: *Xanthomonas oryzae* pv. *oryzae* (Ishiyama 1922) Swings, Van den Mooter, Vauterin, Hoste, Gillis, Mew, and Kersters 1990

Rice bacterial leaf streak: *Xanthomonas oryzae* pv. *oryzicola* (Ishiyama 1922) Swings, Van den Mooter, Vauterin, Hoste, Gillis, Mew, and Kersters 1990

Soybean bacterial blight: *Pseudomonas savastanoi* pv. *glycinea* (Coerper 1919) Gardan, Bollet, Abu Ghorrah, Grimont, and Grimont 1992 (*Pseudomonas syringae* pv. *glycinea* [Coerper 1919] Young, Dye, and Wilkie 1978)*

Tomato bacterial canker: *Clavibacter michiganensis* ssp. *michiganensis* (Smith 1910) Davis, Gillaspie, Vidaver, and Harris 1984

Tomato bacterial speck: *Pseudomonas syringae* pv. *tomato* (Okabe 1933) Young, Dye, and Wilkie 1978

Tomato bacterial spot: *Xanthomonas vesicatoria* (ex Doidge 1920) Vauterin, Hoste, Kersters, and Swings 1995 (*X. campestris* pv. *vesicatoria* [Doidge 1920] Dye 1978)*

Tomato bacterial stem rot and fruit rot: *Erwinia carotovora* ssp.
  carotovora (Jones 1902) Bergey, Harrison, Breed, Hammer, and
  Huntoon 1923
Tomato bacterial wilt: *Ralstonia solanacearum* (Smith 1896) Yabauuchi,
  Kosako, Yano, Hotta, and Nishiuchi 1995

CLASSIFICATION OF PLANT BACTERIAL PATHOGENS

Bacteria have been classified into various genera, families, suborders, orders, subclasses, classes, divisions, and domains. These classifications are based on numerical analyses, serology, membrane protein profiles, and DNA analyses. In contrast to bacterial nomenclature, no "official" classification of bacteria exists. Taxonomy remains a matter of scientific judgment and general agreement. The classification given here is the one widely accepted by microbiologists (Euzeby, 2001). Bacteria are classified into domain or empire (suffix not covered by rules), division (suffix not covered by the rules), class (suffix of the names of class is –ia), subclass (suffix is –idae), order (suffix is –ales), suborder (-ineae), family (-aceae), and tribe (-eae). The taxonomic categories of domain and division are not covered by the rules of bacteriological code. The following 17 families include 28 genera of plant-pathogenic bacteria:

Acetobacteraceae: *Acetobacter, Gluconobacter*
Bacillaceae: *Bacillus*
Burkholderiaceae: *Burkholderia*
Clostridiaceae: *Clostridium*
Comamonadaceae: *Acidovorax*
Corynebacteriaceae: *Corynebacterium*
Enterobacteriaceae: *Erwinia, Pantoea, Pectobacterium,*
    Enterobacter, Serratia
Microbacteriaceae: *Curtobacterium, Clavibacter, Rathayibacter*
Micrococcaceae: *Arthrobacter*
Nocardiaceae: *Nocardia, Rhodococcus*
Pseudomonadaceae: *Pseudomonas, Xylophilus, Rhizobacter*
Ralstoniaceae: *Ralstonia*
Rhizobiaceae: *Agrobacterium*
Sphingomonadaceae: *Rhizomonas*
Spiroplasmataceae: *Spiroplasma*
Streptomycetaceae: *Streptomyces*
Xanthomonadaceae: *Xanthomonas, Xylella*

The following 12 orders contain 17 families consisting of plant-pathogenic bacteria:

Actinomycetales: Corynebacteriaceae, Microbacteriaceae,
    Nocardiaceae, Streptomycetaceae
Bacillales: Bacillaceae
Burkholderiales: Burkholderiaceae, Ralstoniaceae,
    Comamonadaceae
Clostridiales: Clostridiaceae
Enterobacteriales: Enterobacteriaceae
Entomoplasmatales: Spiroplasmataceae
Micrococcales: Micrococcaceae
Pseudomonadales: Pseudomonadaceae
Rhizobiales: Rhizobiaceae
Rhodospirillales: Acetobacteraceae
Sphingomonadales: Sphingomonadaceae
Xanthomonadales: Xanthomonadaceae

The following six classes contain 12 orders consisting of plant bacterial pathogens:

Amoxyphotobacteria: Rhodospirillales
Bacilli: Bacillales
Clostridia: Clostridiales
Mollicutes: Entomoplasmatales
Proteobacteria:
    Alpha subdivision: Sphingomonadales, Rhizobiales
    Beta subdivision: Burkholderiales
    Gamma subdivision: Enterobacteriales, Xanthomonadales,
      Pseudomonadales
Schizomycetes: Actinomycetales, Micrococcales

Two divisions have been recognized to include some classes consisting of bacterial pathogens:

Firmicutes—class Mollicutes (order Entomoplasmatales, family
    Spiroplasmataceae, genus *Spiroplasma*).
Gracilicutes—class Anoxyphotobacteria (order Rhodospirillales,
    family Acetobacteraceae, genus *Gluconobacter*)

Some bacteria have not been described in sufficient detail to warrant establishment of a novel taxon. These bacteria have been included in a new category of indefinite rank called *Candidatus* (Murray and Schleifer, 1994).

Candidatus is used to categorize prokaryotic entities for which more than one sequence is available, but for which characteristics required for description according to the *International Code of Nomenclature of Bacteria* (ICSB, 1992) are lacking. *Candidatus* Liberobacter africanum Jagoueix et al., 1994, and *C. L.* asiaticum are the phloem-limited bacteria causing greening disease of *Citrus*. They are members of the Alpha subdivision of the Proteobacteria.

SYSTEMATIC POSITION OF CROP BACTERIAL PATHOGENS

The following is a list of bacterial genera that cause diseases in crops. Systematic positions and names of important species in each genus are given. The names of many bacterial species were changed recently (Young et al., 1996; Euzeby, 2001). The most important plant pathogenic genus, *Pseudomonas,* was reclassified into five different genera, including *Pseudomonas, Burkholderia, Ralstonia, Acidovorax,* and *Acetobacter.* The genus *Xanthomonas* was reclassified into two genera, *Xanthomonas* and *Xylophilus*. The genus *Erwinia* was reassigned into four different genera: *Erwinia, Pantoea, Enterobacter,* and *Serratia*. The genus *Corynebacterium* was not recognized, but remains a valid synonym of the genera *Curtobacterium, Clavibacter, Erwinia, Rhodococcus, Arthrobacter,* and *Rathayibacter.* The names of various bacterial species as recommended by Young et al. (1996) are provided. Suggested modifications in the names of bacterial species (Schaad et al., 2000) are marked with an asterisk (*). Valid synonyms of some bacterial species are given as = (without asterisk).

Acetobacter—Tribe: Acetobactereae; Family: Acetobacteraceae; Order: Rhodospirillales; Class: Anoxyphotobacteria; Division: Gracilicutes; Domain: Bacteria
> *Acetobacter pasteurianus* (Hansen 1879) Beijerinck and Folpmers 1916 = *Pseudomonas pomi* Cole 1959

Acidovorax—Family: Comamonadaceae; Order: Bukholderiales; Class: Proteobacteria Beta Subdivision; Domain: Bacteria
> *Acidovorax avenae* ssp. *avenae* (Manns 1909) Willems et al. 1992 = *Pseudomonas avenae* ssp. *avenae* 1909
> *A. avenae* ssp. *citrulli* (Schaad et al. 1978) Willems et al. 1992 = *P. avenae* ssp. *citrulli* (Schaad et al. 1978) Hu et al. 1991

Agrobacterium—Family: Rhizobiaceae; Order: Rhizobiales; Class: Proteobacteria Alpha subdivision; Domain: Bacteria
> *Agrobacterium rhizogenes* (Riker et al. 1930) Conn 1942
> *A. rubi* (Hildebrand 1940) Starr and Weiss 1943

A. tumefaciens (Smith and Townsend 1907) Conn 1942
= *A. radiobacter* (Beijerinck and van Delden 1902) Conn 1942
A. vitis Ophel and Kerr 1990

Arthrobacter—Family: Micrococcaceae; Order: Micrococcales; Class: Schizomycetes; Domain: Bacteria
Arthrobacter ilicis (Mandel et al. 1961) Collins et al. 1982 = *Corynebacterium ilicis* Mandel et al. 1961

Bacillus—Family: Bacillaceae; Order: Bacillales; Class: Bacilli; Division: Firmicutes; Domain: Bacteria
Bacillus megaterium pv. *cerealis* Hosford 1982

Burkholderia—Family: Burkholderiaceae; Order: Burkholderiales; Class: Proteobacteria Beta subdivision; Domain: Bacteria
Burkholderia caryophylli (Burkholder 1942) Yabuuchi et al. 1993 = *Pseudomonas caryophylli* (Burkholder 1942) Starr and Burkholder 1942
B. cepacia (ex Burkholder 1950) = *P. cepacia* (ex Burkholder 1950) Palleroni and Holmes 1981
B. gladioli pv. *gladioli* (Severini 1913) Yabuuchi et al. 1992 = *P. gladioli* pv. *gladioli* (Severini 1913)

Candidatus—Class: Proteobacteria Alpha subdivision; Domain: Bacteria (Phloem-limited plant bacteria)
Candidatus Liberobacter africanum Jagoueix et al., 1994
Candidatus Liberobacter asiaticum Jagoueix et al., 1994

Clavibacter—Family: Microbacteriaceae; Suborder: Micrococcineae; Order: Actinomycetales; Subclass: Actinobacteridae; Class: Schizomycetes; Domain: Bacteria
Clavibacter michiganensis ssp. *insidiosus* (McCulloch 1925) Davis et al. 1984 = *Corynebacterium michiganense* ssp. *insidiosum* (McCulloch 1925) Carlson and Vidaver 1982
Clavibacter michiganensis ssp. *sepedonicus* (Spieckermann and Kotthoff 1914) Davis et al. 1984 = *Corynebacterium michiganense* ssp. *sepedonicum* (Spieckermann and Kotthoff 1914) Carlson and Vidaver 1982
Clavibacter xyli Davis et al. 1984

Clostridium—Family: Clostridiaceae; Order: Clostridiales; Class: Clostridia; Domain: Bacteria
Clostridium puniceum Lund et al. 1981

Corynebacterium—Family: Corynebacteriaceae; Suborder: Corynebacterineae; Order: Actinomycetales; Subclass: Actinobacteridae; Class: Schizomycetes; Domain: Bacteria. In the revised nomenclature, the bacteria belonging to the genus

Corynebacterium were assigned to the genera *Curtobacterium, Erwinia, Rhodococcus, Arthrobacter, Clavibacter,* and *Rathayibacter.*
Curtobacterium—Family: Microbacteriaceae; Suborder: Micrococcineae; Order: Actinomycetales; Subclass: Actinobacteridae; Class: Schizomycetes; Domain: Bacteria

    Curtobacterium flaccumfaciens pv. *betae* (Keyworth et al. 1956) Collins and Jones 1983 = *Corynebacterium flaccumfaciens* pv. *betae* (Keyworth et al. 1956) Dye and Kemp 1977

    Curtobacterium flaccumfaciens pv. *flaccumfaciens* (Hedges 1922) Collins and Jones 1983 = *Corynebacterium flaccumfaciens* pv. *flaccumfaciens* (Hedges 1922) Dowson 1942

    Curtobacterium flaccumfaciens pv. *poinsettiae* (Starr and Pirone 1942) Collins and Jones 1983 = *Corynebacterium flaccumfaciens* pv. *poinsettiae* (Starr and Pirone 1942) Dye and Kemp 1977

Enterobacter—Family: Enterobacteriaceae; Order: Enterobacteriales; Class: Proteobacteria Gamma subdivision; Domain: Bacteria

    Enterobacter dissolvens (Rosen 1922) Brenner et al. 1988 = *Erwinia dissolvens* (Rosen 1922) Bukholder 1948

Erwinia—Family: Enterobacteriaceae; Suborder: Erwinieae; Order: Enterobacteriales; Class: Proteobacteria Gamma subdivision; Domain: Bacteria

    Erwinia amylovora (Burrill 1882) Winslow et al. 1920

    E. carotovora ssp. *atroseptica* (van Hall 1902) Dye 1969

    E. carotovora ssp. *betavasculorum* Thomson et al. 1923

    E. carotovora ssp. *carotovora* (Jones 1901) Bergey et al. 1923

    E. chrysanthemi pv. *chrysanthemi* Burkholder et al. 1953 = *Pectobacterium chrysanthemi* (Burkholder et al. 1953) Brenner et al. 1973

    E. chrysanthemi pv. *dianthicola* (Hellmers 1958) Dickey 1979

    E. chrysanthemi pv. *dieffenbachiae* (McFadden 1961) Dye 1978

    E. chrysanthemi pv. *zeae* (Sabet 1954) Victoria et al. 1975

    E. herbicola pv. *gypsophilae* (Brown 1934) Miller et al. 1981

    E. psidii Neto et al. 1988

    E. quercina Hildebrand and Schroth 1967

    E. rhapontici (Millard 1924) Burkholder 1948

    E. tracheiphila (Smith 1895) Bergey et al. 1923

Gluconobacter—Family: Acetobacteraceae; Order: Rhodospirillales; Class: Anoxyphotobacteria; Division: Gracilicutes; Domain: Bacteria

    Gluconobacter oxydans (Henneberg 1897) De Ley 1961

Nocardia—Family: Nocardiaceae; Suborder: Corynebacterineae; Order: Actinomycetales; Subclass: Actinobacteridae; Class: Schizomycetes; Domain: Bacteria

    Nocardia vaccini Demaree and Smith 1952

Pantoea—Family: Enterobacteriaceae; Order: Enterobacteriales; Class: Proteobacteria Gamma subdivision; Domain: Bacteria

    Pantoea agglomerans (Beijerinck 1888) Gavini et al. 1989 = *Erwinia herbicola* (Lohnis 1911) Dye 1964

    P. ananas pv. *ananas* (Serrano 1928) Mergaert et al. 1993 = *E. ananas* pv. *ananas* Serrano 1928

    P. stewartii ssp. *stewartii* (Smith 1898) Mergaert et al. 1993 = *E. stewartii* (Smith 1898) Dye 1963

Pectobacterium—Family: Enterobacteriaceae; Order: Enterobacteriales; Class: Proteobacteria Gamma subdivision; Domain: Bacteria

    Pectobacterium spp. are now synonyms of various *Erwinia* spp.

Pseudomonas—Tribe: Pseudomonadeae; Family: Pseudomonadaceae; Suborder: Pseudomonadineae; Order: Pseudomonadales; Class: Proteobacteria Gamma subdivision; Domain: Bacteria

    Pseudomonas amygdali Psallidas and Panagopoulos 1975

    P. caricapapayae Robbs 1956

    P. cichorii (Swingle 1925) Stapp 1928

    P. corrugata (ex Scarlett et al. 1978) Roberts and Scarlett 1981

    P. fuscovaginae (ex Tanii et al. 1976) Miyajima et al. 1983

    P. marginalis pv. *alfalfae* (Shinde and Lukezic 1974) Young et al. 1978

    P. marginalis pv. *marginalis* (Brown 1918) Stevens 1925

    P. meliae Ogimi 1981

    P. savastanoi pv. *glycinea* (Coerper 1919) Gardan et al. 1992 = *P. syringae* pv. *glycinea* (Coerper 1919) Young et al. 1978*

    P. savastanoi pv. *phaseolicola* (Burkholder 1926) Gardan et al. 1992 = *P. syringae* pv. *phaseolicola* (Burkholder 1926) Young et al. 1978*

    P. savastanoi pv. *savastanoi* (ex Smith 1908) Gardan et al. 1992 = *P. savastanoi**

    P. syringae pv. *antirrrhini* (Takimoto 1920) Young, Dye, and Wilkie 1978

    P. syringae pv. *apii* (Jagger 1921) Young, Dye, and Wilkie 1978

    P. syringae pv. *aptata* (Brown and Jamieson 1913) Young, Dye, and Wilkie 1978

    P. syringae pv. *atrofaciens* (McCulloch 1920) Young, Dye, and Wilkie 1978

    P. syringae pv. *atropurpurea* (Reddy and Godkin 1923) Young, Dye, and Wilkie 1978

    P. syringae pv. *cannabina* (Sutic and Dowson 1959) Young, Dye, and Wilkie 1978

    P. syringae pv. *castaneae* Takanashi and Shimizu 1989

    P. syringae pv. *coronafaciens* (Elliott 1920) Young, Dye, and Wilkie 1978

P. syringae pv. *delphinii* (Smith 1904) Young, Dye, and Wilkie 1978

P. syringae pv. *helianthi* (Kawamura 1934) Young, Dye, and Wilkie 1978

P. syringae pv. *lachrymans* (Smith and Bryan 1915) Young, Dye, and Wilkie 1978

P. syringae pv. *maculicola* (McCulloch 1911) Young, Dye, and Wilkie 1978

P. syringae pv. *mellea* (Johnson 1923) Young, Dye, and Wilkie 1978

P. syringae pv. *mori* (Boyer and Lambert 1893) Young, Dye, and Wilkie 1978

P. syringae pv. *morsprunorum* (Wormald 1931) Young, Dye, and Wilkie 1978

P. syringae pv. *oryzae* (Kuwata 1985) Young et al. 1991

P. syringae pv. *populans* (Rose 1917) Dhanvantari 1977

P. syringae pv. *passiflorae* (Reid 1938) Young, Dye, and Wilkie 1978

P. syringae pv. *persicae* (Prunier et al. 1970) Young, Dye, and Wilkie 1978

P. syringae pv. *pisi* (Sackett 1916) Young, Dye, and Wilkie 1978

P. syringae pv. *sesami* (Malkoff 1906) Young, Dye, and Wilkie 1978

P. syringae pv. *striafaciens* (Elliott 1927) Young, Dye, and Wilkie 1978

P. syringae pv. *syringae* van Hall 1902

P. syringae pv. *tabaci* (Wolf and Foster 1917) Young, Dye, and Wilkie 1978

P. syringae pv. *tagetis* (Hellmers 1955) Young, Dye, and Wilkie 1978

P. syringae pv. *theae* (Hori 1915) Young, Dye, and Wilkie 1978

P. syringae pv. *tomato* (Okabe 1933) Young, Dye, and Wilkie 1978

P. syringae pv. *ulmi* (Sutic and Tesic 1958) Young, Dye, and Wilkie 1978

P. syringae pv. *viburni* (Thornberry and Anderson 1931) Young, Dye, and Wilkie 1978

P. syringae pv. *zizaniae* (ex Bowden and Pereich 1983) Young et al. 1991

P. viridiflava (Burkholder 1930) Dowson 1939

Ralstonia—Family: Ralstoniaceae; Order: Burkholderiales; Class: Proteobacteria Beta subdivision; Domain: Bacteria

Ralstonia solanacearum (Smith 1896) Yabauuchi et al. 1995 = *Pseudomonas solanacearum* (Smith 1896) Smith 1914 = *Burkholderia solanacearum* (Smith 1896) Yabauuchi et al. 1993

Rathayibacter—Family: Microbacteriaceae; Suborder: Micrococcineae; Order: Actinomycetales; Subclass: Actinobacteridae; Class: Schizomycetes; Domain: Bacteria

 Rathayibacter rathayi (Smith 1913) Zgurskaya et al. 1993 = *Clavibacter rathayi* (Smith 1913) Davis et al. 1984 = *Corynebacterium rathayi* (Smith 1913) Dowson 1942

 R. tritici (ex Hutchinson 1917) Zgurskaya et al. 1993

Rhizobacter—Family: Pseudomonadaceae; Suborder: Pseudomonadineae; Order: Pseudomonadales; Class: Proteobacteria Gamma subdivision; Domain: Bacteria

 Rhizobacter dauci Goto and Kuwata 1988

Rhizomonas—Family: Sphingomonadaceae; Order: Sphingomonadales; Class: Proteobacteria Alpha subdivision; Domain: Bacteria

 Rhizomonas suberifaciens van Bruggen, Jochimsen and Brown 1990

Rhodococcus—Family: Nocardiaceae; Suborder: Corynebacterineae; Order: Actinomycetales; Subclass: Actinobacteridae; Class: Schizomycetes; Domain: Bacteria

 Rhodococcus fascians (Tilford 1936) Goodfellow 1984 = *Corynebacterium fascians* (Tilford 1936) Dowson 1942

Serratia—Tribe: Serratiaceae; Family: Enterobacteriaceae; Order: Enterobacteriales; Class: Proteobacteria Gamma subdivision; Domain: Bacteria

 Serratia proteamaculans (Paine and Stansfield 1919) Grimont, Grimont, and Starr 1978 = *Erwinia proteamaculans* (Paine and Stansfield 1919) Dye 1966

Spiroplasma—Family: Spiroplasmataceae; Order: Entomoplastales; Class: Mollicutes; Division: Firmicutes; Domain: Bacteria

 Spiroplasma citri Saglio, L'hospital, Lafleche, Dupont, Bove, Tully, and Freundt 1973

 S. kunkelii Whitcomb et al. 1986

 S. phoeniceum Saillard et al. 1987

Streptomyces—Family: Streptomycetaceae; Order: Actinomycetales; Actinobacteridae; Class: Schizomycetes; Domain: Bacteria

 Streptomyces scabies (ex Thaxter 1892) Lambert and Loria 1989

 S. ipomoeae (Person and Martin 1940) Waksman and Henrici 1948

Xanthomonas—Family: Xanthomonadaceae; Order: Xanthomonadales; Class: Proteobacteria Gamma subdivision; Domain: Bacteria

 Xanthomonas albilineans (Ashby 1929) Dowson 1943

 X. arboricola pv. *juglandis* (Pierce 1901) Vauterin et al. 1995 = *X. juglandis* pv. *juglandis**

 X. arboricola pv. *populi* (ex de Kam 1984) Vauterin et al. 1995 = *X. campestris* pv. *populi* (ex de Kam 1984) Young et al. 1991

X. arboricola pv. *pruni* (Smith 1903) Vauterin et al. 1995 = *X. juglandis* pv. *pruni**

X. arboricola pv. *corylina* (Miller et al. 1940) Vauterin et al. 1995 = *X. juglandis* pv. *corylina**

X. axonopodis pv. *alfalfae* (Riker et al. 1935) Vauterin et al. 1995 = *X. campestris* pv. *alfalfae* ((Riker et al. 1935) Dye 1978*

X. axonopodis pv. *axonopodis* Starr and Garces 1950

X. axonopodis pv. *bauhiniae* (Padhya et al. 1965) Vauterin et al. 1995 = *X. campestris* pv. *bauhiniae* (Padhya et al. 1965) Dye 1978*

X. axonopodis pv. *begoniae* (Takimoto 1934) Vauterin et al. 1995 = *X. campestris* pv. *begoniae* (Takimoto 1934) Dye 1978*

X. axonopodis pv. *cajani* (Kulkarni et al. 1950) Vauterin et al. 1995 = *X. campestris* pv. *cajani* (Kulkarni et al. 1950) Dye 1978*

X. axonopodis pv. *cassiae* (Kulkarni et al. 1951) Vauterin et al. 1995 = *X. campestris* pv. *cassiae* (Kulkarni et al. 1951) Dye 1978*

X. axonopodis pv. *citri* (Hasse 1915) Vauterin et al. 1995 = *X. campestris* pv. *citri* (Hasse 1915) Dye 1978*

X. axonopodis pv. *clitoriae* (Pandit and Kulkarni 1979) Vauterin et al. 1995 = *X. campestris* pv. *clitoriae* (Pandit and Kulkarni 1979) Dye et al. 1980*

X. axonopodis pv. *coracanae* (Desai et al. 1965) Vauterin et al. 1995 = *X. campestris* pv. *coracanae* (Desai et al. 1965) Dye 1978*

X. axonopodis pv. *cyamopsidis* (Patel et al. 1953) Vauterin et al. 1995 = *X. campestris* pv. *cyamopsidis* (Patel et al. 1953) Dye 1978*

X. axonopodis pv. *desmodii* (Patel 1949) Vauterin et al. 1995 = *X. campestris* pv. *desmodii* (Patel 1949) Dye 1978*

X. axonopodis pv. *desmodiilaxiflori* (Pant and Kulkarni 1976) Vauterin et al. 1995 = *X. campestris* pv. *desmodiilaxiflori* Pant and Kulkarni 1976*

X. axonopodis pv. *desmodiirotundifolii* (Deasi and Shah 1960) Vauterin et al. 1995 = *X. campestris* pv. *desmodiirotundifolii* (Deasi and Shah 1960) Dye 1978*

X. axonopodis pv. *dieffenbachiae* (McCulloch and Pirone 1939) Vauterin et al. 1995 = *X. campestris* pv. *dieffenbachiae* (McCulloch and Pirone 1939) Dye 1978*

X. axonopodis pv. *erythrinae* (Patel et al. 1952) Vauterin et al. 1995 = *X. campestris* pv. *erythrinae* (Patel et al. 1952) Dye 1978*

X. axonopodis pv. *glycines* (Nakano 1919) Vauterin et al. 1995 = *X. campestris* pv. *glycines* (Nakano 1919) Dye 1978*

X. axonopodis pv. *lespedezae* (Ayres et al. 1939) Vauterin et al. 1995
= *X. campestris* pv. *lespedezae* (Ayres et al. 1939) Dye 1978*

X. axonopodis pv. *malvacearum* (Smith 1901) Vauterin et al. 1995 =
X. campestris pv. *malvacearum* (Smith 1901) Dye 1978*

X. axonopodis pv. *manihotis* (Bondar 1915) Vauterin et al. 1995 = *X. campestris* pv. *manihotis* (Bondar 1915) Dye 1978*

X. axonopodis pv. *patelii* (Desai and Shah 1959) Vauterin et al. 1995
= *X. campestris* pv. *patelii* (Desai and Shah 1959) Dye 1978*

X. axonopodis pv. *phaseoli* (Smith 1897) Vauterin et al. 1995 = *X. campestris* pv. *phaseoli* (Smith 1897) Dye 1978*

X. axonopodis pv. *phyllanthi* (Sabet et al. 1969) Vauterin et al. 1995
= *X. campestris* pv. *phyllanthi* (Sabet et al. 1969) Dye 1978*

X. axonopodis pv. *poinsettiicola* (Patel et al. 1951) Vauterin et al.
1995 = *X. campestris* pv. *poinsettiicola* (Patel et al. 1951) Dye
1978*

X. axonopodis pv. *rhynchosiae* (Sabet et al. 1969) Vauterin et al.
1995 = *X. campestris* pv. *rhynchosiae* (Sabet et al. 1969) Dye
1978*

X. axonopodis pv. *ricini* (Yoshii and Takimoto 1928) Vauterin et al.
1995 = *X. campestris* pv. *ricini* (Yoshii and Takimoto 1928) Dye
1978*

X. axonopodis pv. *sesbaniae* (Patel et al. 1952) Vauterin et al. 1995
= *X. campestris* pv. *sesbaniae* (Patel et al. 1952) Dye 1978*

X. axonopodis pv. *tamarindi* (Patel et al. 1951) Vauterin et al. 1995
= *X. campestris* pv. *tamarindi* (Patel et al. 1951) Dye 1978*

X. axonopodis pv. *vasculorum* (Cobb 1894) Vauterin et al. 1995 = *X. campestris* pv. *vasculorum* (Cobb 1894) Dye 1978*

X. axonopodis pv. *vignaeradiatae* (Sabet et al. 1969) Vauterin et al.
1995 = *X. campestris* pv. *vignaeradiatae* (Sabet et al. 1969) Dye
1978*

X. axonopodis pv. *vignicola* (Burkholder 1944) Vauterin et al. 1995
= *X. campestris* pv. *vignicola* (Burkholder 1944) Dye 1978*

X. axonopodis pv. *vitians* (Brown 1918) Vauterin et al. 1995 = *X. campestris* pv. *vitians* (Brown 1918) Dye 1978*

X. bromi Vauterin et al. 1995 = *X. campestris* pv. *bromi**

X. campestris pv. *aberrans* (Knosel 1961) Dye 1978

X. campestris pv. *aracearum* (Berniac 1974) Dye 1978

X. campestris pv. *arecae* (Rao and Mohan 1970) Dye 1978

X. campestris pv. *argemones* (Srinivasan et al. 1961) Dye 1978

X. campestris pv. *armoraciae* (McCulloch 1929) Dye 1978

X. campestris pv. *arracaciae* (Pereira et al. 1971) Dye 1978

X. campestris pv. *asclepiadis* Flynn and Vidaver 1990

X. campestris pv. *azadirachtae* (Desai et al. 1966) Dye 1978

X. campestris pv. *betae* Robbs et al. 1981

X. campestris pv. *campestris* (Pammel 1895) Dowson 1939

X. campestris pv. *carissae* (Moniz et al. 1964) Dye 1978

X. campestris pv. *eucalypti* (Truman 1974) Dye 1978

X. campestris pv. *euphorbiae* (Sabet et al. 1969) Dye 1978

X. campestris pv. *fici* (Cavara 1905) Dye 1978

X. campestris pv. *guizotiae* (Yirgou 1964) Dye 1978

X. campestris pv. *incanae* (Kendrick and Baker) Dye 1978

X. campestris pv. *leersiae* (ex Fang et al. 1957) Young et al. 1991

X. campestris pv. *mangiferaeindicae* (Patel et al. 1948) Robbs et al. 1974

X. campestris pv. *musacearum* (Yirgou and Bradbury 1968) Dye 1978

X. campestris pv. *nigromaculans* (Takimoto 1927) Dye 1978

X. campestris pv. *olitorii* (Sabet 1957) Dye 1978

X. campestris pv. *papavericola* (Bryan and McWhorter 1930) Dye 1978

X. campestris pv. *passiflorae* (Pereira 1969) Dye 1978

X. campestris pv. *paulliniae* Robbs et al. 1982

X. campestris pv. *phormiicola* (Takimoto 1933) Dye 1978

X. campestris pv. *plantaginis* (Thornberry and Anderson 1937) Dye 1978

X. campestris pv. *raphani* (White 1930) Dye 1978

X. campestris pv. *sesami* (Sabet and Dowson 1960) Dye 1978

X. campestris pv. *syngonii* Dickey and Zumoff 1987

X. campestris pv. *viticola* (Nayudu 1972) Dye 1978

X. campestris pv. *vitiscarnosae* (Moniz and Patel 1958) Dye 1978

X. campestris pv. *vitistrifoliae* (Padhya et al. 1965) Dye 1978

X. campestris pv. *vitiswoodrowii* (Patel and Kulkarni 1951) Dye 1978

X. campestris pv. *zingibericola* (Ren and Fang 1981) Bradbury 1986

X. campestris pv. *zinniae* (Hopkins and Dowson 1949) Dye 1978

X. cassavae (ex Wiehe and Dowson 1953) Vauterin et al. 1995

X. cucurbitae (ex Bryon 1926) Vauterin et al. 1995

X. exitiosa (Gardner and Kendrick 1921) Schaad et al. 2000

X. fragariae Kennedy and King 1962

X. hederae (Arnaud 1920) Schaad et al. 2000

X. hortorum pv. *hederae* (Arnaud 1920) Vauterin et al. 1995 = *X. hederae* pv. *hederae* Schaad et al. 2000*

X. hortorum pv. *hortorum* Vauterin et al. 1995

X. hortorum pv. *pelargonii* (Brown 1923) Vauterin et al. 1995 = *X. hederae* pv. *pelargonii* Schaad et al. 2000*

X. hyacinthi (Wakker 1883) Vauterin et al. 1995

X. melonis (Neto et al. 1984) Vauterin et al. 1995 = *X. campestris* pv. *melonis* (Neto et al. 1984)*

X. oryzae pv. *oryzae* (Ishiyama 1922) Swings et al. 1990

X. oryzae pv. *oryzicola* (Fang et al. 1957) Swings et al. 1990

X. sacchari Vauterin et al. 1995 = *X. albilineans* Schaad et al. 2000*

X. theicola (Uehara et al. 1980) Vauterin et al. 1995 = *X. campestris* pv. *theicola* (Uehara et al., 1980)*

X. translucens pv. *arrhenatheri* (Egli and Schmidt 1982) Vauterin et al. 1995 = *X. campestris* pv. *arrhenatheri* Egli and Schmidt 1982*

X. translucens pv. *graminis* (Egli et al. 1975) Vauterin et al. 1995 = *X. campestris* pv. *graminis* (Egli et al. 1975) Dye 1978*

X. translucens pv. *phlei* (Egli and Schmidt 1982) Vauterin et al. 1995 = *X. campestris* pv. *phlei* Egli and Schmidt 1982*

X. translucens pv. *phleipratensis* (Wallin and Reddy 1945) Vauterin et al. 1995 = *X. campestris* pv. *phleipratensis* (Wallin and Reddy 1945) Dye 1978*

X. translucens pv. *poae* (Egli and Schmidt 1982) Vauterin et al. 1995 = *X. campestris* pv. *poae* Egli and Schmidt 1982

X. vasicola pv. *holcicola* (Elliott 1930) Vauterin et al. 1995 = *X. campestris* pv. *holcicola* Schaad et al. 2000*

X. vasicola pv. *vasculorum* Vauterin et al. 1995 = *X. campestris* pv. *vasculorum* (Cobb 1894) Dye 1978

X. vesicatoria (ex Doidge 1920) Vauterin et al. 1995 = *X. campestris* pv. *vesicatoria* (ex Doidge 1920) Dye 1978*

Xylella—Family: Xanthomonadaceae; Order: Xanthomonadales; Class: Proteobacteria Gamma subdivision; Domain: Bacteria

Xylella fastidiosa Wells, Raju, Hung, Weisburg, Mandelco–Paul, and Brenner 1987 (Xylem-limited fastidious plant bacteria)

Xylophilus—Family: Pseudomonadaceae; Suborder: Pseudomonadineae; Order: Pseudomonadales; Class: Proteobacteria Gamma subdivision; Domain: Bacteria

Xylophilus ampelinus (Panagopoulos 1969) Willems et al. 1987 = *Xanthomonas ampelina* Panagopoulos 1969

REFERENCES

Euzeby, J. P. (2001). Taxonomic Scheme. Taxa included in the categories above the rank of genus. Web site: <http://www.bacterio.cict.fr/classification.html>.

International Committee on Systematic Bacteriology (ICSB) (1992). *International code of nomenclature of bacteria: Bacteriological code (1990 revision).* Washington, DC: American Society for Microbiology.

Murray, R. G. E. and Schleifer, K. H. (1994). Taxonomic notes: A proposal for recording the properties of putative taxa of prokaryotes. *Int J Syst Bacteriol,* 44:174-176.

Schaad, N. W., Vidaver, A. K., Lacy, G. H., Rudolph, K., and Jones, J. B. (2000). Evaluation of proposed amended names of several pseudomonads and xanthomonads and recommendations. *Phytopathology,* 90:208-213.

Young, J. M., Saddler, G. S., Takikawa, Y., De Boer, S. H., Vauterin, L., Gardan, L., Gvozdyak, R. I., and Stead, D. E. (1996). Names of plant pathogenic bacteria 1864-1995. *Rev Plant Pathol,* 75:721-761.

3

Fungi, Including Chromista and Protozoa

Fungi are the largest group of crop pathogens. Fungi belong to three kingdoms: Fungi, Chromista, and Protozoa. The Kingdom Chromista includes the phylum Oomycota, which contains many plant-pathogenic fungi. The Kingdom Protozoa consists of one phylum, Plasmodiophoromycota, which contains crop pathogens. The Kingdom Fungi consists of four phyla: Ascomycota, Basidiomycota, Zygomycota, and Chytridiomycota. Another group, Mitosporic fungi, includes fungi that have not been correlated with any meiotic states. This section describes the structure and reproduction of all these fungal phyla. It also provides a complete list of crop fungal pathogens and their systematic positions.

THREE KINGDOMS CONTAINING "FUNGI": FUNGI, CHROMISTA, AND PROTOZOA

Fungi are the most important group of plant pathogens. Out of about 56,360 species of fungi, more than 8,000 species are known to cause diseases in plants. Originally all fungi were classified as belonging to the Kingdom Fungi (= Eukaryota). Recently some fungi have been considered not to belong to this kingdom and have been placed under the Kingdoms Chromista and Protozoa. With advances in ultrastructural, biochemical, and especially molecular biology, the treatment of fungi as a single kingdom has become untenable. The organisms so far called fungi are now established as polyphyletic (i.e., with different phylogenies) and have to be referred to three different kingdoms (Hawksworth et al., 1995): Fungi, Chromista, and Protozoa. The Kingdom Fungi contains "true fungi" while the Kingdoms Chromista and Protozoa contain "pseudofungi." The Kingdom Fungi consists exclusively of fungi; the Chromista and Protozoa mainly comprise nonfungal phyla. Chromista consists of three fungal phyla and Protozoa consists of only four fungal phyla.

Fungi are organisms that are eukaryotic and heterotrophic. They develop branching filaments (or more rarely are single-celled), reproduce by spores, and their cell walls contain chitin and β-glucans. They are mostly nonflagellate. When present, the flagella always lack *mastigonemes* (i.e., surfaces of flagella are not covered by hairlike processes). Chromista are organisms that bear flagella with mastigonemes. Their cell walls contain cellulose and glucan rather than chitin. Mycolaminarin is the energy storage molecule found in Chromista. Protozoa are organisms that are predominantly unicellular, plasmodial, or colonial. They are *phagotrophic,* i.e., they feed by ingestion, engulfing food. They are wall-less in the trophic state. They have ciliary hairs that are never rigid or tubular.

OOMYCOTA

Classification of Oomycota

Three fungal (pseudofungal) phyla that belong to Chromista are Hypochytriomycota, Labyrinthulomycota, and Oomycota. Among them, only Oomycota consists of plant pathogens. There are more than 500 species in Oomycota. These include so-called water molds and downy mildews. "Oomycota" means "egg fungi" and refers to the large round oogonia (the structures containing the female gametes). Oomycota are oogamous, producing large nonmotile gametes called "eggs," and smaller gametes called "sperm." Oomycota were classified previously as fungi because of their filamentous growth. However, Oomycota cell walls are not composed of chitin, as in the fungi, but are made up of a mix of cellulosic compounds and glycan. The nuclei within the filaments are diploid, with two sets of genetic information, not haploid as in the fungi. The ultrastructure, biochemistry, and molecular sequences of Oomycota suggest that they belong with Chromista. The free-swimming spores that are produced bear two dissimilar flagella, one with mastigonemes. This feature is common in chromists. Presence of the chemical mycolaminarin in Oomycota is similar to that found in kelps and diatoms. Hence, Oomycota have been placed in the Kingdom Chromista.

The phylum (division) Oomycota consists of nine orders: Olpidiopsidales, Peronosporales, Pythiales, Sclerosporales, Saprolegniales, Leptomidales, Myzocytiopsidales, Rhipidiales, and Salilagenidales. Only the orders Peronosporales, Pythiales, Sclerosporales, and Saprolegniales contain crop pathogens. Oomycota contains many important pathogens that have affected the economies of many countries. In the order Pythiales, the family Pythiaceae contains various *Phytophthora* and *Pythium* species that cause several diseases in various crops. *Phytophthora infestans* causes late blight in potato and tomato. *Phytophthora citrophthora* is the causal organism of gummosis and foot rot diseases in *Citrus. Phytophthora capsici* causes blight in pepper, *P. cactorum* incites crown and root rot in apple and peach,

P. parasitica pv. *nicotianae* causes black shank in tobacco, and *P. sojae* causes root and stem rot of soybean. *Pythium debaryanum, P. ultimum,* and *P. aphanidermatum* cause damping-off of vegetables and fruit trees. *Pythium arrhenomanes, P. graminicola,* and *P. tardicrescens* cause root rot in cereals.

In the order Peronosporales, the family Albuginaceae contains *Albugo candida* (white rust of crucifers) and *Albugo ipomoeae-panduranae* (white rust of sweet potato). In the order Peronosporales, the family Peronosporaceae contains *Plasmopara viticola* (downy mildew of grapevine), *Peronospora parasitica* (downy mildew of crucifers), *Pseudoperonospora cubensis* (downy mildew of cucurbits), and *Bremia lactucae* (downy mildew of lettuce).

In the order Sclerosporales, the family Sclerosporaceae contains *Peronosclerospora sorghi* (downy mildew of corn and sorghum), *Sclerospora graminicola* (green ear and downy mildew of corn), *P. philippinensis* (Philippine downy mildew of corn), and *P. sacchari* (sugarcane downy mildew). In the order Sclerosporales, the family Verrucalvaceae contains *Sclerophthora macrospora* (crazy-top downy mildew of corn) and *S. rayssiae* var. *zeae* (brown-stripe downy mildew of corn). In the order Saprolegniales, the family Saprolegniaceae contains the crucifer black-root pathogen *Aphanomyces raphani* and the pea root-rot pathogen *A. euteiches.*

According to FungalWeb (n.d.) <http://www.fungalweb.com>, Oomycetes is recognized as a class by an international group of mycologists who met in Copenhagen in October 1999. It consists of the orders Leptomitales, Peronosporales, Pythiales, Rhipidales, Saprolegniales, and Sclerosporales. The orders Leptomitales and Rhipidales do not have any plant pathogens. The order Peronosporales consists of the families Peronosporaceae *(Peronospora, Plasmopara, Bremia)* and Albuginaceae *(Albugo).* The order Pythiales consists of the family Pythiaceae *(Phytophthora, Pythium),* the order Saprolegniales consists of the family Saprolegniaceae *(Aphanomyces),* and the order Sclerosporales consists of the families Sclerosporaceae *(Peronosclerospora, Sclerospora)* and Verrucalvaceae *(Sclerophthora).*

Structure of Oomycota

The vegetative body of a fungus is called the mycelium. The *mycelium* is the mass of hyphae that constitutes the thallus. Hyphae are filamentous structures made of thin, transparent, tubular walls filled or lined with a layer of protoplasm. Hyphae of Oomycota do not have septa and are called coenocytic or aseptate. Oomycota produce spores both asexually and sexually. Oomycota produce *sporangium,* a saclike structure, on the somatic hyphae. Entire contents of sporangium are converted into motile *zoospores*. Zoospores bear two dissimilar flagella, one with mastigonemes.

Sexual reproduction in Oomycota results in the union of two different nuclei. The process of sexual reproduction consists of three distinct phases:

1. *Plasmogamy*—The two protoplasts form a union, bringing their nuclei close together within the same cell.
2. *Karyogamy*—The two nuclei fuse.
3. *Meiosis*—Nuclear fusion is followed by meiosis (reduction division).

Meiosis reduces the number of chromosomes to haploid. The sex organs of fungi are called gametangia. The male gametangium is called *antheridium* and the female gametangium is called *oogonium*.

The oogonium contains a multinucleate oosphere surrounded by a layer of periplasm. The antheridium is small and club shaped. During reproduction, the antheridium comes into contact with the oogonium, develops a fertilization tube, and penetrates the oogonial wall and the periplasm. The male nucleus passes through the tube into the oosphere, unites with the female nucleus, and forms a zygote. The zygote is transformed into an oospore by the development of a thick cell wall. Oospores are resting spores that germinate by producing a germ tube. Germ tubes may terminate in a sporangium (as in the case of *Phytophthora* and *Plasmopara*) or the oospores may germinate producing a vesicle (as in the case of *Pythium* and *Albugo*). Zoospores are formed in the sporangium or in the vesicle by meiosis—the division of oospore contents into a large number of uninucleate sections. Each of the sections becomes a biflagellate zoospore.

In some Oomycota (as in the case of *Phytophthora infestans*), sexual reproduction can occur between the oogonium and antheridium from two different, sexually compatible mycelia (thalli). This condition is called *heterothallism*. In many other cases (as in the case of *Pythium*), sexual reproduction occurs between the oogonium and antheridium from the same thallus. This condition is called *homothallism*.

PROTOZOA

Fungal Phyla in Protozoa

The Kingdom Protozoa includes four fungal phyla: Acrasomycota, Dictyosteliomycota, Myxomycota, and Plasmodiophoromycota. Only Plasmodiophoromycota contains important crop pathogens. In this phylum, the order

Plasmodiophorales contains the family Plasmodiophoraceae, in which two important crop pathogens are included: *Plasmodiophora brassicae,* which causes clubroot in various cruciferous plants (including cabbage), and *Spongospora subterranea,* which causes powdery scab disease in potatoes.

Structure of Plasmodiophoromycota

The important character of Plasmodiophoromycota is the presence of plasmodium. *Plasmodium* is a naked (wall-less) motile mass of protoplasm with many nuclei that is bounded by a plasma membrane. *Plasmodium* moves and feeds in amoeboid fashion (i.e., it engulfs food and feeds by ingestion). Plasmodiophoromycota produce resting spores that germinate and produce zoospores. Zoospores swim in water droplets, encyst, penetrate the host tissue, and develop into multinucleate plasmodium. The plasmodium is cleaved into multinucleate portions. Each portion is surrounded by a membrane and develops into zoosporangium. Zoospores are released outside the host and can infect the host. Instead of developing into zoosporangia, some plasmodia produce resting spores that are formed by the formation and copulation of gametes. Plasmogamy, karyogamy, and meiosis may follow and resting spores are formed.

PHYLA IN THE KINGDOM FUNGI

The Kingdom Fungi encompasses large numbers of fungi. It consists of four phyla, 103 orders, 484 families, and 4,970 genera (Hawksworth et al., 1995). Several authors have tried to classify fungi. Barr (1992) and Hawksworth et al. (1995) divided Fungi into Ascomycota, Basidiomycota, Chytridiomycota, and Zygomycota. However, Barr (1992) called the Kingdom Fungi as Eumycota. It is now accepted by FungalWeb (n.d.) that the Kingdom Fungi should have four phyla (divisions): Ascomycota, Basidiomycota, Chytridiomycota, and Zygomycota. FungalWeb was created by an international group of mycologists who met in Copenhagen in October 1999. The Web site <http://www.fungalweb.com> provides a classification for all fungal genera based on a combination of phenotypic and genotypic data. Another group of fungi is called Mitosporic fungi. It was previously recognized as Deuteromycetes or Deuteromycotina, or Fungi Imperfecti. Mitosporic fungi are an artificial assemblage of fungi that have not been correlated with any meiotic states.

ASCOMYCOTA

Structure of Ascomycota

Ascomycota are either single-celled (yeasts), filamentous (hyphal), or both (dimorphic). Yeasts (order: Saccharomycetales) grow by budding or fission. Hyphae of other Ascomycota (Euascomycetes) grow apically and branch laterally. Yeasts are not important plant pathogens. Only *Galactomyces, Geotrichum, Saccharomyces,* and *Zygosaccharomyces* are known to cause some minor disorders in crop plants. The Euascomycetes are characterized by septate mycelium. The cross-walls that divide the hypha into cells are called septa. During certain stages of fungal development, the mycelium becomes organized into loosely or compactly woven tissues, as distinguished from the loose hyphae ordinarily composing a thallus. The loosely woven tissue in which component hyphae lie more or less parallel to one another is called *prosenchyma.* The compact woven fungal tissue consists of closely packed, more or less isodiametric or oval cells resembling the parenchyma cells of higher plants, and this type of fungal tissue is called *pseudoparenchyma.* Both prosenchyma and pseudoparenchyma compose various types of vegetative (somatic) and reproductive structures of Ascomycota.

Stroma is usually made up of prosenchyma, whereas *sclerotium* is made up of pseudoparenchymatous tissue. Both stromata and sclerotia are somatic structures of fungi. Stroma is a compact somatic structure that looks like a mattress. On or in the stroma, fructifications (the structures containing spores) are formed. Sclerotium is a hard and compact vegetative resting structure that is resistant to unfavorable conditions. Fungi may overwinter in the form of sclerotia. The mycelium of some fungi forms thick strands. In such strands, the hyphae lose their individuality and form complex tissues. The strands are called *rhizomorph.* The strands have a thick, hard cortex and a growing tip that looks like a root tip.

Asexual Reproduction in Ascomycota

Ascomycota reproduce both sexually and asexually. Asexual reproduction is repeated several times in the life cycle of Ascomycota, whereas sexual reproduction occurs only once per life cycle. Different types of asexual reproduction have been reported in Ascomycota. The fungi may multiply by fragmentation of hyphae. By this method, the hyphae break up into their component cells, which behave like spores. These spores are called *oidia* or *arthrospores.* When the cells become enveloped in a thick wall before they separate from one another or from other hyphal cells adjoining them, they are called chlamydospores.

Yeasts reproduce by fission or budding. During fission, the cell is split into daughter cells by constriction and formation of a cell wall. In the budding process, a small outgrowth (bud) is produced from a parent cell. The nucleus of the parent cell divides and one daughter nucleus migrates into the bud. The bud increases in size and ultimately breaks off, forming a new individual.

Most Ascomycota reproduce by producing *conidia*. Conidia are produced at the tips or sides of hyphae. The specialized hypha, which bears conidia, is called a *conidiophore*. Conidia are generally borne on conidiophores, which may be produced loosely and indiscriminately by the somatic hyphae or grouped in various types of asexual fruiting bodies. Conidiophores may be simple or branched. They may look like somatic hyphae or they may be provided with *sterigmata* (small hyphal branches that support the conidia) or specialized branches on which they bear conidia.

Conidiophores may be organized into definite fruiting bodies in the case of some fungi. The various fruiting bodies reported in Ascomycota include:

1. *Pycnidium*—A hollow, globose, or flask-shaped structure whose pseudoparenchymatous walls are lined with conidiophores.
2. *Acervulus*—An aggregation (mat) of hyphae (pseudoparenchyma) that is subcuticular, epidermal, or deeper in origin and never entirely superficial. From the mat, short, closely-packed conidiophores arise, forming a bedlike mass.
3. *Sporodochium*—A cushion-shaped stroma covered with short conidiophores that are cemented together. The spore mass is supported by these conidiophores.
4. *Pionnote sporodochium*—A minute sporodochium near the surface of the substratum having no stroma. The spores form a continuous slimy layer.
5. *Synnema*—A more or less compacted group of erect and sometimes fused conidiophores bearing conidia at the apex only or on both the apex and sides. In synnema, conidiophores may be cemented together to form the elongated, spore-bearing structure. This structure may be split in different ways near the apex, sometimes resembling a feather duster.

Sexual Reproduction in Ascomycota

In most species in Ascomycota, the sex organs are ascogonium (the female gametangium) and antheridium (the male gametangium). The male nucleus passes from the antheridium into the ascogonium through a pore developed at the point of contact between the two gametangia. The ascogonium contains a slender structure called the trichogyne, which receives the male nucleus. Male nuclei enter the ascogonium and, in many cases, pair *with* ascogonial nucleus (plasmogamy). Fusion of nuclei (karyokamy) does not take place immediately. The ascogonium produces a number of papillae. The nuclei from the ascogonium begin to pass into these papillae one by one. The papillae elongate into ascogenous hyphae. In this ascogenous hyphae, a leading pair of nuclei appear, followed by a second pair.

The nuclei in the ascogenous hyphae and those still in the ascogonium undergo simultaneous mitosis. Septa are formed and the tip cell of the ascogenous hypha becomes uninucleate. Several other cells become binucleate. One nucleus in each binucleate cell of the ascogenous hyphae is antheridial in origin while the other is ascogonial. One of the binucleate cells of the ascogenous hypha elongates and bends over to form a *crozier* (hook). The two nuclei in this hooked cell divide in such a way that their spindles are oriented more or less vertically and parallel to each other, so that two of the daughter nuclei—one from each spindle and, therefore, of different origin—are close to each other at the end of the hook, while one of the other two nuclei is located at the tip and one near the basal septum of the hook. Two septa form, separating the hook into three cells. The tip and basal cells are uninucleate, one containing an antheridial nucleus and one an ascogonial nucleus. The crook (hooked) cell is binucleate. The crook cell becomes the *ascus* and is called the *ascus mother cell*.

Karyokamy takes place in the ascus mother cell soon after the septa are formed in the hook. The young ascus with its diploid zygote nucleus begins to elongate. The zygote nucleus soon undergoes meiosis, resulting in four haploid nuclei. The nuclei divide mitotically to form eight nuclei. Each nucleus is enveloped by a wall and eight ascospores are formed. Each ascogenous hypha branches and rebranches in various ways and produces a cluster of asci. The saclike structure, which contains the ascospores, is called the *asci*.

In most of Ascomycota, the asci are elongated and either club shaped or cylindrical. The ascus has a single cavity in which the ascospores are formed. Asci may be stalked or sessile. A definite layer of asci, whether naked or enclosed in a fruiting body, is called the *hymenium*. Sterile, elongated hairs called *paraphyses* are found between the asci in the hymenium. Asci are categorized based on the structure of their walls—*unitunicate* or *bitunicate*. The wall of a unitunicate ascus consists of two thin layers. In the bitunicate ascus, two distinct wall layers exist: a rigid outer and an extensible inner wall.

With a few exceptions, Ascomycota produce their asci in fruiting bodies called *ascocarps*. A few Ascomycota produce naked asci without any fruiting body. Different kinds of ascocarps have been reported:

1. *Cleistothecium*—A completely closed ascocarp
2. *Perithecium*—A more or less closed ascocarp. At maturity, the perithecium is provided with a pore through which the ascospores escape
3. *Apothecium*—An open ascocarp
4. *Pseudothecium*—An ascostromatic ascocarp in which asci are formed in numerous unwalled locules (cavities) within the stroma. The stroma itself forms the wall of the ascocarp.

In some Ascomycota, no antheridia are formed. However, nuclei reach ascogonia by means of spermatia, microconidia, or conidia. *Spermatia* are minute, sphaerical or rod-shaped, uninucleate, male sex cells that become attached to the receptive organs (trichogynes or somatic hyphae) and empty their contents into them. Spermatia are detached from the *spermatiophore* (parent hyphae) and are carried by wind, water, or insects to receptive organs. *Microconidia* are minute conidia that behave as spermatia, but are capable of germinating and giving rise to mycelium. Conidia and oidia may also function as spermatia. In some Ascomycota, fusion of somatic hyphae of two compatible mycelia takes place, and the nuclei migrate to the ascogonia through septal perforations.

Classification of Ascomycota

There is considerable variation between the classification systems proposed by several authors (Hawksworth et al., 1995). There is no consensus about categories above order in the classification. According to Hawksworth et al. (1995), the following 46 orders have been recognized in Ascomycota: Arthoniales, Caliciales, Calosphaeriales, Coryneliales, Cyttariales, Diaporthales, Diatrypales, Dothideales, Elaphomycetales, Erysiphales, Eurotiales, Gyalectales, Halosphaeriales, Hypocreales, Laboulbeniales, Lahmiales, Lecanorales, Leotiales, Lichinales, Medeolariales, Meliolales, Microascales, Neolectales, Onygenales, Ophiostomatales, Ostropales, Patellariales, Peltigerales, Pertusariales, Pezizales, Phyllachorales, Pneumocystidales, Protomycetales, Pyrenulales, Rhytismatales, Saccharomycetales, Schizosaccharomycetales, Sordariales, Spathulosporales, Taphrinales, Teloschistales, Triblidiales, Trichosphaeriales, Trichotheliales, Verrucariales, and Xylariales. Among these orders, only a few contain crop pathogens. Important crop pathogens belong to the following orders:

Order Diaporthales, Family Valsaceae: *Diaporthe, Gnomonia, Endothia, Phoma, Valsa*

Order Dothideales, Family Capnodiaceae: *Capnodium*

Order Dothideales, Family Elsinoaceae: *Elsinoe*

Order Dothideales, Family Leptosphaeriaceae: *Leptosphaeria, Ophiobolus*

Order Dothideales, Family Lophiostomataceae: *Trichometasphaeria*

Order Dothideales, Family Mycosphaerellaceae: *Mycosphaerella, Guignardia, Sphaerulina*

Order Dothideales, Family Myriangiaceae: *Myriangium*

Order Dothideales, Family Phaeosphaeriaceae: *Phaeosphaeria*

Order Dothideales, Family Pleosporaceae: *Pleospora, Pyrenophora, Cochliobolus, Leptosphaerulina, Setosphaeria*

Order Dothideales, Family Venturiaceae: *Venturia*

Order Dothideales, Genera Incertae Sedis: *Didymella*

Order Erysiphales, Family Erysiphaceae: *Blumeria, Sphaerotheca, Erysiphe, Uncinula, Leveillula, Phyllactinia, Podosphaera*

Order Eurotiales, Family Trichocomaceae: *Emericella, Eurotium, Talaromyces*

Order Hypocreales, Family Clavicipitaceae: *Balansia, Claviceps*

Order Hypocreales, Family Hypocreaceae: *Nectria, Calonectria, Gibberella*

Order Microascales, Familia Incertae Sedis: *Ceratocystis*

Order Microascales, Family Microascaceae: *Microascus*

Order Ophiostomatales, Family Ophiostomataceae: *Ophiostoma*

Order Phyllachorales, Family Phyllachoraceae: *Phyllachora, Glomerella, Polystigma*

Order Protomycetales, Family Protomycetaceae: *Protomyces*

Order Sordariales, Family Ceratostomataceae: *Arxiomyces* (= *Ceratostoma*), *Scopinella* (= *Ophiostomella*)

Order Taphrinales, Family Taphrinaceae: *Taphrina*

Order Xylariales, Family Xylariaceae: *Hypoxylon, Xylaria, Ustulina, Rosellinia*

Ascomycota, Familia Incertae Sedis:

Familia Incertae Sedis—Hyponectriaceae: *Physalospora, Hyponectria, Monographella*

Familia Incertae Sedis—Magnaporthaceae: *Gaeumannomyces, Magnaporthe*

Familia Incertae Sedis—Phlyctidaceae: *Phlyctis*

Ascomycota, Familia Incertae Sedis, Genera Incertae Sedis: *Leptosphaerella*

According to FungalWeb, the phylum Ascomycota consists of three subphyla (Pezizomycotina, Saccharomycotina, and Taphrinomycotina), several uncertain orders (Ascomycota order uncertain), uncertain families (Ascomycota families uncertain), and uncertain genera (Ascomycota genera uncertain). The subphylum Pezizomycotina consists of several classes such as Arthoniomycetes, Chaetothyriomycetes, Dothideomycetes, Dothideomycetes uncertain, Eurotiomycetes, Lecanoromycetes, Lecanoromycetes uncertain, Leotiomycetes, Pezizomycetes, Sordariomycetes, and Dothidiomycetes et Chaetothyriomycetes uncertain. Several Sordariomycetes orders uncertain, Sordariomycetes families uncertain, and Sordariomycetes genera uncertain are also included under Pezizomycotina. The class Dothideomycetes contains three orders, Dothideales, Patellariales, and Pleosporales, and the important families in these orders containing crop pathogens are given here:

> Order Dothideales, Family Dothideaceae: *Dothidia*
> Order Pleosporales, Family Leptosphaeriaceae: *Leptosphaeria, Ophiobolus*
> Order Pleosporales, Family Lophiostomataceae: *Trichometasphaeria*
> Order Pleosporales, Family Pleosporaceae: *Cochliobolus, Pseudocochliobolus, Pleospora, Macrospora, Pyrenophora, Leptosphaerulina*

The class Dothideomycetes et Chaetothyriomycetes uncertain contains some important pathogens:

> Family Botryosphaeriaceae: *Botyosphaeria*
> Family Elsinoaceae: *Elsinoe*
> Family Mycosphaerellaceae: *Mycosphaerella, Guignardia*
> Family Venturiaceae: *Venturia*

Dothideomycetes uncertain includes the crop pathogens *Didymella* and *Diapleela*.

The class Leotiomycetes consists of the following important powdery mildew fungi and other important pathogens:

> Order Erysiphales, Family Erysiphaceae: *Erysiphe, Uncinula, Sphaerotheca, Blumeria, Leveillula, Phyllactinia, Podosphaera, Microsphaera* (powdery mildew fungi)
> Order Leotiales, Family Dermateaceae: *Diplocarpon, Tapesia*
> Order Leotiales, Family Sclerotiniaceae: *Botryotinia, Sclerotinia*

The class Sordariomycetes contains many crop pathogens as follows:

Order Diaporthales, Family Melanconidaceae: *Schizoparme, Sydowiella*

Order Diaporthales, Family Valsaceae: *Diaporthe, Endothia, Gnomonia, Valsa*

Order Diaporthales Uncertain: *Cryptonectria*

Order Hypocreales, Family Bionectriaceae: *Nectriella*

Order Hypocreales, Family Clavicipitaceae: *Balansia, Claviceps*

Order Hypocreales, Family Nectriaceae: *Nectria, Gibberella, Calonectria*

Order Microascales, Family Microascaceae Uncertain: *Ceratocystis*

Order Ophiostomatales, Family Ophiostomataceae: *Ophiostoma*

Order Sordariales, Family Ceratostomataceae: *Ceratostoma*

Order Sordariales Uncertain: *Monosporascus*

Order Xylariales, Family Clypeosphaeriaceae: *Ceratostomella*

Order Xylariales, Family Diatrypaceae: *Eutypa*

Order Xylariales, Family Hypoxylon: *Rosellinia, Xylaria, Hypoxylon*

Class Sordariomycetes Orders Uncertain, Order Meliolales, Family Meliolaceae: *Meliola*

Class Sordariomycetes Orders Uncertain, Order Phyllachorales, Family Phyllachoraceae: *Phyllachora*

Order Sordariomycetes Families Uncertain, Family Hyponectriaceae: *Hyponectria, Physalospora, Monographella*

Order Sordariomycetes Families Uncertain, Family Magnaporthaceae: *Magnaporthe, Gaeumannomyces*

Order Sordariomycetes Genera Uncertain: *Glomerella*

BASIDIOMYCOTA

Structure and Reproduction in Basidiomycota

The phylum Basidiomycota consists of septate mycelium. The mycelium of most Basidiomycota passes through three distinct stages of development, the primary, the secondary, and the tertiary, before the fungus completes its life cycle. The *basidiospore* germinates and grows into the primary mycelium, which is multinucleate in the initial stage. Septa develop and the primary mycelium becomes septate and uninucleate.

The secondary mycelium originates from the primary mycelium. Its cells are binucleate. Protoplasts of two uninucleate cells fuse without actual karyogamy and binucleate cells develop (plasmogamy). The binucleate cell thus formed produces a branch into which the nuclear pair migrates. The two nuclei divide conjugately and the sister nuclei separate into two daughter cells, thus initiating the binucleate mycelium. In some fungi, the daughter cell formation occurs through clamp connection. *Clamp connection* is a bridgelike connection in the secondary mycelium. The binucleate mycelium gives rise to the basidium in which karyogamy and meiosis occur. It is similar to the ascogenous hyphae from which asci arise.

The tertiary mycelium is an organized tissue that composes the *sporophores*. The cells of the tertiary mycelium are binucleate, with the sporophores actually originating when the secondary mycelium forms complex tissues.

The basidium, dikaryotic mycelium, and the formation of clamp connections are characteristics of Basidiomycota. In addition, septa of secondary mycelium of some Basidiomycota are *dolipore septa*. This septum flares up in the middle portion of the hypha, forming a barrel-shaped structure with open ends.

The sexual structure of Basidiomycota is the *basidium*. The basidium originates as a terminal cell of a binucleate hypha and is separated from the rest of the hypha by a septum over which a clamp connection is generally found. The basidium soon enlarges and becomes broader. The two nuclei within the young basidium fuse (karyogamy). The zygote nucleus soon undergoes meiosis, giving rise to four haploid nuclei. Four sterigmata arise at the top of the basidium and their tips enlarge, forming the basidiospore initials. The four nuclei move through the sterigmata into the young basidiospores. Uninucleate basidiospores develop subsequently. Some Basidiomycota produce their basidia in highly organized fruiting bodies that are called *basidiocarps*. However, the fungi causing rust and smut diseases do not form any basidiocarps.

Asexual reproduction in the Basidiomycota takes place by means of budding, by fragmentation of the mycelium, and by the production of conidia, arthrospores, or oidia. The rusts produce urediniospores, which are conidial in origin and function. Smuts produce chlamydospores.

Classification of Basidiomycota

Hawksworth et al. (1995) recognized three classes in the phylum Basidiomycota: Basidiomycetes, Teliomycetes, and Ustomycetes. The class Basidiomycetes consists of 32 orders. The important orders, which consist of crop

pathogens, are Agaricales, Atractiellales, Boletales, Cantharellales, Ceratobasidiales, Ganodermatales, Hymenochaetales, Lachnocladiales, Poriales,
Schizophyllales, and Stereales. The important plant-pathogenic genera included in these orders are given here:

Order Agaricales, Family Coprinaceae—*Coprinus*
Order Agaricales, Family Strophariaceae—*Psilocybe*
Order Agaricales, Family Tricholomataceae—*Armillaria,*
    Armillariella, Baeospora, Clitocybe, Collybia, Crinipellis,
    Flammulina, Marasmiellus, Marasmius, Mycena
Order Atractiellales, Family Chionosphaeraceae—*Stilbum*
Order Boletales, Family Coniophoraceae—*Serpula*
Order Cantharellales, Family Typhulaceae—*Typhula*
Order Ceratobasidiales, Family Ceratobasidiaceae—
    Ceratobasidium, Oncobasidium, Thanatephorus
Order Ganodermatales, Family Ganodermateaceae—*Ganoderma*
Order Hymenochaetales, Family Hymenochaetaceae—*Fomitiporia,*
    Hymenochaete, Phellinus
Order Lachnocladiales, Family Lachnocladiaceae—*Scytinostroma*
Order Poriales, Family Coriolaceae—*Coriolus, Fomes, Fomitella,*
    Fomitopsis, Gloeophyllum, Heterobasidion, Hexagonia,
    Hirschioporus, Laetiporus, Leptoporus, Oxyporus, Pycnoporus,
    Perenniporia, Phaeolus, Pseudophaeolus, Pycnoporus,
    Rigidoporus, Trichaptum, Trametes
Order Schizophyllales, Family Schizophyllaceae—*Schizophyllum,*
    Solenia
Order Stereales, Family Aleurodiscaceae—*Aleurodiscus*
Order Stereales, Family Atheliaceae—*Athelia, Butlerelfia*
Order Stereales, Family Botryobasidiaceae—*Waitea*
Order Stereales, Family Corticiaceae—*Corticium, Hypochnus,*
    Pellicularia
Order Stereales, Family Peniophoraceae—*Peniophora*
Order Stereales, Family Sistrotremataceae—*Trechispora*
Order Stereales, Family Steccherinaceae—*Steccherinum, Irpex*
Order Stereales, Family Stereaceae—*Stereum*

In the class Teliomycetes, many rust fungi are included. The important
pathogens in this class include:

Order Septobasidiales, Family Septobasidiaceae—*Septobasidium,*
    Uredinella

Order Uredinales, Family Coleosporiaceae—*Coleosporium*
Order Uredinales, Family Cronartiaceae—*Cronartium*
Order Uredinales, Family Melampsoraceae—*Melampsora*
Order Uredinales, Family Mikronegeriaceae—*Mikronegeria*
Order Uredinales, Family Phakopsoraceae—*Phakopsora,*
Cerotelium, Dasturella, Monosporidium, Phragmidiella,
Uredopeltis, Uredostilbe, Angiopsora
Order Uredinales, Family Phragmidiaceae—*Phragmidium,*
Trachyspora, Frommea
Order Uredinales, Family Pileolariaceae—*Uromycladium*
Order Uredinales, Family Pucciniaceae—*Puccinia, Uromyces,*
Gymnosporangium, Chrysella, Chrysopsora, Peridiopsora,
Marvalia, Stereostratum
Order Uredinales, Family Pucciniastraceae—*Pucciniastrum,*
Hyalopsora, Melampsorella, Melampsoridium, Uredinopsis,
Melampsora
Order Uredinales, Family Pucciniosiraceae—*Pucciniosira,*
Trichopsora
Order Uredinales, Family Raveneliaceae—*Ravenelia, Cystomyces,*
Haploravenelia
Order Uredinales, Family Sphaerophragmiaceae—
Sphaerophragmium
Order Uredinales, Family Uropyxidaceae—*Didymopsorella,*
Tranzschelia, Uropyxis
Order Uredinales, Familia Incertae Sedis—*Aecidium, Blastospora,*
Cystopsora, Uredo, Hemileia, Physopella

The class Ustomycetes consists of the orders Exobasidiales, Graphiolales, Platygloeales, Sporidiales, and Ustilaginales, each of which contain crop pathogens. The important crop pathogens in these orders include the following:

Order Exobasidiales, Family Exobasidiaceae—*Exobasidium*
Order Graphiolales, Family Graphiolaceae—*Graphiola,*
Trichodesmium
Order Platygloeales, Family Platygloeacea—*Cystobasidium,*
Helicobasidium, Platycarpa
Order Sporidiales, Family Sporidiobolaceae—*Leucosporidium*
Order Ustilaginales, Family Tilletiaceae—*Entyloma, Burrillia,*
Neovossia, Urocystis, Tilletia, Tilletiella
Order Ustilaginales, Family Ustilaginaceae—*Ustilago, Planetella,*
Sorosporium, Sporisorium, Sphacelotheca, Tolyposporidium,
Tranzscheliella, Ustacystis, Pericladium

According to FungalWeb (n.d.), the phylum Basidiomycota consists of three classes (Hymenomycetes, Teliomycetes, and Ustilaginomycetes) and several families uncertain and genera uncertain. The class Hymenomycetes contains the orders Agaricales, Polyporales, Septobasidiales s.l. (s.l. stands for *sensu lato,* "in the broad sense"), Tremellales, Tulasnellales, and Thelephorales. The important pathogens in this class are:

Order Agaricales, family Marasmiaceae—*Armillaria*
Order Polyporales, Family Fomitaceae—*Fomes*
Order Polyporales, Family Ganodermataceae—*Ganoderma*
Order Septobasidiales s.l., Family Cystobasidiaceae—
    Cystobasidium
Order Tulasnellales, Family Ceratobasidiaceae—*Ceratobasidium*

In the class Teliomycetes, three orders consist of important crop pathogens:

Order Pucciniales, Family Coleosporiaceae—*Coleosporium*
Order Pucciniales, Family Cronartiaceae—*Cronartium*
Order Pucciniales, Family Melamsporaceae—*Melampsora*
Order Pucciniales, Family Phakopsoraceae—*Cerotelium,*
    Phakopsora
Order Pucciniales, Family Phragmidiaceae—*Phragmidium*
Order Pucciniales, Family Pucciniaceae—*Puccinia, Uromyces,*
    Gymnosporangium
Order Pucciniales, Family Pucciniastraceae—*Melampsorella,*
    Melampsoridium
Order Pucciniales, Family Raveneliaceae—*Ravenelia*
Order Pucciniales, Family Sphaerophragmiaceae—
    Sphaerophragmium
Order Pucciniales, Family Uropyxidaceae—*Tranzschelia*
Order Septobasidiales s.str. (s.str. stands for *sensu stricto* = "in the
    strict sense"), Family Septobasidiaceae—*Septobasidium,*
    Uredinella

The class Ustilaginomycetes consists of the orders Entylomatales, Exobasidiales, Tilletiales, Urocystales, Ustilaginales, and Microstromatales, each of which include crop pathogens. The important crop pathogens in this class are provided here:

Order Entylomatales, Family Entylomataceae—*Entyloma*
Order Exobasidiales, Family Exobasidiaceae—*Exobasidium,*
    Graphiola

Order Tilletiales, Family Tilletiaceae—*Tilletia, Neovossia*
Order Urocystales, Family Urocystaceae—*Urocystis*
Order Ustilaginales, Family Ustilaginaceae—*Ustilago,
    Tolyposporium*

In the Basidiomycota families uncertain, one important crop pathogen, *Corticium* is included in the family Corticiaceae. In the Basidiomycota genera uncertain, the crop pathogen *Hemileia* is included.

CHYTRIDIOMYCOTA

The thallus of Chytridiomycota can be coenocytic, holocarpic (i.e., the thallus is entirely converted into one or more reproductive structures), eucarpic (i.e., reproductive structures form on certain portions of the thallus, which continues to perform its somatic functions), monocentric (i.e., the thallus radiates from a single point at which a reproductive organ such as the sporangium is formed), polycentric (i.e., the thallus radiates from many centers at which reproductive organs are formed), or mycelial. At least in hyphal stages, the cell wall consists of chitin. The zoospores are monoflagellate, lacking mastigonemes or scales. An important crop pathogen belonging to the phylum Chytridiomycota is *Synchytrium endobioticum*, which causes wart disease in potato. It possesses uniflagellate zoospores. Zoospores swim in soil moisture and penetrate the epidermal cells of host tissues, shedding their flagellum outside. The unicellular fungus grows in size and the nucleus becomes greatly enlarged. After reaching a certain size, it secretes a thick, golden-brown wall. The well-developed structure is called *prosorus*. The prosorus germinates within the host cell. The cell wall of prosorus ruptures and the proplast is surrounded by a thin hyaline membrane. Repeated mitotic divisions of the nucleus take place, forming about 32 nuclei. Thin hyaline walls are formed in such a way as to divide the prosorus into four to nine multinucleate segments. Nuclear division continues and about 300 nuclei are formed. Each prosorus portion develops into a sporangium or a gametangium, depending on environment. The mass of sporangia is called the *sorus*. If water is abundantly present, zoospores are formed. If drought sets in, the motile cells that are released become *planogametes* (motile gametes) instead of zoospores. The planogametes copulate in pairs (fusion of naked gametes takes place) resulting in the formation of biflagellate zygotes. The zygotes swim in soil moisture until they penetrate epidermal host cells. Inside each host cell the zygote elongates and a heavy wall is formed around it. It is now called *resting sporangium*. From the resting sporangium zoospores are formed.

The phylum Chytridiomycota consists of five orders: Blastocladiales, Chytridiales, Monoblepharidales, Neocallimastigales, and Spizellomycetales (Hawksworth et al., 1995). The plant pathogens of Chytridiomycota include:

> Order Blastocladiales, Family Physodermateaceae—*Physoderma*
> Order Chytridiales, Family Chytridiaceae—*Phlyctidium*
> Order Chytridiales, Family Synchytriaceae—*Synchytrium*
> Order Spizellomycetales, Family Olpidiaceae—*Olpidium*
> Order Spizellomycetales, Family Spizellomycetaceae—
>   *Rhizophlyctis*
> Order Spizellomycetales, Family Urophlyctidaceae—*Urophlyctis*

ZYGOMYCOTA

The mycelium of Zygomycota is *coenocytic* (aseptate). The important character of this phylum is the production of *zygospores*. A zygospore typically results from the fusion of two gamentangia. Zygospores differ from the oospores of Oomycota in that the latter are derived from an oosphere. Asexual reproduction in this phylum is by means of nonmotile *(aplanate)* spores. Zygomycota includes only a few weak pathogens such as *Rhizopus* and *Mucor.*

The mycelium of *Rhizopus* develops rhizoids (a short branch of a thallus that looks like a root) at certain points of the mycelium. Directly above the rhizoids, sporangiophores are produced. The top of the sporangiophores becomes swollen and a sporangium develops. The central portion of the sporangium is a columella (a sterile structure that is an extension of the stalk). The peripheral zone is the spore-bearing portion of the sporangium. Sporangiospores (*aplanospores* = nonmotile spores) are produced inside the sporangium. *Rhizopus* is *heterothallic* (i.e., two different thalli are required for sexual reproduction). When two opposite strains (+ and -) come in contact with each another, *progamentangia* (before gamentagium) are formed. A septum forms near the tip of each progamentagium, separating into two cells (a terminal gamentagium and a *suspensor cell*). The walls of the two contacting gamentagia dissolve at the point of contact and the two protoplasts mix. Fusion of nuclei takes place. The cell wall of the fused gametangia thickens and the resulting structure is called a zygospore. A sporangiophore arises and develops into a sporangium, which is called a *zoosporangium*. Aplanospores develop in the zoosporangium and are released into the ambient water droplets (Vidhyasekaran, 1993).

The phylum Zygomycota consists of two classes, Trichomycetes and Zygomycetes (Hawksworth et al., 1995). Zygomycetes contains plant pathogens and consists of seven orders: Dimargaritales, Endogonales, Entomophthorales, Glomales, Kickxellales, Mucorales, and Zoopagales. Mucorales contains crop pathogens, which include:

Order: Mucorales, Family Mucoraceae—*Rhizopus, Mucor*
Order: Mucorales, Family Cunninghamellaceae—*Cunninghamella*

According to the FungalWeb (n.d.), the order Mucorales does not contain families.

MITOSPORIC FUNGI

Mitosporic fungi is an artificial assemblage of fungi in which sexually produced spores such as ascospores or basidiospores are absent or presumably absent. Mitosporic fungi produce conidia that are formed by mitosis. Mitosporic fungi appears to lack a teleomorph (sexual stage [form]) and has only an anamorph (asexual stage [form]). Mitosporic fungi also include fungi in which both meiotic (sexual) and mitotic (asexual) reproductive structures are absent. Mitosporic fungi cannot be assigned to families in the accepted phyla. However, some of these fungi have been found correlated with teleomorphs in the Ascomycota and Basidiomycota. Such fungi have been placed in families of the corresponding Ascomycota and Basidiomycota phyla as Anamorphic families. For example, the asexual spores (conidia) produced by *Botrytis cinerea* (a Mitosporic fungi which lacks a sexual stage) are similar to the asexual spores produced by *Botryotinia fuckeliana* (a fungus with both sexual and asexual stages, belonging to the family Sclerotiniaceae, class Ascomycota). As such, *Botrytis cinerea* is classified as Anamorphic Sclerotiniaceae. Numerous crop pathogens belong to Mitosporic fungi. Some important pathogens in this group are listed here:

Anamorphic Amphisphaeriaceae—*Pestalotiopsis*
Anamorphic Botryosphaeriaceae—*Diplodia, Fusicoccum*
Anamorphic Capnodiaceae—*Tripospermum*
Anamorphic Clavicipitaceae—*Ephelis*
Anamorphic Corticiaceae—*Rhizoctonia*
Anamorphic Dermateaceae—*Cylindrosporium*
Anamorphic Dothideaceae—*Septoria*
Anamorphic Dothideales—*Ascochyta*
Anamorphic Elsinoaceae—*Sphaceloma*
Anamorphic Erysiphaceae—*Ovulariopsis*

Anamorphic Hypocreaceae—*Fusarium*
Anamorphic Magnaporthaceae—*Pyricularia*
Anamorphic Melanconidaceae—*Melanconium*
Anamorphic Mycosphaerellaceae—*Cercospora, Phyllosticta*
Anamorphic Phyllachoraceae—*Colletotrichum*
Anamorphic Pleosporaceae—*Alternaria, Drechslera, Phoma,*
    Stemphylium, Exserohilum
Anamorphic Sclerotiniaceae—*Botrytis*
Anamorphic Sistrotremataceae—*Phymatotrichopsis*
Anamorphic Trichocomaceae—*Aspergillus, Penicillium*
Anamorphic Uredinales—*Aecidium, Uredo*
Anamorphic Valsaceae—*Phoma, Phomopsis*
Anamorphic Venturiaceae—*Fusicladium*
Anamorphic Xylariaceae—*Dematophora*
Mitosporic Fungi—*Botryodiplodia, Cephalotrichum,*
    Cercosporidium, Chalara, Cordana, Corynespora, Fulvia,
    Gloeosporium, Helminthosporium, Lasiodiplodia,
    Macrophomina, Myrothecium, Pestalotia, Phaeoisariopsis,
    Phyllostictella, Phymatotrichum, Rhynchosporium, Sarocladium,
    Sclerotium, Thielaviopsis, Trichoconis, Verticillium

SYSTEMATIC POSITIONS OF CROP FUNGAL PATHOGENS (INCLUDING CHROMISTA AND PROTOZOA)

The systematic positions of all crop fungal pathogens are given here following the classification of Hawksworth et al. (1995) and Wrobel and Creber (1998). The positions of these fungi according to FungalWeb are given within parentheses. The names of genera of fungi are also provided. The names of species of crop pathogens with authority and the names of diseases caused by them are provided in Chapter 30.

Acanthorhynchus—Hyponectriaceae, Familia Incertae Sedis, Ascomycota
    (Hyponectriaceae, Sordariomycetes Families Uncertain, Ascomycota)
Achlya—Saprolegniaceae, Saprolegniales, Oomycota, Chromista
    (Saprolegniaceae, Saprolegniales, Oomycetes, Chromista)
Acremonium—Mitosporic fungi
Acrocalymma—Anamorphic Lophiostomataceae
Acrocylindrium—Mitosporic fungi
Acrodontium—Mitosporic fungi
Acrophialophora—Mitosporic fungi
Actinonema—Mitosporic fungi

Aecidium—Anamorphic Uredinales (Basidiomycota Genera Uncertain, Basidiomycota Families Uncertain, Basidiomycota)

Akaropeltopsis—Micropeltidaceae, Dothideales, Ascomycota (Micropeltidaceae, Dothideomycetes et Chaetothyriomycetes Uncertain, Ascomycota)

Albugo—Albuginaceae, Peronosporales, Oomycota, Chromista (Albuginaceae, Peronosporales, Oomycetes, Chromista)

Aleurodiscus—Aleurodiscaceae, Stereales, Basidiomycetes, Basidiomycota (Aleurodiscaceae, Hericiales, Hymenomycetes, Basidiomycota)

Allantophomopsis—Mitosporic fungi

Alternaria—Anamorphic Pleosporaceae

Angiopsora—Phakopsoraceae, Uredinales, Teliomycetes, Basidiomycota (Phakopsoraceae, Pucciniales, Teliomycetes, Basidiomycota)

Antennularia—Venturiaceae, Dothideales, Ascomycota (Venturiaceae, Dothideomycetes et Chaetothyriomycetes Uncertain, Ascomycota)

Aphanomyces—Saprolegniaceae, Saprolegniales, Oomycota, Chromista (Saprolegniaceae, Saprolegniales, Oomycetes, Chromista)

Apiognomonia—Valsaceae, Diaporthales, Ascomycota (Valsaceae, Diaporthales, Sordariomycetidae, Sordariomycetes, Ascomycota)

Apiospora—Lasiosphaeriaceae, Sordariales, Ascomycota (Apiosporaceae, Sordariomycetes Families Uncertain, Ascomycota)

Apiosporina—Venturiaceae, Dothideales, Ascomycota (Venturiaceae, Dothideomycetes et Chaetothyriomycetes Uncertain, Ascomycota)

Apostrasseria—Anamorphic Phacidiaceae

Armillaria—Tricholomataceae, Agaricales, Basidiomycetes, Basidiomycota (Marasmiaceae, Agaricales, Hymenomycetes, Basidiomycota)

Armillariella—Tricholomataceae, Agaricales, Basidiomycetes, Basidiomycota (Marasmiaceae, Agaricales, Hymenomycetes, Basidiomycota)

Arthonia—Arthoniaceae, Arthoniales, Ascomycota (Arthoniaceae, Arthoniales, Arthoniomycetes, Pezizomycotina, Ascomycota)

Arthrinium—Anamorphic Lasiosphaeriaceae

Ascochyta—Anamorphic Dothideales, Mycosphaerellaceae, Leptosphaeriaceae

Aspergillus—Anamorphic Trichocomaceae

Asperisporium—Mitosporic fungi

Asteridiella—Meliolaceae, Meliolales, Ascomycota (Meliolaceae, Meliolales, Sordariomycetes Orders Uncertain, Ascomycota)

Asteromella—Anamorphic Mycosphaerellaceae

Athelia—Atheliaceae, Stereales, Basidiomycetes, Basidiomycota
(Atheliaceae, Basidiomycota Families Uncertain, Basidiomycota)

Aureobasidium—Mitosporic fungi

Baeospora—Tricholomataceae, Agaricales, Basidiomycetes,
Basidiomycota (Tricholomataceae, Agaricales, Hymenomycetes,
Basidiomycota)

Balansia—Clavicipitaceae, Hypocreales, Ascomycota (Clavicipitaceae,
Hypocreales, Hypocreomycetidae, Sordariomycetes, Ascomycota)

Beniowskia—Mitosporic fungi

Bipolaris—Anamorphic Pleosporaceae

Biscogniauxia—Xylariaceae, Xylariales, Ascomycota (Xylariaceae,
Xylariales, Xylariomycetidae, Sordariomycetes, Ascomycota)

Bitrimonospora—Sordariales, Ascomycota (Sordariales,
Sordariomycetidae, Sordariomycetes, Ascomycota)

Blumeria—Erysiphaceae, Erysiphales, Ascomycota (Erysiphaceae,
Erysiphales, Leotiomycetes, Ascomycota)

Blumeriella—Dermateaceae, Leotiales, Ascomycota (Dermateaceae,
Leotiales, Leotiomycetes, Ascomycota)

Botryodiplodia—Mitosporic fungi

Botryosphaeria—Botryosphaeriaceae, Dothideales, Ascomycota
(Botryosphaeriaceae, Dothideomycetes et Chaetothyriomycetes Uncertain, Ascomycota)

Botryotinia—Sclerotiniaceae, Leotiales, Ascomycota (Sclerotiniaceae,
Leotiales, Leotiomycetes, Ascomycota)

Botrytis—Anamorphic Sclerotiniaceae

Bremia—Peronosporaceae, Peronosporales, Oomycota, Chromista
(Peronosporaceae, Peronosporales, Oomycetes, Chromista)

Briosia—Mitosporic fungi

Butlerelfia—Atheliaceae, Stereales, Basidiomycetes, Basidiomycota
(Hyphodermataceae, Hymenochaetales s.l., Hymenomycetes,
Basidiomycota)

Byssochlamys—Trichocomaceae, Eurotiales, Ascomycota
(Trichocomaceae, Eurotiales, Eurotiomycetes, Ascomycota)

Calonectria—Hypocreaceae, Hypocreales, Ascomycota (Nectriaceae,
Hypocreales, Hypocreomycetidae, Sordariomycetes, Ascomycota)

Capitorostrum—Mitosporic fungi

Capnodium—Capnodiaceae, Dothideales, Ascomycota (Capnodiaceae,
Dothideomycetes et Chaetothyriomycetes Uncertain, Ascomycota)

Catacauma—Phyllachoraceae, Phyllachorales, Ascomycota
(Phyllachoraceae, Phyllachorales, Sordariomycetes Orders Uncertain,
Ascomycota)

Cenangium—Leotiaceae, Leotiales, Ascomycota (Leotiaceae, Leotiales, Leotiomycetes, Ascomycota)

Centrospora—Mitosporic fungi

Cephalosporium—Mitosporic fungi

Cephalothecium—Mitosporic fungi

Cephalotrichum—Mitosporic fungi

Ceratobasidium—Ceratobasidiaceae, Ceratobasidiales, Basidiomycetes, Basidiomycota (Ceratobasidiaceae, Tulasnellales, Hymenomycetes, Basidiomycota)

Ceratocystis—Familia Incertae Sedis, Microascales, Ascomycota (Microascaceae Uncertain, Microascales, Hypocreomycetidae, Sordariomycetes, Ascomycota)

Ceratomyces—Ceratomycetaceae, Laboulbeniales, Ascomycota (Ceratomycetaceae, Laboulbeniales, Ascomycota Orders Uncertain, Ascomycota)

Ceratophoma—Mitosporic fungi

Cercoseptoria—Mitosporic fungi

Cercospora—Anamorphic Mycosphaerellaceae

Cercosporella—Mitosporic fungi

Cercosporidium—Mitosporic fungi

Cerotelium—Phakopsoraceae, Uredinales, Teliomycetes, Basidiomycota (Phakopsoraceae, Pucciniales, Teliomycetes, Basidiomycota)

Ceuthospora—Mitosporic fungi

Chaetosphaeropsis—Mitosporic fungi

Chalara—Mitosporic fungi

Chalaropsis—Mitosporic fungi

Choanephora—Choanephoraceae, Mucorales, Zygomycetes, Zygomycota (Mucorales, Zygomycota)

Chondrostereum—Meruliaceae, Stereales, Basidiomycetes, Basidiomycota (Meruliaceae, Meruliales s.str., Hymenomycetes, Basidiomycota)

Chrysomyxa—Coleosporiaceae, Uredinales, Teliomycetes, Basidiomycota (Coleosporiaceae, Pucciniales, Teliomycetes, Basidiomycota)

Ciboria—Sclerotiniaceae, Leotiales, Ascomycota (Sclerotiniaceae, Leotiales, Leotiomycetes, Ascomycota)

Cladosporium—Anamorphic Mycosphaerellaceae

Clasterosporium—Anamorphic Magnaporthaceae

Clathridium—Amphisphaeriaceae, Xylariales, Ascomycota (Amphisphaeriaceae, Xylariales, Xylariomycetidae, Sordariomycetes, Ascomycota)

Claviceps—Clavicipitaceae, Hypocreales, Ascomycota (Clavicipitaceae, Hypocreales, Hypocreomycetidae, Sordariomycetes, Ascomycota)

Clitocybe—Tricholomataceae, Agaricales, Basidiomycetes,
  Basidiomycota (Tricholomataceae, Agaricales, Hymenomycetes,
  Basidiomycota)
Clypeoporthe—Valsaceae, Diaporthales, Ascomycota (Valsaceae,
  Diaporthales, Sordariomycetidae, Sordariomycetes, Ascomycota)
Coccomyces—Rhytismataceae, Rhytismatales, Ascomycota
  (Rhytismataceae, Rhytismatales, Leotiomycetes, Ascomycota)
Coccostroma—Phyllachoraceae, Phyllachorales, Ascomycota
  (Phyllachoraceae, Phyllachorales, Sordariomyctes Orders Uncertain,
  Ascomycota)
Cochliobolus—Pleosporaceae, Dothideales, Ascomycota (Pleosporaceae,
  Pleosporales, Dothideomycetes, Ascomycota)
Coleomyces—Mitosporic fungi
Coleosporium—Coleosporiaceae, Uredinales, Teliomycetes, Basidio-
  mycota (Coleosporiaceae, Pucciniales, Teliomycetes, Basidiomycota)
Colletotrichum—Anamorphic Phyllachoraceae
Collybia—Tricholomataceae, Agaricales, Basidiomycetes, Basidiomycota
  (Tricholomataceae, Agaricales, Hymenomycetes, Basidiomycota)
Coniella—Anamorphic Melanconidaceae
Coniothecium—Mitosporic fungi
Coniothyrium—Anamorphic Leptosphaeriaceae
Coprinus—Coprinaceae, Agaricales, Basidiomycetes, Basidiomycota
  (Agaricaceae, Agaricales, Hymenomycetes, Basidiomycota)
Cordana—Mitosporic fungi
Cordyceps—Clavicipitaceae, Hypocreales, Ascomycota (Clavicipitaceae,
  Hypocreales, Hypocreomycetidae, Sordariomycetes, Ascomycota)
Coriolus—Coriolaceae, Poriales, Basidiomycetes, Basidiomycota
  (Coriolaceae, Polyporales, Hymenomycetes, Basidiomycota)
Corticium—Corticiaceae, Stereales, Basidiomycetes, Basidiomycota
  (Corticiaceae, Basidiomycota Families Uncertain, Basidiomycota)
Corynespora—Mitosporic fungi
Coryneum—Anamorphic Melanconidaceae
Crinipellis—Tricholomataceae, Agaricales, Basidiomycetes,
  Basidiomycota (Marasmiaceae, Tricholomataceae, Agaricales,
  Hymenomycetes, Basidiomycota)
Cristulariella—Anamorphic Sclerotiniaceae
Cronartium—Cronartiaceae, Uredinales, Teliomycetes, Basidiomycota
  (Cronartiaceae, Pucciniales, Teliomycetes, Basidiomycota)
Cryphonectria—Valsaceae, Diaporthales, Ascomycota (Valsaceae,
  Diaporthales, Sordariomycetidae, Sordariomycetes, Ascomycota)
Cryptodiaporthe—Valsaceae, Diaporthales, Ascomycota (Valsaceae,
  Diaporthales, Sordariomycetidae, Sordariomycetes, Ascomycota)

Cryptospora—Valsaceae, Diaporthales, Ascomycota (Valsaceae, Diaporthales, Sordariomycetidae, Sordariomycetes, Ascomycota)

Cryptosporella—Valsaceae, Diaporthales, Ascomycota (Valsaceae, Diaporthales, Sordariomycetidae, Sordariomycetes, Ascomycota)

Cryptosporiopsis—Anamorphic Dermateaceae

Cryptosporium –Mitosporic fungi

Cunninghamella—Cunninghamellaceae, Mucorales, Zygomycetes, Zygomycota (Mucorales, Zygomycota)

Curvularia—Anamorphic Pleosporaceae

Cycloconium—Anamorphic Venturiaceae

Cylindrocarpon—Anamorphic Hypocreaceae

Cylindrocladiella—Mitosporic fungi

Cylindrocladium—Anamorphic Hypocreaceae

Cylindrosporium—Anamorphic Dermateaceae

Cystopus—Albuginaceae, Peronosporales, Oomycota, Chromista (Albuginaceae, Peronosporales, Oomycota, Chromista)

Cytospora—Anamorphic Valsaceae

Cytosporina—Mitosporic fungi

Dactuliophora—Mitosporic fungi

Dactylium—Mitosporic fungi

Dasyscypha—Hyaloscyphaceae, Leotiales, Ascomycota (Hyaloscyphaceae, Leotiales, Leotiomycetes, Ascomycota)

Deightoniella—Mitosporic fungi

Dematophora—Anamorphic Xylariaceae

Dendrophoma—Mitosporic fungi

Deuterophoma—Mitosporic fungi

Diachea—Stemonitidaceae, Stemonitales, Myxomycota (Didymiaceae, Physarales, Myxomycetes, Myxomycota)

Diapleela—Dothideales, Ascomycota (Dothideomycetes Uncertain)

Diaporthe—Valsaceae, Diaporthales, Ascomycota (Valsaceae, Diaporthales, Sordariomycetidae, Sordariomycetes, Ascomycota)

Dibotryon—Venturiaceae, Dothideales, Ascomycota (Venturiaceae, Dothideomycetes et Chaetothyriomycetes Uncertain, Ascomycota)

Dictyochaeta—Anamorphic Lasiosphaeriaceae

Didymella—Dothideales Incertae Sedis, Ascomycota (Dothideomycetes uncertain)

Didymosphaeria—Didymosphaeriaceae, Dothideales, Ascomycota (Didymosphaeriaceae, Dothideomycetes et Chaetothyriomycetes Uncertain, Ascomycota)

Dilophospora—Anamorphic Dothideales

Dimeriella—Parodiopsidaceae, Dothideales, Ascomycota (Parodiopsidaceae, Dothideomycetes et Chaetothyriomycetes Uncertain, Ascomycota)

Dinemasporium—Mitosporic fungi

Diplocarpon—Dermateaceae, Leotiales, Ascomycota (Dermateaceae, Leotiales, Leotiomycetes, Ascomycota)

Diplodia—Anamorphic Botryosphaeriaceae

Diplodina—Anamorphic Valsaceae

Discohainesia—Dermateaceae, Leotiales, Ascomycota (Dermateaceae, Leotiales, Leotiomycetes, Ascomycota)

Discostroma—Amphisphaeriaceae, Xylariales, Ascomycota (Amphisphaeriaceae, Xylariales, Xylariomycetidae, Sordariomycetes, Ascomycota)

Doratomyces—Mitosporic fungi

Dothidella—Polystomellaceae, Dothideales, Ascomycota (Polystomellaceae, Dothideomycetes et Chaetothyriomycetes Uncertain, Ascomycota)

Dothiorella—Mitosporic fungi

Drechslera—Anamorphic Pleosporaceae

Elsinoe—Elsinoaceae, Dothideales, Ascomycota (Elsinoaceae, Dothideomycetes et Chaetothyriomycetes Uncertain, Ascomycota)

Entomosprorium—Anamorphic Dermateaceae

Entyloma—Tilletiaceae, Ustilaginales, Ustomycetes, Basidiomycota (Entylomataceae, Entylomatales, Ustilaginomycetes, Basidiomycota)

Ephelis—Anamorphic Clavicipitaceae

Epichloe—Clavicipitaceae, Hypocreales, Ascomycota (Clavicipitaceae, Hypocreales, Hypocreomycetidae, Sordariomycetes, Ascomycota)

Epicoccum—Mitosporic fungi

Erysiphe—Erysiphaceae, Erysiphales, Ascomycota (Erysiphaceae, Erysiphales, Leotiomycetes, Ascomycota)

Erythricium—Hyphodermateaceae, Stereales, Basidiomycota (Meruliaceae, Meruliales s.str., Hymenomycetes, Basidiomycota)

Eupropodella—Dermateaceae, Leotiales, Ascomycota (Dermateaceae, Leotiales, Leotiomycetes, Ascomycota)

Eutypa—Diatrypaceae, Diatrypales, Ascomycota (Diatrypaceae, Xylariales, Xylariomycetidae, Sordariomycetes, Ascomycota)

Exobasidium—Exobasidiaceae, Exobasidiales, Ustomycetes, Basidiomycota (Exobasidiaceae, Exobasidiales, Ustilaginomycetes, Basidiomycota)

Exserohilum—Anamorphic Pleosporaceae

Fabraea—Dermateaceae, Leotiales, Ascomycota (Dermateaceae, Leotiales, Leotiomycetes, Ascomycota)

Flammulina—Tricholomataceae, Agaricales, Basidiomycetes, Basidiomycota (Tricholomataceae, Agaricales, Hymenomycetes, Basidiomycota)

Fomes—Coriolaceae, Poriales, Basidiomycetes, Basidiomycota (Fomitaceae, Polyporales, Hymenomycetes, Basidiomycota)

Fomitella—Coriolaceae, Poriales, Basidiomycetes, Basidiomycota (Fomitaceae, Polyporales, Hymenomycetes, Basidiomycota)

Fomitiporia—Hymenochaetaceae, Hymenochaetales, Basidiomycetes, Basidiomycota (Hymenochaetaceae, Hymenochaetales s.str., Hymenomycetes, Basidiomycota)

Fomitopsis—Coriolaceae, Poriales, Basidiomycetes, Basidiomycota (Fomitopsidaceae, Polyporales, Hymenomycetes, Basidiomycota)

Frommea—Phragmidiaceae, Uredinales, Teliomycetes, Basidiomycota (Phragmidiaceae, Pucciniales, Teliomycetes, Basidiomycota)

Fulvia—Mitosporic fungi

Fusarium—Anamorphic Hypocreaceae

Fusicladium—Anamorphic Venturiaceae

Fusicoccum—Anamorphoic Botryosphaeriaceae

Gaeumannomyces—Magnaporthaceae, Incertae Sedis, Ascomycota (Magnaporthaceae, Sordariomycetes Families Uncertain, Ascomycota)

Galactomyces—Dipodascaceae, Saccharomycetales, Ascomycota (Dipodascaceae, Saccharomycetales, Saccharomycetes, Ascomycota)

Ganoderma—Ganodermateaceae, Ganodermatales, Basidiomycetes, Basidiomycota (Ganodermataceae, Polyporales, Hymenomycetes, Basidiomycota)

Geastrumia –Mitosporic fungi

Geotrichum –Anamorphic Dipodascaceae

Gerlachia—Mitosporic fungi

Gibberella—Hypocreaceae, Hypocreales, Ascomycota (Nectriaceae, Hypocreales, Hypocreomycetidae, Sordariomycetes, Ascomycota)

Gibellina—Phyllachoraceae, Phyllachorales, Ascomycota (Phyllachoraceae, Phyllachorales, Sordariomycetes Orders Uncertain, Ascomycota)

Gilberetella—Gilbertellaceae, Mucorales, Zygomycetes, Zygomycota (Mucorales, Zygomycota)

Gloeocercospora—Mitosporic fungi

Gloeodes –Mitosporic fungi

Gloeophyllum—Coriolaceae, Poriales, Basidiomycetes, Basidiomycota (Fomitopsidaceae, Polyporales, Hymenomycetes, Basidiomycota)

Gloeosporium—Mitosporic fungi

Glomerella—Phyllachoraceae, Phyllachorales, Ascomycota (Sordariomycetes Genera Uncertain, Ascomycota)

Gnomonia—Valsaceae, Diaporthales, Ascomycota (Valsaceae, Diaporthales, Sordariomycetidae, Sordariomycetes, Ascomycota)

Godronia—Leotiaceae, Leotiales, Ascomycota (Leotiaceae, Leotiales, Leotiomycetes, Ascomycota)

Gonatobotrys—Anamorphic Ceratostomataceae

Graphiola—Graphiolaceae, Graphiolales, Ustomycetes, Basidiomycota (Exobasidiaceae, Exobasidiales, Ustilaginomycetes, Basidiomycota)

Graphium—Anamorphic Ophiostomataceae

Greeneria—Mitosporic fungi

Griphosphaeria—Amphisphaeriaceae, Xylariales, Ascomycota (Amphisphaeriaceae, Xylariales, Xylariomycetidae, Sordariomycetes, Ascomycota)

Grovesinia—Sclerotiniaceae, Leotiales, Ascomycota (Sclerotiniaceae, Leotiales, Leotiomycetes, Ascomycota)

Guignardia—Mycosphaerellaceae, Dothideales, Ascomycota (Mycosphaerellaceae, Dothideomycetes et Chaetothyriomycetes Uncertain, Ascomycota)

Gymnoconia—Phragmidiaceae, Uredinales, Teliomycetes, Basidiomycota (Phragmidiaceae, Pucciniales, Teliomycetes, Basidiomycota)

Gymnosporangium—Pucciniaceae, Uredinales, Teliomycetes, Basidiomycota (Pucciniaceae, Pucciniales, Teliomycetes, Basidiomycota)

Hainesia—Anamorphic Leotiaceae

Halosphaeria—Halosphaeriaceae, Halosphaeriales, Ascomycota (Halosphaeriaceae, Halosphaeriales, Hypocreomycetidae, Sordariomycetes, Ascomycota)

Hansenula—Saccharomycetaceae, Saccharomycetales, Ascomycota (Saccharomycetaceae, Saccharomycetales, Saccharomycetes, Saccharomycotina, Ascomycota)

Haplobasidion—Coriolaceae, Poriales, Basidiomycetes, Basidiomycota (Coriolaceae, Polyporales, Hymenomycetes, Basidiomycota)

Helicobasidium—Platygloeaceae, Platygloeales, Ustomycetes, Basidiomycota (Platygloeaceae, Septobasidiales s.l., Hymenomycetes, Basidiomycota)

Helminthosporium—Mitosporic fungi

Hemileia—Incertae Sedis, Uredinales, Teliomycetes, Basidiomycota (Basidiomycota Genera Uncertain, Basidiomycota Families Uncertain, Basidiomycota)

Hendersonia—Mitosporic fungi

Hendersonula—Mitosporic fungi

Heterobasidion—Coriolaceae, Poriales, Basidiomycetes, Basidiomycota (Echinodontiaceae, Hericiales, Hymenomycetes, Basidiomycota)

Hetrosporium—Mitosporic fungi

Hexagonia—Coriolaceae, Poriales, Basidiomycetes, Basidiomycota
(Coriolaceae, Polyporales, Hymenomycetes, Basidiomycota)

Hirschioporus—Coriolaceae, Poriales, Basidiomycetes, Basidiomycota
(Coriolaceae, Polyporales, Hymenomycetes, Basidiomycota)

Hormodendrum—Mitosporic fungi

Hyalothyridium—Mitosporic fungi

Hymenella—Mitosporic fungi

Hymenochaete—Hymenochaetaceae, Hymenochaetales, Basidiomycota
(Hymenochaetaceae, Hymenochaetales s.str., Hymenomycetes,
Basidiomycota)

Hymenula—Mitosporic fungi

Hypochnus—Arthoniaceae, Arthoniomycetes, Ascomycota
(Arthoniaceae, Arthoniomycetes, Pezizomycotina, Ascomycota)

Hypocrea—Hypocreaceae, Hypocreales, Ascomycota (Hypocreaceae,
Hypocreales, Hypocreomycetidae, Sordariomycetes, Ascomycota)

Hypomyces—Hypocreaceae, Hypocreales, Ascomycota (Hypocreaceae,
Hypocreales, Hypocreomycetidae, Sordariomycetes, Ascomycota)

Hyponectria—Hyponectriaceae, Familia Incertae Sedis, Ascomycota
(Hyponectriaceae, Sordariomycetes Families Uncertain, Ascomycota)

Hypoxylon—Xylariaceae, Xylariales, Ascomycota (Xylariaceae,
Xylariales, Xylariomycetidae, Sordariomycetes, Ascomycota)

Idriella –Mitosporic fungi

Irpex—Steccherinaceae, Stereales, Basidiomycetes, Basidiomycota
(Steccherinaceae, Hymenochaetales s.l., Hymenomycetes,
Basidiomycota)

Isariopsis—Mitosporic fungi

Itersonilia—Mitosporic fungi

Johncouchia—Anamorphic Septobasidiaceae

Junghuhnia—Steccherinaceae, Stereales, Basidiomycota
(Steccherinaceae, Hymenochaetales s.l. Hymenomycetes,
Basidiomycota)

Kabatiella—Mitosporic fungi

Kalmusia—Dothideales, Ascomycota (Dothideomycetes Uncertain,
Pezizomycotina, Ascomycota)

Khuskia—Ascomycota Incertae Sedis (Sordariomycetes Genera Uncertain, Pezizomycotina, Ascomycota)

Kuehneola—Phragmidiaceae, Uredinales, Teliomycetes, Basidiomycota
(Phragmidiaceae, Pucciniales, Teliomycetes, Basidiomycota)

Kunkelia—Uredinales, Incertae Sedis

Kutilakesa—Mitosporic fungi

Laetiporus—Coriolaceae, Poriales, Basidiomycetes, Basidiomycota (Phaeolaceae, Polyporales, Hymenomycetes, Basidiomycota)

Lasiodiplodia—Mitosporic fungi

Leandria—Mitosporic fungi

Lepteutypa—Amphisphaeriaceae, Xylariales, Ascomycota (Amphisphaeriaceae, Xylariales, Xylariomycetidae, Sordariomycetes, Ascomycota)

Leptodontium—Mitosporic fungi

Leptoporus—Coriolaceae, Poriales, Basidiomycetes, Basidiomycota (Phaeolaceae, Polyporales, Hymenomycetes, Basidiomycota)

Leptosphaeria—Leptosphaeriaceae, Dothideales, Ascomycota (Leptosphaeriaceae, Pleosporales, Dothideomycetes, Ascomycota)

Leptosphaerulina—Pleosporaceae, Dothideales, Ascomycota (Pleosporaceae, Pleosporales, Dothideomycetes, Ascomycota)

Leptothyrium—Mitosporic fungi

Leptotrochila—Dermateaceae, Leotiales, Ascomycota (Dermateaceae, Leotiales, Leotiomycetes, Ascomycota)

Leucocytospora—Mitosporic fungi

Leucostoma—Valsaceae, Diaporthales, Ascomycota (Valsaceae, Diaporthales, Sordariomycetidae, Sordariomycetes, Ascomycota)

Leucothallia- Erysiphaceae, Erysiphales, Ascomycota (Erysiphaceae, Erysiphales, Leotiomycetes, Ascomycota)

Leveillula—Erysiphaceae, Erysiphales, Ascomycota (Erysiphaceae, Erysiphales, Leotiomycetes, Ascomycota)

Libertella—Anamorphic Diatrypaceae, Xylariaceae

Limacinula—Coccodiniaceae, Dothideales, Ascomycota (Coccodiniaceae, Dothideales, Dothideomycetes, Ascomycota)

Lophodermium—Rhytimataceae, Rhytismatales, Ascomycota (Rhytimataceae, Rhytismatales, Leotiomycetes, Ascomycota)

Macrophoma—Mitosporic fungi

Macrophomina—Mitosporic fungi

Macrosporium—Mitosporic fungi

Magnaporthe—Magnaporthaceae, Incertae Sedis, Ascomycota (Magnaporthaceae, Sordariomycetes Families Uncertain, Ascomycota)

Marasmiellus—Tricholomataceae, Agaricales, Basidiomycetes, Basidiomycota (Marasmiaceae, Agaricales, Hymenomycetes, Basidiomycota)

Marasmius—Tricholomataceae, Agaricales, Basidiomycetes, Basidiomycota (Marasmiaceae, Agaricales, Hymenomycetes, Basidiomycota)

Marssonina—Anamorphic Dermateaceae

Massarina—Lophiostomataceae, Dothideales, Ascomycota
(Lophiostomataceae, Pleosporales, Dothideomycetes, Ascomycota)
Mauginiella—Mitosporic fungi
Melampsora—Melampsoraceae, Uredinales, Teliomycetes,
Basidiomycota (Melampsoraceae, Pucciniales, Teliomycetes,
Basidiomycota)
Melampsorella—Pucciniastraceae, Uredinales, Teliomycetes,
Basidiomycota (Pucciniastraceae, Pucciniales, Teliomycetes,
Basidiomycota)
Melanconium—Anamorphic Melanconidaceae
Melanospora—Ceratostomataceae, Sordariales, Ascomycota
(Ceratostomataceae, Sordariales, Sordariomycetidae, Sordariomycetes,
Ascomycota)
Meliola—Meliolaceae, Meliolales, Ascomycota (Meliolaceae, Meliolales,
Sordariomycetes Orders Uncertain, Ascomycota)
Meria—Mitosporic fungi
Microdochium—Anamorphic Hyponectriaceae
Micronectriella—Mycosphaerellaceae, Dothideales, Ascomycota
(Mycosphaerellaceae, Dothideomycetes et Chaetothyriomycetes Un-
certain, Ascomycota)
Microsphaera—Erysiphaceae, Erysiphales, Ascomycota (Erysiphaceae,
Erysiphales, Leotiomycetes, Ascomycota)
Moesziomyces—Ustilaginaceae, Ustilaginales, Ustomycetes,
Basidiomycota (Ustilaginaceae, Ustilaginales, Ustilaginomycetes,
Basidiomycota)
Monilia—Mitosporic fungi
Monilinia—Sclerotiniaceae, Leotiales, Ascomycota (Sclerotiniaceae,
Leotiales, Leotiomycetes, Ascomycota)
Moniliophthora—Mitosporic fungi
Monilochaetes—Mitosporic fungi
Monochaetia—Mitosporic fungi
Monographella—Hyponectriaceae, Familia Incertae Sedis, Ascomycota
(Hyponectriaceae, Sordariomycetes Families Uncertain, Ascomycota)
Monosporascus—Sordariales Incertae Sedis, Ascomycota (Sordariales
Uncertain, Sordariomycetidae, Sordariomycetes, Ascomycota)
Mucor—Mucoraceae, Mucorales, Zygomycetes, Zycomycota (Mucorales,
Zygomycota)
Myceliophthora—Anamorphic Arthrodermataceae
Mycena—Tricholomataceae, Agaricales, Basidiomycetes, Basidiomycota
(Favolaschiaceae, Agaricales, Hymenomycetes, Basidiomycota)
Mycocentrospora—Mitosporic fungi

Mycoleptodiscus—Anamorphic Magnaporthaceae
Mycosphaerella—Mycosphaerellaceae, Dothideales, Ascomycota
  (Mycosphaerellaceae, Dothideomycetes et Chaetothyriomycetes Un-
  certain, Ascomycota)
Mycovellosiella –Mitosporic fungi
Myriogenospora—Clavicipitaceae, Hypocreales, Ascomycota
  (Clavicipitaceae, Hypocreales, Hypocreomycetidae, Sordariomycetes,
  Ascomycota)
Myriosclerotinia—Sclerotiniaceae, Leotiales, Ascomycota
Myrothecium—Mitosporic fungi
Mystrosporium—Mitosporic fungi
Naevia—Dermateaceae, Leotiales, Ascomycota (Dermateaceae,
  Leotiales, Leotiomycetes, Ascomycota)
Nakataea—Mitosporic fungi
Nattrassia—Mitosporic fungi
Necator—Anamorphic Corticiaceae
Nectria—Hypocreaceae, Hypocreales, Ascomycota (Nectriaceae,
  Hypocreales, Hypocreomycetidae, Sordariomycetes, Ascomycota)
Nectriella—Hypocreaceae, Hypocreales, Ascomycota (Bionectriaceae,
  Hypocreales, Hypocreomycetidae, Sordariomycetes, Ascomycota)
Nematospora—Metschnikowiaceae, Saccharomycetales, Ascomycota
  (Eremotheciaceae, Saccharomycetales, Saccharomycetes, Saccharo-
  mycotina, Ascomycota)
Neocosmospora—Hypocreaceae, Hypocreales, Ascomycota (Nectriaceae,
  Hypocreales, Hypocreomycetidae, Sordariomycetes, Ascomycota)
Neofabrae—Dermateaceae, Leotiales, Ascomycota (Dermateaceae,
  Leotiales, Leotiomycetes, Ascomycota)
Neovossia—Tilletiaceae, Ustilaginales, Ustomycetes, Basidiomycota
  (Tilletiaceae, Tilletiales, Ustilaginomycetes, Basidiomycota)
Nigrospora—Anamorphic Trichosphaeriales
Nummularia—Xylariaceae, Xylariales, Ascomycota (Xylariaceae,
  Xylariales, Xylariomycetidae, Sordariomycetes, Ascomycota)
Oidiopsis—Anamorphic Erysiphaceae
Oidium—Anamorphic Erysiphaceae
Olpidium—Olpidiaceae, Spizellomycetales, Chytridiomycetes,
  Chytridiomycota (Olpidiacae, Spizellomycetales, Chytridiomycota)
Omnidemptus—Magnaporthaceae, Incertae Sedis, Ascomycota
  (Magnaporthaceae, Sordariomycetes Families Uncertain, Ascomycota)
Omphalina—Tricholomataceae, Agaricales, Basidiomycetes,
  Basidiomycota (Tricholomataceae, Agaricales, Hymenomycetes,
  Basidiomycota)

Oncobasidium—Ceratobasidiaceae, Ceratobasidiales, Basidiomycota
  (Ceratobasidiaceae, Tulasnellales, Hymenomycetes, Basidiomycota)
Oospora—Mitosporic fungi
Operculella—Mitosporic fungi
Ophiobolus—Leptosphaeriaceae, Dothideales, Ascomycota
  (Leptosphaeriaceae, Pleosporales, Dothideomycetes, Ascomycota)
Ophiosphaerella—Phaeosphaeriaceae, Dothideales, Ascomycota
  (Phaeosphaeriaceae, Dothideomycetes et Chaetothyriomycetes Uncertain, Ascomycota)
Ophiostoma—Ophiostomataceae, Ophiostomatales, Ascomycota
  (Ophiostomataceae, Ophiostomatales, Sordariomycetidae,
  Sordariomycetes, Ascomycota)
Ovulariopsis—Anamorphic Erysiphaceae
Oxyporus—Coriolaceae, Poriales, Basidiomycetes, Basidiomycota
  (Rigidiporaceae, Hymenochaetales s.l., Hymenomycetes,
  Basidiomycota)
Ozonium—Mitosporic fungi
Paecilomyces—Anamorphic Trichocomaceae
Paracercospora- Anamorphic Mycosphaerellaceae
Paraphaeosphaeria—Phaeosphaeriaceae, Dothideales, Ascomycota
  (Phaeosphaeriaceae, Dothideomycetes et Chaetothyriomycetes Uncertain, Ascomycota)
Patellaria—Patellariaceae, Patellariales, Ascomycota (Patellariaceae,
  Patellariales, Dothideomycetes, Ascomycota)
Patellina—Mitosporic fungi
Pellicularia—Corticiaceae, Stereales, Basidiomycetes, Basidiomycota
  (Corticiaceae, Basidiomycota Families Uncertain, Basidiomycota)
Peltaster—Mitosporic fungi
Penicillium—Anamorphic Trichocomaceae
Peniophora—Peniophoraceae, Stereales, Basidiomycetes, Basidiomycota
  (Peniophoraceae, Basidiomycota Families Uncertain, Basidiomycota)
Perenniporia—Coriolaceae, Poriales, Basidiomycetes, Basidiomycota
  (Perenniporiaceae, Polyporales, Hymenomycetes, Basidiomycota)
Periconia—Mitosporic fungi
Periconiella—Mitosporic fungi
Peronosclerospora—Sclerosporaceae, Sclerosporales, Oomycota,
  Chromista (Sclerosporaceae, Sclerosporales, Oomycetes, Chromista)
Peronospora—Peronosporaceae, Peronosporales, Oomycota, Chromista
  (Peronosporaceae, Peronosporales, Oomycetes, Chromista)
Pestalotia—Mitosporic fungi
Pestalotiopsis—Anamorphic Amphisphaeriaceae
Pestalozzia—Mitosporic fungi

Pezicula—Dermateaceae, Leotiales, Ascomycota (Dermateaceae, Leotiales, Leotiomycetes, Ascomycota)

Pezizella—Leotiaceae, Leotiales, Ascomycota (Thelebolaceae, Erysiphales, Leotiomycetes, Ascomycota)

Phacidiopycnis—Anamorphic Cryptomycetaceae

Phacidium—Phacidaceae, Leotiales, Ascomycota (Phacidaceae, Leotiales, Leotiomycetes, Ascomycota)

Phaeoacremonium—Mitosporic fungi

Phaeocytosporella—Mitosporic fungi

Phaeocytostroma—Mitosporic fungi

Phaeoisariopsis—Mitosporic fungi

Phaeolus—Coriolaceae, Poriales, Basidiomycetes, Basidiomycota (Phaeolaceae, Polyporales, Hymenomycetes, Basidiomycota)

Phaeoramularia—Mitosporic fungi

Phaeoseptoria—Mitosporic fungi

Phaeosphaerella—Venturiaceae, Dothideales, Ascomycota (Venturiaceae, Dothideomycetes et Chaetothyriomycetes Uncertain, Ascomycota)

Phaeosphaeria—Phaeosphaeriaceae, Dothideales, Ascomycota (Phaeosphaeriaceae, Dothideomycetes et Chaetothyriomycetes Uncertain, Ascomycota)

Phaeosphaerulina—Dothioraceae, Dothideales, Ascomycota (Dothioraceae, Dothideomycetes et Chaetothyriomycetes Uncertain, Ascomycota)

Phakopsora—Phakopsoraceae, Uredinales, Teliomycetes, Basidiomycota (Phakopsoraceae, Pucciniales, Teliomycetes, Basidiomycota)

Phellinus—Hymenochaetaceae, Hymenochaetales, Basidiomycetes, Basidiomycota (Hymenochaetaceae, Hymenochaetales s.str., Hymenomycetes, Basidiomycota)

Phialophora—Anamorphic Magnaporthaceae

Phoma—Anamorphic Pleosporaceae

Phomopsis—Anamorphic Valsaceae

Phragmidium—Phragmidiaceae, Uredinales, Teliomycetes, Basidiomycota (Phragmidiaceae, Pucciniales, Teliomycetes, Basidiomycota)

Phyllachora—Phyllachoraceae, Phyllachorales, Ascomycota (Phyllachoraceae, Phyllachorales, Sordariomyctes Orders Uncertain, Ascomycota)

Phyllactinia—Erysiphaceae, Erysiphales, Ascomycota (Erysiphaceae, Erysiphales, Leotiomycetes, Ascomycota)

Phyllosticta—Anamorphic Mycosphaerellaceae

Phyllostictella—Mitosporic fungi

Phymatotrichopsis—Anamorphic Sistrotremataceae

Phymatotrichum—Mitosporic fungi

Physalospora—Hyponectriaceae, Familia Incertae Sedis, Ascomycota (Hyponectriaceae, Sordariomycetes Families Uncertain, Ascomycota)

Physarum—Physaraceae, Physarales, Myxomycota (Physaraceae, Physarales, Myxomycetes, Myxomycota)

Physoderma—Physodermataceae, Blastocladiales, Chytridiomycota (Physodermataceae, Blastocladiales, Chytridiomycota)

Physopella—Uredinales Incertae Sedis, Basidiomycota (Basidiomycota Genera Uncertain, Basidiomycota Families Uncertain, Basidiomycota)

Phytophthora—Pythiaceae, Pythiales, Oomycota, Chromista (Pythiaceae, Pythiales, Oomycetes, Chromista)

Pichia—Saccharomycetaceae, Saccharomycetales, Ascomycota (Saccharomycetaceae, Saccharomycetales, Saccharomycetes, Ascomycota)

Pilidiella—Mitosporic fungi

Pithomyces—Anamorphic Pleosporaceae

Plasmodiophora—Plasmodiophoraceae, Plasmodiophorales, Plasmodiophoromycota, Protozoa (Plasmodiophoraceae, Plasmodiophorales, Plasmodiophoromycota, Protozoa)

Plasmopara—Peronosporaceae, Peronosporales, Oomycota, Chromista (Peronosporaceae, Peronosporales, Oomycetes, Chromista)

Platyspora—Hysteriaceae, Dothideales, Ascomycota (Diademaceae, Dothideomycetes et Chaetothyriomycetes Uncertain, Ascomycota)

Plenodomas—Mitosporic fungi

Pleocyta—Mitosporic fungi

Pleosphaerulina—Dothioraceae, Dothideales, Ascomycota (Dothioraceae, Dothideomycetes et Chaetothyriomycetes Uncertain, Ascomycota)

Pleospora—Pleosporaceae, Dothideales, Ascomycota (Pleosporaceae, Pleosporales, Dothideomycetes, Ascomycota)

Podosphaera—Erysiphaceae, Erysiphales, Ascomycota (Erysiphaceae, Erysiphales, Leotiomycetes, Ascomycota)

Polyporus—Polyporaceae, Poriales, Basidiomycetes, Basidiomycota (Polyporaceae, Polyporales, Hymenomycetes, Basidiomycota)

Polyscytalum—Mitosporic fungi

Polystigma—Phyllachoraceae, Phyllachorales, Ascomycota (Phyllachoraceae, Phyllachorales, Sordariomyctes Orders Uncertain, Ascomycota)

Potebniamyces—Cryptomycetaceae, Rhytismatales, Ascomycota (Cryptomycetaceae, Rhytismatales, Leotiomycetes, Ascomycota)

Proventuria—Venturiaceae, Dothideales, Ascomycota (Venturiaceae, Dothideomycetes et Chaetothyriomycetes Uncertain, Ascomycota)

Pseudocercospora—Mitosporic fungi

Pseudocercosporella—Anamorphic Mycospherellaceae

Pseudocochliobolus—Pleosporaceae, Dothideales, Ascomycota (Pleosporaceae, Pleosporales, Dothideomycetes, Ascomycota)

Pseudoepicoccum—Mitosporic fungi

Pseudoperonospora—Peronosporaceae, Peronosporales, Oomycota, Chromista (Peronosporaceae, Peronosporales, Oomycetes, Chromista)

Pseudopezicula—Leotiaceae, Leotiales, Ascomycota (Leotiaceae, Leotiales, Leotiomycetes, Ascomycota)

Pseudopeziza—Dermateaceae, Leotiales, Ascomycota (Dermateaceae, Leotiales, Leotiomycetes, Ascomycota)

Pseudophaeolus—Coriolaceae, Poriales, Basidiomycetes, Basidiomycota (Phaeolaceae, Polyporales, Hymenomycetes, Basidiomycota)

Pseudoseptoria—Mitosporic fungi

Psilocybe—Strophariaceae, Agaricales, Basidiomycetes, Basidiomycota (Strophariaceae, Agaricales, Hymenomycetes, Basidiomycota)

Puccinia—Pucciniaceae, Uredinales, Teliomycetes, Basidiomycota (Pucciniaceae, Pucciniales, Teliomycetes, Basidiomycota)

Pucciniastrum- Pucciniastraceae, Uredinales, Teliomycetes, Basidiomycota (Pucciniastraceae, Pucciniales, Teliomycetes, Basidiomycota)

Pycnoporus—Coriolaceae, Poriales, Basidiomycetes, Basidiomycota (Coriolaceae, Polyporales, Hymenomycetes, Basidiomycota)

Pyrenobotrys—Venturiaceae, Dothideales, Ascomycota (Venturiaceae, Dothideomycetes et Chaetothyriomycetes Uncertain, Ascomycota)

Pyrenochaeta—Anamorphic Lophiostomataceae

Pyrenopeziza—Dermateaceae, Leotiales, Ascomycota (Dermateaceae, Leotiales, Leotiomycetes, Ascomycota)

Pyrenophora—Pleosporaceae, Dothideales, Ascomycota (Pleosporaceae, Pleosporales, Dothideomycetes, Ascomycota)

Pyricularia—Anamorphic Magnaporthaceae

Pythium—Pythiaceae, Pythiales, Oomycota, Chromista (Pythiaceae, Pythiales, Oomycetes, Chromista)

Ramichloridium—Mitosporic fungi

Ramularia—Anamorphic Dothideaceae

Ramulispora—Mitosporic fungi

Rhinocladium—Mitosporic fungi

Rhizoctonia—Anamorphic Corticiaceae, Ceratobasidiaceae, Otideaceae

Rhizomorpha—Fungi, Incertae Sedis

Rhizopus—Mucoraceae, Mucorales, Zygomycetes, Zygomycota (Mucorales, Zygomycota)

Rhizosphaera—Anamorphic Venturiaceae

Rhopographus—Phaeosphaeriaceae, Dothideales, Ascomycota
  (Phaeosphaeriaceae, Dothideomycetes et Chaetothyriomycetes Uncertain, Ascomycota)
Rhynchosporium—Mitosporic fungi
Rhytisma—Rhytismataceae, Rhytismatales, Ascomycota
  (Rhytismataceae, Rhytismatales, Leotiomycetes, Ascomycota)
Rigidoporus—Coriolaceae, Poriales, Basidiomycetes, Basidiomycota
  (Rigidiporaceae, Hymenochaetales s.l., Hymenomycetes,
  Basidiomycota)
Roesleria—Caliciales Incertae Sedis, Ascomycota (Ascomycota Genera
  Uncertain)
Rosellinia—Xylariaceae, Xylariales, Ascomycota (Xylariaceae,
  Xylariales, Xylariomycetidae, Sordariomycetes, Ascomycota)
Saccharomyces—Saccharomycetaceae, Saccharomycetales, Ascomycota
  (Saccharomycetaceae, Saccharomycetales, Saccharomycetes,
  Saccharomycotina, Ascomycota)
Salmonia—Erysiphaceae, Erysiphales, Ascomycota (Erysiphaceae,
  Erysiphales, Leotiomycetes, Ascomycota)
Sarocladium—Mitosporic fungi
Schizoparme—Melanconidaceae, Diaporthales, Ascomycota
  (Melanconidaceae, Diaporthales, Sordariomycetidae, Sordariomycetes,
  Ascomycota)
Schizophyllum—Schizophyllaceae, Schizophyllales, Basidiomycetes,
  Basidiomycota (Schizophyllaceae, Agaricales, Hymenomycetes,
  Basidiomycota)
Schizothyrium—Schizothyriaceae, Dothideales, Ascomycota
  (Schizothyriaceae, Dothideomycetes et Chaetothyriomycetes Uncertain, Ascomycota)
Sclerophthora—Verrucalvaceae, Sclerosporales, Oomycota, Chromista
  (Verrucalvaceae, Sclerosporales, Oomycetes, Chromista)
Sclerospora—Sclerosporaceae, Sclerosporales, Oomycota, Chromista
  (Sclerosporaceae, Sclerosporales, Oomycetes, Chromista)
Sclerotinia—Sclerotiniaceae, Leotiales, Ascomycota (Sclerotiniaceae,
  Leotiales, Leotiomycetes, Ascomycota)
Sclerotium—Mitosporic fungi
Scolecosporiella—Mitosporic fungi
Scolicotrichum—Mitosporic fungi
Scopulariopsis—Anamorphic Microascaceae
Scytalidium—Mitosporic fungi
Scytinostroma—Lachnocladiaceae, Lachnocladiales, Basidiomycetes,
  Basidiomycota (Lachnocladiaceae, Lachnocladiales, Hymenomycetes,
  Basidiomycota)

Seimatosporium—Anamorphic Amphisphaeriaceae

Selenophoma—Mitosporic fungi

Septobasidium—Septobasidiaceae, Septobasidiales, Teliomycetes, Basidiomycota (Septobasidiaceae, Septobasidiales s.str., Teliomycetes, Basidiomycota)

Septocylindrium—Mitosporic fungi

Septogloeum—Mitosporic fungi

Septoria—Anamorphic Dothideaceae

Serpula—Coniophoraceae, Boletales, Basidiomycota (Boletaceae, Boletales, Hymenomycetes, Basidiomycota)

Setosphaeria—Pleosporaceae, Dothideales, Ascomycota (Pleosporaceae, Pleosporales, Dothideomycetes, Ascomycota)

Sitosporium—Mitosporic fungi

Solenia—Schizophyllaceae, Schizopyllales, Basidiomycota (Schizophyllaceae, Agaricales, Hymenomycetes, Basidiomycota)

Sphacelia—Anamorphic Clavicipitaceae

Sphaceloma—Anamorphic Elsinoaceae

Sphacelotheca—Ustilaginaceae, Ustilaginales, Ustomycetes, Basidiomycota (Ustilaginaceae, Ustilaginales, Ustilaginomycetes. Basidiomycota)

Sphaerodothis—Phyllachoraceae, Phyllachorales, Ascomycota (Phyllachoraceae, Phyllachorales, Sordariomycetes Orders Uncertain, Ascomycota)

Sphaeropsis—Mitosporic fungi

Sphaerostilbe—Hypocreaceae, Hypocreales, Ascomycota (Nectriaceae, Hypocreales, Hypocreomycetidae, Sordariomycetes, Ascomycota)

Sphaerotheca—Erysiphaceae, Erysiphales, Ascomycota (Erysiphaceae, Erysiphales, Leotiomycetes, Ascomycota)

Sphaerulina—Mycospherellaceae, Dothideales, Ascomycota (Mycosphaerellaceae, Dothideomycetes et Chaetothyriomycetes Uncertain, Ascomycota)

Spilocaea—Anamorphic Venturiaceae

Spondylocladium—Mitosporic fungi

Spongospora—Plasmodiophoraceae, Plasmodiophorales, Plasmodiophoromycota, Protozoa (Plasmodiophoraceae, Plasmodiophorales, Plasmodiophoromycota, Protozoa)

Sporendonema—Mitosporic fungi

Sporisorium—Ustilaginaceae, Ustilaginales, Ustomycetes, Basidiomycota (Ustilaginaceae, Ustilaginales, Ustilaginomycetes, Basidiomycota)

Sporobolomyces—Mitosporic fungi

Sporonema—Mitosporic fungi

Sporotrichum—Mitosporic fungi

Stagonospora—Mitosporic fungi

Stagonosporopsis—Mitosporic fungi

Steccherinum—Steccherinaceae, Stereales, Basidiomycota
(Steccherinaceae, Hymenochaetales s.l., Hymenomycetes,
Basidiomycota)

Stemphylium—Anamorphic Pleosporaceae

Stenocarpella—Mitosporic fungi

Stereum—Stereaceae, Stereales, Basidiomycota (Stereaceae, Hericiales,
Hymenomycetes, Basidiomycota)

Stigmatea—Ascomycota, Incertae Sedis (Ascomycota Genera Uncertain)

Stigmina—Anamorphic Dothideales

Stilbum—Chionosphaeraceae, Atractiellales, Basidiomycetes,
Basidiomycota (Chionosphaeraceae, Basidiomycota Families Uncertain, Basidiomycota)

Synchronoblastia—Mitosporic fungi

Synchytrium—Synchytriaceae, Chytridiales, Chytridiomycota
(Synchytriaceae, Synchytridiales, Chytridiomycota)

Tapesia—Dermateaceae, Leotiales, Ascomycota (Dermateaceae,
Leotiales, Leotiomycetes, Ascomycota)

Taphrina—Taphrinaceae, Taphrinales, Ascomycota (Taphrinaceae,
Taphrinales, Taphrinomycetes, Taphrinomycotina, Ascomycota)

Thanatephorus—Ceratobasidiaceae, Ceratobasidiales, Basidiomycetes,
Basidiomycota (Ceratobasidiaceae, Tulasnellales, Hymenomycetes,
Basidiomycota)

Thielaviopsis—Mitosporic fungi

Thryospora—Mitosporic fungi

Thyrostroma—Mitosporic fungi

Tilletia—Tilletiaceae, Ustilaginales, Ustomycetes, Basidiomycota
(Tilletiaceae, Tilletiales, Ustilaginomycetes, Basidiomycota)

Tolyposporium—Ustilaginaceae, Ustilaginales, Ustomycetes,
Basidiomycota (Ustilaginaceae, Ustilaginales, Ustilaginomycetes.
Basidiomycota)

Trachysphaera—Pythiaceae, Pythiales, Oomycota, Chromista
(Pythiaceae, Pythiales, Oomycetes, Chromista)

Trametes—Coriolaceae, Poriales, Basidiomycetes, Basidiomycota
(Coriolaceae, Polyporales, Hymenomycetes, Basidiomycota)

Tranzschelia—Uropyxidaceae, Uredinales, Teliomycetes, Basidiomycota
(Uropyxidaceae, Pucciniales, Teliomycetes, Basidiomycota)

Trechispora—Sistrotremataceae, Stereales, Basidiomycetes,
Basidiomycota (Sistrotremataceae, Basidiomycota Families Uncertain,
Basidiomycota)

Trichaptum—Coriolaceae, Poriales, Basidiomycetes, Basidiomycota (Steccherinaceae, Hymenochaetales s.l., Hymenomycetes, Basidiomycota)

Trichoconis—Mitosporic fungi

Trichoconium—Mitosporic fungi

Trichoderma—Anamorphic Hypocreaceae

Trichometasphaeria—Lophiostomataceae, Dothideales, Ascomycota (Lophiostomataceae, Pleosporales, Dothideomycetes, Ascomycota)

Trichothecium—Mitosporic fungi

Tripospermum—Anamorphic Capnodiaceae

Tubercularia—Anamorphic Hypocreaceae

Tunstallia—Ascomycota Incertae Sedis

Typhula—Typhulaceae, Cantharellales, Basidiomycetes, Basidiomycota (Basidiomycota Genera Uncertain, Basidiomycota Families Uncertain, Basidiomycota)

Ulocladium—Mitosporic fungi

Uncinula—Erysiphaceae, Erysiphales, Ascomycota (Erysiphaceae, Erysiphales, Leotiomycetes, Ascomycota)

Uredo—Anamorphic Uredinales (Basidiomycota Genera Uncertain, Basidiomycota Families Uncertain, Basidiomycota)

Urocystis—Tilletiaceae, Ustilaginales, Ustomycetes, Basidiomycota (Urocystaceae, Urocystales, Ustilaginomycetes, Basidiomycota)

Uromyces—Pucciniaceae, Uredinales, Teliomycetes, Basidiomycota (Pucciniaceae, Pucciniales, Teliomycetes, Basidiomycota)

Urophlyctis—Urophlyctidiaceae, Spizellomycetales, Chytridiomycetes, Chytridiomycota (Urophlyctidiaceae, Spizellomycetales, Chytridiomycota)

Ustilago—Ustilaginaceae, Ustilaginales, Ustomycetes, Basidiomycota (Ustilaginaceae, Ustilaginales, Ustilaginomycetes. Basidiomycota)

Ustulilaginoidea—Mitosporic fungi

Ustulina—Xylariaceae, Xylariales, Ascomycota (Xylariaceae, Xylariales, Xylariomycetidae, Sordariomycetes, Ascomycota)

Valsa—Valsaceae, Diaporthales, Ascomycota (Valsaceae, Diaporthales, Sordariomycetidae, Sordariomycetes, Ascomycota)

Venturia—Venturiaceae, Dothideales, Ascomycota (Venturiaceae, Dothideomycetes et Chaetothyriomycetes Uncertain, Ascomycota)

Veronaea—Mitosporic fungi

Verticillium—Mitosporic fungi

Waitea—Botryobasidiaceae, Stereales, Basidiomycetes, Basidiomycota (Ceratobasidiaceae, Tulasnellales, Hymenomycetes, Basidiomycota)

Whetzelinia –Sclerotiniaceae, Leotiales, Ascomycota (Sclerotiniaceae, Leotiales, Leotiomycetes, Ascomycota)

Wilsonomyces—Mitosporic fungi

Xylaria—Xylariaceae, Xylariales, Ascomycota (Xylariaceae, Xylariales, Xylariomycetidae, Sordariomycetes, Ascomycota)

Zimmermaniella—Phyllachoraceae, Phyllachorales, Ascomycota (Phyllachoraceae, Phyllachorales, Sordariomycetes Orders Uncertain, Ascomycota)

Zopfia—Zopfiaceae, Dothideales, Ascomycota (Zopfiaceae, Dothideomycetes et Chaetothyriomycetes Uncertain, Ascomycota)

Zygophiala—Mitosporic fungi

Zygosaccharomyces—Saccharomycetaceae, Saccharomycetales, Ascomycota (Saccharomycetaceae, Saccharomycetales, Saccharomycetes, Saccharomycotina, Ascomycota)

Zythia—Mitosporic fungi

REFERENCES

Barr, D. J. S. (1992). Evolution and kingdoms of organisms from the perspective of a mycologist. *Mycologia,* 84:1-11.

FungalWeb (n.d.). <http://www.fungalweb.com>.

Hawksworth, D. L., Kirk, P. M., Sutton, B. C., and Pegler, D. N. (1995). *Ainsworth and Bisby's Dictionary of the Fungi.* CAB International, U.K.

Vidhyasekaran, P. (1993). *Principles of Plant Pathology.* CBS Publishers, Delhi, India.

Wrobel, W. and Creber, G. (1998). *Elsevier's Dictionary of Fungi and Fungal Plant Diseases.* Elsevier, Amsterdam, Holland.

4

Parasitic Flagellate Protozoa

In addition to pathogens, plant diseases are also caused by parasites. Pathogens cause harmful deviation from normal functioning of physiological processes, whereas parasites get their food from the plant. A parasite is defined as an organism that lives in or on the cells of another organism and obtains its food from the latter without necessarily causing disease symptoms (Strobel and Barash, 1990). However, they can cause disease symptoms, probably by depleting host nutrients. The flagellated protozoa of the genus *Phytomonas* and some *Herpetomonas* spp. are parasites of plants. The parasitism of protozoa occurs without any apparent pathogenicity (Dutra et al., 2000). The parasites live mostly in the phloem and laticifers of infected plants. Such parasites can also cause diseases of economic significance in plantations of coconut, oil palm, cassava, and coffee (Dollet, 1984; Camargo, 1990). Parasites have been detected in edible fruits such as pomegranates, peaches, guavas, and tangerines (Dutra et al., 2000). *Phytomonas* is the important genus causing diseases. The important diseases caused by *Phytomonas* are described here.

PHLOEM NECROSIS OF COFFEE

Phytomonas parasitizes coffee plants and causes phloem necrosis. The parasitized coffee plants show a reduction in their starch reserves. The oldest leaves turn yellow and prematurely fall. The new leaves are fewer and smaller than on a healthy tree. They get paler until they yellow and then fall, leaving bare branches. The diseased tree dies in three to 12 months. In the acute form of the disease, only a few old leaves fall; the others lose turgescence and hang limp without falling (Dollet, 1984). After two to three weeks, the leaves turn brownish and necrotic. The roots also turn brown and perish. Cytological examinations show hyperplasia of the phloem, producing sieve tubes three times smaller than normal. Browning of cells and deposition of callose are seen. The vector of the disease is not known. Scale insects may be involved in the transmission of the disease.

HARTROT OF COCONUT

Hartrot of coconut is otherwise called lethal yellowing, bronze wilt, or Coronie wilt. The earliest symptom is yellowing of the oldest leaves, followed by yellowing of the younger leaves. The unripe coconuts may fall. These nuts show internal browning of the husk and blackening of the endocarp. The petioles of the oldest leaves break and necrosis starts in the spear. The apical region of the crown rots, producing a foul odor. Two Pentatomidae insects, *Lincus croupius* and *L. styliger,* are believed to transmit the disease (Desmier de Chenon et al., 1983).

MARCHITEZ DISEASE OF OIL PALM

Early symptoms of Marchitez disease include browning of the leaflet tips. The root system deteriorates, growth slows down, and fruit bunches become dull and rot or fall. A Pentatomidae insect of genus *Lincus* may be the vector of the disease. Insecticides reduce the spread of the disease.

Not much work has been done on *Phytomonas,* most likely because it does not affect major crops and does not cause any significant economic losses.

REFERENCES

Camargo, E. P. (1990). *Phytomonas* and other trypanosomatid parasites of plants and fruit. *Adv Parasitol,* 42:29-112.

Desmier de Chenon, R., Merlan, E., Genty, P., Morin, J. P., and Doller, M. (1983). Research on the genus *Lincus,* Pentatomidae Discocephalidae and its possible role in the transmission of the "Marchitez" of oil palm and Hart rot of coconut. Presented in 4a Reun Com Tecn Reg San Veg SARAH—IICA, Cancun, Mexico.

Dollet, M. (1984). Plant diseases caused by Flagellate protozoa *(Phytomonas). Annu Rev Phytopathol,* 22:115-132.

Dutra, P. M. L., Rodrigues, C. O., Romeiro, A., Grillo, L. A. M., Dias, F. A., Attias, M., De Souza, W., Lopes, A. H. C. S., and Meyer-Fernandes, J. R. (2000). Characterization of ectophosphatase activities in trypanosomatid parasites of plants. *Phytopathology,* 90:1032-1038.

Strobel, G. A. and Barash, I. (1990). Microbial phytotoxins and plant diseases. In P. Vidhyasekaran (Ed.), *Basic Research for Crop Disease Management,* Daya Publishing House, Delhi, pp. 65-73.

5

Parasitic Green Algae

Algae that are parasitic on plants are known only among the phylum Chlorophyta. Of those, only the genus *Cephaleuros* (family Trentepohliaceae order Trentepohliales) is known to cause diseases in economically important crops. The diseases caused by *Cephaleuros* spp. are commonly called red rusts or algal spots.

ECONOMIC IMPORTANCE

Cephaleuros virescens (= *C. mycoidea*) is known to cause diseases in *Citrus* spp. (e.g., limes, lemons, oranges, grapefruit), mango, papaya, pecan, avocado, cacao, tea, coffee, pepper, oil palms, vanilla, litchi, sapota, and guava. *Cephaleuros virescens* is prevalent in India, Indonesia, China, Japan, Malaysia, Australia, the United States, Brazil, West Indies, and Africa. In general, the alga affects perennial trees, not annual crops. It does not cause economical losses in many crops. However, considerable losses in tea, pepper, and *Citrus* spp. due to the alga have been reported.

DISEASE SYMPTOMS

Cephaleuros virescens infects the leaves, stems, and fruit of the trees. Leaf infection can be seen on the lamina, veins, and petiole. Yellow-green pinpoint specks appear on the upper leaf surface. Occasionally the specks occur on the lower leaf surface. The algal thallus grows mostly subcuticular and the cuticle imparts a glistening appearance to the early developmental stage. As growth of the disc progresses, sporophores and sterile hairs develop, both of which contain the orange pigment hematochrome. The algal colony becomes velvety in texture and its color changes from green to orange brown. The thallus often becomes slightly raised. Green islands around the infection foci are common. The cell layers nearest to the alga die and assume a corklike appearance. They may serve as a barrier to further algal penetration. Only a vigorous host is capable of producing an effective barrier. The alga takes up water and minerals from the host by osmosis.

In tea leaves, an intercellular type of *Cephaleuros* infection is often seen. A small purple translucent spot gradually extends from the upper to the lower surface. On such lesions, fructifications of the alga can be seen on both surfaces. Ultimately, the affected region dries and drops out leaving holes in the affected leaves (shot hole pattern). In coffee leaves, the spongy mesophyll cells enlarge, divide, and undergo considerable cell wall thickening. The alga appears to be completely isolated from the healthy tissue by a continuous barrier of thick-walled cells (Joubert and Rijkenberg, 1971).

Stem infection by the algal parasite results in the appearance of dark gray or purple lesions. These lesions assume the typical rusty red appearance similar to those found in leaf infections. The bark becomes scaly and cracks. Localized swelling is common at the site of infection. Lesions coalesce, often girdling the stem completely for a long distance. Young wood may die due to infection. Affected branches are often stunted and bear fewer leaves, which typically show signs of chlorosis and variegation. Fruit infections are common on guava, citrus, tea, and avocado and coffee berries. Slightly raised, irregularly shaped, dark green, brown, or black lesions appear on fruits. The alga penetrates up to the superficial cell layers of the fruit only.

LIFE CYCLE

Cephaleuros spp. have a disclike thallus composed of symmetrically arranged cells that radiate dichotomously from the center to the periphery (Joubert and Rijkenberg, 1971). The cells are elongated and sometimes barrel shaped. The thallus cells at the margin are monolayered, whereas several layers develop in the center of older discs. The algal filaments usually extend between the cuticle and the epidermis of host leaves or, more rarely, between the epidermis and palisade cells. Sometimes, the parasite penetrates between the mesophyll cells of the leaf. The alga appears to extend by the mechanical force of its expansion rather than by enzymatic dissolution of host structures. A large airspace is found between the algal disc and the necrotic epidermal cells of the host. Rhizoids arise from the thallus and extend through the airspace to the underlying host cells. The rhizoids serve to anchor the thallus and furnish it with water and other substances from the host (Vidhyasekaran, 1993).

Large, sessile, flask-shaped cells known as gametangia arise in the thallus. In the presence of free water, these cells open by fissure and release swarm spores. Eight to 32, but most commonly 16, biflagellate isogametes per gametangium are found. Copulation takes place between two gametes either from the same gametangium or from different gamentangia. The resulting zygote produces a dwarf sporophyte that bears a small, dehiscent microsporangium. After meiotic division, four quardiflagellate microzoospores are produced (Joubert and Rijkenberg, 1971). These microzoospores rarely infect the host. The older algal discs produce hairlike initials at the end of some of the radiating rows of cells. After rupturing the cuticle or epidermis of the host, the hairlike initials differentiate into either setae or sporangiophores, both of which are produced abundantly on the same thallus. The sporangiophores terminate in a swollen apical cell from which many lateral protrusions are subtended. Each protrusion develops into a zoosporangium attached to the apex by a pedicel. Under favorable conditions, sporangia release zoospores through an ostiole. Zoospores are biflagellate, and about 30 are produced from each sporangium. When the zoospore reaches the host cell, it comes to rest. A supracuticular "primary disc" develops as a result of repeated cross-wall formation in the zoospore after it has come to rest. Buds develop from the undersurface of the primary disc cells, penetrate the host leaf, and form a "secondary disc" in the plant tissue. There are also reports that nearly all superficial thalli perish and that only those zoospores that lodge in a crevice or abrasion succeed in infecting the host (Wolf, 1930). The infection may be initiated by zoospores washed into the stomata by rain. Zoosporangia are disseminated by the wind. Rain water running over the surfaces of lesions and splashing onto new leaves may rapidly spread the zoospores. Hence, at the end of rainy season the disease becomes severe (Mann and Hutchinson, 1907).

DISEASE MANAGEMENT

Host debility is usually accompanied by an increased incidence and severity of the disease, which can be significantly reduced by the application of nitrogen and potassium. Disease incidence also can be reduced by providing proper soil drainage. All badly diseased or dead wood should be removed. Pruning is recommended for citrus, tea, and cacao trees. Shade trees and shelterbelts may serve as the primary sources of inoculum. Hence, shade trees selected should not be alternate hosts for the algal pathogen. Irrigation may reduce the disease incidence by improving the vigor of the trees. Fungicide sprays may reduce the algal infection. Bordeaux mixture is used extensively in the control of algae. Copper oxychloride and cuprous oxide can be sprayed to control the disease.

REFERENCES

Joubert, J. J. and Rijkenberg, F. H. J. (1971). Parasitic green algae. *Annu Rev Phytopathol*, 9:45-64.

Mann, H. H. and Hutchinson, C. M. (1907). *Cephaleuros virescens* Kunze the "red rust" of tea. *Mem Dep Agr India Bot Ser*, 1(6):1-33.

Vidhyasekaran, P. (1993). *Principles of Plant Pathology*. CBS Publishers, Delhi.

Wolf, F. A. (1930). A parasitic alga, *Cephaleuros virescens* Kunze, on *Citrus* and certain other plants. *J Elisha Mitchell Sci Soc*, 45:187-205.

6

Parasitic Higher Plants

Some plants are parasitic to other plants, depending upon them for food and/or water. Such plants cause disease by taking away the nutrients of the host. The important parasitic higher plants are *Striga, Orobanche, Loranthus, Cassytha,* and *Cuscuta.*

STRIGA

Several *Striga* species parasitize crop plants. More than 30 species of *Striga* have been reported in Africa (Berner et al., 1995). Three of these species are endemic to Australia. Two species, *Striga asiatica* and *S. gesnerioides,* have been detected in the United States. *Striga* spp., which are also called witchweed, affect many crops, including corn, sorghum, sugarcane, rice, cowpea, and tobacco. *Striga* are characterized by opposite leaves and irregular flowers with a pronounced bend in the corolla tube. They are obligate parasites and will not develop without hosts. They are also root parasites that produce haustoria, food-absorbing outgrowths that graft with the roots of the host. *Striga* spp. lack typical root hairs and root caps (Vidhyasekaran, 1993).

The seeds of *Striga* spp. are extremely small and require afterripening or postharvest ripening before germination. Germination of seeds requires exposure to an exogenous germination stimulant after an environmental conditioning period in which the seeds imbibe water. Usually this stimulant is a host-root exudate, but some nonhost-root exudates and synthetic compounds can also stimulate germination of *Striga* seeds. The stimulant in the root exudate has been identified as strigol. Strigol promotes the germination of *S. asiatica* seeds at concentrations as low as 10^{-16} M. Ethylene, gibberellins, cytokinins, and coumarins also stimulate the germination of *Striga* seeds. After germination, endosperm nutrients can sustain the seedlings for three to seven days in the absence of a host. If the seedlings do not attach to a host and successfully establish a parasitic link within this period, they will die. If a host root is in close proximity (2 to 3 mm) to a germinated *Striga* seedling, chemical signals are exchanged that direct the seedling's radicle to the host root, initiate haustorium induction, and result in the successful attachment and establishment of xylem-to-xylem connections between the parasite and host (Berner et al., 1995). One of these signals has been identified as 2,6-dimethoxy-*p*-benzoquinone, which may be the product of enzymatic degradation of the host root responsible for stimulating formation of the parasite haustorium. Haustoria of parasitic plants are specialized organs developed from parasite radicles prior to penetration. After successful attachment, developing *Striga* plants grow underground for four to seven weeks prior to emergence. Numerous parasitic attachments occur on the same plant.

Striga absorb water and foodstuffs through their haustoria. Carbohydrate and nitrogen of the host are utilized by the parasites. This process results in sugar sink and nitrogen starvation in the host. Differential absorption of phosphorous, potassium, sulfur, and iron by the infected host has also been reported. Infected sorghum plants show 90 to 95 percent less cytokinins and 30 to 80 percent less gibberellins. A strong transpirational pull from host to parasite is observed. High humidity inhibits the growth of *Striga* due to a reduced flow of materials from host to parasite. For the same reason, *Striga* survive more often in dry soils. *Striga* strengthen the roots that are parasitized and stimulate root production of the host (Vidhyasekaran, 1993).

Symptoms of parasitism resemble drought stress, nutrient deficiency, and vascular disease. Infected plants become stunted with heavy yield production. Parasites blossom and form fruits within three or four weeks of emergence, producing several hundred thousand minute seeds per plant. The seeds are dispersed by the wind, finally settling on the soil. Such seeds may be viable for up to 14 years in soil.

Biological strains of *Striga* have been reported. *S. gesnerioides* isolates from tobacco do not infect cowpea and the isolates from cowpea do not infect tobacco. The *Sorghum* strain of *S. hermonthica* does not infect corn and the corn strain of *S. hermonthica* does not infect sorghum. Cowpea cultivar-specific races of *S. gesnerioides* have been reported also. *Striga* seeds are often found mixed with crop seeds, and the contaminated crop seeds may transmit the disease into a new area of the field. Hence, seeds should be obtained only from *Striga*-free fields. Seed treatment by soaking seeds in an aqueous solution with low amounts of imazaquin (an acetolactate synthase [ALS] inhibitor that selectively interferes with inhibition of amino acid biosynthesis by the parasite) effectively contols *Striga* infection in cowpea (Berner et al., 1994). Crop varieties resistant to the parasites have been identified. Four cowpea cultivars, Gorom Local (SUVITA-2) from Burkina Faso, 58-57 from Senegal, B301 from Botswana, and IT82D-849 from Nigeria, have been identified as resistant to *S. gesnerioides*. However, several races of *Striga* are known in different countries and some of these resistant cowpea cultivars are susceptible to new races of *S. gesnerioides* (Berner et al., 1995).

Some cultural practices may reduce parasite incidence. Soil mulching with polyethylene film can be useful in controlling *Striga* seed germination. Soil fumigation with methyl bromide kills *Striga* seeds. Soil injection of ethylene gas is used in the United States to induce *Striga* seed germination in the absence of a host crop. Trap crops, which stimulate germination but do not support parasitism, can be grown to remove *Striga* from the infested field. Cotton and peanut are trap crops for *S. asiatica,* and sudangrass is a trap crop for *S. hermonthica.* Parasites should be pulled out and destroyed before their seeds set. Weedicides such as 2,4-D and MCPA can be useful in controlling *Striga.*

OROBANCHE

Orobanche is commonly called broomrape. Several species of *Orobanche* attack crops, and at least seven species attack tobacco. *Orobanche aegyptica* and *O. cernua* affect tobacco, sunflowers, melons, tomatoes, and broad-beans. Tobacco is seriously affected by *Orobanche* in India, Pakistan, Russia, Mexico, Italy, Australia, New Zealand, and many Eastern European countries. *Orobanche* is a minor parasite in the United States. It is an annual fleshy flower plant, growing up to a height of about 12 inches. It has a cylindrical, thick, fleshy, whitish to purple stem. The stem bears scaly leaves that end in spikes. It bears many tiny flowers that vary from purple to white in color.

It is an obligate parasite that carries no chlorophyll. The single stem is attached to a tobacco root, ten or more of which can be carried by one tobacco plant. *Orobanche* produces abundant minute seeds. The seeds remain dormant in the soil for more than 20 years. Germination occurs only when the seeds are close to the roots of plants that are able to provide necessary stimulation. Stimulants in the root exudates have been identified as benzopyran derivatives. Several growth regulators, such as ethylene, gibberellins, cytokinins, and coumarins, also induce seed germination. *Orobanche* emerges five to six weeks after tobacco is planted and as many as 50 shoots may be found around a single plant. Affected plants are stunted and chlorosis of the leaves is predominant. *Orobanche* shoots develop singly or in bunches, and they are attached to host roots in a fragile manner by swollen cloves. They grow quickly, sometimes rising above the height of the host plant. Some cultural practices, such as hand weeding, may reduce parasite incidence. *Carum ajown,* a trap crop for *O. cernua,* can be exploited for the control of *Orobanche.* Methyl bromide fumigation and metham-sodium applied as a preplanting drench can reduce parasite incidence. Allyl alcohol, dazomet, and methyl isothiocyanate are also effective in reducing the emergence of *Orobanche* in the field.

LORANTHUS

Loranthus is a stem parasite that attacks orange, mango, guava, pomegranate, mulberry, rubber, teak, sandalwood, rosewood, kapok, and casuarina trees. *Loranthus* has green leaves and small berries that are attractive to birds. Birds eat the fruit and the seeds are defecated. The defecated seeds adhere to the branches due to their sticking properties. The seed coat is gelatinous and able to absorb water from rain, mist, or dew. This property prevents the seed from perishing. The natural fall of berries from higher to lower levels of the same host tree ensures a continuous infestation. The seeds sprout in due course, growing into the host tissue by producing a holdfast. The holdfast may become as large as a tennis ball, completely surrounding the branch. The tops of every infected branch die. Very few leaves are formed, and losses ranging from 20 to 30 percent of the harvest have been reported. For effective control, parasitized branches should be cut. To kill the parasite, one hole is made by drilling at the bottom of the *Loranthus* and a mixture of 7 g of copper sulfate and 1 g of 2,4-D is applied. Refined diesel oil is also used. Forty percent emulsion of the diesel oil is prepared in water containing 0.005 percent washing soap used as an emulsifying agent, which is then sprayed on *Loranthus*.

CASSYTHA

The common species is *Cassytha filiformis,* which attacks several plants. It is a slender, threadlike, cylindrical herbaceous vine. Leaves are reduced to mere scales. *Cassytha* covers other vegetation like a mantle. It attaches itself to the host by means of haustoria. It has a spicy fragrance and produces fruit, globose drupes that are one to two inches in diameter and covered by a fleshy receptacle. Birds eat these fleshy fruits and seeds and distribute them. Orange trees are often affected by this parasite. *Cassytha* should be removed manually before it seeds.

CUSCUTA

Cuscuta is commonly called dodder. Dodder affects many crops, including alfalfa, clover, and tobacco. Dodder is threadlike, with leafless stems that are yellow, orange, or greenish in color. Minute leaves, which are functionless, are also seen in some species. Dodder vines usually appear as a tangled mass of yellowish threads. Dodder seeds can germinate without external root stimulation. The dodder first produces a rudimentary root system of its own. However, if the aerial parts are unsuccessful in finding a host plant within a few weeks, the vine dies. Dodder obtains water and nutrients from the host plant through haustoria that are embedded in the stems of the host plant. A close orientation of the vascular elements of the host and the parasite is seen. Some dodders have traces of chlorophyll, and stomata can be seen on the stem.

Dodder plants produce small white flowers and abundant seeds. The seeds fall on the soil and germinate, producing tiny colorless seedlings that evidence a spiral growth at the uppermost portions. These seedlings will survive only for a short time if a host is not available. When a host is present, the seedlings will attach to the host. After contact between the host and the parasite is established, the portion of the seedling embedded in the soil gradually shrivels and disappears. Parasitic plants can be manually removed. Fumigation of the tobacco seedbed with methyl bromide is recommended to kill dodder seeds in tobacco cultivation.

REFERENCES

Berner, D. K., Awad, A. E., and Algbokhan, E. I. (1994). Potential of imazaquin seed treatment for control of *Striga gesnerioides* and *Alectra vogelii* in cowpea *(Vigna unguiculata)*. *Plant Dis*, 78:18-23.

Berner, D. K., Kling, J. G., and B. B. Singh (1995). *Striga* research and control—A perspective from Africa. *Plant Dis*, 79:652-660.

Vidhyasekaran, P. (1993). *Principles of Plant Pathology*. CBS Publishers, Delhi.

7

Parasitic Nematodes

Nematodes are a large group of invertebrates living in soil and water. Some feed on higher plants and can cause diseases. Some plant parasitic nematodes cause heavy crop losses. Cereal cyst nematode *Heterodera avenae,* potato cyst nematode *Globodora rostochiensis,* and the root-knot nematodes *Meloidogyne javanica, M. incognita,* and *M. arenaria* severely affect several cereals and vegetables. Some nematodes cause damage in association, with fungi such as *Fusarium, Verticillium, Rhizoctonia,* and *Phytophthora* spp. causing wilt and root rots in several crops. In association with bacterial pathogens such as *Clavibacter tritici,* the nematode *Anguina tritici* causes diseases such as spike blight of wheat. Some nematodes such as *Xiphinema, Trichodorus,* and *Paratrichodorus* serve as vectors of viruses causing serious diseases in tomato, peach, strawberry, raspberry, grapevine, pea, cowpea, and tobacco. Several nematicides have been developed, and these broad-spectrum nematicides were highly effective in management of nematode-related diseases. However, several effective nematicides have been removed from the marketplace as the result of federal deregulation. Research work has been intensified in developing nematode-resistant plants exploiting single-gene resistance. Cultivars with vertical or horizontal resistance have been developed, with more success with vertical resistance. Several biocontrol agents have been identified, but their practical uses in the field are yet to be demonstrated. All these aspects are described in this chapter.

STRUCTURE OF NEMATODES

Most plant-parasitic nematodes are minute, vermiform animals. Their bodies are elongated and threadlike (*nema* in Greek means "thread") without any segments. They are cylindrical, tapering at each end especially toward the tail. Nematode females may swell to become spherical. The size of plant-parasitic nematodes ranges from 0.2 to 10 mm and is commonly 0.5 to 1.5 mm. The basic body structure of nematodes consists of a flexible body wall that is composed of cuticle, hypodermis, and somatic muscles. The wall surrounds a tubelike gut, which is made up of esophagus and intestine. The body cavity between the gut and the body wall is usually regarded as a pseudocoelom containing a pseudocoelomic fluid (Bird, 1981; Vidhyasekaran, 1993).

The nerve system consists of a nerve ring that encircles the gut usually in the region of the esophageal isthmus. Several nerves extend anteriorly and posteriorly. An excretory duct opens to the exterior via an excretory pore. The gut is an internal tube beginning at the oral opening and ending at the ventrally placed anus in juveniles and females and at the cloaca in males (Southey, 1982).

Females usually have a vulva in the midbody region with two reproductive tracts (didelphic) that are often opposed to each other (amphidelphic). A single reproductive tract (monodelphic) is usually directed anterior to the vulva (prodelphic), and the vulva is often located more posteriorly. When only one fully formed reproductive tract is present, it may be opposed by a pre- or postvulval sac (Southey, 1982). Males usually have a single or paired testis, seminal vesicle, and vas deferens. The vas deferens opens into the cloaca, which has a ventrally placed opening. Males usually have a pair of copulatory spicules that lie in an invagination of the cloacal wall. Sclerotized structures such as a gubernaculum or lateral accessory pieces guide the spicules.

The females lay eggs after copulation. The young ones are called larvae. Nematodes typically moult four times to reach the adult stage (Vidhyasekaran, 1993).

PLANT PARASITIC NEMATODES

The important parasitic nematodes are listed here. The crops affected by these nematodes are given within parentheses.

Awl Nematodes

Dolichodorus spp. (Crops: celery, corn, crucifers, sorghum)
Dolichodorus heterocephalus Cobb (Crops: bean, celery, corn)

Bulb and Stem Nematode, Leaf and Stem Nematode

Ditylenchus dipsaci (Kühn) Filipjev (Crops: alfalfa, bean, beet, corn, oats, potato, rice, rye, tobacco)

Burrowing Nematode

>   *Radopholus similis* (Cobb) Thorne (Crop: corn)

Citrus Nematode

>   *Tylenchulus semipenetrans* Cobb (Crops: citrus, grape)

Crimp Nematode

>   *Aphelenchoides besseyi* Christie (Crop: rice)

Cyst Nematodes

>   *Globdera pallida* (Stone) Mulvey and Stone (Crop: potato)
>   *G. rostochiensis* (Wollenweber) Mulvey and Stone (Crop: potato)
>   *G. solanacearum* (Miller and Gray) Behrens = *G. virginiae* (Miller
>   and Gray) Behrens (Crop: tobacco)
>   *G. tabacum* (Lownsbery and Lownsbery) Behrens (Crop: tobacco)
>   *Heterodera* spp. (Crops: celery, wheat)
>   *H. avenae* Wollenweber (Crops: barley, corn, oats, rye, wheat)
>   *H. carotae* Jones (Crop: carrot)
>   *H. cruciferae* Franklin (Crop: crucifers)
>   *H. filipjevi* (Madzhidov) Stelter (Crop: barley)
>   *H. glycines* Ichinohe (Crops: bean, soybean)
>   *H. hordecalis* Andersson (Crop: oat, wheat)
>   *H. latipons* Franklin (Crops: barley, oats, wheat)
>   *H. schachtii* Schmidt (Crops: beet, cotton, crucifers)
>   *H. trifolii* Goffart (Crops: alfalfa, beet)
>   *H. zeae* Koshy et al. (Crop: corn)
>   *Punctodera chalcoensis* Stone et al. (Crop: corn, oat, wheat)

Dagger Nematodes

>   *Xiphinema* spp. (Crops: citrus, grape, strawberry)
>   *X. americanum* Cobb (Crops: alfalfa, almond, apple, apricot, bean,
>   corn, cotton, cucurbit, grape, oat, peach, pear, sorghum, straw-
>   berry, tobacco, wheat)
>   *X. brevicolle* Lordello and daCosta (Crop: mango)
>   *X. diversicaudatum* (Micoletzky) Thorne (Crop: rose)

X. index Thorne and Allen (Crop: grape)

X. mediterraneum Lima (Crop: corn)

X. rivesi Dalmasso (Crops: almond, apple, apricot, peach, pear)

X. vuittenezi Luc et al. (Crop: apple, pear)

False Root-Knot Nematodes

Nacobbus aberrans (Thorne and Schuster) Sher (Crops: beet, potato)

N. dorsalis Thorne and Allen (Crop: corn)

Foliar Nematode

Aphelenchoides ritzemabosi (Schwartz) Steiner (Crops: tobacco)

Grass-Root Gall Nematode

Subanguina radiciola (Greeff) Paramonov (Crop: wheat)

Lance Nematodes

Hoplolaimus spp. (Crops: bean, corn)

H. columbus Sher (Crops: soybean, corn, cotton, mango)

H. galeatus (Cobb) Thorne (Crops: corn, soybean)

H. uniformis (Crop: carrot)

Needle Nematodes

Longidorus spp. (Crops: alfalfa, beet, corn, citrus, grape)

L. africans Merny (Crops: cotton, sorghum)

L. breviannulatus Norton and Hoffman (Crop: corn)

Rotylenchulus spp. (Crop: alfalfa)

Pin Nematodes

Paratylenchus spp. (Crops: alfalfa, apple, bean, celery, crucifers, cucurbits, peanut, sorghum)

Paratylenchus hamatus Thorne and Allen (Crops: alfalfa, celery, cotton)

Pod Lesion Nematode

Tylenchorhynchus brevilineatus Williams = *T. brevicadatus* Hopper (Crop: peanut)

Potato Rot Nematode

Ditylenchus destructor Thorne (Crops: beet, potato)

Reniform Nematodes

Rotylenchulus spp. (Crop: sorghum)
Rotylenchulus reniformis Linford and Oliveira (Crop: bean, cotton, cucurbits, soybean, tobacco)

Ring Nematodes

Criconemella spp. (Crops: corn, cotton, cucurbit, grape, oat, sorghum, wheat)
C. axesta (Fassuliostis and Williams) Luc and Raski (Crop: rose)
C. ornata (Raski) Luc and Raski (Crop: corn, peanut, soybean)
C. xenoplax (Raski) Luc and Raski (Crops: almond, apple, apricot, peach)
Criconemoides ovantus Raski (Crop: bean)
Nothocriconemella mutabilis (Taylor) Ebsary (Crop: oats)

Root Nematode

Hirschmaniella oryzae (Soltwedel) Lucand Goodey (Crop: rice)

Root Gall Nematode

Subanguina radicicola (Greeff) Paramonov (Crops: barley, rye)

Root-Knot Nematodes

Meloidogyne spp. (Crops: alfalfa, apple, barley, beet, citrus, coffee, corn, crucifer, cucurbit, grape, oat, pear, potato, rice, rye, sorghum, strawberry, sugarcane, tomato, wheat)
M. arenaria (Neal) Chitwood (Crops: almond, alfalfa, apricot, banana, bean, peanut, peach, soybean, tobacco)

M. artiellia Franklin (Crop: barley)

M. chitwoodi Golden et al. (Crops: alfalfa, barley, corn, oat, wheat)

M. hapla Chitwood (Crops: alfalfa, bean, carrot, peanut, potato, rose, soybean, tobacco, strawberry)

M. incognita (Kofoid and White) Chitwood (Crops: alfalfa, almond, apricot, banana, bean, corn, cotton, peach, potato, soybean, tobacco)

M. javanica (Treub) Chitwood (Crops: alfalfa, almond, apricot, banana, bean, peach, peanut, potato, soybean, tobacco, corn)

M. naasi Franklin (Crops: barley, oat, wheat)

Root Lesion Nematodes

Pratylenchus spp. (Crops: alfalfa, apple, barley, bean, beet, carrot, celery, citrus, corn, cotton, cucurbit, oat, pear, potato, grape, sorghum, soybean, sugarcane, tobacco, wheat)

P. brachyurus (Godfrey) Filipjev and Schuurmans-Stekhoven (Crops: banana, corn, peanut, potato, tobacco)

P. coffeae (Zimmerman) Schuurmans-Stekhoven (Crops: banana, peanut, strawberry)

P. crenatus Loof (Crop: corn)

P. goodeyi Sher and Allen (Crop: banana)

P. hexincisus Taylor and Jenkins (Crop: corn)

P. minyus Sher and Allen (Crop: wheat)

P. neglectus (Rensch) Filipjev and Schuurmans-Stekhoven (Crops: alfalfa, corn)

P. penetrans (Cobb) Filipjev and Schuurmans-Stekhoven (Crops: alfalfa, apple, bean, carrot, corn, peach, pear, potato, rose, strawberry, tobacco)

P. pratensis (de Man) Filipjev (Crops: crucifer, strawberry)

P. reniformia Linford and Oliveira (Crop: banana)

P. scribneri Steiner (Crops: corn, strawberry)

P. thornei Sher and Allen (Crops: corn, oat, wheat)

P. vulnus Allen and Jensen (Crops: almond, apricot, grape, peach, rose)

P. zeae Graham (Crop: corn)

Seed and Pod Nematode

Ditylenchus destructor Thorne (Crop: peanut)

Seed Gall Nematode

Anguina tritici (Steinbuch) Chitwood (Crops: rye, wheat)

Sheath Nematode

Hemicycliophora spp. (Crops: oats, soybean)
Hemicycliophora arenaria Raski (Crop: citrus)

Sheathoid Nematode

Hemicriconemoides mangiferae Siddiqi (Crop: mango)

Spiral Nematodes

Helicotylenchus spp. (Crops: alfalfa, corn, cotton, cucurbit, oat, sor-
ghum, soybean, sugarcane, tobacco, wheat)
H. dihystera Cobb (Crops: banana, bean)
H. multicinctus Cobb (Crop: banana)
Helicotylenchus nannus Steiner (Crop: rose)
Rotylenchus spp. (Crops: rose, sugarcane)
Scutellonema spp. (Crops: cotton, sugarcane)
S. cavenessi Sher (Crops: peanut)

Stem Nematode

Ditylenchus angustus (Butler) Filipjev (Crop: rice)

Sting Nematodes

Belonolaimus spp. (Crops: celery, corn, crucifers)
B. gracilis Steiner (Crop: peanut, soybean, strawberry)
B. longicaudatus Rau (Crops: bean, carrot, citrus, corn, cotton, cucurbit, oats, peanut, potato, sorghum, soybean, strawberry, tomato)

Stubby-Root Nematodes

Paratrichodorus spp. (Crops: alfalfa, beet, celery, citrus, corn, cotton, potato, sorghum, tobacco, tomato, wheat)
P. christiei (Allen) Siddiqi (Crops: bean, corn)

P. minor (Colbran) Siddiqi (Crops: corn, cucurbit, oat, sorghum)
Trichodorus spp. (Crops: beet, corn, potato, tobacco, tomato)

Stunt Nematodes

Merlinius spp. (Crops: bean, cotton, oat, tobacco)
M. brevidens (Allen) Siddiqi (Crops: barley, wheat)
Quinisulcius acutus (Allen) Siddiqi (Crops: corn, bean, rose)
Tylenchorhynchus spp. (Crops: alfalfa, bean, citrus, cotton, oat, rose, sorghum, tobacco)
T. claytoni Steiner (Crop: cucurbit)
T. dubius (Butschli) Filipjef (Crops: barley, corn)
T. maximus Allen (Crop: barley)

Testa Nematode

Aphelenchoides arachidis Bos (Crop: peanut)

DISEASE SYMPTOMS

Most nematodes attack the roots of plants. They may be endoparasites (i.e., they enter the root and lose contact with soil or a large part of their body will be inside the root tissue) or ectoparasites (i.e., they are free living in soil and intermittently feed on the epidermis of roots and root hairs near the root tip). *Meloidogyne, Heterodera, Globodera, Pratylenchus, Radopholous, Rotylenchulus, Helicotylenchus, Criconemella,* and *Paratylenchus* are endoparasites. *Paratrichodorus, Trichodorus,* and *Belonolaimus* are ectoparasites. Some nematodes, called foliage parasites, primarily infect and damage the above-ground plant parts. These nematodes may survive in the soil and/or host-plant residues. Under favorable conditions, the parasites may crawl up the plant and attack young seedlings or mature plants. The stem nematode *Ditylenchus,* the seed gall nematode *Anguina,* and the foliar nematode *Aphelenchoides* belong to this category.

Symptoms of nematode-infested plants are mostly nonspecific. Patchy yellow appearance, dieback symptoms, and slow decline are frequently observed symptoms in the nematode-infested field. Above-ground symptoms due to root parasites include stunting (poor growth and lack of vigor), yellowing (probably due to nutritional deficiency), wilting, and reduced yield. Foliage parasites may cause crinkled and distorted stems and leaves. The growing points of seedlings may be killed by the nematodes (e.g., *Anguina tritici*). The infected flower primordium develops into a gall, as in the case of seed galls caused by *A. tritici*. The stem-and-bulb nematode *Ditylenchus dipsaci* causes necrosis and discoloration in stem and leaves. Lesions and leaf spots appear due to feeding of *Aphelenchoides*. Leaf and stem galls are produced by *Anguina millefolii*.

Some characteristic below-ground symptoms are induced by root-parasitic nematodes. These include the following:

Root Knots

A root knot is the formation of a hard mass of wood in the root. Galls (swellings) may also be produced in the infested root. *Meloidogyne* forms characterisic galls on the roots of many plants. *Ditylenchus, Longidorus,* and *Xiphinema* may cause gall-like swellings on roots.

Stubby Root

Paratrichodorus induces stubby (short and broad) root, with elongated swellings on root endings.

Root Lesions

These are typical injuries caused by the penetration and movement of nematodes through the cortex of the root. The lesion nematode *Pratylenchus* and the burrowing nematode *Radopholus* cause root lesions. Small lesions are sometimes induced by the ring nematode *Criconemoides*.

Root Rots

Some nematodes may cause extensive damage to fleshy root tissues. This may be followed by rotting, probably due to a secondary invasion by fungi and bacteria. *Ditylenchus destructor* causes rotting of potato and beet roots.

Hairy Root

During feeding, some nematodes kill distal parts of the root by girdling. The remaining portion of the root that is alive usually develops branch rootlets. This abnormality is called hairy root, bearding, or witches'-broom. *Pratylenchus* spp., *Naccobus,* and *Meloidogyne hapla* cause this type of symptom.

Root Surface Necrosis

Some ectoparasites may kill the surface cells of large areas, causing superficial discoloration. *Paratrichodorus* may cause this symptom.

Devitalized Roots

Due to nematode feeding, the growth of root tips stops even though the cells are not actually killed. These roots may produce branches, which in turn stop further growth. *Trichodorus* causes the production of devitalized roots.

Curly Tip

Sometimes a root injury caused by nematodes may retard elongation of root tip and cause it to curl. Dagger nematode *(Xiphinema)* and root-knot nematode *(Meloidogyne)* induce curly tip symptoms.

Coarse Root

Nematode infestation may inhibit the growth of lateral roots, resulting in an open root system that has main roots without branches. This type of symptom is called coarse root and is caused by *Paratrichodorus* and *Belanolaimus*.

MANAGEMENT OF PARASITIC NEMATODES

Quarantine

Parasitic nematodes can be excluded from nonendemic areas that extend beyond governmental boundaries where it is possible for quarantine personnel to examine them at a few entry points. Domestic quarantines have been established in the United States to prevent the spread of the potato cyst (golden) nematodes since 1941. All movement of plants, soil, etc., from land still under cultivation on Long Island was rigidly controlled. The United States Department of Agriculture (U.S.D.A.) Golden Nematode Control Project prescribed stringent import controls to restrict the spread of the nematode (Southey, 1982). Most European countries legislated to prevent the spread of the potato cyst nematodes to apparently clean areas and to prevent multiplication to damaging levels. In the Netherlands, *Ditylenchus dipsaci* in flower bulbs was excluded by strict quarantine. The Dutch plant health regulations require inspection of all bulb crops, and nematode-infested stocks must be given a prescribed hot-water treatment or be destroyed.

Sanitation

Nematodes do not move more than a few inches per year of their own accord, and most spread is passive. Infested farm implements may spread the nematodes to neighboring fields. Farm implements should be cleaned before taking them to new fields.

Use of Clean Seed

Some seeds may be contaminated with nematodes. *Anguina tritici* galls are often found mixed with wheat seeds. Such seeds should be cleaned and the galls should be eliminated.

Selection of Healthy Propagating Material

In plants that are vegetatively propagated, it is possible to eliminate nematodes by selecting uninfected plants or parts of plants for propagation. The banana rhizome rot nematode *Radopholus similis* spreads through suckers. Hence, healthy suckers alone should be used for planting. In Florida, a state-operated certifying program exists. It certifies the citrus planting material as free from the citrus nematode *Tylenchulus semipenetrans* to exclude the nematode from new citrus planting areas of the state (Duncan and Cohn, 1990).

Crop Rotation

Crop rotation may be beneficial in eliminating or reducing some nematode populations in the field. *Heterodera, Globodera,* and *Meloidogyne* do not survive in soil for long, and therefore can be controlled by proper crop rotation. *Aphelenchoides* disappears in soil in about two months and hence no crop rotation is needed to eliminate it. On the other hand, *Ditylenchus dipsaci* persists in heavy soil for a long time and crop rotation does little to reduce its population.

Soil Amendments

Soil amendments are commonly used to control nematodes. Manures, compost, and oil cakes affect the population levels of plant-parasitic nematodes. Application of organic manures may enhance nematophagus fungi and antagonistic bacteria, and hence reduce the nematode infestation (Oka and Yermiyahu, 2002). Application of chitin can also reduce nematode population. A commercial product (Clandosan) containing chitin and urea has been registered in the United States. Chitin amendments may release ammonia upon degradation at rates that can be nematicidal. Chitin application may result in a buildup of chitinolytic organisms that may contribute to the mortality of nematode eggs that contain chitin (Duncan, 1991).

Plants Inhibitory to Nematodes

Root exudates of some plants are inhibitory to nematodes. *Tagetes erecta* reduces *Pratylenchus* infection through nematicidal action. Asparagus reduces *Paratrichodorus christiei* population. Crotolaria reduces *Meloidogyne* population. White mustard reduces the potato cyst nematode. These plants can be grown as an alternate crop or as an intercrop. Intercropping will be more beneficial.

Trap Cropping

Trap cropping is one of the important components in the integrated nematode management system. It is useful to control *Globodera* and *Heterodera* spp. In this system, a crop that causes the nematodes to hatch is planted; if the crop is a host plant it must be destroyed before the nematodes mature, but if the larvae do not develop, the crop can be allowed to mature normally. Oat is an efficient trap crop for *Heterodera avenae,* a nematode that attacks wheat and barley. Oat should be ploughed in before the nematodes mature.

Biological Control

Several antagonistic organisms have been reported to be present in soil. Several *Pseudomonas* spp. are antagonistic to the nematodes. Predacious nematodes are common in soil. A large amount of organic matter is essential to activate these biological agents for control of nematode infection. The bacterium *Pasteuria penetrans,* nematophagus fungi *Dactylella, Dactylaria,* and *Arthrobotrys,* and egg parasites such as *Paecilomyces lilacinus* are used for the control of nematodes (Sayre and Walter, 1991). Mixtures of biocontrol agents enhance biological control of *Meloidogyne javanica* in tomato (Siddiqui and Shukat, 2002).

Physical Methods

Infested soil can be sterilized with steam. Potato cyst nematodes are killed by passing the steam into the soil through perforated pipes. Hot-water treatment can be beneficial to control cyst nematodes in potato tubers. Solarization using plastic mulches may help to reduce the nemetode population (Salch et al., 1988).

Host Plant Resistance

Several resistant crop varieties have been developed against *Ditylenchus, Meloidogyne, Heterodera,* and *Globodera* spp. using both dominant major gene (qualitative) resistance and polygenic (quantitative) resistance. Resistance in most of the developed resistant varieties is conferred by dominant major genes. Major gene resistance may be more durable against nematodes than against fungal and bacterial pathogens because nematodes disperse slowly and reproduce at relatively low numbers. Hence, several nematode-resistant varieties have been developed incorporating a single resistance gene (Duncan, 1991). Some genes confer resistance to more than one species of nematode. A gene in tomato, *Mi,* confers resistance to *Meloidogyne incognita, M. javanica,* and *M. arenaria.* This gene has been incorporated into a large number of tomato varieties with good agronomic characteristics. Growers in California choose to use these resistant varieties rather than other means to manage the nematodes (Duncan, 1991). The resistance gene *Mi* in tomato has been cloned (Rossi et al., 1998), and another gene, *Mi-9,* has been characterized (Ammiraju et al., 2003).

Chemical Control

Many fumigants are available to eradicate nematodes. Methyl bromide, dichloropropene, dibromochloropropane, and metham-sodium can be injected into the soil to eradicate nematodes. Some of the granules are known to reduce the nematode population. Aldicarb, oxamyl, and carbofuran can also be useful to reduce the nematode infestation. The following are the nematicides available for management of nematodes.

Aldicarb (2-methyl-2-(methylthio) propionaldehyde *O*-methylcarbamoyloxime); Temik (trade name); granules; used to control free-living nematodes

Chloropicrin (trichloronitromethane); Pic-Clor, Chlor-O-Pic (trade names); preplant soil fumigant

Carbofuran (2,3-dihydro-2,2-dimethylbenzofuran-7-yl-methylcarbamate); Furadan (trade name); granules and flowable formulations; soil application

D-D (1,2-dichloropropane with 1,3-dichloropropene); Vidden D (trade name); soil-injected nematicide; mixed formulations: D-D + methyl isothiocyanate, D-D + chloropicrin

Dazomet (tetrahydro-3.5-dimethyl-1,3,5-thiadiazine-2 thione); Basamid, Mylone (trade names); soil sterilant; used as preplant treatment in tobacco, turf seedbeds

Diazinon (*O,O*-diethyl *O*-2-isopropyl-6-methylpyrimidin-4-yl phosphorothioate); Diazol, Basudin, Neocidol, Exodin, Diazitol, Spectracide, Sarolex, Diagran (trade names); granules; effective against many nematodes

Dibromochloropropane (1,2-dibromochloropropane); Nemagon, Fumazone (trade names); emulsifiable concentrate and nonemulsifiable concentrate; nematicidal soil sterilant used on citrus, grapes, peanuts, deciduous fruit, cotton, and vegetables

Dichlofenthion (*O*-(2,4-dichlorophenyl) *O,O*-diethyl phosphorothioate); Mobilawn, VC 13 Nemacide (trade names); nonsystemic nematicide; controls noncyst-forming nematodes on turf and ornamentals; mixed formulation: dichlofenthion + thiram

Dichloropropene (1,3-dichloropropene); Telone (trade name); liquid formulation; nematicidal control in fruit, flower, and vegetable crops, cotton, soybean, and peanut; effective against potato cyst nematode, root-knot nematode in cucumbers and tomato, and stem nematode in strawberry

Ethoprophos (*O*-ethyl S, S-dipropyl phosphorodithioate); Mocap, Prophos (trade names); granules and emulsifiable concentrate formulations; nonsystemic nematicide; controls nematodes in potato

Ethylene dibromide (1,2-dibromoethane); Dowfume, Bromofume, Soilbrom, Soilfume (trade names); liquid formulation; mixed formulations: Ethylene dibromide + chloropicrin, ethylene + methyl bromide

Fenamiphos (ethyl-4-methylthio-*m*-tolyl isopropylphosphoramidate); Nemacur (trade name); granules and emulsifiable concentrate; systemic nematicidal action; absorbed through leaves and roots and controls ecto- and endo-parasitic nematodes; applied broadcast, in band, in-the-row, by drench before or at planting time or to established plants

Fensulfothion (*O,O*-diethyl *O*-4-methylsulphinylphenyl phosphorothioate); Dasanit, Terracur P (trade names); systemic and contact action; soil application for nematicidal control of parasitic, sedentary, and free-living nematodes

Metham-sodium (sodium methyldithiocarbamate); Vapam, VPM,
Trimaton, Maposol, Sistan (trade names); soil fumigant applied prior to
planting edible crops
Methyl bromide (bromomethane); Haltox, Terabol, Meth-O-Gas,
Dowfume, Brom-O-Gaz, Bercema (trade names); soil fumigant
Methyl isothiocyanate (isothiocyanatomethane); Trapex (trade name); soil
fumigant
Oxamyl (*N,N*-dimethyl-2-methylcarbamoyloxyimino-2-(methylthio)
acetamide); Vydate (trade name); granules and water-soluble liquid
formulations; systemic contact nematicide; uptake through leaves and
roots, transported in plants downward to the root
Terbufos (S-*tert*-butylthiomethyl *O,O*-diethyl phosphorodithionate);
Counter (trade name); granules
Thionazin (*O-O*-diethyl-*O*-pyrazin-2 yl-phosphorothionate); Nemafos,
Zinophos, Cynem, Nemaphos (trade names); granules, emulsifiable
concentrate formulations; controls free-living and plant-parasitic nematodes

Some nematodes act as vectors of viruses that cause serious diseases in
crops. They are described in Chapter 10.

REFERENCES

Ammiraju, J. S. S., Veremis, J. C., Huang, X., Roberts, P. A., and Kaloshian, I.
(2003). The heat-stable root-knot nematode resistance gene *Mi-9* from *Lycoper-
sicon peruvianum* gene is localized on the short arm of chromosome 6. *Theor
Appl Genet,* 106:478-484.
Bird, A. F. (1981). *The Structure of Nematodes.* Academic Press, London.
Duncan, L. W. (1991). Current options for nematode management. *Annu Rev
Phytopathol,* 29:469-490.
Duncan, L. W. and Cohn, E. (1990). Nematode parasites of citrus. In M. Luc, R. A.
Sikora, and J. Bridge (Eds.), *Plant Parasitic Nematodes in Subtropical and
Tropical Agriculture,* CAB Institute, Wallingford, pp. 321-346.
Oka, Y. and Yermiyahu, U. (2002). Suppressive effects of composts against the
root-knot nematode *Meloidogyne javanica* on tomato. *Nematology,* 4:891-898.
Rossi, M., Goggin, F. L., Milligan, S. B., Kolashian, I., Ullman, D. E., and William-
son, V. M. (1998). The nematode resistance gene *Mi* of tomato confers resistance against the potato aphid. *Proc Natl Acad Sci USA,* 95:9750-9754.
Salch, H., Abu-Gharbich, W. I., and Al-Banna, L. (1988). Effect of solarization
combined with solar-heated water on *Meloidogyne javanica. Nematologica,*
34:290-291.
Sayre, R. M. and Walter, D. E. (1991). Factors affecting the efficacy of natural ene-
mies of nematodes. *Annu Rev Phytopathol,* 29:149-166.

Siddiqui, I. A. and Shukat, S. S. (2002). Mixtures of plant disease suppressive bacteria enhance biological control of multiple tomato pathogens. *Biol Fertil Soils,* 36:260-268.

Southey, J. F. (1982). *Plant Nematology.* Her Majesty's Stationery Office, London.

Vidhyasekaran, P. (1993). *Principles of Plant Pathology.* CBS Publishers.

Williams, T. D. (1984). Plant parasitic nematodes. In The Commonwealth Mycological Institute (compilation), *Plant Pathologist's Pocketbook.* Common Mycological Institute, Kew, England, pp. 119-136.

8

Phytoplasmas and Spiroplasmas

Phytoplasmas and spiroplasmas are bacteria that lack rigid cell walls. They belong to the class Mollicutes. Phytoplasmas cannot be cultured, whereas some spiroplasmas can be cultured. Both mollicutes are phloem limited in plants and can be transmitted by leafhoppers. Spiroplasmas are helical in shape. The structure of phytoplasmas and spiroplasmas, classification of phytoplasmas based on their 16S rRNA gene sequences, and diseases caused by them are described.

PHYTOPLASMAS

What Are Phytoplasmas?

Phytoplasmas are minute bacteria without cell walls. They inhabit phloem sieve elements in infected plants. They cannot be cultured in artificial media to date. They can pass through a bacteria-proof filter. They have been associated with diseases in cereals, vegetables, fruit crops, ornamental plants, and timber and shade trees.

Yellows are an important group of diseases and were once considered to be caused by viruses. One disease, aster yellows, was first reported in 1902, and it was considered a virus disease until 1967. Doi et al. (1967) reported that particles in ultrathin sections of the phloem of plants affected by yellows diseases, including aster yellows, resembled animal and human mycoplasmas. The particles lacked rigid cell walls, were surrounded by a single unit membrane, and were sensitive to the antibiotic, tetracycline. The term *mycoplasmalike organisms* (MLOs) was used to refer to such causal organisms of yellows diseases from 1967 to 1994. The name "phytoplasma" was introduced by the Phytoplasma Working Team at the Tenth Congress of the International Organization for Mycoplasmology (Lee et al., 2000).

Classification of Phytoplasmas

Despite several attempts during the past three decades, phytoplasmas have not been cultured. Hence, phytoplasmas could not be classified based on the traditional tests applied to cultured prokaryotes. Woese et al. (1980) distinguished phytoplasmas (mycoplasmas) by analyzing highly conserved rRNA gene sequences in prokaryotes. It has been suggested that the phyto-pathogenic, mycoplasmalike organisms belong to the class Mollicutes. Four major phylogenetic groups (clades) have been identified in the class Mollicutes: *Mycoplasma hominis*, *M. pneumoniae*, *Spiroplasma*, and *Anaeroplasma*. Phylogenetic analyses of 16S rRNA and *rp* (ribosomal proteins) gene operon sequences showed that phytoplasmas formed a large, discrete, monophyletic clade within the expanded Anaeroplasma clade (Gundersen et al., 1994). Within the phytoplasma clade, several distinct subclades (monophyletic groups or taxa) have been identified based on the restriction fragment length polymorphism (RFLP) of 16S rRNA and ribosomal protein sequences (Lee et al., 2000). It has been suggested that the phytoplasma clade should be distinguished at the taxonomic level of a genus, and each subclade (corresponding 16S rRNA sequence) should represent a species. However, the naming of new species in the class Mollicutes requires descriptions of the species in pure culture. Phytoplasmas cannot be isolated in pure culture, and the phenotypic characteristics used to describe mollicute species are unattainable for uncultured phytoplasmas. Therefore, a provisional classification system using the *Candidatus* category has been developed (Lee et al., 2000). Five *Candidatus* Phytoplasma species have been recognized. The following is the classification of phytoplasmas based on RFLP analysis of 16S rRNA and ribosomal protein sequences (Seemuller et al., 1998; Lee et al., 2000):

Aster Yellows Group (16 SrI)

> Subgroup 16SrI-A, 16SrI-A (rp-A): aster yellows (in China aster), lettuce yellows, periwinkle little leaf, tomato big bud (Arkansas), gladiolus virescence (Italy)
> Subgroup 16SrI-B, 16SrI-B (rp-B): aster yellows (in potato, carrot, celery, clover), broccoli phyllody, chrysanthemum yellows, cabbage witches'-broom, onion virescence (yellows), mulberry dwarf, eggplant dwarf, turnip virescence
> Subgroup 16SrI-B, 16Sr-B (rp-K): hydrangea phyllody
> Subgroup 16SrI-B, 16Sr-B (rp-L): maize bushy stunt
> Subgroup 16SrI-C: strawberry green petal, clover phyllody, olive witches'-broom

Subgroup 16SrI-D: paulownia witches'-broom
Subgroup 16SrI-E: blueberry stunt
Subgroup 16SrI-F: apricot chlorotic leaf roll
Subgroup 16SrI-K: strawberry multiplier

Peanut Witches'-Broom Group (16SrIII)

Subgroup 16SrII-A: peanut witches'-broom, sweet potato witches'-broom, sunn hemp witches'-broom
Subgroup 16SrII-B (*Candidatus* Phytoplasma aurantifolia): lime witches'-broom
Subgroup 16SrII-C: cotton phyllody, soybean phyllody, *faba* bean phyllody
Subgroup 16SrII-D: sweet potato little leaf
Subgroup 16SrII-E (*Candidatus* Phytoplasma australasia): papaya mosaic, papaya yellow crinkle, tomato big bud

X-Disease Group (16SrIII)

Subgroup 16SrIII-A: peach X-disease, cherry X-disease
Subgroup 16SrIII-B: clover yellow edge (Canada), italian clover phyllody
Subgroup 16SrIII-C: pecan bunch
Subgroup 16SrIII-D: goldenrod yellows
Subgroup 16SrIII-E: spirea stunt
Subgroup 16SrIII-F: milkweed yellows
Subgroup 16SrIII-G: walnut witches'-broom
Subgroup 16SrIII-H: poinsettia branching inducing

Coconut Lethal Yellows Group (16SrIV)

Subgroup 16SrIV-A: coconut lethal yellows
Subgroup 16SrIV-B: tanzanian coconut lethal decline

Elm Yellows Group (16SrV)

Subgroup 16SrV-A, 16SrV-A (rp-A): alder yellows, elm yellows, elm witches'-broom, rubus stunt
Subgroup 16SrV-B, 16SrV-B (rp-B): cherry lethal yellows
Subgroup 16SrV-B, 16SrV-B (rp-C): jujube witches'-broom
Subgroup 16SrV-C, 16SrV-A (rp-D): flavescence doree (grapevine)

Clover Proliferation Group (16SrVI)

> Subgroup 16SrVI-A: potato witches'-broom, alfalfa witches'-broom, tomato big bud (California), clover proliferation
> Subgroup 16SrVI-B: strawberry multiplier (Canada)
> Unclassified: brinjal little leaf, willow witches'-broom

Ash Yellows Group (16SrVII)

> Subgroup 16SrVII-A (*Candidatus* Phytoplasma fraxini): ash yellows, lilac witches'-broom

Loofah Witches'-Broom Group (16SrVIII)

> Subgroup 16SrVIII-A: loofah witches'-broom

Pigeon Pea Witches'-Broom Group (16SrIX)

> Subgroup 16SrIX-A: pigeon pea witches'-broom
> Unclassified: gliricidia little leaf, picris phyllody

Apple Proliferation Group (16SrX)

> Subgroup 16SrX-A: apple proliferation, hazel decline
> Subgroup 16SrX-B: apricot chlorotic leaf roll, plum leptonecrosis, European stone fruit yellows
> Subgroup 16SrX-C: pear decline, peach yellow leaf roll

Rice Yellow Dwarf Group (16SrXI)

> Subgroup 16SrXI-A: rice yellow dwarf
> Subgroup 16SrXI-B: sugarcane grassy shoot, sugarcane whiteleaf

Stolbur Group (16SrXII)

> Subgroup 16SrXII-A: pepper stolbur, tomato stolbur, grapevine yellows (bois noir), celery yellows
> Subgroup 16SrXII-B (*Candidatus* Phytoplasma australiense): Australian grapevine yellows, papaya dieback)
> *Candidatus* Phytoplasma japonicum: Japanese hydrangea phyllody

Mexican Periwinkle Virescence Group (16SrXIII)

Subgroup 16SrXIII-A: Mexican periwinkle virescence
Subgroup 16SrXIII-B: strawberry green petal

Bermudagrass White Leaf Group (16SrXIV)

Subgroup 16SrXIV-A: bermudagrass white leaf

Symptoms of Phytoplasma Diseases

Phytoplasma diseases are characterized by several types of symptoms. The important symptoms are:

Phyllody—floral parts transformed into green leafy structures
Virescence—the development of green flowers and loss of normal flower pigments
Witches'-broom—proliferation (mass outgrowths) of the branches of woody plants; proliferation of auxiliary shoots
Yellows—uniform yellow to almost white discoloration of leaves; yellowing will be the conspicuous symptom in the diseased plant
Stunting—retardation of plant growth, small flowers, and leaves; shortened internodes
Sterility—suppressed development of reproductive structures
Leaf curling—the distortion, puffing, and crinkling of a leaf resulting from the unequal growth of its two sides
Bushy stunt—diseased plant severely stunted; its shoots are crowded, giving bushy appearance
Little leaf—in the diseased plants, the leaves are malformed into tiny chlorotic structures
Leaf roll—plant leaves tend to curl
Big bud—the fruit bearing shoots become thick, dark green structures; fruits formed in infected plants become hard and woody
Dieback—death of twigs from the tip backward
Slender shoots—abnormal elongation of internodes resulting in slender shoots
Reddening—predomination of red pigments
Blackening—blackening of tissues due to intensive necrosis

Transmission and Mode of Spread

Phytoplasmas are phloem-limited pathogens that are found in phloem sieve elements. Phytoplasmas cannot be transmitted mechanically. Sap-sucking insects such as leafhoppers and planthoppers can transmit phytoplasmas. These insects feed on phloem tissues, where they acquire phytoplasmas and transmit them from plant to plant. Phytoplasmas may overwinter in infected vectors. They may survive in perennial plants. No seed-borne phytoplasmas have been reported. However, phytoplasmas are spread by vegetative propagation through cuttings, storage tubers, rhizomes, or bulbs. They can be transmitted also through grafts.

Disease Diagnosis

Because phytoplasmas cannot be cultured, their detection is difficult. The presence of characteristic symptoms in diseased plants and subsequent observation of phytoplasma bodies in ultrathin sections were used to diagnose phytoplasma diseases. Recently, several molecular probes have been developed to diagnose these diseases. Enzyme-linked immunosorbent assays, immunofluorescence microscopy, DNA probes, Southern hybridization and RFLP analysis of phytoplasma genomic DNA, and PCR assays using PCR primers based on sequences of cloned phytoplasma DNA fragments have been used to detect phytoplasma diseases (see Chapter 13 for details about these techniques). The most useful method is the PCR method, which employs phytoplasma universal (generic) or phytoplasma group-specific oligonucleotide primers that are based on highly conserved 16S rRNA gene sequences.

Phytoplasma Disease Management

Disease-free planting material should be used for planting. Disease-resistant varieties are available for protection against some diseases. Insect vectors should also be controlled. Genetic engineering technology has been exploited to develop transgenic plants that are resistant to phytoplasma. The expression of antibodies in transgenic plants confers resistance to phytoplasma diseases (Chen and Chen, 1998).

SPIROPLASMAS

Spiroplasmas are wall-less prokaryotes that belong to the class Mollicutes (Kirkpatrick and Smart, 1995). They are helical prokaryotes that lack a rigid cell wall. Unlike phytoplasmas, some spiroplasmas can be cultured (Daniels, 1983). Three plant-pathogenic spiroplasmas have been identified: *Spiroplasma citri,* the causal agent of citrus stubborn and horseradish brittleroot; *S. kunkelii,* the causal agent of corn stunt disease; and *S. phoen-*

icium, a spiroplasma isolated from diseased periwinkle plants (Kirkpatrick and Smart, 1995). *Spiroplasma citri* can be cultured, whereas *S. kunkelli* cannot. Spiroplasmas, in common with phytoplasmas, are found only in phloem sieve tubes (Bove, 1984). *Spiroplasma citri* has a genome size close to 10^9 Daltons (Bove, 1984). The spiroplasma has been shown to contain three DNA polymerases. Spiroplasmas carry only one (16S-rRNA) or two rRNA operons (16S-23S-5 S rRNA), whereas bacteria such as *Escherichia coli* and *Bacillus subtilis* have seven to ten rRNA operons.

Spiroplasma citri may be transmitted by the beet leafhopper *Circulifer tenellus. Spiroplasma kunkelii* is transmitted by the corn leafhopper *Dalbulus maidis* and certain other *Dalbulus* and *Baldulus* species. Spiroplasmas can infect and multiply within the insects. Spiroplasmas can be detected in many organs and tissues of infected insects. Spiroplasmas pass from the insect gut lumen into the blood stream, passing through the endoplasmic reticulum. However, the transmission of spiroplasmas by insects has not been convincingly demonstrated.

Important Diseases Caused by Spiroplasmas

Citrus stubborn: *Spiroplasma citri* has a wide host range. At least 15 families of wild dicotyledonous and one family of monocotyledonous plants have been recorded as hosts (Daniels, 1983).

Corn stunt: *Spiroplasma kunkelli* causes dwarfing of the plant, chlorotic, distorted leaf margins, loss of apical dominance, and reduced flower size. However, infected corn plants do not wilt and die.

REFERENCES

Bove, J. M. (1984). Wall-less prokaryotes of plants. *Annu Rev Phytopathol,* 22:361-396.

Chen, Y. D. and Chen, T-A. (1998). Expression of engineered antibodies in plants: A possible tool for spiroplasma and phytoplasma disease control. *Phytopathology,* 88:1367-1371.

Daniels, M. J. (1983). Mechanisms of spiroplasma pathogenicity. *Annu Rev Phytopathol,* 21:29-43.

Doi, Y. M., Teranaka, M., Yora, K., and Asuyama, H. (1967). Mycoplasma or PLT-group-like microorganisms found in the phloem elements of plants infected with mulberry dwarf, potato witches' broom, aster yellows, or paulownia witches' broom. *Ann Phytopathol Soc Jpn,* 33:259-266.

Gundersen, D. E., Lee, I.-M., Rehner, S. A., Davis, R. E., and Kingsbury, D. T. (1994). Phylogeny of mycoplasmalike organisms (phytoplasmas): A basis for their classification. *J Bacteriol,* 176:5244-5254.

Kirkpatrick, B. C. and Smart, C. D. (1995). Phytoplasmas: Can phylogeny provide the means to understand pathogenicity? *Adv Bot Res,* 21:188-212.

Lee, I-M., Davis, R. E., Gundersen-Rindal, D. E. (2000). Phytoplasma: Phytopathogenic mollicutes. *Annu Rev Microbiol,* 54:221-255.

Seemuller, E., Marcone, C., Lauer, U., Ragozzino, A., and Goschl, M. (1998). Current status of molecular classification of the phytoplasmas. *J Plant Pathol,* 80: 3-26.

Woese, C. R., Maniloff, J., and Zablen, L. B. (1980). Phylogenetic analysis of the mycoplasmas. *Proc Natl Acad Sci USA,* 77:494-498.

9

Viroids

Viroids are covalently closed, circular RNA molecules (Hammond and Zhao, 2000). Viroids were the first circular RNAs to be discovered in nature. These are the smallest known infectious agents (Elena et al., 2001). Potato spindle tuber viroid was the first viroid reported, and it is widely prevalent in different potato growing areas. Citrus exocortis viroid is widespread in citrus production areas where trifoliate orange *(Poncirus trifoliate)* is used as root stock (Guo et al., 2002). Hop stunt viroid has a wide range of hosts. Mechanism of viroid pathogenesis in plants has been elucidated recently. This chapter describes the viroids and viroid-like satellite RNAs (virusoids).

STRUCTURE OF VIROIDS

Viroids are nucleic acids that exist naturally with no protein coat. They consist of ribonucleic acid (RNA). These miniviruses are the smallest known causal organisms of infectious diseases. They are subviral, and their size ranges from 246 to 388 nucleotides in length (Symons, 1991). The RNA structure of viroids is different from transfer RNA (tRNA), ribosomal RNA (rRNA), and messenger RNA (mRNA). Viroids were the first circular RNAs to be discovered in nature; this feature differentiates them from other RNAs (Reisner, 1991).

Important Viroids Causing Diseases

The following are viroids that cause diseases in important crops:

>   Apple scar skin viroid
>   Australian grapevine viroid
>   Avocado sunblotch viroid
>   Chrysanthemum chlorotic mottle viroid
>   Chrysanthemum stunt viroid
>   Citrus exocortis viroid
>   Coconut cadang-cadang viroid
>   Coconut tinangaja viroid

Columnea pale fruit viroid
Cucumber pale fruit viroid
Grapevine viroid 1B
Grapevine yellow speckle viroid
Hop latent viroid
Hop stunt viroid
Potato spindle tuber viroid
Tomato apical stunt viroid
Tomato planta macho viroid

VIROID INFECTION PROCESS AND MANAGEMENT

Infection Process

The viroid RNA does not code for any genes. Viroid replication and pathogenesis may depend completely on the enzyme systems of the host. The viroid RNA is dependent upon the host for its replication as well as intraplant movement. The functions necessary for propagation of the viroids are derived completely from the host. The viroids are associated with and replicate in either the nucleus or the chloroplasts of the plants (Zaitlin and Palukaitis, 2000). They are replicated by host-encoded RNA polymerases (Fels et al., 2001). They do not encode proteins. Viroids replicate within the nucleus of infected cells without a helper virus. Viroids are transported into the plant nucleus, and typically potato spindle tuber viroid (PSTV) possesses a sequence and/or structural motif for nuclear transport (Woo et al., 1999). Phloem proteins may be involved in systemic transport of viroids in the plants (Gomez and Pallas, 2001). Phloem protein 2, a dimeric lectin, is the abundant component of phloem exudates of cucumber. This protein interacts with the viroid RNA and facilitates the systemic movement of hop stunt viroid (Owens et al., 2001).

Symptoms

Infection with viroids does not result in obvious macroscopic symptoms. Common symptoms of viroid diseases include retardation of plant growth and stunting. Potato plants infected with the potato spindle tuber viroid are smaller than healthy plants. However, tuber symptoms are prominent. The diseased tubers are spindle shaped. Citrus trees infected with the *Citrus exocortis* viroid are stunted. Symptoms of the disease include scaling of the bark below the graft union. Stunted trees crop well for their size, and the fruit is normal (Diener, 1979). Stunting is the important symptom of tomato plants affected by the tomato bunchy top viroid, hop plants infected by the hop stunt viroid, and chrysanthemum plants infected by the chrysanthemum stunt viroid and the chrysanthemum chlorotic mottle viroid.

Mode of Spread

Viroids are highly contagious and mechanically transmitted. They are spread by leaf contact. Viroids are spread also by contaminated planting and cultivating equipments. They may be disseminated mostly as a result of cultural operations through contaminated knives, tools, and hands. Some reports indicate that viroids are transmitted by insects. Potato spindle tuber viroid has been reported to be transmitted by aphids (*Myzus persicae* and *Macrosiphum euphorbiae*), grasshoppers (*Melanoplus* spp.), flea beetles (*Epitrix cucumeris* and *Systena taeniata*), tarnished plant bugs *(Lygus pratensis),* larvae of the Colorado potato beetle *(Leptinotarsa decemlineata),* and the leaf beetle *(Disonycha triangularis).* However, the transmission of viroids by insects is negligible, and mechanical transmission is more important.

Viroid Disease Management

Because viroids are spread mechanically, disease-free planting materials should be used for planting. Cutting knives and all planting and field equipment should be cleaned scrupulously. Commercial cultivars with high resistance to the diseases are lacking.

VIRUSOIDS (ENCAPSIDATED, VIROIDLIKE, SATELLITE RNAs)

Some viruses contain a viroidlike satellite RNA in addition to a linear, single-stranded molecule of genomic RNA. Such viroid-like satellite RNAs are called *virusoids*. They show little sequence homology with viroids, but they do show significant homology with the linear satellite RNA associated with *Tobacco ringspot virus*. The virusoids in infected plants exist almost solely as circular molecules, either free or encapsidated within virion of the helper virus. This is in contrast to similarly sized satellite RNA of the *Tobacco ringspot virus* that is found as both linear and circular single-stranded molecules in vivo, but only the linear form is encapsidated (Symons, 1991). Virusoids do not code for any polypeptides.

Only four virusoids have been discovered (Symons, 1991). They are found with the helper viruses, *Lucerne transient streak virus* (virusoid name abbreviation: vLTSV, 324 nucleotides in length), *Solanum nodiflorum mottle virus* (vSNMV, 377 nucleotides in length), *Subterranean clover mottle virus* (vSCMoV, 332 and 388 nucleotides in length), and *Velvet tobacco mottle virus* (vVTMoV, 365 and 366 nucleotides in length).

REFERENCES

Diener, T. (1979). *Viroids and Viroid Diseases.* John Wiley & Sons, New York.

Elena, S. F., Dopazo, J., Pena, M. de la., Flores, R., Diener, T. O., and Moya, A. (2001). Phylogenetic analysis of viroid-like satellite RNAs from plants: A reassessment. *J Mol Evol,* 53:155-159.

Fels, A., Hu, K. H., and Riesner, D. (2001). Transcription of potato spindle tuber viroid by RNA polymerase II starts predominantly at two specific sites. *Nucleic Acids Res,* 29:4589-4597.

Gomez, G. and Pallas, V. (2001). Identification of an in vitro ribonucleoprotein complex between viroid RNA and a phloem protein from cucumber plants. *Mol Plant-Microbe Interact,* 14:910-913.

Guo, W. W., Cheng, Y. J., and Deng, X. X. (2002). Regeneration and molecular characterization of intergeneric somatic hybrids between *Citrus reticulata* and *Poncirus trifoliata. Plant Cell Rep,* 20:829-834.

Hammond, R. W. and Zhao, Y. (2000). Characterization of tomato protein kinase gene induced by infection by potato spindle tuber viroid. *Mol Plant-Microbe Interact,* 13:903-910.

Owens, R. A., Blackburn, M., and Ding, B. (2001). Possible involvement of the phloem lectin in long-distance viroid movement. *Mol Plant-Microbe Interact,* 14:905-909.

Reisner, D. (1991). Viroids: From thermodynamics to cellular structure and function. *Mol Plant-Microbe Interact,* 4:122-131.

Symons, R. H. (1991). The intriguing viroids and virusoids: What is their information content and how did they evolve? *Mol Plant-Microbe Interact,* 4:111-121.

Woo, Y. M., Itaya, A., Owens, R. A., Tang, L., Hammond, R. W., Chou, H. C., Lai, M. M. C., and Ding, B. (1999). Characterization of nuclear import of potato spindle tuber viroid RNA in permeabilized protoplasts. *Plant J,* 17:627-635.

Zaitlin, M. and Palukaitis, P. (2000). Advances in understanding plant viruses and virus diseases. *Annu Rev Phytopathol,* 38:117-143.

– 10 –

Viruses

Viruses are important groups of plant pathogens. Several viruses are known to cause crop diseases, and 977 plant virus species have been officially or provisionally recognized (Fauquet and Mayo, 1999). Rice tungro virus disease causes heavy losses in rice grown in different countries in Asia (Azzam and Chancellor, 2002). Citrus tristeza virus disease has caused the decline and death of close to 40 million trees grafted on sour orange rootstocks in Spain and several million trees in other growing areas in the Americas, Australia, and South Africa (Terrada et al., 2000). Tomato spotted wilt virus causes significant yield losses in many food and ornamental crops (Garcia et al., 2000). It was difficult to classify plant viruses into class, order, family, genus, and species due to lack of knowledge of genetic relationships between viruses. Recently viruses have been grouped into genera, based on molecular biological studies. A few families and an order have been created. The virus infection process is now studied at the molecular level, and the viral genes involved in the pathogenicity have been characterized. Insects, mites, fungi, protozoa, and nematodes serve as vectors of viruses, and studies of their interaction help to develop integrated virus disease management technology. All these aspects are described in this chapter.

STRUCTURE OF VIRUSES

Viruses are infectious agents that are not visible under the microscope (i.e., they are submicroscopic) and are small enough to pass through a bacterial filter. Viruses do not have a metabolism of their own and depend upon living host cells for multiplication. They have only one type of nucleic acid, either ribonucleic acid or deoxyribonucleic acid. Nucleic acids may be single stranded (ss) or double stranded (ds), and may be single segmented or with a few segments. The nucleic acids are enclosed in a coat of protein. Some viruses may have an extra coat (envelope). The number of coat protein polypeptides may range from one to seven. The virus particles are of different shapes and sizes. They may be elongated (rod shaped), as in the case of *Tobacco mosaic virus;* isometric *(Cowpea mosaic virus);* bacilliform *(Alfalfa mosaic virus, Rice tungro bacilliform virus);* spherical *(Rice tungro spherical virus);* or geminate (i.e., arranged in pairs). Some viruses have different particle types that differ in size or nucleic acids. Multipartite (multicomponent, multigenome segments) viruses consist of two to four single-stranded (ss) RNAs. Two RNA pieces occur in *Tobacco rattle virus, Bean pod mottle virus,* and *Pea enation mosaic virus;* three RNA pieces occur in *Cucumber mosaic virus* and *Prune dwarf virus;* and four RNA pieces occur in *Alfalfa mosaic virus* and *Tomato spotted wilt virus.* All RNA pieces are needed for multicomponent virus infection.

The particle shape and size of important characterized viruses, number of genome segments, nature of RNA or DNA, and number of coat protein polypeptides of these viruses are provided here:

Alfalfa mosaic virus (AMV)—Bacilliform, 28-58 nm × 18 nm, 4 genome
    segments, linear ssRNA, 1 coat protein polypeptide
Andean potato latent virus (APLV)—Isometric, about 29 nm, 1 genome
    segment, linear ssRNA, 1 coat protein polypeptide
Andean potato mottle virus (APMV)—Isometric, about 30 nm, 2 genome
    segments, linear ssRNA, 2 coat protein polypeptides
Arabis mosaic virus (ArMV)—Isometric, about 30 nm, 2 genome
    segments, linear ssRNA, 2 coat protein polypeptides
Barley stripe mosaic virus (BSMV)—Rigid rod, 100-150 nm × 20 nm,
    3 genome segments, linear ssRNA, 1 coat protein polypeptide
Barley yellow dwarf virus (BYDV)—Polyhedral, about 30 nm, 1 genome
    segment, linear ssRNA, 1 coat protein polypeptide
Barley yellow mosaic virus (BaYMV)—Flexuous rod, 680-900 nm × 11
    nm, 1 genome segment, linear ssRNA, 1 coat protein polypeptide
Barley yellow striate mosaic virus (BYSMV)—Bacilliform, 330 × 45 nm,
    1 genome segment, linear ssRNA, 5 coat protein polypeptides
Bean common mosaic virus (BCMV)—Flexuous rod, 680-900 nm × 11 nm,
    1 genome segment, linear ssRNA, 1 coat protein polypeptide
Bean golden mosaic virus (BGMV)—Geminate (arranged in pairs),
    18 × 30 nm
Bean pod mottle virus (BPMV)—Isometric, about 30 nm, 2 genome segments, linear ssRNA, 2 coat protein polypeptides
Bean rugose mosaic virus (BRMV)—Isometric, about 30 nm, 2 genome
    segments, linear ssRNA, 2 coat protein polypeptides
Bean yellow mosaic virus (BYMV)—Flexuous rod, 680-900 nm × 11 nm,
    1 genome segment, linear ssRNA, 1 coat protein polypeptide
Beet leaf curl virus (BLCV)—Bacilliform, 225-350 nm × 45 nm,
    1 genome segment, linear ssRNA, 5 coat protein polypeptides

Beet mosaic virus (BtMV)—Flexuous rod, 680-900 nm × 11 nm, 1 genome segment, linear ssRNA, 1 coat protein polypeptide

Beet western yellows virus (BWYV)—Polyhedral, about 30 nm, 1 genome segment, linear ssRNA, 1 coat protein polypeptide

Beet yellow stunt virus (BYSV)—Flexuous rod, 600-2,000 nm × 12 nm, 1 genome segment, linear ssRNA, 1 coat protein polypeptide

Beet yellows virus (BYV)—Flexuous rod, 600-2000 nm × 12 nm, 1 genome segment, linear ssRNA, 1 coat protein polypeptide

Black raspberry latent virus (BRLV)—Isometric, 26-35 nm, 3 genome segments, linear ssRNA, 1 coat protein polypeptide

Broad bean stain virus (BBSV)—Isometric, about 30 nm, 2 genome segments, linear ssRNA, 2 coat protein polypeptides

Broad bean true mosaic virus (BBTMV)—Isometric, about 30 nm, 2 genome segments, linear ssRNA, 2 coat protein polypeptides

Brome mosaic virus (BMV)—Isometric, about 29 nm, 3 genome segments, linear ssRNA, 1 coat protein polypeptide

Carnation Italian ringspot virus (CIRSV)—Isometric, about 34 nm, 1 genome segment, linear ssRNA, 1 coat protein polypeptide

Carnation latent virus (CLV)—Flexuous rod, 600-700 nm × 13 nm, 1 genome segment, linear ssRNA, 1 coat protein polypeptide

Carnation necrotic fleck virus (CNFV)—Flexuous rod, 600-2000 nm × 12 nm, 1 genome segment, linear ssRNA, 1 coat protein polypeptide

Carnation ringspot virus (CRSV)—Isometric, about 34 nm, 2 genome segments, linear ssRNA, 1 coat protein polypeptide

Carnation vein mottle virus (CVMV)—Flexuous rod, 680-900 nm × 11 nm, 1 genome segment, linear ssRNA, 1 coat protein polypeptide

Carrot latent virus (CaLV)—Bacilliform, 220 × 70 nm, 1 genome segment, linear ssRNA, 5 coat protein polypeptides

Carrot red leaf virus (CaRLV)—Polyhedral, 25-30 nm, 1 genome segment, linear ssRNA, 1 coat protein polypeptide

Carrot thin leaf virus (CTLV)—Flexuous rod, 680-900 nm × 11 nm, 1 genome segment, linear ssRNA, 1 coat protein polypeptide

Carrot yellow leaf virus (CYLV)—Flexuous rod, 600-2000 nm × 12 nm, 1 genome segment, linear ssRNA, 1 coat protein polypeptide

Cauliflower mosaic virus (CaMV)—Isometric, 42-50 nm, 1 circular genome segment, dsDNA, 1 coat protein polypeptide

Celery mosaic virus (CeMV)—Flexuous rod, 680-900 nm × 11 nm, 1 genome segment, linear ssRNA, 1 coat protein polypeptide

Cherry leafroll virus (CLRV)—Isometric, about 30 nm, 2 genome segments, linear ssRNA, 2 coat protein polypeptides

Chloris striate mosaic virus (CSMV)—Geminate (arranged in pairs), 18 × 30 nm

Citrus leaf rugose virus (CiLRV)—Isometric, 26-35 nm, 3 genome
  segments, linear ssRNA, 1 coat protein polypeptide
Citrus tristeza virus (CTV)—Flexuous rod, 600-2000 nm × 12 nm,
  1 genome segment, linear ssRNA, 1 coat protein polypeptide
Citrus variegation virus (CVV)—Isometric, 26-35 nm, 3 genome
  segments, linear ssRNA, 1 coat protein polypeptide
Clover yellow mosaic virus (ClYMV)—Flexuous rod, 480-580 nm ×
  13 nm, 1 genome segment, linear ssRNA, 1 coat protein polypeptide
Clover yellow vein virus (ClYVV)—Flexuous rod, 680-900 nm × 11 nm,
  1 genome segment, linear ssRNA, 1 coat protein polypeptide
Clover yellows virus (CYV)—Flexuous rod, 600-2000 nm × 12 nm,
  1 genome segment, linear ssRNA, 1 coat protein polypeptide
Cowpea aphid-borne mosaic virus (CABMV)—Flexuous rod, 680-900
  nm × 11 nm, 1 genome segment, linear ssRNA, 1 coat protein
  polypeptide
Cowpea mosaic virus (CPMV)—Isometric, about 30 nm, 2 genome
  segments, linear ssRNA, 2 coat protein polypeptides
Cucumber mosaic virus (CMV)—Isometric, about 29 nm, 3 genome
  segments, linear ssRNA, 1 coat protein polypeptide
Cymbidium mosaic virus (CybMV)—Flexuous rod, 480-580 nm × 13 nm,
  1 genome segment, linear ssRNA, 1 coat protein polypeptide
Glycine mottle virus (GmoV)—Isometric, about 34 nm, 1 genome
  segment, linear ssRNA, 1 coat protein polypeptide
Grapevine fan leaf virus (GFLV)—Isometric, about 30 nm, 2 genome
  segments, linear ssRNA, 2 coat protein polypeptides
Lettuce mosaic virus (LMV)—Flexuous rod, 680-900 nm × 11 nm,
  1 genome segment, linear ssRNA, 1 coat protein polypeptide
Lettuce necrotic yellows virus (LNYV)—Bacilliform (enveloped), 360 ×
  52 nm, 1 genome segment, linear ssRNA, 5 coat protein polypeptides
Maize chlorotic dwarf virus (MCDV)—Isometric, about 30 nm, 1 linear
  genome segment, ssRNA
Maize mosaic virus (MMV)—Bacilliform, 240 × 48 nm, 1 genome
  segment, linear ssRNA, 5 coat protein polypeptides
Maize streak virus (MSV)—Geminate, 18 × 30 nm, 2 circular genome
  segments, ssDNA, 1 coat protein polypeptide
Muskmelon vein necrosis virus (MkVNV)—Flexuous rod, 600-700 nm
  × 13 nm, 1 genome segment, linear ssRNA, 1 coat protein polypeptide
Narcissus yellow stripe virus (NYSV)—Flexuous rod, 680-900 nm ×
  11 nm, 1 genome segment, linear ssRNA, 1 coat protein polypeptide
Oat blue dwarf virus (OBDV)—Polyhedral, 30 nm, 1 genome segment,
  ssRNA

Oat mosaic virus (OMV)—Flexuous rod, 680-900 nm × 11 nm, 1 genome segment, linear ssRNA, 1 coat protein polypeptide

Oat striate mosaic virus (OSMV)—Bacilliform, 400 × 100 nm, 1 genome segment, linear ssRNA, 5 coat protein polypeptides

Onion yellow dwarf virus (OYDV)—Flexuous rod, 680-900 nm × 11 nm, 1 genome segment, linear ssRNA, 1 coat protein polypeptide

Papaya mosaic virus (PapMV)—Flexuous rod, 480-580 nm × 13 nm, 1 genome segment, linear ssRNA, 1 coat protein polypeptide

Papaya ringspot virus (PRSV)—Flexuous rod, 680-900 nm × 11 nm, 1 genome segment, linear ssRNA, 1 coat protein polypeptide

Parsnip mosaic virus (ParMV)—Flexuous rod, 680-900 nm × 11 nm, 1 genome segment, linear ssRNA, 1 coat protein polypeptide

Pea early browning virus (PEBV)—Rigid rod, 180-215 nm × 22 nm and 46-114 nm × 22 nm, 2 genome segments, 2 linear ssRNA, 1 coat protein polypeptide

Pea enation mosaic virus (PEMV)—Polyhedral, about 30 nm, 2 genome segments, linear ssRNA, 1 coat protein polypeptide

Pea leafroll virus (PeLRV)—Polyhedral, 25-30 nm, 1 genome segment, linear ssRNA, 1 coat protein polypeptide

Pea seedborne mosaic virus (PSMV)—Flexuous rod, 680-900 nm × 11 nm, 1 genome segment, linear ssRNA, 1 coat protein polypeptide

Pea streak virus (PeSV)—Flexuous rod, 600-700 nm × 13 nm, 1 genome segment, linear ssRNA, 1 coat protein polypeptide

Peach rosette mosaic virus (PRMV)—Isometric, about 30 nm, 2 genome segments, linear ssRNA, 2 coat protein polypeptides

Peanut mottle virus (PeMoV)—Flexuous rod, 680-900 nm × 11 nm, 1 genome segment, linear ssRNA, 1 coat protein polypeptide

Peanut stunt virus (PSV)—Isometric, about 29 nm, 3 genome segments, linear ssRNA, 1 coat protein polypeptide

Peanut yellow mottle virus (PYMV)—isometric, about 29 nm, 1 genome segment, linear ssRNA, 1 coat protein polypeptide

Pepino latent virus (PeLV)—Flexuous rod, 600-700 nm × 13 nm, 1 genome segment, linear ssRNA, 1 coat protein polypeptide

Pepper mottle virus (PepMoV)—Flexuous rod, 680-900 nm × 11 nm, 1 genome segment, linear ssRNA, 1 coat protein polypeptide

Pepper veinal mottle virus (PVMV)—Flexuous rod, 680-900 nm × 11 nm, 1 genome segment, linear ssRNA, 1 coat protein polypeptide

Plum pox virus (PPV)—Flexuous rod, 680-900 nm × 11 nm, 1 genome segment, linear ssRNA, 1 coat protein polypeptide

Potato black ringspot virus (PBRV)—Isometric, about 30 nm, 2 genome segments, linear ssRNA, 2 coat protein polypeptides

Potato leafroll virus (PLRV)—Polyhedral, 25-30 nm, 1 genome segment, linear ssRNA, 1 coat protein polypeptide

Potato virus A (PVA)—Flexuous rod, 680-900 nm × 11 nm, 1 genome segment, linear ssRNA, 1 coat protein polypeptide

Potato virus M (PVM)—Flexuous rod, 600-700 nm × 13 nm, 1 genome segment, linear ssRNA, 1 coat protein polypeptide

Potato virus S (PVS)—Flexuous rod, 600-700 nm × 13 nm, 1 genome segment, linear ssRNA, 1 coat protein polypeptide

Potato virus X (PVX)—Flexuous rod, 480-580 nm × 13 nm, 1 genome segment, linear ssRNA, 1 coat protein polypeptide

Potato virus Y (PVY)—Flexuous rod, 680-900 nm × 11 nm, 1 genome segment, linear ssRNA, 1 coat protein polypeptide

Potato yellow dwarf virus (PYDV)—Bacilliform, 380 × 75 nm, 1 genome segment, linear ssRNA, 5 coat protein polypeptides

Prune dwarf virus (PDV)—Isometric, 26-35 nm, 3 genome segments, linear ssRNA, 1 coat protein polypeptide

Prunus necrotic ringspot virus (PNRSV)—Isometric, 26-35 nm, 3 genome segments, linear ssRNA, 1 coat protein polypeptide

Raspberry ringspot virus (RRV)—Isometric, about 30 nm, 2 genome segments, linear ssRNA, 2 coat protein polypeptides

Red clover necrotic mosaic virus (RCNMV)—Isometric, about 34 nm, 2 genome segments, linear ssRNA, 1 coat protein polypeptide

Red clover vein mosaic virus (RCVMV)—Flexuous rod, 600-700 nm × 13 nm, 1 genome segment, linear ssRNA, 1 coat protein polypeptide

Rice black-streaked dwarf virus (RBSDV)—Polyhedral, about 80 nm in diameter, 10 genome segments, dsRNA, 5 coat protein polypeptides

Rice dwarf virus (RDV)—Polyhedral, about 70 nm in diameter, 12 genome segments, dsRNA, 7 coat protein polypeptides

Rice necrosis mosaic virus (RNMV)—Flexuous rod, 680-900 nm × 11 nm, 1 genome segment, linear ssRNA, 1 coat protein polypeptide

Rice ragged stunt virus (RRSV)—Polyhedral, about 50 nm in diameter, 10 genome segments, dsRNA, 5 coat protein polypeptides

Rice transitory yellowing virus (RTYV)—Bullet shaped, 180-210 nm × 9 nm, 1 genome segment, ssRNA, 4 or 5 coat protein polypeptides

Rice tungro bacilliform virus (RTBV)—Bacilliform, 100-300 nm in length and 30-35 nm in width, 1 genome segment, circular dsDNA, 1 coat protein polypeptide

Rice tungro spherical virus (RTSV)—Polyhedral, about 30 nm in diameter, 1 genome segment, 3 coat protein polypeptides

Rice yellow mottle virus (RYMV)—Polyhedral, about 30 nm, 1 genome segment, linear ssRNA, 1 coat protein polypeptide

Shallot latent virus (SLV)—Flexuous rod, 600-700 nm × 13 nm,
1 genome segment, linear ssRNA, 1 coat protein polypeptide

Southern bean mosaic virus (SBMV)—Polyhedral, about 30 nm,
1 genome segment, linear ssRNA, 1 coat protein polypeptide

Sowbane mosaic virus (SoMV)—Polyhedral, about 30 nm, 1 genome
segment, linear ssRNA, 1 coat protein polypeptide

Soybean dwarf virus (SoyDV)—Polyhedral, 25-30 nm, 1 genome
segment, linear ssRNA, 1 coat protein polypeptide

Soybean mosaic virus (SoyMV)—Flexuous rod, 680-900 nm × 11 nm,
1 genome segment, linear ssRNA, 1 coat protein polypeptide

Subterranean clover mottle virus (SCMoV)—Polyhedral, about 30 nm,
1 genome segment, linear ssRNA, 1 coat protein polypeptide

Subterranean clover red leaf virus (SCRLV)—Polyhedral, about 30 nm,
1 genome segment, linear ssRNA, 1 coat protein polypeptide

Sugarcane mosaic virus (SCMV)—Flexuous rod, 680-900 nm × 11 nm,
1 genome segment, linear ssRNA, 1 coat protein polypeptide

Sweet clover necrotic mosaic virus (SCNMV)—Isometric, about 34 nm,
2 genome segments, linear ssRNA, 1 coat protein polypeptide

Tobacco etch virus (TEV)—Flexuous rod, 680-900 nm × 11 nm,
1 genome segment, linear ssRNA, 1 coat protein polypeptide

Tobacco mosaic virus (TMV)—Rigid rod, about 300 nm × 18 nm,
1 genome segment, linear ssRNA, 1 coat protein polypeptide

Tobacco necrosis virus (TNV)—Polyhedral, about 30 nm, 1 genome
segment, linear ssRNA, 1 coat protein polypeptide

Tobacco necrotic dwarf virus (TNDV)—Polyhedral, 25-30 nm, 1 genome
segment, linear ssRNA, 1 coat protein polypeptide

Tobacco rattle virus (TRV)—Rigid rod, 180-215 nm × 22 nm and
46-114 nm × 22 nm, 2 genome segments, 2 linear ssRNA, 1 coat
protein polypeptide

Tobacco ringspot virus (TRSV)—Isometric, about 30 nm, 2 genome
segment, linear ssRNA, 1 coat protein polypeptide

Tobacco streak virus (TSV)—Isometric, 26-35 nm, 3 genome segments,
linear ssRNA, 1 coat protein polypeptide

Tobacco vein mottling virus (TVMV)—Flexuous rod, 680-900 nm × 11 nm,
1 genome segment, linear ssRNA, 1 coat protein polypeptide

Tomato aspermy virus (TAV)—Isometric, about 29 nm, 3 genome
segments, linear ssRNA, 1 coat protein polypeptide

Tomato black ring virus (TBRV)—Isometric, about 30 nm, 2 genome
segments, linear ssRNA, 2 coat protein polypeptides

Tomato bushy stunt virus (TBSV)—Isometric, about 34 nm, 1 genome
segment, linear ssRNA, 1 coat protein polypeptide

Tomato golden mosaic virus (ToGMV)—Geminate, 18 × 30 nm, 2 circular genome segments, ssDNA, 1 coat protein polypeptide

Tomato mosaic virus (ToMV)—Rigid rod, about 300 nm × 18 nm, 1 genome segment, linear ssRNA, 1 coat protein polypeptide

Tomato ringspot virus (ToRSV)—Isometric, about 30 nm, 2 genome segments, linear ssRNA, 2 coat protein polypeptides

Tomato spotted wilt virus (TSWV)—Spherical, about 85 nm, 4 genome segments, linear ssRNA, 4 coat protein polypeptides

Tulip breaking virus (TBV)—Flexuous rod, 680-900 nm × 11 nm, 1 genome segment, linear ssRNA, 1 coat protein polypeptide

Turnip mosaic virus (TuMV)—Flexuous rod, 680-900 nm × 11 nm, 1 genome segment, linear ssRNA, 1 coat protein polypeptide

Turnip rosette virus (TRoV)—Polyhedral, about 30 nm, 1 genome segment, linear ssRNA, 1 coat protein polypeptide

Turnip yellow mosaic virus (TYMV)—Isometric, about 29 nm, 1 genome segment, linear ssRNA, 1 coat protein polypeptide

Watermelon mosaic virus (WMV)—Flexuous rod, 680-900 nm × 11 nm, 1 genome segment, linear ssRNA, 1 coat protein polypeptide

Wheat chlorotic streak virus (WCSV)—Bacilliform, 355 × 55 nm, 1 genome segment, linear ssRNA, 5 coat protein polypeptides

Wheat yellow leaf virus (WYLV)—Flexuous rod, 600-2000 nm × 12 nm, 1 genome segment, linear ssRNA, 1 coat protein polypeptide

White clover mosaic virus (WClMV)—Flexuous rod, 480-580 nm × 13 nm, 1 genome segment, linear ssRNA, 1 coat protein polypeptide

MECHANISM OF PLANT VIRUS INFECTION

Viruses depend completely upon host cells for the supply of precursors, energy, enzymes, and the structural machinery—in fact for all the infrastructure except the coded message (virus nucleic acid)—for their replication. Thus, viruses are parasites at the genetic level. The infection process consists of viral attachment and ingress of the viral genome into the host cell. Plant viruses possess no capacity of their own to push through the plant cell wall barrier in order to enter the cell. They can come in contact with the cytoplasm or reach the interior of the plant cell only through wounds caused mechanically or by vectors (Zaitlin and Hull, 1987). Wounding damages the cuticle and cell wall and/or breaks the trichomes to expose specific attachment sites that differ from nonspecific attachment sites found on unabraded cell walls (De Zoeten, 1995). The former exposes the cell membrane, whereas the latter exposes the plasmodesmata. Plasmodesmata are abundant in the cell walls of hairs and also between hair cells and underlying epidermal cells. In this way, virus particles become attached to the cell membrane. The virus may enter the protoplasts by pinocytosis or endocytosis, the process by which substances enter the cell from the exterior and pass into the cytoplasm. Attachment of virus particles to plasmalemma induces the plasmalemma's invagination at the point of attachment. The neck of the invagination closes, and the virus particle is imprisoned within an intracytoplasmic vesicle.

The virus may enter the cell by any one of the possible preexisting cell mechanisms of nutrient uptake. Many invaginations of various sizes are normally present on the surface of plasmalemma of protoplasts, and the adsorption processes are most effective in these invaginations. The adsorbed virus particles may enter the protoplast by closing the walls of invagination. The resultant vesicle later may disintegrate to release virus particles into the cytoplasm. The adsorbed virus particles may also penetrate the protoplast by the same type of electroosmotic forces responsible for transmembrane transport of ions into the protoplast through plasmalemma.

Within a few seconds to minutes after entry of the virions (the complete and infectious nucleoprotein particle of the virus) into the plant cell, uncoating of virus particles (i.e., removal of the envelope and coat protein and release of the viral genome) occurs. TMV particles often uncoat within 15 to 30 min after inoculation. Uncoating of *Tobacco necrosis virus* particles commences immediately after end-on attachment to the cell walls of plants. It has been shown by several researchers that uncoating of virus particles occurs on the plant cell wall. Uncoating appears to be a nonspecific event. Uncoating of the virus particles occurs not only on host plants but also with equal efficiency on nonhosts. It has also been suggested that the uncoating process may take place on the plasma membrane of pinocytic vesicles. The virus particles uncoat and disappear from view from pinocytic vesicles within a few minutes after their entry so that the viral genome, rather than complete virus particles, enters the cytosol. According to some reports, the uncoating process occurs intracellularly (i.e., in the cytoplasm). However, this process has not been confirmed. It is most likely that this process occurs outside of the cell (i.e., at an extracellular site).

The mechanism of uncoating is not yet fully understood (De Zoeten, 1995). It is suggested that more or less complete virus particles pass through the plasmalemma. During their passage, these particles may encounter low ionic (Ca^{2+}) strength and Ca^{2+} ion gradients, the hydrophobic environment of the plasmalemma, intracellular phospholipid membrane, or subcellular low pH compartments. These conditions cause the virions to swell or lose some protein subunits from the 5' end leader sequence so that partially encapsidated virions enter the cytosol. There the ribosomes take over and result in cotranslational disassembly. During translation, the ribosomes may extract or release RNA from the spherical capsid without disrupting the capsid completely. The RNA may escape through a hole formed in the capsid by removal of some capsid protein units.

RNA-dependent RNA polymerase has been detected in many plants, and it may be involved in viral replication in infected host cells. This enzyme is activated by virus infection. However, host polymerase may not be the complete replicase needed for replication of a specific virus RNA (De Zoeten, 1995). A virus-coded polypeptide may associate with the host-specific enzyme to form a fully competent replicase (holoenzyme), which also possesses the needed specificity for viral RNA. Thus, a virus-encoded protein is an integral component of the viral replicase. It is suggested that the complete virus-specific replicase is constituted by two units, a host-coded part and a virus-coded part. The virus-coded part is a virally coded polypeptide that functions in template selection. For example, *Turnip yellow mosaic virus* RNA replicase is made up of two major subunits: A 115-kilodalton (kDa) molecular weight subunit encoded by the viral genome and a host-encoded 45 kDa unit that is different from the host's RNA-dependent RNA replicase.

Immediately after partial or complete uncoating, viral RNA directs the synthesis of some early virus-specific proteins. One of these proteins is the specific virus-encoded subunit of the polymerase complex, which is synthesized within 1 to 10 minutes after inoculation. Replication of viral RNA involves transcription of the virus (+) RNA into the complementary (−) chain by the enzymes of the replicase system. The replication process starts from the 3' end. Viral RNA acts as a template for the synthesis of complementary strand. Thus viral RNA now exists as a double-stranded structure, the replicative form (RF) of RNA. RF RNA is a full-length double-stranded structure composed entirely of base pairs that is never infectious. The minus strands of the initially dsRNA (RF RNA) are then released and act as templates for the synthesis of complementary plus strands of viral RNA (progeny RNA) by again generating dsRNA structures, the replicative intermediate (RI). Progeny plus strands generated may further enter into any one of the three routes of the virus replication cycle: act as mRNA and code for coat protein synthesis (the translation step), be converted to double-stranded RF RNA to provide more templates for synthesis of viral plus RNA, or be encapsidated to produce virions. The assembly of nucleic acid and coat protein into virions is almost a spontaneous process when viral nucleic acid and specific proteins meet.

The replication process may vary slightly from virus to virus. For viruses with (−) strand RNA (such as *Rhabdovirus*), special mRNA has first to be transcribed. In the case of double-stranded RNA viruses such as *Reovirus,* each of the ten to 12 segments may produce its own mRNA. DNA viruses such as *Caulimovirus* use host polymerases for replication. All viruses have to code for their specific replicase. Viral (+) RNA and complementary RNA strands formed after transcription of the viral genome of some plant viruses (*Reovirus* and *Rhabdovirus*) act as mRNA. This leads to the synthesis of virus-specific capsids and other proteins by translation of this mRNA. Two types of ribosomes are present in plant cells: 80S cytoplasmic ribosomes and 70S chloroplast and mitochondrial ribosomes. Plant viral mRNAs are translated only on the 80S cytoplasmic ribosomes in several cases. Viral proteins are formed on 80S ribosomes. Viral RNA may encode its specific proteins in the midst of continuing synthesis of host proteins. Viruses have only a few genes. *Tobacco mosaic virus* RNA carries information for at least four proteins. The coat protein is encoded by the coat protein subgenomic RNA. A second subgenomic RNA, I_2 RNA, encodes the 30 kDa protein. Another subgenomic RNA, I_1 RNA, encodes a 50 kDa protein. These three subgenomic RNAs occur on ribosomes and act as mRNAs (Mandahar, 1991).

Rice tungro bacilliform virus, a double-stranded DNA genome, has been shown to have four open reading frames (ORFs). ORF I includes the gene *P24,* which codes for a 24 kDa protein that may be associated with particle assembly. The gene *P12* in ORF II codes for a 12 kDa protein. The amino acid sequence of the *Rice tungro bacilliform virus* ORF III product *(P194)* has motifs suggestive of viral coat protein, aspartate protease, reverse transcriptase, and ribonuclease H. *P194* is processed to give several products, such as two coat proteins (37 kDa, 33 kDa), aspartate protease (21 kDa), reverse transcriptase (62 kDa), and ribonuclease H (55 kDa). ORF IV consists of the gene *P46* and may be involved in controlling expression of the RTBV genome (Hull, 1996).

The plant virus genome encodes a wide range of proteins. All viruses produce replicases and coat protein(s), and most probably encode one or more proteins that potentiate virus movement from cell to cell and long distances in the plant. Some viruses, which are transmitted by insects, fungi, or nematodes, often produce helper proteins that aid in the acquisition of the virus by the vector. Some produce proteases that cleave the polyprotein. *Cauliflower mosaic virus* encodes a multifunctional protein from *gene VI,* a major component of the inclusion bodies that accumulates in infected cells. *Gene VI* may play a role in virus assembly in the determination of the host range of the virus and the severity of the symptoms (Zaitlin and Palukaitis, 2000).

Self-assembly of viral particles involves recognition of the specific genomic RNA by specific capsid protein subunits and the assembly of viral RNA with its specific proteins to form complete virus particles. Empty capsids are also formed in some cases. Extensive virus synthesis occurs, and complete virus particles gradually diffuse throughout the cytoplasm and ultimately lead to the formation of large virus aggregates or virus crystals.

Virus particles are synthesized in infected host cells and they spread from cell to cell as well as long distances in host tissues. The molecular basis of viral systemic movement has been studied in detail. Specific viral movement proteins are involved in the transfer of viral infectious transcripts through plasmodesmata (Lucas, 1993). While competency for viruses to replicate is produced by wounding, competency to transport is induced by virus replication itself. Movement proteins are produced during virus replication. These proteins open the gates, i.e., the plasmodesmata, for transport of the virus (Deom et al., 1992).

Plasmodesma is a membrane-lined pore (approximately 60 nm in diameter) through which passes a modified form of the endoplasmic reticulum, termed the *desmotubule* or *axial component.* The axial component is continuous with the endoplasmic reticulum of the neighboring cells, but does not have an associated lumen. It exists as a lipidic cylinder. In such a situation, cytoplasmic continuity would occur through the cytoplasmic annulus, or sleeve, that is located between the plasma membrane and the axial component. Void spaces within the cytoplasmic sleeve may have an effective diameter of 1.5-2.0 nm. Because the particles of most plant viruses are either icosahedral with a diameter of 18-80 nm, helical or filamentous rods (rigid or flexuous with diameters ranging from 10-25 nm and lengths of up to 2.5 μm), their physical dimensions preclude their movement through unmodified plasmodesmata (Lucas, 1993). Numerous researchers have provided electron microscopic evidence that plant viruses have the capability to cause dramatic changes in plasmodesmatal structure within systemically infected tissues (Esau, 1967; Roberts and Lucas, 1990; Lucas, 1993). Virions have often been identified within the enlarged cytoplasmic "sleeve" of these modified plasmodesmata. Their presence suggests that a viral encoded product is involved in effecting these structural changes in the plasmodesmata. This product has been identified as a protein called "movement protein." A 30-kDa protein has been identified as the movement protein in *Tobacco mosaic virus.* There are also reports that the virus coat protein is involved in long-distance transport (De Zoeten, 1995).

INCLUSION BODIES

Inclusion bodies are commonly found in virus-infected plants. They are either amorphous, crystalline, or granular structures found in the cytoplasm or in the nucleus. They differ in structure and composition, according to virus. Some contain high concentrations of virus particles embedded in amorphous material (e.g., *Cauliflower mosaic virus*). In cases of infection with viruses such as *Tobacco mosaic virus,* the inclusion bodies consist of pure virus in regular crystalline array. In cases of infection with *Clover yellow mosaic virus,* the pure virus inclusion bodies are in banded array. Crystalline inclusion bodies are formed due to infection from some viruses belonging to the genus *Potyvirus* (potyviruses). These bodies consist of protein formed as a result of virus infection. Crystalline inclusion bodies consisting of small, noninfectious nucleoprotein particles are found in plants infected with *Red clover vein mosaic virus.* Granular inclusions caused by *Bean yellow mosaic virus* and some other potyviruses contain peculiar pinwheels, scrolls, and rolls, consisting of newly formed protein together with abnormal organelles. Thus, inclusion bodies consist of pure virus particles (viroplasm), virus-induced protein, or other material. These bodies can be detected by taking freehand sections, staining the infected leaves, and viewing under a light microscope.

SATELLITE VIRUSES, SATELLITE RNA, AND DEFECTIVE INTERFERING VIRUSES

Satellite viruses were first reported by Kassanis in 1962. Satellite viruses have a coat but a small genome. They are dependent on other viruses to supply replicase and other enzymes necessary for replication. They act as parasites of plant-parasitic viruses. A satellite virus has been reported as being associated with *Tobacco necrosis virus* (Kassanis, 1962). The viruses are not serologically related. TNV is often found alone and can multiply indefinitely without causing the production of a satellite virus. However, the satellite virus is entirely dependent on TNV for its multiplication. Both viruses occur in the roots of apparently normal plants and are transmitted among roots by the fungus *Olpidium brassicae.*

The first satellite RNA described was the satellite RNA of *Tobacco ringspot virus* (Schneider, 1969). Satellite RNA has no coat protein. It completely depends upon those viruses for multiplication. Viruses that help multiplication of satellite RNA are called "helper viruses." Satellite RNA aggravates or reduces the symptom development. This type of association has also been reported in *Cucumber mosaic virus* (Tien and Wu, 1991). Sometimes satellite viruses also may have satellite RNAs. The satellite of *Tobacco necrosis virus* has been shown to contain a small satellite RNA that is dependent on *Tobacco necrosis virus* for replication and on the satellite virus for encapsulation.

Defective interfering viruses were first described in 1970 with an animal virus (Zaitlin and Palukaitis, 2000). Several plant viruses in the family *Tombusviridae* are now known to generate such RNAs during replication (Rubio et al., 1999). In many cases, the viral symptoms are ameliorated (Roux et al., 1991). However, in the case of *Turnip crinkle virus,* the defective interfering viruses are known to intensify viral symptoms.

CLASSIFICATION OF VIRUSES

Classification of plant viruses into class, order, family, genus, and species has not been achieved, mostly due to the lack of knowledge of genetic relationships among the viruses. It is now known that viruses are biological entities that possess genes, replicate, interact with hosts, and are exposed to selection pressure, thus specializing and evolving. Recent molecular biological studies have facilitated grouping the viruses into genera. A few families have also been created to include related viruses. The formation of orders is very much in preliminary stage, and only one order has been created thus far. The classification of plant viruses into orders, families, and genera was formulated and approved by the International Committee on Taxonomy of Viruses (ICTV) in 1995 and is being updated continually. The Latinized binomial system is not being followed for virus classification, unlike the classification of fungi and bacteria. At present, only groups of viruses have been recognized and these groups are provisionally given the status of the possible genera without establishing families. Each genus, regardless of whether it is a member of a family, will have its own name. The virus species is a class of viruses consisting of a replicating lineage and occupying a particular niche. This definition was approved by Executive Committee of International Committee on Taxonomy of Viruses in 1991 at Atlanta. Species names are actually the vernacular name of the virus, indicating the common name of the plant and type of disease symptoms induced by the virus. Virus names do not include the genus or group names. Van Regenmortel (1999) provided guidelines for writing the names of species, genera, and families of various viruses. The names of species, genera, and families should be written in italics. The first word and proper noun in the name of the species should be capitalized. Subsequent reference to the same virus should be by the accepted acronym, which is not italicized. The ICTV's list of approved virus genera and important virus species included in each genus (Mayo and Horzinek, 1998; Van Regenmortel et al., 1999) is given here. Many viruses remain unclassified, and their names are also given.

Genera of Plant Viruses and Names of Important Virus Species in These Genera

Genus *Alfamovirus: Alfalfa mosaic virus*

Genus *Alphacryptovirus : White clover cryptic virus 1, Beet 1 virus, Beet 2 virus*

Genus *Badnavirus: Banana streak virus, Cacao swollen shoot virus, Piper yellow mottle virus, Rice tungro bacilliform virus, Sugarcane bacilliform virus, Sweet potato leaf curl virus*

Genus *Betacryptovirus: White clover cryptic virus 2*

Genus *Bigeminivirus: Bean golden mosaic virus, Cassava African mosaic virus, Cassava Indian mosaic virus, Cotton leaf crumple virus, Cotton leaf curl virus, Dolichos yellow mosaic virus, Mungbean yellow mosaic virus, Okra (bhendi) yellow vein mosaic virus, Pepper hausteco virus, Potato yellow mosaic virus, Soybean crinkle leaf virus, Squash leaf curl virus, Tobacco leaf curl virus, Tomato yellow leaf curl virus, Tomato golden mosaic virus, Tomato Indian leaf curl virus, Tomato mottle virus, Watermelon chlorotic stunt virus*

Genus *Bromovirus: Broad bean mottle virus, Cowpea chlorotic mottle virus*

Genus *Bymovirus: Barley mild mosaic virus, Barley yellow mosaic virus, Oat mosaic virus, Rice necrosis mosaic virus, Wheat yellow mosaic virus, Wheat spindle streak virus, Barley mild mottle virus*

Genus *Capillovirus: Apple stem grooving virus, Citrus tatter virus, Cherry A virus*

Genus *Carlavirus: Blueberry scorch virus, Carnation latent virus, Chrysanthemum B virus, Hop mosaic virus, Cole latent virus, Cowpea mild mottle virus, Poplar mosaic virus, Potato M virus, Potato S virus, Red clover vein mosaic virus, Shallot latent virus, Passiflora latent virus, Hop latent virus*

Genus *Carmovirus: Carnation mottle virus, Cowpeas mottle virus, Melon necrotic spot virus, Pelargonium flower break virus, Turnip crinkle virus, Hibiscus chlorotic ringspot virus, Pelargonium line pattern virus, Blackgram mottle virus, Narcissus tip necrosis virus*

Genus *Caulimovirus: Cassava vein mosaic virus, Cauliflower mosaic virus, Peanut chlorotic streak virus, Petunia vein-clearing virus, Strawberry vein-banding virus, Blueberry red ringspot virus, Carnation etched ring virus, Figwort mosaic virus, Dahlia mosaic virus, Soybean chlorotic mottle virus*

Genus *Closterovirus: Beet pseudo-yellows virus, Beet yellows virus, Citrus tristeza virus, Cucurbit yellow stunting disorder virus, Grapevine corky bark-associated virus, Grapevine leafroll-associated virus, Grapevine stem pitting-associated virus, Lettuce infectious yellows virus, Little cherry virus, Pineapple wilt-associated virus, Sweet potato infectious chlorosis virus, Tomato infectious chlorosis virus, Beet yellow stunt virus, Sweet potato sunken vein virus*

Genus *Comovirus: Bean pod mottle virus, Bean rugose mosaic virus, Bean severe mosaic virus, Broad bean stain virus, Cowpea mosaic virus, Cowpea severe mosaic virus, Potato Andean mottle virus, Red clover mottle virus, Squash mosaic virus*

Genus *Crinivirus: Sweet potato chlorotic stunt virus*

Genus *Cucumovirus: Cucumber mosaic virus, Peanut stunt virus, Tomato aspermy virus*

Genus *Cytorhabdovirus: American wheat striate mosaic virus, Barley yellow striate mosaic virus, Festuca leaf streak virus, Lettuce necrotic yellows virus, Strawberry crinkle virus, Cereal northern mosaic virus*

Genus *Dianthovirus: Carnation ringspot virus, Red clover necrotic mosaic virus*

Genus *Enamovirus: Pea enation mosaic virus*

Genus *Fabavirus: Broad bean wilt virus*

Genus *Fijivirus: Maize rough dwarf virus, Oat sterile dwarf virus, Rice black-streaked dwarf virus, Sugarcane Fiji disease virus*

Genus *Foveavirus: Apple stem pitting virus*

Genus *Furovirus: Beet necrotic yellow vein virus, Beet soilborne virus, Broad bean necrosis virus, Oat golden stripe virus, Potato mop-top virus, Wheat soilborne mosaic virus, Peanut clump virus, Rice stripe necrosis virus*

Genus *Geminivirus: Taino tomato mottle virus, Tomato yellow vein streak virus, Sinaloa tomato leaf curl virus, Tomato greenhouse whiteflyborne virus, Horse radish curly top virus, Tomato leaf curl virus*

Genus *Hordeivirus: Barley stripe mosaic virus*

Genus *Hybrigeminivirus: Beet curly top virus, Tomato pseudo-curly top virus*

Genus *Idaeovirus: Raspberry bushy dwarf virus*

Genus *Ilarvirus: Apple mosaic virus, Asparagus 2 virus, Prune dwarf virus, Prunus necrotic ringspot virus, Spinach latent virus, Tobacco streak virus, Citrus leaf rugose virus, Citrus variegation virus, Tomato 1 virus*

Genus *Ipomovirus: Sweet potato mild mottle virus*

Genus *Luteovirus: Barley yellow dwarf virus, Bean yellow dwarf virus, Beet western yellows virus, Carrot red leaf virus, Cucurbit aphidborne yellows virus, Groundnut rosette assistor virus, Potato leafroll virus, Soybean dwarf virus, Subterranean clover red leaf virus, Beet mild yel-*

lowing virus, Bean leaf roll virus, Pepper vein yellows virus, Soybean Indonesian dwarf virus, Strawberry mild yellow edge virus

Genus *Machlomovirus: Maize chlorotic mottle virus*

Genus *Macluravirus: Narcissus latent virus, Maclura mosaic virus*

Genus *Marafivirus: Maize rayado fino virus, Oat blue dwarf virus*

Genus *Monogeminivirus: Chickpea chlorotic dwarf virus, Maize streak virus, Wheat dwarf virus, Tobacco yellow dwarf virus, Sugarcane streak virus*

Genus *Nanavirus: Banana bunchy top virus, Coconut foliar decay virus, Faba bean necrotic yellows virus, Subterranean clover stunt virus*

Genus *Necrovirus: Tobacco necrosis virus*

Genus *Nepovirus: Arabis mosaic virus, Artichoke Italian latent virus, Cherry leafroll virus, Cherry rosette virus, Grapevine chrome mosaic virus, Grapevine fanleaf virus, Hibiscus latent ringspot virus, Raspberry ringspot virus, Satsuma dwarf virus, Strawberry latent ringspot virus, Tobacco black ring virus, Tomato ringspot virus, Tomato black ring virus, Cherry rasp leaf virus, Blueberry leaf mottle virus, Peach rosette mosaic virus*

Genus *Nucleorhabdovirus: Maize mosaic virus, Raspberry vein chlorosis virus, Sorghum stunt mosaic virus, Eggplant mottled dwarf virus*

Genus *Oleavirus: Olive latent 2 virus, Epirus cherry virus, Melon ourmia virus*

Genus *Oryzavirus: Rice ragged stunt virus*

Genus *Phytoreovirus: Rice dwarf virus, Clover wound tumor virus*

Genus *Potexvirus: Bamboo mosaic virus, Cactus X virus, Cassava common mosaic virus, Clover yellow mosaic virus, Cymbidium mosaic virus, Foxtail mosaic virus, Hosta X virus, Lily X virus, Narcissus mosaic virus, Papaya mosaic virus, Patchouli X virus, Potato aucuba mosaic virus, Potato X virus, Strawberry mild yellow edge-associated virus, Tulip X virus, Viola mottle virus, White clover mosaic virus*

Genus *Potyvirus: Abaca mosaic virus, Amazon lily mosaic virus, Artichoke latent virus, Asian prunus latent virus, Azuki bean mosaic virus, Banana bract mosaic virus, Bean common mosaic virus, Bean yellow mosaic virus, Beet mosaic virus, Blackeye cowpea mosaic virus, Cardamom mosaic virus, Carnation vein mottle virus, Cassava brown streak virus, Celery mosaic virus, Chilli veinal mottle virus, Clover yellow vein virus, Cowpea aphidborne mosaic virus, Dasheen mosaic virus, Datura Colombian virus, Garlic yellow streak virus, Johnsongrass mosaic virus, Kalanchoe mosaic virus, Leek yellow stripe virus, Lettuce mosaic virus, Maize dwarf mosaic virus, Narcissus yellow stripe virus, Onion yellow dwarf virus, Palm mosaic virus, Papaya leaf-distortion mosaic virus, Papaya ringspot virus, Passion fruit woodiness* virus, *Pea seedborne mosaic virus, Peanut chlorotic blotch virus, Peanut mottle virus, Peanut stripe virus, Pepper severe mosaic virus, Pepper mottle virus, Pepper veinal mottle virus, Plum pox virus, Potato A virus, Potato V virus, Potato Y virus, Radish vein clearing virus, Sorghum mosaic virus, Soybean mosaic virus, Sugarcane mosaic virus, Sweet potato feathery mottle virus, Sweet potato latent virus, Sweet potato mild speckling virus, Tamarillo mosaic virus, Taro feathery mottle virus, Tobacco etch virus, Tobacco vein mottling virus, Tulip breaking virus, Turnip mosaic virus, Vanilla necrosis virus, Watermelon Moroccan mosaic virus, Watermelon mosaic 1 virus, Watermelon mosaic 2 virus, Welsh onion yellow stripe virus, Yam mosaic virus, Zucchini yellow fleck virus, Zucchini yellow mosaic virus*

Genus *Rhabdovirus*: *Beet leaf curl virus, Citrus leprosis virus, Orchid fleck virus*

Genus *Rymovirus*: *Agropyron mosaic virus, Brome streak mosaic virus, Garlic miteborne mosaic virus, Hordeum mosaic virus, Ryegrass mosaic virus, Wheat leaf streak virus, Wheat streak mosaic virus*

Genus *Satellivirus*: *Panicum mosaic satellivirus, Tobacco mosaic satellivirus, Tobacco necrosis satellivirus*

Genus *Sequivirus*: *Dandelion yellow mosaic virus, Parsnip yellow fleck virus*

Genus *Sobemovirus*: *Bean southern mosaic virus, Blueberry shoestring virus, Clover mottle virus, Lucerne transient streak virus, Olive latent 1 virus, Panicum mosaic virus, Rice yellow mottle virus, Sowbane mosaic virus, Subterranean clover mottle virus*

Genus *Tenuivirus*: *Maize stripe virus, Rice hoja blanca virus, Rice grassy stunt virus, Rice stripe virus, Wheat Iranian stripe virus, Echinochloa hoja blanca virus, Maize yellow stripe virus*

Genus *Tobamovirus*: *Cucumber green mottle mosaic virus, Paprika mild mottle virus, Pepper mild mottle virus, Sunnhemp mosaic virus, Tomato mosaic virus, Tobacco mosaic virus, Tobacco mild green mosaic virus, Turnip vein-clearing virus*

Genus *Tobravirus*: *Pea early browning virus, Tobacco rattle virus, Tobacco ringspot virus*

Genus *Tombusvirus*: *Tomato bushy stunt virus, Carnation Italian ringspot virus, Oat chlorotic stunt virus, Artichoke mottled crinkle virus, Cymbidium ringspot virus, Petunia asteroid mosaic virus, Cucumber necrosis virus*

Genus *Tospovirus*: *Groundnut ringspot virus, Peanut bud necrosis virus, Peanut yellow spot virus, Tomato spotted wilt virus, Tomato chlorotic spot virus, Impatiens necrotic spot virus, Iris yellow spot virus*

Genus *Trichovirus: Apple chlorotic leaf spot virus, Cherry mottle leaf virus, Grapevine berry inner necrosis virus*

Genus *Tymovirus: Turnip yellow mosaic virus, Melon rugose mosaic virus, Physalis mottle virus, Eggplant mosaic virus, Belladonna mottle virus, Calopogonium yellow vein virus, Okra mosaic virus, Poinsettia mosaic virus, Potato Andean latent virus, Sesbania mosaic virus*

Genus *Umbravirus: Carrot mottle virus, Groundnut rosette virus, Sunflower yellow blotch virus*

Genus *Varicosavirus: Lettuce big-vein virus*

Genus *Vitivirus: Grapevine A virus, Grapevine B virus*

Genus *Waikavirus: Rice tungro spherical virus, Maize chlorotic dwarf virus*

Unclassified viruses: *Banana dieback virus, Banana mosaic virus, Bean line pattern mosaic virus, Beet soilborne mosaic virus, Beet virus Q, Black currant reversion associated virus, Black raspberry necrosis virus, Broad bean bushy dwarf virus, Citrus infectious variegation virus, Citrus mosaic virus, Citrus ringspot virus, Citrus psorosis virus, Citrus vein enation virus, East African cassava mosaic virus, Garlic A virus, Garlic latent virus, Garlic X virus, Garlic yellow stripe virus, Grapevine fleck virus, Groundnut bud necrosis virus, Havana tomato virus, Ipomoea crinkle leaf curl virus, Lettuce chlorosis virus, Papaya leaf curl virus, Papaya lethal yellowing virus, Peach mosaic virus, Peanut chlorotic fan-spot virus, Pepper vein banding virus, Pigeonpea sterility mosaic virus, Raspberry leaf spot virus, Raspberry leaf mottle virus, Rice yellow stunt virus, Sour cherry green ring mottle virus, Squash yellow leaf curl virus, Sugarcane mild mosaic virus, Sugarcane yellow leaf virus, Sunflower mosaic virus, Tomato chlorosis virus, Urd bean leaf crinkle virus, Yam mild mosaic virus*

Families of Plant Viruses

A few genera have been grouped into families. The recognized families containing plant viruses are as follows:

1. Family: *Bromoviridae*
   Genera: *Alfamovirus, Bromovirus, Cucumovirus, Ilarvirus, Oleavirus*
2. Family: *Bunyaviridae*
   Genera: *Tospovirus* (Other genera such as *Bunyavirus, Nairovirus, Hantavirus,* and *Phlebovirus* are not plant viruses)
3. Family: *Caulimoviridae*
   Genera: *Badnavirus, Caulimovirus*

4. Family: *Comoviridae*
   Genera: *Comovirus, Fabavirus, Nepovirus*
5. Family: *Geminiviridae*
   Genera: *Mastrevirus* (Subgroup I *Geminivirus*)
            Curtovirus (Subgroup II *Geminivirus*)
            Begomovirus (Subgroup III *Geminivirus*)
6. Family: *Partitiviridae*
   Genera: *Alphacryptovirus, Betacryptovirus*
7. Family: *Potyviride*
   Genera: *Bymovirus, Potyvirus, Rymovirus*
8. Family: *Reoviridae*
   Genera: *Fijivirus, Oryzavirus, Phytoreovirus*
9. Family: *Rhabdoviridae*
   Genera: *Cytorhabdovirus, Nucleorhabdovirus*
10. Family: *Sequiviridae*
    Genera: *Sequivirus, Waikavirus*
11. Family: *Tombusviridae*
    Genera: *Carmovirus, Tombusvirus*

An Order That Contains Plant Viruses

Order: *Mononegavirales*
Family: *Rhabdoviridae*
Genera: *Cytorhabdovirus, Nucleorhabdovirus*

SYSTEMATIC POSITION OF IMPORTANT VIRAL PATHOGENS

Abaca mosaic virus—genus *Potyvirus*, family *Potyviridae*
Agropyron mosaic virus—genus *Rymovirus*, family *Potyviridae*
Alfalfa cryptic virus—unclassified
Alfalfa mosaic virus—genus *Alfamovirus*, family *Bromoviridae*
Amazon lily mosaic virus—genus *Potyvirus*, family *Potyviridae*
American wheat striate mosaic virus—genus *Cytorhabdovirus*, family
     Rhabdoviridae, order *Mononegavirales*
Apple chlorotic leaf spot virus—genus *Trichovirus*
Apple mosaic virus—genus *Ilarvirus*, family *Bromoviridae*
Apple stem grooving virus—genus *Capillovirus*
Apple stem pitting virus—genus *Foveavirus*
Arabis mosaic virus—genus *Nepovirus*, family *Comoviridae*
Artichoke Italian latent virus—genus *Nepovirus*, family *Comoviridae*

Artichoke latent virus—genus *Potyvirus,* family *Potyviridae*
Artichoke mottled crinkle virus—genus *Tombusvirus*
Asian prunus latent virus—genus *Potyvirus,* family *Potyviridae*
Asparagus 2 virus—genus *Ilarvirus,* family *Bromoviridae*
Azuki bean mosaic virus—genus *Potyvirus,* family *Potyviridae*
Bamboo mosaic virus—genus *Potexvirus*
Banana bract mosaic virus—genus *Potyvirus,* family *Potyviridae*
Banana bunchy top virus—genus *Nanavirus*
Banana dieback virus—unclassified
Banana mosaic virus—unclassified
Banana streak virus—genus *Badnavirus,* family *Caulimoviridae*
Barley mild mosaic virus—genus *Bymovirus,* family *Potyviridae*
Barley mild mottle virus—genus *Bymovirus,* family *Potyviridae*
Barley stripe mosaic virus—genus *Hordeivirus*
Barley yellow dwarf virus—genus *Luteovirus*
Barley yellow mosaic virus—genus *Bymovirus,* family *Potyviridae*
Barley yellow striate mosaic virus—genus *Cytorhabdovirus,* family
   Rhabdoviridae, order *Mononegavirales*
Bean common mosaic virus—genus *Potyvirus,* family *Potyviridae*
Bean curly dwarf mosaic virus—genus *Comovirus,* family *Comoviridae*
Bean golden mosaic virus—genus *Bigeminivirus*
Bean leafroll virus—genus *Luteovirus*
Bean line pattern mosaic virus—unclassified
Bean mild mosaic virus—unclassified
Bean pod mottle virus—genus *Comovirus,* family *Comoviridae*
Bean rugose mosaic virus—genus *Comovirus,* family *Comoviridae*
Bean severe mosaic virus—genus *Comovirus,* family *Comoviridae*
Bean southern mosaic virus—genus *Sobemovirus*
Bean yellow dwarf virus—genus *Luteovirus*
Bean yellow mosaic virus—genus *Potyvirus,* family *Potyviridae*
Beet 1 virus—genus *Alphacryptovirus,* family *Partitiviridae*
Beet 2 virus—genus *Alphacryptovirus,* family *Partitiviridae*
Beet curly top virus—genus *Hybrigeminivirus*
Beet leaf curl virus—genus *Rhabdovirus*
Beet mild yellowing virus—genus *Luteovirus*
Beet mosaic virus—genus *Potyvirus,* family *Potyviridae*
Beet necrotic yellow vein virus—genus *Furovirus*
Beet pseudo-yellows virus—genus *Closterovirus*
Beet soilborne mosaic virus—unclassified
Beet soilborne virus—genus *Furovirus*
Beet virus Q—unclassified
Beet western yellows virus—genus *Luteovirus*

Beet yellow stunt virus—genus *Closterovirus*
Beet yellows virus—genus *Closterovirus*
Belladonna mottle virus—genus *Tymovirus*
Black currant reversion associated virus—unclassified
Black raspberry necrosis virus—unclassified
Blackeye cowpea mosaic virus—genus *Potyvirus*, family *Potyviridae*
Blackgram mottle virus—genus *Carmovirus*, family *Tombusviridae*
Blueberry leaf mottle virus—genus *Nepovirus*, family *Comoviridae*
Blueberry red ringspot virus—genus *Caulimovirus*, family
    Caulimoviridae
Blueberry scorch virus—genus *Carlavirus*
Blueberry shoestring virus—genus *Sobemovirus*
Broad bean bushy dwarf virus—unclassified
Broad bean mottle virus—genus *Bromovirus*, family *Bromoviridae*
Broad bean necrosis virus—genus *Furovirus*
Broad bean stain virus—genus *Comovirus*, family *Comoviridae*
Broad bean wilt virus—genus *Fabavirus*, family *Comoviridae*
Brome streak mosaic virus—genus *Rymovirus*, family *Potyviridae*
Cacao swollen shoot virus—genus *Badnavirus*, family *Caulimoviridae*
Cactus X virus—genus *Potexvirus*
Calopogonium yellow vein virus—genus *Tymovirus*
Cardamom mosaic virus—genus *Potyvirus*, family *Potyviridae*
Carnation etched ring virus—genus *Caulimovirus*, family *Caulimoviridae*
Carnation Italian ringspot virus—genus *Tombusvirus*
Carnation latent virus—genus *Carlavirus*
Carnation mottle virus—genus *Carmovirus*, family *Tombusviridae*
Carnation necrotic fleck virus—genus *Closterovirus*
Carnation ringspot virus—genus *Dianthovirus*
Carnation vein mottle virus—genus *Potyvirus*, family *Potyviridae*
Carrot latent virus—genus *Nucleorhabdovirus*
Carrot mottle virus—genus *Umbravirus*
Carrot red leaf virus—genus *Luteovirus*
Cassava African mosaic virus—genus *Bigeminivirus*
Cassava brown streak virus—genus *Potyvirus*, family *Potyviridae*
Cassava common mosaic virus—genus *Potexvirus*
Cassava Indian mosaic virus—genus *Bigeminivirus*
Cassava vein mosaic virus—genus *Caulimovirus*, family *Caulimoviridae*
Cauliflower mosaic virus—genus *Caulimovirus*, family *Caulimoviridae*
Celery mosaic virus—genus *Potyvirus*, family *Potyviridae*
Cereal northern mosaic virus—genus *Cytorhabdovirus*, family
    Rhabdoviridae, order *Mononegavirales*
Cherry A virus—genus *Capillovirus*

Cherry leafroll virus—genus *Nepovirus*, family *Comoviridae*
Cherry mottle leaf virus—genus *Trichovirus*
Cherry rasp leaf virus—genus *Nepovirus*, family *Comoviridae*
Cherry rosette virus—genus *Nepovirus*, family *Comoviridae*
Chickpea chlorotic dwarf virus—genus *Monogeminivirus*
Chickpea virus—genus *Luteovirus*
Chilli veinal mottle virus—genus *Potyvirus*, family *Potyviridae*
Chrysanthemum B virus—genus *Carlavirus*
Citrus infectious variegation virus—unclassified
Citrus leaf rugose virus—genus *Ilarvirus*, family *Bromoviridae*
Citrus leprosis virus—genus *Rhabdovirus*
Citrus mosaic virus—unclassified
Citrus psorosis virus—unclassified
Citrus ringspot virus—unclassified
Citrus tatter virus—genus *Capillovirus*
Citrus tristeza virus—genus *Closterovirus*
Citrus variegation virus—genus *Ilarvirus*, family *Bromoviridae*
Citrus vein enation virus—unclassified
Clover mottle virus—genus *Sobemovirus*
Clover wound tumor virus—genus *Phytoreovirus*, family *Reoviridae*
Clover yellow mosaic virus—genus *Potexvirus*
Clover yellow vein virus—genus *Potyvirus*, family *Potyviridae*
Coconut foliar decay virus—genus *Nanavirus*
Cole latent virus—genus *Carlavirus*
Cotton leaf crumple virus—genus *Bigeminivirus*
Cotton leaf curl virus—genus *Bigeminivirus*
Cowpea aphidborne mosaic virus—genus *Potyvirus*, family *Potyviridae*
Cowpea chlorotic mottle virus—genus *Bromovirus*, family *Bromoviridae*
Cowpea mild mottle virus—genus *Carlavirus*
Cowpea mosaic virus—genus *Comovirus*, family *Comoviridae*
Cowpea severe mosaic virus—genus *Comovirus*, family *Comoviridae*
Cowpeas mottle virus—genus *Carmovirus*, family *Tombusviridae*
Cucumber green mottle mosaic virus—genus *Tobamovirus*
Cucumber mosaic virus—genus *Cucumovirus*, family *Bromoviridae*
Cucumber necrosis virus—genus *Tombusvirus*
Cucurbit aphidborne yellows virus—genus *Luteovirus*
Cucurbit yellow stunting disorder virus—genus *Closterovirus*
Cymbidium mosaic virus—genus *Potexvirus*
Cymbidium ringspot virus—genus *Tombusvirus*
Dahlia mosaic virus—genus *Caulimovirus*, family *Caulimoviridae*
Dandelion yellow mosaic virus—genus *Sequivirus*, family *Sequiviridae*
Dasheen mosaic virus—genus *Potyvirus*, family *Potyviridae*

Datura Colombian virus—genus *Potyvirus*, family *Potyviridae*
Dolichos yellow mosaic virus—genus *Bigeminivirus*
East African cassava mosaic virus—unclassified
Echinochloa hoja blanca virus—genus *Tenuivirus*
Eggplant mosaic virus—genus *Tymovirus*
Eggplant mottled dwarf virus—genus *Nucleorhabdovirus*, family *Rhabdoviridae*, order *Mononegavirales*
Epirus cherry virus—genus *Oleavirus*, family *Bromoviridae*
Faba bean necrotic yellows virus—genus *Nanavirus*
Festuca leaf streak virus—genus *Cytorhabdovirus*, family *Rhabdoviridae*, order *Mononegavirales*
Figwort mosaic virus—genus *Caulimovirus*, family *Caulimoviridae*
Foxtail mosaic virus—genus *Potexvirus*
Garlic A virus—unclassified
Garlic latent virus—unclassified
Garlic miteborne mosaic virus—genus *Rymovirus*, family *Potyviridae*
Garlic X virus—unclassified
Garlic yellow streak virus—genus *Potyvirus*, family *Potyviridae*
Garlic yellow stripe virus—unclassified
Grapevine A virus—genus *Vitivirus*
Grapevine B virus—genus *Vitivirus*
Grapevine berry inner necrosis virus—genus *Trichovirus*
Grapevine chrome mosaic virus—genus *Nepovirus*, family *Comoviridae*
Grapevine corky bark-associated virus—genus *Closterovirus*
Grapevine fanleaf virus—genus *Nepovirus*, family *Comoviridae*
Grapevine fleck virus—unclassified
Grapevine leafroll-associated virus—genus *Closterovirus*
Grapevine stem pitting-associated virus—genus *Closterovirus*
Groundnut bud necrosis virus—unclassified
Groundnut ringspot virus—genus *Tospovirus*
Groundnut rosette assistor virus—genus *Luteovirus*
Groundnut rosette virus—genus *Umbravirus*
Havana tomato virus—unclassified
Hibiscus chlorotic ringspot virus—genus *Carmovirus*, family *Tombusviridae*
Hibiscus latent ringspot virus—genus *Nepovirus*, family *Comoviridae*
Hop latent virus—genus *Carlavirus*
Hop mosaic virus—genus *Carlavirus*
Hordeum mosaic virus—genus *Rymovirus*, family *Potyviridae*
Horse radish curly top virus—genus *Geminivirus*, family *Geminiviridae*
Hosta X virus—genus *Potexvirus*
Impatiens necrotic spot virus—genus *Tospovirus*, family *Bunyaviridae*

Ipomoea crinkle leaf curl virus—unclassified
Iris yellow spot virus—genus *Tospovirus*
Johnsongrass mosaic virus—genus *Potyvirus,* family *Potyviridae*
Kalanchoe mosaic virus—genus *Potyvirus,* family *Potyviridae*
Leek yellow stripe virus—genus *Potyvirus,* family *Potyviridae*
Lettuce big-vein virus—genus *Varicosavirus*
Lettuce chlorosis virus—unclassified
Lettuce infectious yellows virus—genus *Closterovirus*
Lettuce mosaic virus—genus *Potyvirus,* family *Potyviridae*
Lettuce necrotic yellows virus—genus *Cytorhabdovirus,* family
    Rhabdoviridae, order *Mononegavirales*
Lettuce speckles mottle virus—genus *Umbravirus*
Lily X virus—genus *Potexvirus*
Little cherry virus—genus *Closterovirus*
Lucerne transient streak virus—genus *Sobemovirus*
Maclura mosaic virus—genus *Macluravirus*
Maize chlorotic dwarf virus—genus *Waikavirus*
Maize chlorotic mottle virus—genus *Machlomovirus*
Maize dwarf mosaic virus—genus *Potyvirus,* family *Potyviridae*
Maize mosaic virus—genus *Nucleorhabdovirus,* family *Rhabdoviridae,*
    order *Mononegavirales*
Maize rayado fino virus—genus *Marafivirus*
Maize rough dwarf virus—genus *Fijivirus,* family *Reoviridae*
Maize streak virus—genus *Monogeminivirus*
Maize stripe virus—genus *Tenuivirus*
Maize yellow stripe virus—genus *Tenuivirus*
Melon necrotic spot virus—genus *Carmovirus,* family *Tombusviridae*
Melon ourmia virus—genus *Oleavirus,* family *Bromoviridae*
Melon rugose mosaic virus—genus *Tymovirus*
Mungbean yellow mosaic virus—genus *Bigeminivirus*
Narcissus latent virus—genus *Macluravirus*
Narcissus mosaic virus—genus *Potexvirus*
Narcissus tip necrosis virus—genus *Carmovirus,* family *Tombusviridae*
Narcissus yellow stripe virus—genus *Potyvirus,* family *Potyviridae*
Oat blue dwarf virus—genus *Marafivirus*
Oat chlorotic stunt virus—genus *Tombusvirus*
Oat golden stripe virus—genus *Furovirus*
Oat mosaic virus—genus *Bymovirus,* family *Potyviridae*
Oat sterile dwarf virus—genus *Fijivirus,* family *Reoviridae*
Okra (bhendi) yellow vein mosaic virus—genus *Bigeminivirus*
Okra mosaic virus—genus *Tymovirus*
Olive latent 1 virus—genus *Sobemovirus*

Olive latent 2 virus—genus *Oleavirus,* family *Bromoviridae*
Onion yellow dwarf virus—genus *Potyvirus,* family *Potyviridae*
Orchid fleck virus—genus *Rhabdovirus*
Palm mosaic virus—genus *Potyvirus,* family *Potyviridae*
Panicum mosaic satellivirus—genus *Satellivirus*
Panicum mosaic virus—genus *Sobemovirus*
Papaya apical necrosis virus—genus *Rhabdovirus*
Papaya droopy necrosis virus—genus *Rhabdovirus*
Papaya leaf curl virus—unclassified
Papaya leaf-distortion mosaic virus—genus *Potyvirus,* family *Potyviridae*
Papaya lethal yellowing virus—unclassified
Papaya mosaic virus—genus *Potexvirus*
Papaya ringspot virus—genus *Potyvirus,* family *Potyviridae*
Paprika mild mottle virus—genus *Tobamovirus*
Parsnip yellow fleck virus—genus *Sequivirus,* family *Sequiviridae*
Passiflora latent virus—genus *Carlavirus*
Passion fruit woodiness virus—genus *Potyvirus,* family *Potyviridae*
Patchouli X virus—genus *Potexvirus*
Pea early browning virus—genus *Tobravirus*
Pea enation mosaic virus—genus *Enamovirus*
Pea seedborne mosaic virus—genus *Potyvirus,* family *Potyviridae*
Pea streak virus—genus *Carlavirus*
Peach mosaic virus—unclassified
Peach rosette mosaic virus—genus *Nepovirus,* family *Comoviridae*
Peanut bud necrosis virus—genus *Tospovirus,* family *Bunyaviridae*
Peanut chlorotic blotch virus—genus *Potyvirus,* family *Potyviridae*
Peanut chlorotic fan-spot virus—unclassified
Peanut chlorotic streak virus—genus *Caulimovirus,* family
    Caulimoviridae
Peanut clump virus—genus *Furovirus*
Peanut mottle virus—genus *Potyvirus,* family *Potyviridae*
Peanut stripe virus—genus *Potyvirus,* family *Potyviridae*
Peanut stunt virus—genus *Cucumovirus,* family *Bromoviridae*
Peanut yellow spot virus—genus *Tospovirus,* family *Bunyaviridae*
Pelargonium flower break virus—genus *Carmovirus,* family
    Tombusviridae
Pelargonium line pattern virus—genus *Carmovirus,* family *Tombusviridae*
Pepper hausteco virus—genus *Bigeminivirus*
Pepper mild mottle virus—genus *Tobamovirus*
Pepper mottle virus—genus *Potyvirus,* family *Potyviridae*
Pepper severe mosaic virus—genus *Potyvirus,* family *Potyviridae*
Pepper vein banding virus—unclassified

Pepper vein yellows virus—genus *Luteovirus*
Pepper veinal mottle virus—genus *Potyvirus*, family *Potyviridae*
Petunia asteroid mosaic virus—genus *Tombusvirus*
Petunia vein-clearing virus—genus *Caulimovirus*, family *Caulimoviridae*
Physalis mottle virus—genus *Tymovirus*
Pigeonpea sterility mosaic virus—unclassified
Pineapple wilt-associated virus—genus *Closterovirus*
Piper yellow mottle virus—genus *Badnavirus*, family *Caulimoviridae*
Plum pox virus—genus *Potyvirus*, family *Potyviridae*
Poinsettia mosaic virus—genus *Tymovirus*
Poplar mosaic virus—genus *Carlavirus*
Potato A virus—genus *Potyvirus*, family *Potyviridae*
Potato Andean latent virus—genus *Tymovirus*
Potato Andean mottle virus—genus *Comovirus*, family *Comoviridae*
Potato aucuba mosaic virus—genus *Potexvirus*
Potato leafroll virus—genus *Luteovirus*
Potato M virus—genus *Carlavirus*
Potato mop-top virus—genus *Furovirus*
Potato S virus—genus *Carlavirus*
Potato T virus—genus *Trichovirus*
Potato V virus—genus *Potyvirus*, family *Potyviridae*
Potato X virus—genus *Potexvirus*
Potato Y virus—genus *Potyvirus*, family *Potyviridae*
Potato yellow mosaic virus—genus *Bigeminivirus*
Prune dwarf virus—genus *Ilarvirus*, family *Bromoviridae*
Prunus necrotic ringspot virus—genus *Ilarvirus*, family *Bromoviridae*
Radish vein clearing virus—genus *Potyvirus*, family *Potyviridae*
Raspberry bushy dwarf virus—genus *Idaeovirus*
Raspberry leaf mottle virus—unclassified
Raspberry leaf spot virus—unclassified
Raspberry ringspot virus—genus *Nepovirus*, family *Comoviridae*
Raspberry vein chlorosis virus—genus *Nucleorhabdovirus*, family
    Rhabdoviridae, order *Mononegavirales*
Red clover mottle virus—genus *Comovirus*, family *Comoviridae*
Red clover necrotic mosaic virus—genus *Dianthovirus*
Red clover vein mosaic virus—genus *Carlavirus*
Rice black-streaked dwarf virus—genus *Fijivirus*, family *Reoviridae*
Rice dwarf virus—genus *Phytoreovirus*, family *Reoviridae*
Rice gall dwarf virus—genus *Phytoreovirus*, family *Reoviridae*
Rice grassy stunt virus—genus *Tenuivirus*
Rice hoja blanca virus—genus *Tenuivirus*
Rice necrosis mosaic virus—genus *Bymovirus*, family *Potyviridae*

Rice ragged stunt virus—genus *Oryzavirus*, family *Reoviridae*
Rice stripe necrosis virus—genus *Furovirus*
Rice stripe virus—genus *Tenuivirus*
Rice tungro bacilliform virus—genus *Badnavirus*, family *Caulimoviridae*
Rice tungro spherical virus—genus *Waikavirus*, family *Sequiviridae*
Rice yellow mottle virus—genus *Sobemovirus*
Rice yellow stunt virus—unclassified
Ryegrass mosaic virus—genus *Rymovirus*, family *Potyviridae*
Satsuma dwarf virus—genus *Nepovirus*, family *Comoviridae*
Sesbania mosaic virus—genus *Tymovirus*
Shallot latent virus—genus *Carlavirus*
Sinaloa tomato leaf curl virus—genus *Geminivirus*, family *Geminiviridae*
Sorghum mosaic virus—genus *Potyvirus*, family *Potyviridae*
Sorghum stunt mosaic virus—genus *Nucleorhabdovirus*, family *Rhabdoviridae*, order *Mononegavirales*
Sour cherry green ring mottle virus—unclassified
Sowbane mosaic virus—genus *Sobemovirus*
Soybean chlorotic mottle virus—genus *Caulimovirus*, family *Caulimoviridae*
Soybean crinkle leaf virus—genus *Bigeminivirus*
Soybean dwarf virus—genus *Luteovirus*
Soybean Indonesian dwarf virus—genus *Luteovirus*
Soybean mosaic virus—genus *Potyvirus*, family *Potyviridae*
Soybean virus—genus *Rhabdovirus*
Spinach latent virus—genus *Ilarvirus*, family *Bromoviridae*
Squash leaf curl virus—genus *Bigeminivirus*
Squash mosaic virus—genus *Comovirus*, family *Comoviridae*
Squash yellow leaf curl virus—unclassified
Strawberry crinkle virus—genus *Cytorhabdovirus*, family *Rhabdoviridae*, order *Mononegavirales*
Strawberry latent ringspot virus—genus *Nepovirus*, family *Comoviridae*
Strawberry mild yellow edge virus—genus *Luteovirus*
Strawberry mild yellow edge-associated virus—genus *Potexvirus*
Strawberry vein banding virus—genus *Caulimovirus*, family *Caulimoviridae*
Subterranean clover mottle virus—genus *Sobemovirus*
Subterranean clover red leaf virus—genus *Luteovirus*
Subterranean clover stunt virus—genus *Nanavirus*
Sugarcane bacilliform virus—genus *Badnavirus*, family *Caulimoviridae*
Sugarcane Fiji disease virus—genus *Fijivirus*, family *Reoviridae*
Sugarcane mild mosaic virus—unclassified
Sugarcane mosaic virus—genus *Potyvirus*, family *Potyviridae*

Sugarcane streak virus—genus *Monogeminivirus*
Sugarcane yellow leaf virus—unclassified
Sunflower mosaic virus—unclassified
Sunflower yellow blotch virus—genus *Umbravirus*
Sunnhemp mosaic virus—genus *Tobamovirus*
Sweet potato chlorotic stunt virus—genus *Crinivirus*
Sweet potato feathery mottle virus—genus *Potyvirus*, family *Potyviridae*
Sweet potato infectious chlorosis virus—genus *Closterovirus*
Sweet potato latent virus—genus *Potyvirus*, family *Potyviridae*
Sweet potato leaf curl virus—genus *Badnavirus*, family *Caulimoviridae*
Sweet potato mild mottle virus—genus *Ipomovirus*
Sweet potato mild speckling virus—genus *Potyvirus*, family *Potyviridae*
Sweet potato sunken vein virus—genus *Closterovirus*
Taino tomato mottle virus—genus *Geminivirus*, family *Geminiviridae*
Tamarillo mosaic virus—genus *Potyvirus*, family *Potyviridae*
Taro feathery mottle virus—genus *Potyvirus*, family *Potyviridae*
Tobacco black ring virus—genus *Nepovirus*, family *Comoviridae*
Tobacco etch virus—genus *Potyvirus*, family *Potyviridae*
Tobacco leaf curl virus—genus *Bigeminivirus*
Tobacco mild green mosaic virus—genus *Tobamovirus*
Tobacco mosaic satellivirus—genus *Satellivirus*
Tobacco mosaic virus—genus *Tobamovirus*
Tobacco necrosis satellivirus—genus *Satellivirus*
Tobacco necrosis virus—genus *Necrovirus*
Tobacco rattle virus—genus *Tobravirus*
Tobacco ringspot virus—genus *Tobravirus*
Tobacco streak virus—genus *Ilarvirus*, family *Bromoviridae*
Tobacco vein mottling virus—genus *Potyvirus*, family *Potyviridae*
Tobacco yellow dwarf virus—genus *Monogeminivirus*
Tomato 1 virus—genus *Ilarvirus*, family *Bromoviridae*
Tomato aspermy virus—genus *Cucumovirus*, family *Bromoviridae*
Tomato black ring virus—genus *Nepovirus,* family *Comoviridae*
Tomato bushy stunt virus—genus *Tombusvirus*
Tomato chlorosis virus—unclassified
Tomato chlorotic spot virus—genus *Tospovirus*
Tomato golden mosaic virus—genus *Bigeminivirus*
Tomato greenhouse whitefly-borne virus—genus *Geminivirus*, family
   Geminiviridae
Tomato Indian leaf curl virus—genus *Bigeminivirus*
Tomato infectious chlorosis virus—genus *Closterovirus*
Tomato leaf curl virus—genus *Geminivirus*, family *Geminiviridae*
Tomato mosaic virus—genus *Tobamovirus*

Tomato mottle virus—genus *Bigeminivirus*
Tomato pseudo-curly top virus—genus *Hybrigeminivirus*
Tomato ringspot virus—genus *Nepovirus,* family *Comoviridae*
Tomato spotted wilt virus—genus *Tospovirus,* family *Bunyaviridae*
Tomato yellow leaf curl virus—genus *Bigeminivirus*
Tomato yellow vein streak virus—genus *Geminivirus,* family *Geminiviridae*
Tulip breaking virus—genus *Potyvirus,* family *Potyviridae*
Tulip X virus—genus *Potexvirus*
Turnip crinkle virus—genus *Carmovirus,* family *Tombusviridae*
Turnip mosaic virus—genus *Potyvirus,* family *Potyviridae*
Turnip vein-clearing virus—genus *Tobamovirus*
Turnip yellow mosaic virus—genus *Tymovirus*
Urd bean leaf crinkle virus—unclassified
Vanilla necrosis virus—genus *Potyvirus,* family *Potyviridae*
Viola mottle virus—genus *Potexvirus*
Watermelon chlorotic stunt virus—genus *Bigeminivirus*
Watermelon Moroccan mosaic virus—genus *Potyvirus,* family *Potyviridae*
Watermelon mosaic 1 virus—genus *Potyvirus,* family *Potyviridae*
Watermelon mosaic 2 virus—genus *Potyvirus,* family *Potyviridae*
Watermelon silvery mottle virus—genus *Tospovirus,* family *Bunyaviridae*
Welsh onion yellow stripe virus—genus *Potyvirus,* family *Potyviridae*
Wheat dwarf virus—genus *Monogeminivirus*
Wheat Iranian stripe virus—genus *Tenuivirus*
Wheat leaf streak virus—genus *Rymovirus,* family *Potyviridae*
Wheat soilborne mosaic virus—genus *Furovirus*
Wheat spindle streak virus—genus *Bymovirus,* family *Potyviridae*
Wheat streak mosaic virus—genus *Rymovirus,* family *Potyviridae*
Wheat yellow mosaic virus—genus *Bymovirus,* family *Potyviridae*
White clover cryptic virus 1—genus *Alphacryptovirus,* family *Partitiviridae*
White clover cryptic virus 2—genus *Betacryptovirus,* family *Partitiviridae*
White clover mosaic virus—genus *Potexvirus*
Yam mild mosaic virus—unclassified
Yam mosaic virus—genus *Potyvirus,* family *Potyviridae*
Zucchini yellow fleck virus—genus *Potyvirus,* family *Potyviridae*
Zucchini yellow mosaic virus—genus *Potyvirus,* family *Potyviridae*

SYMPTOMS OF CROP VIRUS DISEASES

Virus diseases are recognized by their symptoms. The important symptoms are: mosaic, chlorosis, mottle, yellows, ring spot, streak, stripe, vein band, fleck, line pattern, vein chlorosis, vein clearing, necrosis, reddening, browning, blackening, wilt, etch, leaf rolling, leaf curling, curly top, leaf crinkling, leaf distortion, fern-leaf, malformations, rugose, enation, big vein, rosette, tumors, bud blight, flower color-breaking, dwarf, stunt, and sterility. These symptoms are described in Chapter 11.

MODE OF TRANSMISSION OF VIRUS DISEASES

Mechanical Transmission

Some viruses pass directly from plant to plant when their leaves rub together. Mechanical transmission will occur if virus concentrations in plant sap and virus stability are high. *Tobacco mosaic virus, Tomato mosaic virus, Cucumber green mottle virus, Tobacco necrosis virus,* and *Potato virus X* are easily transmitted by mechanical contact. Viruses also may spread by sticking on the clothes and hands of workers and equipment. For example, mowing machines spread *White clover mosaic virus* and *Red clover mottle virus.*

Transmission by Grafting

Grafting is a common practice in horticultural crops. Viruses may pass through grafting. *Citrus tristeza virus* is transmitted by this method.

Transmission Through Vegetative Propagation Material

Viruses normally invade the host systemically and hence all plant parts contain the virus. Rhizomes, tubers, corms, bud woods, and stem cuttings are the common vegetative propagation materials and they may serve as sources for virus infection. *Banana bunchy top virus* is transmitted mostly through suckers. *Potato virus Y, Potato virus X,* and *Potato virus T* are transmitted through potato seed tubers.

Transmission Through Seed

Several viruses are found in immature seed coats, which consist of the integuments and nucellar remnants, and in the periplasm. Because the seed coat and periplasm are part of the mother plant, they may become infected due to systemic invasion by the viruses. Viruses do not survive desiccation, and during maturation of the seed coat, they perish. Because there is no vascular contact between the embryo and mother plant, viruses cannot move to the embryo from the seed coat. However, some viruses, such as *Tobacco mosaic virus,* are highly stable, and the seed coat contamination with these viruses may lead to seed transmission. However, this is rare.

Most seed transmission of viruses occurs through the embryo only. Both the embryo and endosperm are formed within the embryo sac after fertilization and lack direct connection and cellular contact with the mother plant through plasmodesmata. Hence, a virus from the mother plant cannot move to the embryo after fertilization. Embryo infection can occur only when the mother plant is infected before the production of gametes or before cytoplasmic separation of embryonic tissue. Infection after flowering does not lead to transmission through seed or pollen. Viruses that cannot pass beyond the phloem (e.g., viruses that are transmitted by phloem-feeding insects) will not be seedborne. The important viruses that are transmitted by seed (Jackson et al., 1989; Mink, 1993; Johansen et al., 1994) include the following:

Barley—*Barley stripe mosaic virus*
Bean—*Bean common mosaic virus, Bean pod mottle virus, Bean southern mosaic virus, Bean yellow mosaic virus*
Broad bean—*Broad bean mottle virus, Broad bean stain virus, Broad bean wilt virus*
Corn—*Maize chlorotic dwarf virus, Maize dwarf mosaic virus, Maize mosaic virus*
Cowpea—*Blackeye cowpea mosaic virus, Cowpea aphidborne mosaic virus, Cowpea mild mottle virus, Cowpea mosaic virus, Cowpea severe mosaic virus*
Lettuce—*Lettuce mosaic virus*
Lucerne—*Lucerne latent virus, Lucerne transient streak virus*
Melon—*Melon necrotic spot virus*
Mungbean—*Mungbean mosaic virus*
Muskmelon—*Muskmelon necrotic ringspot virus*
Oat—*Oat mosaic virus*
Pea—*Pea early-browning virus, Pea enation mosaic virus, Pea seedborne mosaic virus*
Peanut—*Peanut clump virus, Peanut mottle virus, Peanut stripe virus, Peanut stunt virus*
Soybean—*Soybean mosaic virus*
Sunflower—*Sunflower mosaic virus*
Tobacco—*Tobacco etch virus, Tobacco mosaic virus, Tobacco rattle virus, Tobacco ringspot virus, Tobacco streak virus*

Tomato—*Tomato aspermy virus, Tomato black ring virus, Tomato
    bushy stunt virus, Tomato spotted wilt virus, Tomato ringspot virus*
Urdbean—*Urdbean leaf crinkle virus*
Watermelon—*Watermelon mosaic virus*
Wheat—*Wheat streak mosaic virus, Wheat striate mosaic virus*

Transmission Through Pollen

Some viruses are known to be transmitted through pollens. *Prunus necrotic ringspot virus* and *Prune dwarf virus* in stone fruits, *Blueberry shock
virus* and *Blueberry leaf mottle virus* in blueberry, *Cherry leafroll virus* in
walnut, *Raspberry bushy dwarf virus* in raspberry, and *Tobacco streak virus*
in many crops are transmitted from plant to plant through pollens. Insects
may move virus-contaminated pollen to flowers on healthy plants. Pollen-
feeding insects, such as thrips which thrive in the pollen receptors of flowers, may create wounds needed for the mechanical transmission of viruses
to nongametophytic tissues (Mink, 1993).

Transmission by Insects

Several insects transmit virus diseases. The transfer is not merely mechanical, and the relationship between virus and vector is complex. Some
mechanically transmissible viruses are not transmitted by insects. Mechanically transmitted viruses such as *Tobacco mosaic virus, Potato virus X,*
and *White clover mosaic virus* are not transmitted by insects. Virus transmission by insects is highly specific. Viruses transmitted by leafhoppers are
not transmitted by aphids. Strains of one virus may have different insect
species for their transmission. Insects have different kinds of mouthparts
and ways of feeding. Insects such as beetles and caterpillars have biting and
chewing mouthparts. They eat away the leaf pieces and damage the plants
severely. Cells that are touched by the insects' mouthparts are destroyed. Viruses coming through these mouthparts cannot establish on the leaves, and
hence transmission is rare among the biters and chewers. However, some viruses, such as *Southern bean mosaic virus, Broad bean stain virus, Radish
mosaic virus, Turnip yellow mosaic virus,* and *Cowpea mosaic virus,* are
transmitted by beetles.

The majority of insects that transmit viruses have piercing and sucking
mouthparts. Aphids and leafhoppers, for example, have these kinds of
mouthparts. Aphids make initial, brief sap-sampling probes between epidermal cells or into them. However, aphids take up food from phloem vessels only. They need several minutes to hours to penetrate between plant

cells rather than through cells to finally reach the phloem. The mouthparts of leafhoppers are most robust compared to aphids. They penetrate the plant cells and reach the phloem within a few minutes. In the phloem, sap flows under pressure and will pass up into the insect's food canal. Hence, leafhoppers transmit viruses found in the phloem of infected plants. Thrips have rasping and sucking mouthparts. They feed by grasping open the epidermal cells or by rasping them and sucking up their contents. Thus, thrips can transmit viruses found in epidermal cells also.

The virus-vector relationship varies widely depending upon the duration of the virus in the vector (persistence). In the case of persistent viruses, the virus may simply circulate through the body of the vector or it may propagate. Hence, this relationship can be classified as (1) noncirculative nonpersistent, (2) noncirculative semipersistent, (3) circulative nonpropagative, and (4) circulative propagative transmission.

Noncirculative Nonpersistent Transmission

Aphids transmit viruses mostly from and into the host's parenchyma. The virus is acquired and inoculated during a brief feeding period of a few seconds to some minutes, mostly during probes. Aphids become infective immediately after virus uptake. Persistence in the vector is very brief. In exceptional cases, infectivity persists up to 40 hours when the vector has no access to plants after acquiring the virus. These viruses are "styletborne." Hence, the transmission is noncirculative nonpersistent. The transmission is of low specificity. Aphids, which are not pests and do not colonize on the crop, but merely probe the surface for a palatable host, can also transmit the viruses efficiently (Vidhyasekaran, 1993).

Circulative Nonpropagative (Persistent) Transmission

Viruses taken up by the vector enter the alimentary canal, pass through the gut wall, circulate in body fluid (haemolymph), and contaminate the saliva. After a latent period, the insect becomes infective at the next feeding. Infectivity is lost at moulting. Such uptake and inoculation happens only after long feeding times, usually after 15 minutes. Viruses are acquired from the phloem of vascular bundles. Persistent transmission is associated with highly specific virus-vector relationships. Certain viruses are spread by a single vector species or a biotype, and even individuals may differ in efficiency as vectors. Leafhoppers generally transmit viruses in a persistent manner.

Circulative Propagative (Persistent) Transmission

Certain viruses are propagated inside the insect body and may last until death. Even transovarial transfer of the viruses through insect eggs is found in leafhoppers. *Wheat striate mosaic virus, Rice dwarf virus,* and *Rice stripe virus* are transmitted in this manner. Such viruses can persist in populations without access to susceptible hosts. Once infective, the insects transmit the virus to healthy plants throughout their life period.

Semipersistent Viruses

Semipersistent viruses can be transmitted mechanically with some diffi-culty. Viruses are taken up from the phloem by insects as they feed. The viruses are adsorbed and gradually eluted from the line of pharynx. They are then transmitted through ejection or egression at the start of subsequent feeding. *Beet yellows virus* is transmitted this way.

Infection of Insect Vectors

Harmful effects of some plant viruses on insect vectors have also been re-ported. The planthopper *Laodelphax striatellus,* which transmits *Rice stripe virus,* is infected by the virus itself. Individuals from eggs laid by infective planthoppers die prematurely, if they hatch at all. The mortality of nymphs is also high, particularly in the first and second instars.

The following is a list of viruses transmitted by various insects:

Aphids

> *Banana bunchy top virus—Pentalonia nigronervosa*
> *Bean yellow mosaic virus—Myzus persicae*
> *Beet yellow net virus—Myzus persicae*
> *Beet yellow stunt virus—Nasonova lactucae*
> *Beet yellows virus—Myzus persicae, Aphis fabae*
> *Bean common mosaic virus—Myzus persicae*
> *Broccoli necrotic yellows virus—Brevicoryne brassicae*
> *Carnation necrotic fleck virus—Myzus persicae*
> *Carrot latent virus—Semiaphis heraclei*
> *Carrot yellow leaf virus—Semiaphis heraclei*
> *Celery yellow spot virus—Hyadaphis foeniculi*
> *Citrus tristeza virus—Toxoptera citricidus, Aphis gossypii*
> *Clover yellows virus—Aphis craccivora*

Cotton anthocyanosis virus—Aphis gossypii
Cucumber mosaic virus—Myzus persicae
Groundnut rosette assistor virus—Aphis craccivora
Lettuce necrotic yellows virus—Hyperomyzus lactucae
Lucerne enation virus—Aphis craccivora
Pea enation mosaic virus—Acrythosiphon pisum, Aulacorthum solani
Pea seedborne mosaic virus—Macrosiphum euphorbiae, Myzus persicae, Acyrthosiphon pisum, Aphis craccivora, Rhophalosiphum padi, Aphis fabae, Macrosiphum rosae, Aulacorthum solani, Phorodon cannabis, Semiaphis dauci, Brevicoryne brassicae
Peanut mottle virus—Myzus persicae
Peanut stunt virus—Myzus persicae
Physalis mild chlorosis virus—Myzus persicae
Raspberry leaf curl virus—Aphis rubicola
Raspberry vein chlorosis virus—Aphis idaei
Sonchus yellow net virus—Aphis coreopsidis
Sowthistle yellow vein virus—Hyperomyzus lactucae
Soybean mosaic virus—Aphis craccivora, Myzus persicae
Strawberry crinkle virus—Chaetosiphon fragaefolii
Strawberry mild yellow edge virus—Chaetosiphon fragaefolii
Subterranean clover stunt virus—Aphis craccivora
Tobacco vein distorting virus—Myzus persicae
Tobacco yellow net virus—Myzus persicae
Tobacco yellow vein assistor virus—Myzus persicae
Tomato yellow net virus—Myzus persicae
Wheat yellow leaf virus—Rhopalosiphum padi

Leafhoppers

American wheat striate mosaic virus—Endria inimica
Cereal chlorotic mottle virus—Nesoclutha pallida
Maize streak virus—Cicadulina mbila
Oat striate mosaic virus—Graminella nigrifrons
Potato yellow dwarf virus—Aceratogallia sanguinolenta, Agallia constricta
Rice transitory yellowing virus—Nephotettix apicalis
Tobacco yellow dwarf virus—Orosius argentatus
Wheat dwarf virus—Psammotettix alienus

Planthoppers

Barley yellow striate mosaic virus—Laodelphax striatellus
Colocasia bobone disease virus—Tarophagus proserpina
European wheat striate mosaic virus—Javesella dubia, J. pellucida
Maize mosaic virus—Peregrinus maidis
Maize rough dwarf virus—Peregrinus maidis
Maize sterile stunt virus—Sogatella kolophon, Peregrinus maidis
Northern cereal mosaic virus—Terthron albovittatus, Ukanodes
 sapporona, U. albifascia, Muellerianella fairmairei, Laodelphax
 striatellus
Oat sterile dwarf virus—Dicranotropis hamata, Javesella discolor,
 J. dubia, J. obscurella, J. pellucida
Rice black-streaked dwarf virus—Ukanodes sapporona,
 U. albifascia, Laodelphax striatellus
Rice grassy stunt virus—Nilaparvata lugens
Rice hoja blanca virus—Sogatodes cubanus, S. oryzicola
Rice stripe virus—Ukanodes sapporona, U. albifascia, Laodelphax
 striatellus
Sugarcane Fiji disease virus—Perkinsiella saccharicida, P. vastatrix

Thrips

Impatiens necrotic spot virus—Thrips tabaci
Peanut bud necrosis virus—Thrips tabaci
Peanut yellow spot virus—Thrips tabaci
Tomato spotted wilt virus—Thrips tabaci, Frankliniella spp.
Watermelon silvery mottle virus—Thrips tabaci

Whiteflies

Bean golden mosaic virus—Bemisia tabaci
Cucumber yellows virus—Trialeurodes vaporariorum
Mungbean yellow mosaic virus—Bemisia tabaci
Tomato golden mosaic virus—Bemisia tabaci
Tobacco leaf curl virus—Bemisia tabaci
Tomato yellow dwarf virus—Bemisia tabaci
Tomato yellow leaf curl virus—Bemisia tabaci
Tomato yellow mosaic virus—Bemisia tabaci

Mealybugs

Cacao swollen shoot virus—Delococcus tafoensis, Dysmicoccus brevipes, Ferrisia virgata, Maconellicoccus ugandae, Paracoccus sp., *Paraputo anomalus, Planococcoides njalensis, Planococcus* sp., *Planococcus citri, Planococcus keyae, Pseudococcus concavocerarii, Pseudococcus hargreavesi, Pseudococcus longispinus, Tylococcus westwoodi*
Grapevine virus A—Pseudococcus longispinus

Beetles

Andean potato latent virus—Epitrix sp.
Bean mild mosaic virus—Diabrotica undecimpunctata, Epilachna varivestis
Bean pod mottle virus—Ceratoma trifurcata, Diabrotica balteata, D. undecimpunctata, Colaspis lata, Epicanta vittata
Bean rugose mosaic virus—Cerotoma ruficornis, Diabrotica balteata, D. adelpha
Broad bean mottle virus—Acalymma trivittata, Diabrotica undecimpunctata, Colaspis flavida
Broad bean stain virus—Apion vorax, A. aestivum, A. aethiops, Sitona lineatus
Cowpea mosaic virus—Epilachna varivestis, Ceratoma arcuata, C. atrofasciata, C. variegata, Diabrotica adelpha, Diabrotica virgifera, Acalymma vittatum. Systena sp., *Diphaulaca meridae*
Maize chlorotic mottle virus—Oulema melanopa, Chaetocnema pulicaria, Systena frontalis, Diabrotica undecimpunctata, D. longicornis, D. virgifera
Radish mosaic virus—Epitrix hirtipennis, Diabrotica undecimpunctata, Phyllotreta striolata
Rice yellow mottle virus—Oulema dunbrodiensis, Cryptocephalus chalybeipennis, Cryptocephalus nigrum, Monolepta flaveola, Monolepta hamatura, Sesselia pusilla, Chaetocnema abyssinica, Chaetocnema kenyensis, Chaetocnema pulla, Dactylispa bayoni, Dicladispa paucispina, Dicladispa virdicyanae, Trichispa sericea
Southern bean mosaic virus—Cerotoma trifurcata, Epilachna varivestis, Atrachya menetris, Acalymma vittata
Squash mosaic virus—Diabrotica balteata, D. undecimpunctata, D. longicornis, D. virgifera, Acalymma vittata, A. thiemei, A. trivittata, Epilachna chrysomelina
Turnip yellow mosaic virus—Phyllotreta spp., *Psylliodes* spp., *Phaedon cochleariae*

Lace Bugs

    Beet leaf curl virus—Piesma quadratum

Plant Bugs

    Spinach blight virus—Lygus protensis

Earwigs

    Turnip yellow mosaic virus—Foficula auricularia

Lygacid Bugs

    Centrosema mosaic virus—Nysius sp.

Flies (Leafminer)

    Sowbane mosaic virus—Liriomyza langei (mechanical transmission)
    Tobacco mosaic virus—Liriomyza langei (mechanical transmission)

Grasshoppers

    Potato virus X virus—Melanoplus differentialis

Treehoppers

    Tomato pseudo-curly top virus—Micrutalis malleifera

Transmission by Mites

Mites belong to the class Arachnida and order Acarina. Mites have puncturing and sucking mouthparts. Eryophid mites commonly transmit viruses. Eryophid mites are elongate and tiny, about 0.2 mm long. They are invisible to the naked eye and often hide between hairs of leaves and in buds. They do not have much movement and are carried by the wind. They have slender styles with which they puncture epidermal cells. The two pads at the tip of the rostrum help in bringing saliva to the stylets and in sucking up plant sap.

Spider mites (*Brevipalpus* spp.) also transmit viruses. The important virus diseases transmitted by mites include:

> *Agopyron mosaic virus—Abacarus hystrix*
> *Cherry mottle leaf virus—Phytoptus inaequalis*
> *Citrus leprosies virus—Brevipalpus obovatus*
> *Coffee ringspot virus—Brevipalpus phoenicis*
> *Fig mosaic virus—Eriophyes ficus*
> *Onion mosaic virus—Eriophyes tulipae*
> *Peach mosaic virus—Phytoptus insidiosus*
> *Ryegrass mosaic virus—Abacarus hystrix*
> *Wheat spot mosaic virus—Eriophyes tulipae*
> *Wheat streak mosaic virus—Eriophyes tulipae*

Transmission by Nematodes

Some nematodes belonging to the genera *Longidorus, Xiphinema, Trichodorus,* and *Paratrichodorus* are known to transmit viruses. Nematode-borne viruses are retained in the vector, including the guide-sheath lining of the odontostyle in *Longidorus,* the lumen of the odontophore, the esophagus in *Xiphinema,* and the entire pharynx and esophagus in *Trichodorus.* Nematodes, which have life spans of about two years, retain virus for long periods. The following are viruses transmitted by nematodes:

> *Arabis mosaic virus—Xiphinema diversicaudatum, X. coxi,*
>   *X. bakeri*
> *Brome mosaic virus—Xiphinema diversicaudatum*
> *Carnation ringspot virus—Xiphinema diversicaudatum*
> *Cowpea mosaic virus—Xiphinema basiri*
> *Grapevine fanleaf virus—Xiphinema index, X. italae*
> *Pea early-browning virus—Paratrichodorus pachydermus, P. teres,*
>   *P. anemones, Trichodorous viruliferus*
> *Peach rosette mosaic virus—Xiphinema americanum,*
>   *Criconemoides xenoplax*
> *Prunus necrotic ringspot virus—Longidorus macrosoma*
> *Raspberry ringspot virus—Longidorus elongatus, L. macrosoma,*
>   *L. elongates*
> *Strawberry latent ringspot virus—Xiphinema diversicaudatum,*
>   *X. coxi*
> *Tobacco rattle virus—Paratrichodorus anemones, P. nanus,*
>   *P. pachydermus, P. teres, P. cylindricus, Trichodorus sparsus,*
>   *T. viruliferus*

Tobacco ringspot virus—Xiphinema americanum
Tomato black ring virus—Longidorus attenuatus, L. elongatus
Tomato ringspot virus—Xiphinema americanum, X. brevicolle

Transmission by Fungi (Including Protozoa)

Some soilborne fungal (including Protozoa) pathogens transmit plant viruses. *Olpidium* (belonging to Chytridiomycota, Fungi), *Polymyxa,* and *Spongospora* (belonging to Plasmodiophoromycota, Protozoa) are pathogens of crop plants. The zoospores of these fungi are released from the host, carry the virus, and transmit it to susceptible hosts during the infection process. In some cases, plant viruses are carried on the outside of the fungi. Some examples include *Tobacco necrosis virus* and *Cucumber necrosis virus.* Viruses such as *Lettuce big vein virus* are found inside the zoospores. These viruses persist for years in viable resting sporangia. Transmission by fungi can be of two types: nonpersistent and persistent transmission. The following are viruses transmitted by fungi, including protozoa:

Barley yellow mosaic virus—Polymyxa graminis
Beet necrotic yellow vein virus—Polymyxa betae
Cucumber necrosis virus—Olpidium radicale
   (= *O. cucurbitacearum*)
Lettuce big-vein virus—Olpidium brassicae
Oat mosaic virus—Polymyxa graminis
Potato mop-top virus—Spongospora subterranea
Rice necrosis mosaic virus—Polymyxa graminis
Soilborne wheat mosaic virus—Polymyxa graminis
Tobacco necrosis virus—Olpidium brassicae
Tobacco stunt virus—Olpidium brassicae
Wheat mosaic virus—Polymyxa graminis
Wheat spindle streak mosaic virus—Polymyxa graminis
Wheat yellow mosaic virus—Polymyxa graminis

Transmission Through Soil and Water

Some viruses, such as *Tobacco mosaic virus* and *Tomato mosaic virus,* are highly stable and retain infectivity within soil for a long time. They may survive in diseased plant debris and infect plants grown in infested soil. Similarly, *Cucumber green mottle mosaic virus* survives in water. These viruses are transmitted mechanically (Vidhyasekaran, 1993).

Transmission Through Dodder

Dodder (*Cuscuta* sp.) is a chlorophyll-less parasitic plant. It grows and entwines its yellowish stem around the stems of host plants. It produces haustoria, penetrates the host's vascular bundles, and forms natural bridges between diseased and healthy plants. Viruses from diseased plants pass through dodder stems and enter healthy plants. Most viruses can be transmitted through dodder plants, although a natural transmission of virus diseases has not been reported.

VIRUS DISEASE DIAGNOSIS

Virus disease diagnosis is important to plan for effective disease management. Several immunological methods have been developed to diagnose virus diseases, including agglutination tests, precipitation tests, immunodiffusion tests, enzyme-linked immunosorbent assays (ELISA), double antibody sandwich assays (DAS-ELISA), imunofluorescent tests, and immunosorbent electron microscopy (ISEM). Many diagnostic methods based on the nucleic acids of viral pathogens are also widely used, including restriction fragment-length polymorphisms (RFLPs), polymerase chain reactions (PCR), nucleic acid hybridization dot-blot tests, cDNA probes, reverse transcription-PCR (RT-PCR), inverse PCR, arbitrarily primed PCR (AP-PCR), and DNA amplification fingerprinting (DAF) tests. For more information about these tests, see Chapter 13.

IMPORTANT CROP DISEASES CAUSED BY VIRUSES

The following are economically important virus diseases of major crops:

Banana bunchy top: genus *Nanavirus, Banana bunchy top virus* (BBTV)

Barley yellow dwarf: genus *Luteovirus, Barley yellow dwarf virus* (BYDV)

Barley yellow mosaic: genus *Bymovirus, Barley yellow mosaic virus* (BaYMV)

Citrus tristeza: genus *Closterovirus, Citrus tristeza virus* (CTV)

Cucumber mosaic: genus *Cucumovirus, Cucumber mosaic virus* (CMV)

Pea seedborne mosaic: genus *Potyvirus, Pea seedborne mosaic virus* (PSbMV)

Peach necrotic ring spot: genus *Ilarvirus, Prunus necrotic ringspot virus* (PNRSV)

Plum pox (= Sharka): genus *Potyvirus, Plum pox virus* (PPV)

Potato leafroll mosaic: genus *Luteovirus, Potato leafroll virus* (PLRV)

Potato leafrolling: genus *Carlavirus, Potato virus M* (PVM)

Potato mild mosaic: genus *Potyvirus, Potato virus A* (PVA)

Potato mop-top: genus *Furovirus, Potato mop-top virus* (PMTV)

Potato rugose mosaic: genus *Potyvirus, Potato virus Y* (PVY, strains O, N, and C)

Potato stem mottle (spraing disease): genus *Tobravirus, Tobacco rattle virus* (TRV)

Potato virus X: genus *Potexvirus, Potato virus X* (PVX)

Rice grassy stunt: genus *Tenuivirus, Rice grassy stunt virus* (RGSV)

Rice tungro: genus *Badnavirus, Rice tungro bacilliform virus* (RTBV), and genus *Waikavirus, Rice tungro spherical virus* (RTSV)

Tobacco mosaic: genus *Tobamovirus, Tobacco mosaic virus* (TMV)

Tomato mottle: genus *Bigeminivirus, Tomato mottle virus* (ToMoV)

Tomato spotted wilt: genus *Tospovirus, Tomato spotted wilt virus* (TSWV)

REFERENCES

Azzam, O. and Chancellor, T. C. B. (2002). The biology, epidemiology, and management of rice tungro disease in Asia. *Plant Dis,* 86:88-100.

De Zoeten, G. A. (1995). Plant virus infection: Another point of view. *Adv Bot Res,* 21:105-124.

Deom, C. M., Lapidot, M., and Beachy, R. N. (1992). Plant virus movement proteins. *Cell,* 69:221-224.

Esau, A. W. and Lucas, W. J. (1990). *Viruses in Plant Hosts.* University of Wisconsin Press, Madison, WI.

Esau, K. (1967). Anatomy of plant virus infections. *Annu Rev Phytopathol* 5:45-76.

Fauquet, C. M. and Mayo, M. A. (1999). Updated ICTV list of names and abbreviations of viruses, viroids, and satellites infecting plants. *Arch Virol,* 144:1249-1273.

Garcia, L. E., Brandenburg, R. L., and Bailey, J. E. (2000). Incidence of *Tomato spotted wilt virus (Bunyaviride)* and tobacco thrips in Virginia-type peanuts in North Carolina. *Plant Dis,* 84:459-469.

Hull, R. (1996). Molecular biology of rice tungro viruses. *Annu Rev Phytopathol,* 34:275-297.

Jackson, A. O., Hunter, B. G., and Gustafson, G. D. (1989). Hordeivirus relationships and genome organization. *Annu Rev Phytopathol,* 27:95-121.

Johansen, E., Edwards, M. C., and Hampton, R. O. (1994). Seed transmission of viruses: Current perspectives. *Annu Rev Phytopathol,* 32:363-386.

Kassanis, B. (1962). Properties and behaviour of a virus depending for its multiplication on another. *J Gen Virol,* 27:477-488.

Lucas, W. J. (1993). Movement protein gene for virus disease management. In P. Vidhyasekaran (Ed.), *Genetic Engineering, Molecular Biology, and Tissue Culture for Crop Pest and Disease Management.* Daya Publishing House, Delhi, pp. 2-15.

Mandahar, C. L. (1991). *Plant Viruses. Vol. I. Structure and Replication.* CRC Press, Boca Raton, FL.

Mayo, M. A. and Horzinek, M. (1998). A revised version of the international code of virus classification and nomenclature. *Arch Virol,* 143:1645-1654.

Mink, G. I. (1993). Pollen- and seed-transmitted viruses and viroids. *Annu Rev Phytopathol,* 31:375-402.

Roberts, A. W. and Lucas, W. J. (1990). Plasmodesmata. *Annu Rev Plant Physiol Plant Mol Biol,* 41:369-419.

Roux, L., Simon, A. E., and Holland, J. J. (1991). Effects of defective interfering viruses on virus replication and pathogenesis in vitro and in vivo. *Adv Virus Res,* 40:181-211.

Rubio, T., Borja, M., Scholthof, H. B., Feldstein, P. A., Morris, T. J., and Jackson, A. O. (1999). Broad-spectrum protection against tombusviruses elicited by defective interfering RNAs in transgenic plants. *J Gen Virol,* 73:5070-5078.

Schneider, I. R. (1969). Satellite-like particle of tobacco ringspot virus that resembles tobacco ringspot virus. *Science,* 166:1627-1629.

Terrada, E., Kerschbaumer, R. J., Giunta, G., Galeffi, P., Himmler, G., and Cambra, M. (2000). Fully "recombinant enzyme-linked immunosorbent assay" using genetically engineered single-chain antibody fusion proteins for detection of *Citrus tristeza virus. Phytopathology,* 90:1337-1344.

Tien, P. and Wu, G. (1991). Satellite RNA for the biocontrol of plant disease. *Adv Virus Res,* 39:321-339.

Van Regenmortel, M. H. V. (1999). How to write the names of virus species. *Arch Virol,* 144:1041-1042.

Van Regenmortel, M. H. V., Fauquet, C. M., Bishop, D. H. L., Carstens, E., Estes, M., Lemon, S., McGeoch, D., Wickner, R. B., Mayo, M. A., Pringle, C. R., and Maniloff, J. (1999). *Virus Taxonomy, Seventh Report of the International Committee for the Taxonomy of Viruses.* Academic Press, New York.

Vidhyasekaran, P. (1993). *Principles of Plant Pathology.* CBS Publishers, Delhi.

Zaitlin, M. and Hull, R. (1987). Plant virus-host interactions. *Annu Rev Plant Physiol,* 38:291-315.

Zaitlin, M. and Palukaitis, P. (2000). Advances in understanding plant viruses and virus diseases. *Annu Rev Phytopathol,* 38:117-143.

COMMON DISEASE SYMPTOMS, ASSESSMENT, AND DIAGNOSIS

– 11 –

Common Disease Symptoms

Common names of diseases are always based on their key symptoms. These symptoms are described here.

anthracnose: A characteristic lesion, which is a circular to angular, sometimes irregular sunken spot with grayish-black center and yellow margin in leaves, stems, and fruit (e.g., grapevine anthracnose).

areolate mildew: Mildew growth in the area between veins on a leaf (e.g., cotton areolate mildew).

big vein: A condition in which veins become enlarged (e.g., *Lettuce big vein virus* disease).

black scurf: Black flaky or scaly matter adhering to the surface of a plant part (e.g., black scurf of potato).

blackening: Intensive necrosis that leads to blackening of tissues (e.g., *Potato black ringspot virus* disease).

blast: A disease that kills plants suddenly (e.g., rice blast).

blight: A plant disease characterized by withering and shriveling without rotting (e.g., late blight of potato).

blister blight: A bubblelike elevation on the surface of a diseased leaf that results in withering and shriveling of the leaf (e.g., tea blister blight).

boll rot: Decay of boll, a fruit of plants such as cotton. Boll consists of a rounded capsule containing the seeds (e.g., cotton boll rot).

brown rot: A condition in which decaying tissues turn brownish (e.g., potato brown rot).

browning: Cell death leads to necrosis, which leads to the browning of tissues (e.g., *Pea early browning virus* disease).

bud blight: Necrosis of buds (*Groundnut bud blight virus* disease).

bud rot: Decay of buds (e.g., bud rot of coconut).

bunchy top: Leaves arise in clusters, giving a rosette appearance at the top (e.g., *Banana bunchy top virus* disease).

bushy stunt: The diseased plant is severely stunted and its shoots are crowded, giving a bushy appearance (e.g., *Tomato bushy stunt virus* disease).

canker: A corky outgrowth formed on leaves, twigs, and fruit (e.g., citrus canker).

charcoal rot: Decaying tissues turn charcoal (black) in color (e.g., corn charcoal rot).

chlorosis: Yellowing or whitening may be distributed in the entire plant due to a partial failure of chlorophyll development in leaves and other plant parts, causing more or less uniform discoloration (e.g., *Tomato chlorosis virus* disease, *Lettuce chlorosis virus* disease).

club root: Roots are malformed into clublike structures due to a thick, fleshy growth of roots. The root tips are malformed, leaving the basal portions of the root mostly normal (e.g., club root of cabbage).

collar rot: Decay of the collar region of seedlings at the postemergence stage (e.g., tobacco collar rot).

crown gall: Abnormal outgrowth or swelling produced due to hyperplasia and hypertrophy of host cells (e.g., crown gall of several crops, such as apple, peach, and pear).

crown rot: Rotting of the crown that may spread to the root (e.g., oat crown rot).

curly top: Inward rolling of the leaves with puckering and blisterlike elevation (e.g., *Beet curly top virus* disease).

damping-off: Rotting and collapse of seedlings at soil level or prevention of seedling emergence (e.g., damping-off of vegetables and tobacco).

dieback: Death of diseased plant organs, especially stem or branches, from the tip backward (e.g., eutypa dieback of grape).

downy mildew: Fungal growth that appears as a coating of soft, fine hair on the surface of the host (e.g., downy mildew of grapevine).

dry rot: Putrefactive decomposition of tissues (e.g., dry root rot of citrus and peanut).

dwarf: Plant growth is stunted due to disease (e.g., rice yellow dwarf disease).

ear rot: Decay of the ear of a cereal plant that contains the seeds, grains, or kernels (e.g., corn diplodia ear rot).

enation: Small outgrowths on leaves, especially on veins and stems (e.g., *Pea enation mosaic virus* disease).

etch: Desiccation of superficial tissue (epidermal cells) (e.g., *Tobacco etch virus* disease).

false smut: Individual grains of the panicle are transformed into greenish spore balls. At first, the spore balls are smooth and yellow and are covered by a membrane. When the membrane bursts, the color of the ball becomes orange, yellowish green, or greenish black. The surface of the ball cracks subsequently. When the balls are cut open, the innermost layer is yellowish, the next layer is orange, and the outermost layer is greenish (e.g., false smut of rice).

fern-leaf: Reduced development of the leaf blade in proportion to the midrib leads to fern-leaf development (e.g., *Asparagus fern-leaf virus* disease).

fire blight: Diseased blossoms turn brown or black and exhibit a burnt appearance (e.g., fire blight of apple).

fleck: Small discolored parts are sharply bordered but circular (e.g., *Parsnip yellow fleck virus* disease).

flower color-breaking: Mosaic or variegation of petals of flowers (e.g., *Tulip color-breaking virus* disease).

foot rot: Decay of the basal portion of a plant (e.g., rice foot rot).

freckle: A small brownish spot or small area of discoloration; a freckle is smaller than a spot (e.g., banana freckle).

fruit rot: Decay of fruit (e.g., strawberry alternaria fruit rot).

gray mold: A condition in which infected blossoms and fruit become coated with the fine gray fruiting stalks and spores of the fungal pathogen (e.g., gray mold *[Botrytis cinerea]* of strawberry).

green ear: The floral parts of the ear of a cereal plant are transformed into a green leafy structure; in the spikelet, the glumes, lemmae, paleae, stamens, and pistil are all transformed into green leafy structures (e.g., green ear of pearl millet).

gummosis: Secretion of gum as a symptom of the disease (e.g., citrus gummosis).

head rot: Decay of the head of plants such as sunflowers, which consists of fertilized inflorescence with developed seeds and grains (e.g., sunflower head rot).

kernel rot: Decay of kernels (e.g., corn fusarium kernel rot).

leaf blight: The entire leaf may be blighted (e.g., northern corn leaf blight).

leaf blotch: Irregular, large, discolored, usually black spots (e.g., leaf blotch of barley).

leaf crinkle: Wrinkles form in the infected leaf (e.g., *Turnip crinkle virus* disease).

leaf crumple: The infected leaf appears crushed, forming wrinkles or creases (similar to marks produced by folding) (e.g., cotton leaf crumple).

leaf curl: Leaf margins turn downward and come together at the bottom, exposing the middle upper surface of the leaf blade; the disease results in the downward curling of leaves; the distortion, puffing, and crinkling of a leaf resulting from the unequal growth of its two sides (e.g., *Tobacco leaf curl virus* disease).

leaf roll: A condition in which the leaf tends to roll (e.g., *Potato leafroll virus* disease).

leaf spots: Several types of spots are seen in infected leaves. Water-soaked spots are common in bacteria-infected leaves; circular, cylindrical, irregular, or angular spots are common in fungus and bacteria-infected leaves (e.g., sigatoka leaf spot of banana).

line pattern: Infected leaves show an arrangement of repeated lines (e.g., *Pelargonium line pattern virus* disease).

little leaf: In diseased plants, the leaves are malformed into tiny chlorotic structures (e.g., little leaf of eggplant).

mosaic: Infected leaves show various shades of green and yellow areas that are usually irregularly angular, but sharply delimited. Mosaic is characterized by a patchy variation of normal green color (e.g., *Tobacco mosaic virus* disease).

mottle: Diffusely bordered variegation (differently colored spots), i.e., a pattern of white patches in leaves and other plant parts, that results from the failure of chlorophyll to develop in certain cells. Mottle is an arrangement of spots or confluent blotches of different shades, as of the surface of marble (e.g., *Tomato mottle virus* disease, *Bean pod mottle virus* disease).

necrosis: A symptom in which cells die in patches (e.g., *Rice necrosis virus* disease).

phyllody: The floral parts are transformed into green leafy structures (e.g., phyllody of sesamum).

powdery mildew: Fungal growth seen as a powdery growth on the host's surface (e.g., wheat powdery mildew).

Pox: Infected leaves show various types of mottling, which consist of light green to yellowish-green blotches. The fruits are poxed (i.e., pustules form on the skin) (e.g., *Plum pox virus* disease).

red rot: Decaying tissues appear reddish in color (e.g., sugarcane red rot).

reddening: Red pigments predominate (e.g., *Carrot red leaf virus* disease).

rhizome rot: Decay of rhizome (e.g., turmeric rhizome rot).

ringspot: Infected leaves show ringlike circular spots (e.g., *Groundnut ringspot virus* disease, *Carnation ringspot virus* disease, *Citrus ringspot virus* disease).

root rot: Root decay leads to wilting in plants. In the field, a sudden and complete wilting of the plants is seen. The major difference between wilt and root rot is the discoloration of the stem. When the bark of plants is removed, black streaks can be seen extending upward to the branches and downward to the lateral roots. Root rot, which also leads to wilting, is characterized by root decay (e.g., corn pythium root rot, cotton black root rot).

rosette: Rosetting (decoration resembling a rose) is formed due to impeded internodal expansion at the stem tips; crowding of the foliage excessively in the form of a rosette (e.g., *Groundnut rosette virus* disease).

rot diseases: Decay of tissues. Several types of rot symptoms are seen in crops.

rugose: The growth of veinal tissue is retarded (e.g., *Bean rugose mosaic virus* disease).

rust: Rustlike pustules (small elevated spots resembling pimples). Several types of rust diseases are known and they are referred to by the color of their pustules: **Brown rust** (brown [leaf] rust of wheat), **black rust** (black [stem] rust of wheat), **yellow rust** (yellow [stripe] rust of wheat), and **white rust** (white rust of crucifers). When pustules are arranged in linear rows between the veins of the leaf, the rust is called **stripe rust** (stripe rust of wheat). In the case of leaf rust, rust pustules are seen mostly on leaves (leaf rust of wheat). In the case of **stem rust,** rust pustules are seen mostly on stems (stem rust of wheat).

scab: The surface of affected tissues becomes rough and the affected surface is raised due to an abnormal proliferation of cells in the epidermis. The surface has a rusty appearance that is deeply pitted with corky wounds (e.g., apple scab).

scald: Affected plants show lesions, which appear to be similar to the scalding caused by hot water. Such lesions are mainly bleached and may be partly translucent (e.g., barley scald).

shot hole: Leaf spots in which the necrotic (spot) regions drop out, leaving holes in the affected leaves (e.g., peach *Stigmina* shot hole).

silver leaf: A condition in which infected leaves show a metallic luster (e.g., plum silver leaf).

smut: Black sooty masses of fungal spores that cover the affected plant parts. Several types of smuts are seen, including **covered kernel smut:** smut sori replacing grains of plants (e.g., covered kernel smut of sorghum); **long smut:** smut sori are covered by a fairly thick membrane and are much longer than the other smuts; they are cylindrical in shape (e.g., long smut of pearl millet); **loose smut:** smut sori are covered with a fragile membrane, which breaks easily at the time of spike emergence from the host, exposing a powdery mass of spores (e.g., loose smut of wheat); **covered smut:** smut sori are covered with a thick membrane, which resists easy rupturing (covered smut of oats); **flag smut:** smut sori occur on the leaf blade, leaf sheath, and culm (flag smut of wheat); **head smut:** the entire earhead is converted into smut sori (head smut of corn); **kernel smut:** symptoms appear only on mature grains; minute black pustules or streaks bursting through the glumes are seen; sometimes the entire grain is replaced by a powdery, black mass of smut spores (e.g., kernel smut of rice); and **leaf smut:** smut sori are formed in infected leaves (leaf smut of rice).

soft rot: Soft, water-soaked, irregular lesions appear on tubers, rhizomes, fruits, vegetables, and other storage organs. These lesions are more or less superficial, but soon spread and cover the inner tissues. Lesions lead to rotting of storage organs (e.g., soft rot of potato).

sooty mold: Masses of fungal black spores that stick to the leaf surface, making the foliage appear black and ugly (e.g., sooty mold of citrus, mango, sapota, and guava).

spindle tuber: A condition in which the tubers of affected plants are elongated (spindle shaped) (e.g., *Potato spindle tuber viroid* disease).

spotted wilt: Bronze markings appear on the upper surface of young leaves, and the markings extend from the leaf blade down to the petiole and stem, resulting in wilting of the stem (e.g., tomato spotted wilt).

stalk rot: Decay of plant stalks (e.g., corn bacterial stalk rot).

stem bleeding: Exudation of sap or resin from an infected stem (e.g., coconut stem bleeding).

stem gall: Abnormal outgrowth in the stem (coriander Protomyces stem gall).

stem rot: Decay of stem tissues (e.g., rice stem rot, crucifers sclerotinia stem rot).

stenosis: In diseased plants, the leaves are highly reduced in size and clustered along the stem (e.g., cotton stenosis).

sterility: Suppression of development of reproductive structures. Diseased plants do not produce seeds, fruit, stamens, or pistils (e.g., pigeonpea sterility mosaic disease).

storage rot: Decay of storage organs (e.g., corn aspergillus ear and storage rot).

streak: Mosaic in leaves along veins, which looks like a streak (e.g., *Tobacco streak virus* disease).

stripe: A long band of mosaic pattern along the parallel veins of a leaf (e.g., *Peanut stripe virus* disease).

stunt: Retardation of plant growth due to disease (e.g., *Peanut stunt virus* disease).

tumors: Swellings on stems or roots (e.g., *Sweet clover root tumor virus* disease).

vein band: A broad dark-green band along the veins. The rest of the lamina surface shows chlorosis (e.g., *Strawberry vein banding virus* disease).

vein chlorosis: Chlorosis is restricted to tissues adjoining the veins in leaves (e.g., *Raspberry vein chlorosis virus* disease).

vein clearing: A clearing or chlorosis of the tissue in or immediately adjacent to the vein (e.g., *Petunia vein-clearing virus* disease).

wilt: In the early stages of wilting, yellowing of leaves is seen. A flaccid or drooping condition of the plant due to disease, wilting may be due to a shortage of water, an impeded supply of nutrients, or excessive transpiration (e.g., cotton fusarium wilt, banana fusarium wilt, crucifers verticillium wilt).

witches'-broom: Abnormal proliferation (mass outgrowths) of the branches of woody plants due to disease (e.g., witches'-broom of potato).

yellows: Uniform yellow to almost-white discoloration of leaves; yellowing will be the conspicuous symptom in the diseased plant. Yellow pigments may predominate in infected plants (e.g., crucifers aster yellows, *Beet yellows virus* disease).

– 12 –

Crop Disease Assessment

Crop disease assessment is otherwise called phytopathometry. It involves the measurement and quantification of crop diseases. Accurate disease assessment will help in predicting the development of epidemics and in developing a decision support system for timing the application of fungicides to control diseases. Diseases are assessed by different methods, based on the type of disease symptoms and their relationship with yield loss. Remote sensing and image analysis are important tools in disease assessment.

DISEASE INCIDENCE ASSESSMENT

A disease is assessed either as a percentage of disease incidence or as disease severity. Disease incidence is calculated as follows:

Percentage disease incidence =
(number of infected plants/total number of plants assessed) × 100

Assessment of disease incidence will be useful for measuring systemic infections, which may result in total plant loss. Virus diseases such as rice tungro, barley yellow dwarf, and banana bunchy top, and fungal diseases such as loose smut of wheat and barley and sugarcane smut are assessed by estimating disease incidence. Some fungal pathogens may not cause systemic infection, but may cause total crop loss. Wilt diseases such as Fusarium wilt of tomato, Panama wilt of banana, and Fusarium wilt of chickpea cause total losses and these diseases are assessed as percentage of disease incidence. Monosporascus wilt of melons is assessed as percentage of wilt incidence (Cohen et al., 2000).

Percentage of disease incidence is calculated also for some leaf spot and fruit spot/rot diseases. Brown spot of pear, which is caused by *Stemphylium vesicatarium,* is assessed by recording the number of leaves that show leaf spots out of ten leaves of four shoots per tree (Llorente et al., 2000). The brown spot disease incidence was also recorded as a percentage pear fruit infection (Llorente et al., 2000). Diseases such as damping-off and some root rots in which total losses occur are assessed by percentage of disease incidence.

DISEASE SEVERITY ASSESSMENT

Disease severity is generally calculated as follows (Cooke, 1998):

Disease severity = (area of diseased tissue/total tissue area) × 100

Disease severity is assessed by allotting a grade value (category value or scale value) depending upon the area of infection. The severity scale is fixed, mostly based on the disease's effect on yield loss. Disease severity is usually assessed using either descriptive or pictorial keys. Standard diagrams illustrate the developmental stages of a disease on simple units (e.g., leaves, fruits) or on large composite units such as branches or whole plants. Such standard diagrams are derived from a series of disease symptom pictures, which may be in the form of line drawings, photographs, or preserved specimens (Cooke, 1998). Pictorial disease assessment keys are available for measuring disease severity on a range of hosts using the principle of standard area diagrams (Cooke, 1998). Standard area diagrams are prepared using graph paper outlines. Planimeters, electronic scanners, and image analyzers are used to assess the area of infection. Lesion areas can be determined by computer-assisted image analysis (Timmer et al., 2000).

Disease severity is assessed by using arbitrary categories. Horsfall and Barratt (1945) proposed a logarithmic scale for the measurement of plant disease severity, in which 12 grades were allotted according to the leaf area diseased, keeping 50 percent as a midpoint. These categories were developed taking into account the fact that the human eye apparently assessed diseased areas in logarithmic steps, as stated by the Weber-Fechner law for visual acuity (for appropriate stimuli, visual response is proportional to the logarithm of the stimulus). This law has several steps in the scale around 50 percent infected area and decreasing steps at both ends of the scale. However, many modifications have been made to this scoring system, mostly based on the possible yield loss due to each grade and each type of symptom. Grading systems consisting of five grades (1, 2, 3, 4, and 5 or 0, 1, 2, 3, and 4) or nine grades (1, 3, 5, 7, and 9 or 0 to 9) are commonly used. Citrus greasy spot caused by *Mycosphaerella citri* was assessed using a 0-to-5 scale (Timmer et al., 2000). Greasy symptoms on the top ten leaves of each plant were rated on a scale of

0 = none
1 = 1 to 5 percent
2 = 6 to 10 percent
3 = 11 to 15 percent
4 = 16 to 20 percent
5 = more than 20 percent

of the leaf surface area affected by the disease (Timmer et al., 2000). Monosporascus disease in cucumber was assessed using a 1-to-5 scale (Bruton et al., 2000) as follows: 1 = healthy with no lesions or discoloration on hypocotyls, 2 = slight discoloration, 3 = moderate discoloration and/or with lesions, 4 = moderate maceration, and 5 = severe maceration. Bacterial spot (*Xanthomonas campestris* pv. *vesicatoria*) of bell pepper was assessed using a 0-to-9 rating scale in which

0 = no diseased leaves
1 = <1 percent leaf area diseased
2 = 2 to 10 percent leaf area diseased or defoliated
3 = 11 to 20 percent leaf area diseased or defoliated
4 = 21 to 35 percent leaf area diseased or defoliated
5 = 36 to 50 percent leaf area diseased or defoliated
6 = 51 to 65 percent leaf area diseased or defoliated
7 = 66 to 80 percent leaf area diseased or defoliated
8 = 81 to 99 percent leaf area diseased and very few leaves (one to three) remaining on plant
9 = complete defoliation (plant dying or dead) (Romero et al., 2001)

Foliar symptoms of sudden death syndrome of soybean caused by *Fusarium solani* f. sp. *glycines* are assessed on a 0-to-9 scale based on (different types of symptoms) the percentage of leaf area that is chlorotic, necrotic, or defoliated as follows (Luo et al., 2000):

0 = no detectable leaf symptoms
1 = 1 to 10 percent chlorotic or 1 to 5 percent necrotic
2 = 10 to 20 percent chlorotic or less than 10 percent necrotic
3 = 20 to 40 percent chlorotic or 10 to 20 percent necrotic

4 = 40 to 60 percent chlorotic or 20 to 40 percent necrotic
5 = greater than 60 percent chlorotic or greater than 40 percent
     necrotic
6 = up to one-third premature defoliation
7 = one-third to two-thirds premature defoliation
8 = greater than two-thirds premature defoliation
9 = plants are prematurely dead

Brown spot of pear was measured by calculating the severity of infection. Measurements were performed on ten leaves of four shoots per tree. Disease severity per shoot was calculated according to a scale index based on an approximate lesion number per leaf corresponding to 0 (none), 1 (1 to 5 lesions), 2 (6 to 25 lesions), and 3 (more than 25 lesions) (Llorente et al., 2000).

Shaner and Buechley (1995) developed a leaf blotch severity scale based on the position of leaves of wheat plants. The top leaves, particularly the top two leaves, contribute to grain yield in cereals. The scoring system was based on corresponding yield loss. Hence, Shaner and Buechley (1995) scored the top four leaves (flag leaf, flag leaf-1 [the leaf immediately below the flag leaf], flag leaf-2 [the leaf below flag leaf-1], and flag leaf-3 [the leaf below flag leaf-2]) of tillers into different category values. Flag-3 leaves with 0 to 5 percent severity were given the scale value of 1, 5 to 20 percent (scale value 2), 20 to 40 percent (scale value 3), 40 to 70 percent (scale value 4), 70 to 90 percent (5), 90 to 100 percent (scale value 6), and 100 percent (scale value 7). Contrarily, flag-2 leaves with 1 to 10 percent disease severity were given the scale value of 4, 10 to 25 percent (scale value 5), 25 to 75 percent (scale value 6), 75 to 100 percent (scale value 7), and 100 percent (scale value 8). Flag-1 leaves with 0 to 1 percent disease severity were given scale value of 5, 1 to 10 percent (scale value 6), 10 to 50 percent (scale value 7), 50 to 90 percent (scale value 8), 90 to 100 percent (scale value 9), and 100 percent (scale value 9.5), and flag leaves with 1 to 20 percent disease severity were given a scale value of 8, 20 to 90 percent (scale value 9), and 90 to 100 percent (scale value 9.5). Mean severity was calculated from the average of the four leaves of each tiller, based on midpoint values for each range. Mean severity (P) is calculated from the scale value (S) according to:
$P = -0.38253 - 0.69435\ S + 1.17499\ S^2\ (R^2 = 0.999)$.

Disease severity is also calculated by measuring the height up to which infection spreads in an infected plant (vertical disease progress). Rice sheath blight *(Rhizoctonia solani)* disease intensity is measured using a relative lesion height percentage as follows: The highest point a lesion is seen in a plant (cm)/plant height × 100 (Vidhyasekaran et al., 1997). In this case, it has been shown that the relative lesion height corresponds to yield loss.

DISEASE INDEX

Data from both percentage of disease incidence and disease severity with different scale values are often used to assess the disease index (Llorente et al., 2000; Luo et al., 2000). Disease index is calculated as follows: Disease incidence percentage × Disease severity scale ÷ the maximum severity scale (Luo et al., 2000). Disease index is also calculated as the mean grade value (totaling grade values of all the examined plants and dividing them by number of the examined plants) × 100 ÷ maximum grade value (Cooke, 1998). A brown spot of pear disease index was calculated by adding the scale index of each leaf in a tree and dividing the sum obtained by 3 (the maximum level of severity) and the number of leaves assessed per shoot (Llorente et al., 2000). The disease scale can be recalculated to a percentage severity value called the disease index (Nilsson, 1995).

AREA UNDER DISEASE PROGRESS CURVE (AUDPC)

Disease severity is also measured by calculating the area under disease progress curve (Romero et al., 2001). Bacterial spot (*Xanthomonas axonopodis* pv. *vesicatoria*) of bell pepper was assessed by using a 0-to-9 rating scale. The AUDPC was then calculated (Romero et al., 2001). For the procedure to calculate AUDPC, see Chapter 14.

REMOTE SENSING

Another disease assessment method involves aerial photography and photogrammetry using infrared film or color filter combinations. Remote sensing and digital image analysis are methods of acquisition and interpretation of measurements of an object without physical contact between the measuring device and the object. The term *remote sensing* is restricted to instruments that measure electromagnetic radiation reflected or emitted from an object (Nilsson, 1995). The instruments record radiation in various parts of the electromagnetic spectrum: Ultraviolet (UV, 10-390 nm), visible (390-700 nm), near infrared (NIR, 770-1300 nm), infrared (IR, 1300-2500 nm), etc. The human eye records three visual spectral ranges: Red, green, and blue. However, sensitivity to red over 650-700 nm is only slight. The human eye cannot detect the infrared spectrum. The amount of reflected light (radiance) as a percentage of incoming light (irradiance) is called the reflectance factor. If the radiance from a healthy leaf is measured by a radiometer, substantially high reflectance can be seen in NIR at 750-1100 nm. NIR reflectance decreases greatly when a diseased leaf is viewed due to the reduction in chlorophyll and xanthophylls caused by infection (Nilsson, 1995). A relative decrease in NIR reflectance may indicate the disease severity. Aerial photography using infrared film helps to detect disease incidence.

Remote sensing for detecting and estimating the severity of plant diseases is used at three levels above the crop canopy (Cooke, 1998). Handheld multispectral radiometers or multiple waveband video cameras are used at the lowest altitude (within 1.5 to 2.0 meters above crop height). At 75 to 1500 m, aerial photography is used. At the highest altitude, satellite imagery is employed utilizing satellites orbiting 650 to 850 km above the Earth's surface. Handheld multispectral radiometers or multiple waveband video cameras are useful to assess diseases in the greenhouse and in the field. Aerial infrared photography is useful at field level. Satellite imagery is used for disease surveillance in large areas or regions of the Earth. Aerial IR photography was successfully used in the 1970s for surveillance of southern corn blight *(Helminthosporium maydis)* epidemic in the United States.

DIGITAL IMAGE ANALYSIS

Digital image analysis is being used to assess various plant diseases. Digital image analysis includes analysis of satellite images, aerial photographs and videographs, nucleic magnetic resonance images, and images in electron microscopy. Image processing reduces the total information to a manageable amount. Aerial IR color photographs can be enhanced and color-coded to facilitate visual interpretation of the distribution and severity of diseases in fields (Nilsson, 1995). Blazquez and Hedley (1986) made computer-aided spectrophotometric measurements of 35 mm color IR film of late blight infection in tomato fields. This technology has vast potential for making rapid assessments of plant diseases.

REFERENCES

Blazquez, C. H. and Hedley, L. E. (1986). Late blight detection in tomato fields with 35 mm color infrared aerial photography. *Phytopathology,* 76:1093.
Bruton, B. D., Garcia-Jimenez, J., Armengol, J., and Popham, T. W. (2000). Assessment of virulence of *Acremonium cucurbitacearum* and *Monosporascus cannonballus* on *Cucumis melo. Plant Dis,* 84:907-913.

Cohen, R., Edelstein, M., Pivonia, S., Gamliel, A., Burger, Y., and Katan, J. (2000). Toward management of *Monosporascus* wilt of melons in Israel. *Plant Dis,* 84:496-505.

Cooke, B. M. (1998). Disease assessment and yield loss. In D. G. Jones (Ed.), *The Epidemiology of Plant Diseases.* Kluwer Publishers, Dordrecht, The Netherlands, pp. 42-71.

Horsfall, J. G. and Barratt, R. W. (1945). An improved grading system for measuring plant disease. *Phytopathology,* 35:655.

Llorente, I., Vilardell, P., Bugiani, R., Gherardi, I., and Montesinos, E. (2000). Evaluation of BSPcast disease warning system in reduced fungicide use programs for management of brown spot of pear. *Plant Dis,* 84:631-637.

Luo, Y., Hildebrand, K., Chong, S. K., Myers, O., and Russin, J. S. (2000). Soybean yield loss to sudden death syndrome in relation to symptom expression and root colonization by *Fusarium solani* f. sp. *glycines*. *Plant Dis,* 84:914-920.

Nilsson, H.-E. (1995). Remote sensing and image analysis in plant pathology. *Annu Rev Phytopathol,* 15:489-527.

Romero, A. M., Kousik, C. S., and Ritchie, D. F. (2001). Resistance to bacterial spot in bell pepper induced by acibenzolar-*S*-methyl. *Plant Dis,* 85:189-194.

Shaner, G. and Buechley, G. (1995). Epidemiology of leaf blotch of soft red winter wheat caused by *Septoria tritici* and *Stagonospora nodorum*. *Plant Dis,* 79:928-938.

Timmer, L. W., Roberts, P. D., Darhower, H. M., Bushong, P. M., Stover, E. W., Peever, T. L., and Ibanez, A. M. (2000). Epidemiology and control of citrus greasy spot in different citrus growing areas in Florida. *Plant Dis,* 84:1294-1298.

Vidhyasekaran, P., Ruby Ponmalar, T., Samiyappan, R., Velazhahan, R., Vimala, R., Ramanathan, A., Paranidharan, V., and Muthukrishnan, S. (1997). Host-specific toxin production by *Rhizoctonia solani,* the rice sheath blight pathogen. *Phytopathology,* 87:1258-1263.

Waliyar, F., Adamou, M., and Traore, A. (2000). Rational use of fungicide applications to maximize peanut yield under foliar disease pressure in West Africa. *Plant Dis,* 84:1203-1211.

13

Crop Disease Diagnosis

Disease diagnosis is very important for developing effective strategies for disease management. Without diagnosis, there can be no disease management. Crop disease diagnosis is an art as well as a science (Holmes et al., 2000). The diagnostic process involves the recognition of symptoms (which are associated with disease) and signs (which are not outwardly observable) and requires intuitive judgment as well as the use of scientific methods. Several conventional techniques are followed to diagnose disease incidence. These techniques include visual inspection and recognition of symptoms and isolation and examination of crop pathogens using microscopy. Such techniques are time-consuming and may not be able to detect latent infections. Several diagnostic assays have been developed for early and rapid diagnosis. These include immunoassays, nucleic acid probe-based methods, and PCR-based techniques. The use of these techniques in the diagnosis of fungal, bacterial, viral, viroid, and phytoplasma diseases is described here.

DIAGNOSIS BASED ON DISEASE SYMPTOMS

Some pathogens produce characteristic symptoms that can be easily recognized in the field. Fungal diseases such as powdery mildews, rusts, downy mildews, and smuts show characteristic symptoms. Some bacterial diseases such as canker and crown gall can be recognized based on symptoms. A few virus diseases produce characteristic symptoms such as bunchy top, rosette, witches'-broom, phyllody, and flower color-breaking. Diseases should be diagnosed at early stages of their development, so that epidemics can be effectively prevented. The early stages of many diseases are inconspicuous and it may not be feasible to make a rapid visual assessment until the level of disease is sufficiently high. Further, even at advanced stages of symptom development, some diseases cannot be recognized with certainty. For example, symptoms such as chlorosis, mosaic, leaf drooping, yellowing, dwarfing, stunting, necrosis, root rots, wilts, fruit rot, dieback, leaf blight, and bud rot can be caused by several types of fungal, bacterial, viral and phytoplasma pathogens. In these cases, the appearance of fruiting bodies of the pathogens on leaf, stem, and fruit surfaces must be checked. However, sporulation of the pathogen occurs only at a very late stage of infection. Sometimes saprophytes may develop on these lesions, making diagnosis a difficult task. Hence, the pathogen has to be isolated in pure culture and identified using a microscope. Sometimes selective media may be required to isolate pathogens and induce sporulation for identification. Many fungal pathogens grow slowly in media and may be lost too readily for identification. If the pathogen is a bacterium, virus, or phytoplasma, more complicated tests have to be conducted to identify them. Several biochemical tests must be conducted to identify bacterium. Routines for extraction and purification of viruses require professional proficiency and skill. In fact, conventional plant-pathological techniques need high expertise for routine identification. In the case of latent infections in vegetative planting materials, seeds, and fruit, conventional methods may not be useful to diagnose infection.

DIAGNOSIS BY IMMUNOLOGICAL TECHNIQUES

Immunodiagnostic assays provide a fast method of confirming visible symptoms as well as detecting pathogens that cannot be easily identified by other methods. They permit early detection of plant pathogens and accurate identification of pathogens. Because many fungicides are specific only to certain pathogens or groups of pathogens, immunodiagnosis will be useful in the selection of the most appropriate treatment. Viruses, bacteria, and fungi (especially those spreading as a sterile mycelium) can be readily detected by these methods.

Immunoassays depend on the development of antibodies specific to the particular pathogen. Cells of living animals, particularly mammals, have the ability to recognize binding sites on proteins, glycoproteins, lipopolysaccharides, and carbohydrate molecules that are not present in their bodies (i.e., foreign to that animal). Such molecules, known as antigens, stimulate the immune system of the animal and this leads to the production of specific antibodies, each of which specifically recognizes and binds to its complementary antigen. The role of an immunoassay is to reveal the presence of specific complexes between the antibody and an antigen that are unique to the pathogen (Fox, 1998). The antibodies produced in an animal body can recognize the microbial antigen, which is present on cell walls or found attached with them. In other words, the antibodies can recognize the plant pathogen by recognizing the antigen specific to the pathogen. In principle, immunoassays are based on the fact that antibodies react specifically with the homologous antigen. However, the reaction is not easy to detect. Several techniques have been developed to exploit this reaction in immunoassays.

Production of Polyclonal Antibodies

Rabbits, rats, mice, sheep, or goats are used to produce antibodies. For the production of antibodies to fungal pathogens, mycelia (obtained from sterile cultures grown in the laboratory) are freeze dried. They are then finely ground with a pestle and mortar and used as antigens. Surface washings from fungal mycelia are also used as antigens. These antigens are emulsified with an equal volume of an adjuvant, such as Freund's complete adjuvant for the first injection and Freund's incomplete adjuvant for the second or subsequent injections. Other types of adjuvants such as Quil A are also used. Bacterial antigens can be prepared by first killing the bacteria with heat or formalin and then suspending them in phosphate-buffered saline. The bacterial cells are disrupted by grinding, sonication, or enzyme action to produce soluble cytoplasmic antigens. Virus antigens can be prepared from infected host tissues by differential centrifugation or precipitation with ammonium sulfate.

These antigens (otherwise called immunogens) are then injected into rabbits (or other animals) to induce complementary antibodies. The antigen emulsion is injected under the skin at the back of the neck of the animals. One or more booster injections with the emulsion are given several weeks after first injection. Antiserum is obtained by taking blood from the ears of rabbits or from the tail veins of mice.

Antibodies have a basic common structure, which consists of two heavy and two light chains held together by disulfide bonds. However, they differ in the ability of their tips to bind to sites on different "foreign" protein, glycoprotein, lipoprotein, lipopolysaccharide, and carbohydrate molecules (antigenic substances). Such antigenic substances stimulate the animals' immune systems to produce an assortment of specific antibodies, each of which specifically recognizes and binds almost exclusively to its complementary epitope (antibody reaction site) on the antigen. This interaction between an antibody and an antigen depends on pairing of an individual epitope of the antigen in three dimensions between the contours of the tips of the side chains of a particular antibody. Several antigens have been detected in viral, bacterial, and fungal pathogen preparations. The antisera raised against these antigens may contain a mixture of antibodies directed toward various epitopes of the pathogens. These antisera, which contain polyclonal antibodies, are called "polyclonal antisera."

Production of Monoclonal Antibodies

Kohler and Milstein (1975) were the first to produce monoclonal antibodies. Monoclonal antibodies (MCAs) recognize a single epitope only and can therefore differentiate between related pathogens. Monoclonal antibodies that are specific at genus, species, pathovar, and strain levels are available. MCAs are homogenous and therefore free from the variability common to polyclonal antisera. Once a hybridoma cell line is obtained, an endless supply of the same MCA can be produced. The method of preparation of monoclonal antibodies is given here:

An antigen (immunogen) is prepared from cultures of fungal and bacterial pathogens or from the sap of virus-infected plants as described for the preparation of polyclonal antibodies. Mice, usually females of the BALB/c strain, are immunized by intraperitoneal injection with the antigen preparation. The preparation is given at about two week intervals until antiserum from trial bleeds gives a high titer against the immunogen. The final booster injection is given without an adjuvant about three to four days before fusion. The mice are killed and their spleens are removed. β-Lymphocytes are isolated from the spleens of immunized mice. Meanwhile, a potentially immortal line of myeloma cells is obtained from an animal similar to the animal used for immunization (mostly mouse), subcultured, and grown for fusion with the spleen cells to form hybridoma cells. Viable spleen cells are mixed at a ratio of 1:10 with viable myeloma cells. This cell mixture is washed by centrifugation at 60 g for 10 min and the pellet loosened by tapping the tube. One ml of polyethylene glycol 4000 (PEG 4000) is then added dropwise over a period of 1 minute to the pellet, stirred for another minute, and slowly diluted by the addition of a tissue culture medium. Dulbecco's modified Eagle's medium is generally used as a tissue culture medium. Penicillin-streptomycin solution, L-glutamine (200 mM), phenol-red solution, and controlled processed serum replacement are added to the medium. The cells are washed and resuspended in fresh HAT (hypoxanthine, aminopterin, thymidine) medium (a selective medium made by supplementing the complete tissue culture medium with 0.01 M hypoxanthine, 0.0016 M thymidine, and 0.00004 M aminopterin) and dispensed into 96-well tissue culture plates. The plates are kept in an incubator at 37°C with 5 percent CO_2. After ten to 16 days, aliquots of the supernatant are removed from the wells containing clones and screened by ELISA (enzyme-linked immunosorbent assay). Only hybridomas producing antibodies specific to the antigen are saved and grown on a large scale in vitro. Selected hybridoma cells are grown in bulk. The antibodies are purified on a sepharose column. Once a useful MCA has been generated, the selected hybridoma cells can be immortalized by storing them in liquid nitrogen. Whenever needed, MCAs can be produced indefinitely and in unlimited quantities.

Both polyclonal and monoclonal antibodies are used in various immuno-diagnosis techniques.

Agglutination Test

This test can be carried out in slides or in test tubes. In the slide agglutination test, drops of antigen and diluted antiserum containing antibodies are mixed together on a glass microscope slide. Agglutination is observed by eye or microscope (if ambiguous). In the test-tube agglutination test, the antigens are mixed with antibodies in test tubes, and the aggregation of antigens and antibodies is monitored with a binocular microscope.

Precipitation Test

In this test, aliquot dilutions of antigen are layered over equal volumes of antiserum diluted in normal serum in capillary, or other small tubes. The test is regarded as positive if there is precipitation at the interface (Fox, 1998). When the antigens are layered over the antibodies, the antigens are precipitated out of solution by the antiserum when antigen and antibodies are related.

Latex-Agglutination Test

In this test, the antibodies are first adsorbed onto the much larger latex particles and mixed with the antigen preparation (sap of diseased tissues). An opalescent suspension containing antigen particles will settle out, producing a clear supernatant by granulation or flocculation in a positive test.

Ouchterlony Immunodiffusion Test

In this test, a petri dish containing 1 percent agar is used. A central well (usually 0.5 cm in diameter), surrounded by several satellite wells, is cut into the agar. The antibody is placed on the central well, whereas the antigen is placed on surrounding wells. The antibodies and antigens are allowed to approach each other by diffusion in agar. In the positive test, a precipitin band, usually visible to the naked eye, forms at the leading edge where diffusing antigen and antibody molecules meet.

Immunoelectrophoresis

By this method, mixtures of antigens are separated before immuno-diffusion. A narrow trough is cut in a layer of thin gel parallel to an electric current that passes close to the antigens along the length of the gel. Each antigen moves in a separate wave at a characteristic rate according to its distinct charge. As a result, proteins separate into bands. Once the proteins have separated sufficiently, the current is switched off and antiserum is added to the trough cut in the gel. Precipitin arcs composed of complexes of antibodies and antigens form where the individual electrophoresced antigens have reached.

Antigen-Capture Indirect ELISA

Infected plant sap is used as an antigen. Many variations of ELISA exist, such as antigen-capture indirect ELISA, direct sandwich ELISA, double antibody sandwich ELISA (DAS-ELISA), triple antibody sandwich ELISA, and $F(ab')_2$-based ELISA.

In the antigen-capture indirect ELISA method, the wells are coated with the antigen. The remaining sites are blocked by a blocking agent, and a specific primary antibody is added. A secondary antibody-enzyme conjugate binds to the primary antibody. This is followed by the addition of an enzyme substrate leading to a colored product in proportion to the concentration of pathogen. This methodology involves coating immunoplates with the antigen and then centrifuging them at 600 g for 5 min at 4°C. The plates are post-fixed by incubation of 100 μL per well of 0.25 percent glutraldehyde solution for 5 to 7 min at room temperature. The fluid is flicked off by inverting the plate, and the plates are washed three times with either tris-buffered saline (TBS) Tween (TBST) or phosphate-buffered saline (PBS) Tween (PBST) followed by a short period of drying by laying the plates on a paper tissue. About 200 μg blocking solution (2 percent casein in either TBST or PBST) is absorbed to the remaining sites in the well for 1 h at room temperature. After one washing, the antibody in TBST or PBST is added at 50-100 μL per well. After 1 to 2 h incubation at 37°C and three further washings, antimouse immunoglobulin IgG alkaline phosphatase conjugate diluted in TBST or PBST containing casein (2 percent) is incubated for 1 to 2 h at 37°C (50-100 μL per well). The excess fluid is removed by flicking the plate, followed by washing with TBST or PBST for three to five times. Phosphatase tablets are freshly dissolved in diethanolamine (DEA) buffer, pH 9.6, and this substrate solution is pipetted into the wells. Substrate conversion is quantified by measuring absorbance at 405 nm with a plate reader (after at least 30 min incubation in the dark at 37°C).

Direct Sandwich ELISA

In the direct sandwich ELISA method, 96-well immunoplates are coated with the specific antibody (polyclonal or preferably monoclonal antibody) and incubated successively with the antigen containing sample followed by a second enzyme-labeled specific antibody that is directly conjugated with an enzyme. This leads to a colored product in proportion to concentration of pathogen.

Double Antibody Sandwich ELISA

In the double antibody sandwich ELISA method, a specific capture antibody is immobilized onto a solid surface (Huttinga, 1996), such as the wells of a microtiter plate. The infected plant tissue sample is added, and unbound material is washed away. Bound antigen is detected by the addition of a detecting antibody that has been conjugated with an enzyme, and unbound material is again washed away. The presence of the detecting antibody is determined through the addition of a substrate for the enzyme. The amount of color that develops is proportional to the amount of antigen present in the sample. The intensity of the color can be recorded by automated equipment (Miller, 1996).

Triple Antibody Sandwich ELISA

In the triple antibody sandwich ELISA method, a specific antibody produced in one animal species is bound to the solid substrate (96-well plates), while a second specific antibody produced in another animal species binds to the bound antigen. The "sandwich" is detected by an antibody-enzyme conjugate that binds to the second antibody (Miller, 1996).

F(ab')$_2$-Based ELISA

In F(ab')$_2$-based ELISA, F(ab')$_2$ fragments from target-specific antibodies are used as the capture reagent. A specific second antibody produced in the same animal species binds to antigen already bound to the F(ab')$_2$ fragments. The sandwich in turn is detected by the addition of a general detecting antibody-enzyme conjugate that reacts specifically with the Fc portion of the second antibody (Miller, 1996).

Dipstick ELISA

ELISAs can be carried out on membranes made of nitrocellulose, nylon, or other materials. In the dipstick assay, nitrocellulose dipsticks surface-coated with a capture antibody are used. Dipsticks on which specific MCAs have been bound are dipped into the test specimen, allowing ELISA reactions to occur in situ. Flexible plastic dipsticks are quick and easy to use in the field to test sap squeezed onto the coated surface. Any pathogen (particularly viruses) present is entangled and detected following incubation with the antibody-enzyme conjugate, washing, and then the substrate, further washing, and finally the stopping solution. If the specified pathogen is present, the dipstick becomes colored, whereas if free of pathogen it remains colorless.

Dot-Blot ELISA

In this assay system, ELISA reactions are carried out on nitrocellulose membranes. A drop containing the specific monoclonal antibody is absorbed as a "dot," onto which a drop of the test sample is later added and blotted (Fox, 1998).

Direct Tissue Blot Immunoassay

This assay is also called tissue print ELISA, immunoprinting, or direct tissue print immunoassay. In this assay, stems and leaf petioles are cut with a razor blade, and the cut surface is pressed gently and evenly to the nitrocellulose membrane. These blots are allowed to dry for 10 to 30 min, incubated with the monoclonal antibodies for 2 h, and rinsed with PBST buffer for 10 min. The blots are labeled with an enzyme conjugate, goat antimouse Ig (H + L), for 1 to 2 h at 37°C, rinsed with PBST-polyvinylpyrrolidone (PVP) buffer, and rinsed again with TTBS (20mM tris (hydroxymethyl) aminomethane, 500 mM sodium chloride, and 0.05 percent Tween 20), each for 10 min. The blots are incubated with freshly prepared NBT-BCIP (nitroblue tetrazolium, 5-bromo-4-chloro-3-indolyl phosphate in sodium carbonate buffer) substrate for 5 to 20 min. After stopping the reaction by putting the blots in water in a petri dish, the blots are observed under a light microscope. Development of an area with an intense purple color located at phloem tissue cells is considered a positive reaction (Lin et al., 2000).

In Situ Immunoassay (ISIA)

In this assay, sections 100 to 200 μM thick are cut from stems, petioles, or veins of diseased plants. The sections are transferred to 24-well plastic plates and fixed with 70 percent ethanol for 5 to 10 min. After the alcohol is removed using a pipette, the sections are incubated with specific mono-clonal antibodies for about 30 min. The sections are washed with PBST-PVP for 5 to 10 min, and incubated with enzyme-labeled secondary anti-bodies (alkaline phosphatase conjugated goat antirabbit IgG) at 37°C for 30 to 60 min. The sections are washed again with PBST-PVP and with TTBS buffer. Sections are then incubated with freshly prepared NBT-BCIP sub-strate mixture for 5 to 15 min. After stopping the reaction by removing the substrate solution, the sections are transferred to a glass slide and observed under a light microscope. The development of a purple color in the phloem tissue cells is considered as a positive reaction (Lin et al., 2000).

Recombinant Enzyme-Linked Immunosorbent Assay

In this assay, monoclonal antibodies are subcloned from hybridoma cell lines and expressed in *Escherichia coli* as single-chain variable fragment antibody (scFv) fusion proteins. The scFv technology allows the cloning of variable genes from preexisting monoclonal antibodies' cell lines, linking them with a flexible peptide as an scFv. These constructs are expressed in bacteria as soluble proteins or fused with the capsid proteins of filamentous bacteriophages. In addition, antibody genes are expressed fused with other proteins such as alkaline phosphatase or with amphipathic helices (Terrada et al., 2000).

Phage Display Technology

Recombinant antibody technology, in combination with phage display technology, provides a useful approach for the production of target-specific antibodies without the use of laboratory animals and time-consuming im-munization protocols (Griep et al., 2000). Recombinant antibodies are raised from a human combinatorial antibody library. Helper phages, which contain the entire phage genome but lack an efficient packaging signal, are used to "rescue" phagemids from the combinatorial library. When both the helper phage and single-chain variable fragment encoding the phagemid vector are present within the same bacterium (e.g., *Escherichia coli*), phages are assembled that carry scFv-antibodies (phage antibodies [PhAbs]) on their surface and contain the scFv-encoding phagemid vector. Consequently, within PhAbs, the genotype and phenotype are linked. To select for antigen specificity, PhAbs rescued from the combinatorial antibody library are allowed to bind to immobilized target antigens (panning). Washing removes PhAbs that lack affinity for the target antigens. Bound PhAbs are eluted, and selected PhAbs are applied to three sequential rounds of panning to further enhance the percentage of target-reactive PhAbs (Griep et al., 2000). PhAbs are selected to major structural proteins of the test virus. The scFv-encoding genes are retrieved and expressed in *E. coli* as ready to use antibody-enzyme fusion proteins. After subcloning, the encoding DNA sequences in the expression vector pSKAP/S, which allowed the scFvs to be expressed as alkaline phosphatase fusion proteins. The antibodies are used as coating and detecting reagents in a DAS-ELISA (Griep et al., 2000).

Latex-Protein A Agglutination Assay

In this assay, inert particles such as latex beads are used to detect antigen/antibody complexes. Protein A, a cell surface protein from *Staphylococcus aureus* that binds nonspecifically to the heavy chains of mouse and rabbit IgG antibodies, is used to link the particles coated with the specific antibodies by shaking the two together. Samples of plant extracts are tested by incubation with coated particles. If the test pathogen is present in the sample, the antigens form bridges linking the particles together, resulting in agglutination.

Immunofluorescence

Two methods of immunofluorescence are used to diagnose plant diseases. In the direct immunofluorescence method, specific antibodies bound to their target antigens are detected by using second antibodies conjugated with fluorescent dyes such as fluorescein isothiocyanate (FITC) or rhodamine isothiocyanate. Fluorescence, indicating the presence of the target antigen, is visualized microscopically. The microscope should have a special device for fluorescence using ultraviolet light (fluorescence microscopy). In direct immunofluorescence, the fluorescent dye is conjugated directly to the specific antibody (Salinas and Schots, 1994). Usual protocols involve extraction of the pathogen fraction from the tissue of individual or composite plants that is fixed to a well of a multiwell glass microscope slide by flaming or acetone treatment. Fixed preparations are stained directly by conjugated antibody or by the indirect procedure in which the primary antibody is not conjugated but a secondary conjugated antibody is bound to the primary one (De Boer et al., 1996).

Immunosorbent Electron Microscopy

This assay system is mostly used for the diagnosis of virus diseases. Electron microscope grids coated with carbon strongly adsorb protein, and when they are floated on a drop of antiserum containing antibodies to the pathogen, the antibodies become attached. The grids are then floated on a drop of the sap of an infected plant. After staining, the pathogen (particularly virus particles) adsorbed to the antibodies can be seen under a transmission electron microscope.

Immunosorbent Dilution-Plating Technique

In this method, bacterial pathogens are trapped on antibody-coated petri dishes or inoculation rods so they can be incubated on selective media. Any unbound bacteria are removed by washing. The colonies that form are detected by drying the agar and staining by immunofluorescence.

Western Blotting

This technique involves the transfer of proteins or glycoproteins on polyacrylamide gels electrophoretically onto a membrane or solid phase and the probing of such membrane-bound antigens with a specific antibody solution, based on the covalent binding of the antibody and its antigen (Fox, 1998). In the electrophoresis, different proteins move apart to form distinct bands. The larger molecules migrate more slowly than the smaller molecules. The protein in these bands can be transferred by blotting to a strip of porous nitrocellulose material. Afterward, specific polyclonal or monoclonal antibodies are used to bathe nitrocellulose for about an hour. An enzyme/substrate system (phosphatase or peroxidase conjugate), similar to an ELISA reaction, is also used to locate the position of bound antibodies (Fox, 1998).

Immunogold Assay

In the immunogold assay, following the electrophoresis of proteins on SDS polyacrylamide gel, proteins are transferred to a nitrocellulose membrane. Monoclonal antibody specific to a protein of the test pathogen is conjugated to gold particles and applied on the nitrocellulose membrane. The signal appears as a band and can be visualized on the membrane directly.

Radioimmunoassay

In the radioimmunoassay method, the protein is transferred to a nitrocellulose filter after electrophoresis, and a monoclonal antibody is added to it. MCA binds to a specific protein, and a radioactive detector antibody is added. The band is visualized indirectly by preparing an autoradiograph.

NUCLEIC ACID PROBE-BASED METHODS

Both DNA and RNA probes are used for crop disease diagnosis. DNA is double-stranded and the two strands are held together via specific hydrogen bondings between the base pairs. Pairing occurs with great specificity: Adenine (A) pairs only with thymine (T), and guanine (G) only with cytosine (C). The interaction of the two single-stranded molecules is known as hybridization. Several physical conditions can denature the DNA into single strands. High temperature and strongly alkaline pH denature double-stranded DNA. By manipulating the physical conditions, a strand can anneal with another strand of the target DNA. Nucleic acid hybridization, or reassociation, is a process by which complementary single-stranded nucleic acid anneals to double-stranded nucleic acid. This process occurs because of the hydrogen bonds between the two strands of the DNA. The pairing can also take place between DNA and RNA or RNA and RNA, and the pairing can occur between adenine (A) and uracil (U) in RNA instead of thymine (T) in DNA and guanine (G) and cytosine (C). Single-stranded DNA and RNA can act as probes. A diagnostic nucleic acid probe may be defined as a nucleotide sequence, labeled with a reporter molecule, that is able to identify a target pathogen within a test sample by selectively hybridizing to the complementary sequence present within a microorganism's DNA (De Boer et al., 1996). Nucleic acid probes are sequences of nucleic acids that are labeled with a marker and used to detect complementary nucleic acid sequences in a sample. Probes, which can be either DNA or RNA, range in size from 15 to several thousand base pairs (bp). Hybridization protocols usually require the probe or target sequence to be immobilized on a solid surface, but the reaction can also occur in solution or in situ. The probe-target hybrids are visualized using autoradiography, colorimetric assays, or chemiluminescence.

Hybridization Formats

Nucleic acid hybridizations may be performed in solution or on a solid support. In a solid-support format, the target nucleic acid from the pathogen to be detected is immobilized on a nitrocellulose or nylon membrane. DNA is first cut into fragments by a restriction enzyme, often followed by gel electrophoresis and denaturation (Southern, 1975). RNA is usually transferred by a similar technique known as Northern blotting. After it has been blotted onto the support, the target nucleic acid is fixed by baking or cross-linking under UV light. The labeled probe hybridizes to the target DNA. After the excess unbound probe has been removed by washing, the hybrid is detected by a suitable assay (Fox, 1998). The diagnostic probes are labeled with radioisotopes, primarily ^{32}P. Nonradioactive methods are also available for labeling DNA probes. Nonradioactive probes may be labeled enzymatically or chemically. Biotin, a vitamin, is commonly used for labeling, and the modified nucleotide is biotin-11-dUTP. Probes attached to biotin bind to avidin (a protein) extremely tightly. As a result, a more effective complex of many enzyme molecules can be produced by binding enzymes first to biotin and then to avidin, around which they cluster (Fox, 1998). The steroid digoxigenin may be employed in a similar way with antidigoxigenin enzymes for labeling. Sulfonation of cytosine residues by sulfite was one of the first hapten-based labeling systems (De Boer et al., 1996). The probe is detected using a monoclonal antisulfonate antibody followed by an alkaline phosphatase-conjugated antimouse IgG antibody. Another method of chemical labeling involves photobiotin, a compound in which biotin is bound to aryl azide by a linker arm. In the presence of light, aryl azide is converted into an aryl nitrene moiety that reacts with nucleic acid. Oligonucleotides are most efficiently labeled chemically by a reactive amine. A detectable group, such as alkaline phosphatase or horseradish peroxidase, is added to the amino-oligonucleotide after synthesis (De Boer et al., 1996).

Many commercial nucleic acid-based tests employ the sandwich assay, in which adjacent sections of the target nucleic acid are hybridized by two probes. One is the capture probe, which by hybridization links the target to the solid support to which this probe is already bound. The other probe is labeled with a reporter group, which may be a fluorescent molecule, an enzyme, or a radioactive atom that hybridizes with an adjacent section of the target nucleic acid. The reporter group with a fluorescent molecule can be readily distinguished, even in minute amounts (Fox, 1998).

Dot-Blot Diagnosis

When an extract or cell sap believed to contain the target nucleic acid is dotted on a filter, the method is called dot-blot. Dot-blot is a routine spot hybridization method. In this method, the DNA is not cut by restriction enzymes and fractioned. The DNA is applied as a small drop of sap extracted from an infected plant directly onto a nitrocellulose sheet. The DNA is then dried and hybridized with the probe in a sealed plastic bag using a buffer extract (Maule et al., 1983). The nucleic acid is loaded into a multisample vacuum manifold, which is used to spot it onto the membrane.

Colony Blot

Colony blot can be useful to detect bacteria in infected tissues. Dilutions of the bacteria from an infected tissue are grown on nitrocellulose filters placed on suitable agar media. After incubation, the nucleic acids are simultaneously released and denatured with alkali or by incubation in a microwave oven. The filter is then washed and baked prior to hybridization as a colony.

In Situ Hybridization Assay

In this assay, host tissue sections are fixed to a microscope slide and exposed to probe DNA. In another type of assay, sections of cut plant tissue are pressed against the nitrocellulose filter, which is then treated with alkali to denature the DNA and the tissue print is fixed on the filter. Hybridization reveals the presence of the pathogen.

PCR-Based Methodology

The polymerase chain reaction provides a powerful and rapid technique to exponentially amplify specific DNA sequences by in vitro DNA synthesis (Henson and French, 1993). Three essential steps to a PCR include (1) melting the target DNA, (2) annealing two oligonucleotide primers to the denatured DNA strands, and (3) extending the primer via a thermostable DNA polymerase. Newly synthesized DNA strands serve as targets for subsequent DNA synthesis since the three steps are repeated up to 50 times. The specificity of the method derives from the synthetic oligonucleotide primers, which base-pair to and define each end of the target sequence to be amplified (Henson and French, 1993).

PCR uses a thermostable *Thermus aquaticus (Taq)* DNA polymerase to synthesize DNA from oligonucleotide primers and template DNA. The template DNA may be genomic, first-strand cDNA, or cloned sequences. Primers are designed to anneal to complementary strands of the template such that DNA synthesis initiated at each primer results in replication of the template region between the primers.

The PCR involves three distinct steps governed by temperature. DNA, primers, deoxynucleotides, buffer, and *Taq* polymerase are combined in a microcentrifuge and overlaid with mineral oil. The tube is placed in a thermocycler programmed to repeat a set of short incubations at predetermined temperatures. In the first step, the template DNA is denatured to separate the complementary strands. This is done at 95°C for 5 minutes. In the second step, the mixture is held at an annealing temperature to allow the primers to hybridize to their complementary sequences. This is done at 55°C for 1 min. A PCR primer may comprise two regions, a 3' (priming) region and a 5' (variable) region. The most important region in determining the efficiency of annealing and subsequent DNA synthesis during the PCR is the 3' region, which should be perfectly complementary to the template sequence. The priming region should normally be 20 to 25 bases long. The *Taq* polymerase stabilizes these base-paired structures and initiates DNA synthesis. In the last step, the reaction is heated to about 72°C for 1 to 5 minutes. This process leads to a *Taq* polymerase-directed DNA synthesis. The cycle is repeated by keeping the reaction tubes in a thermal cycler for more than 20 times. In the first cycle each template gives rise to a newly synthesized complement. Thus, the number of copies of the target region is doubled. Similarly, in each subsequent cycle, the DNA concentration corresponding to the target region is almost doubled. About 20 cycles of PCR would produce 10^6-fold amplification of the target DNA.

Initially, very low numbers of DNA molecules can be multiplied enormously by PCR. Only a few nanograms of the initial template DNA is necessary for amplification. Both dsDNA and ssDNA can be amplified by PCR. It is possible to amplify RNA by reverse transcription (RT) into a cDNA copy by RT-PCR. Synthetic oligonucleotides (primers) that are complementary to the end sequences can be produced. Because the PCR process does not depend on the use of purified DNA, the host tissue extract can also been used for diagnosis.

The PCR product is analyzed by agarose gel electrophoresis. The PCR product from a defined band can be recovered from agarose gel. The DNA generated in a PCR can be reamplified and used for sequencing. DNA sequencing reactions are performed using commercially available kits of T7 polymerase or Sequenase.

RAPD (random amplified polymorphic DNA) PCR is an important diagnostic tool. Oligonucleotides of arbitrary sequence are used as primers for amplification. Shorter sequences on the order of 100 to 1,000 base pairs are most efficiently amplified and easily resolved by agarose electrophoresis. Hence, DNA sequences within a few hundred base pairs are usually chosen as primer annealing sites. In these cases, pairs of sequences complementary to the primer may be close to one another and arranged with 3' ends pointing toward one another. Under these circumstances, annealing of the primer to the target genome will result in the production of an amplified fragment after appropriate thermal cycling. RAPD PCR with DNA of infected plant tissues is performed as follows: In a microcentrifuge tube, genomic DNA (25 to 50 ng), deoxynucleoside triphosphate (dNTP) (200µm), primer (200 pmol), PCR buffer (100 mM Tris-HCl [pH 8.3], 500 mM KCl, 15 mM $MgCl_2$, 1 mg/mL gelatin, 0.1 percent Tween-20, 0.1 percent NP), and 1 unit *Taq* polymerase are mixed, spun briefly, and overlaid with 50 µL mineral oil. Forty-five PCR cycles are performed at 94°C for 1 min, 35°C for 1 min, and 72°C for 2 min. The reaction mixture (10 µL) is run through a 1 percent agarose gel in tris borate ethylene diamine tetra-acetic acid (TBE) buffer (89mM Tris, 89 mM boric acid (pH 8.3), 2.5 mM EDTA [ethylene diamine tetra-acetic acid]).

Any DNA or RNA sequence that is specific for a particular organism can be used for detection of that organism. Potential targets specific to pathogens include specific sequences from mitochondrial DNA fragment, pathogen-specific plasmid sequences, and short, interspersed repetitive elements present in bacteria. Ribosomal genes and the spacers between them possess conserved as well as variable sequences, and can be amplified and sequenced with universal primers based on their conserved sequences. Genes for 5.8S, 18S, and 26S (28S) fungal nuclear ribosomal RNA genes (rDNA) are used as primers (Henson and French, 1993).

Considerably greater sequence variation is found in the internal transcribed spacer (ITS) regions between the rRNA genes within an rRNA repeat unit (rDNA). Even more sequence differences are in the nontranscribed spacer (NTS) regions between the rDNA repeat units, and still more are in the intergenic spacer regions or noncoding sequences that occur within the rDNA repeat unit of some fungi (Henson and French, 1993). Any organism that has rDNA repeats may be specifically detected by selecting primer sequences based on variable spacer regions.

Immuno-PCR and Immunocapture PCR

Disease diagnosis methods that combine antibody binding and PCR are highly sensitive and can detect microbial antigens in addition to their nucleic acids. In the immuno-PCR method, a DNA fragment is molecularly linked to antigen-antibody complexes. Protein A and streptavidin portions of the linker molecule bind the antibody and DNA, respectively. Antigen present in a sample binds the specific antibody, which, in turn, binds the linker molecule. The latter is bound to a nonspecific, biotin-labeled DNA sequence that is subsequently amplified by PCR. Immuno-PCR is 10^5 times as sensitive as the enzyme-linked immunosorbent assay for the detection of antigen (Sano et al., 1992).

Immunocapture PCR is another sensitive diagnostic assay. Whereas immuno-PCR requires the antigen-specific antibody only, immunocapture PCR requires the antigen-specific antibody and nucleic acid sequence information from the microbe being detected.

Immunocapture-Reverse Transcription-Polymerase Chain Reaction (IC-RT-PCR)

In a typical IC-RT-PCR assay, PCR tubes are rinsed in PBST, soaked in ethanol for 15 min, and air dried. The tubes are coated with the specific virus antiserum in a coating buffer (15 mM Na_2CO_3; 35 mM $NaHCO_3$; 3 mM NaN_3; pH 9.6) for 3 h at 37°C. The tubes are washed three times with PBST and blotted dry on a tissue. They are placed at –70°C for at least 10 min and the contents thawed at 94°C for 2 min to disassemble the antibody-bound virus and free the viral RNA from the protein coat. A total of 50 μL of RT-PCR mix (4 μL of 5× first-strand reverse transcription buffer; 2 μL of DTT; 2.6 μL of dNTP; 0.1 μL of RNasin RNase inhibitor; 0.25 μL of SuperScript RNase H- Reverse Transcriptase; 3 μL of 10 × PCR buffer; 4.2 μL of 25 mM $MgCl_2$; 1 μL of reverse primer; 0.5 μL of forward primer; 0.2 μL of *Taq* polymerase; and 32.75 μL of nuclease-free water) as added to each tube. The tubes are then treated as follows: 37°C for 1 h, 94°C for 2 min (35 cycles of 94°C for 30 s, 50°C for 30 s, and 72°C for 60 s), and 72°C for 10 min. Following RT-PCR, the products are assessed by electrophoresis in 1.5 percent agarose gels and stained with ethidium bromide (Gillaspie et al., 2000).

Repetitive-Sequence-Based Polymerase Chain Reaction (Rep-PCR)

Rep-PCR is based on PCR-mediated amplification of DNA sequences located between specific interspersed repeated sequences in microbial genomes. These repeated elements are termed BOX, REP, and ERIC elements. Amplification of the DNA sequences between primers based on these repeated elements generates an array of different-sized DNA fragments from the genomes of individual strains. The separation of these fragments on agarose gels yields highly specific DNA fingerprints that can be either visually compared or subjected to computer-assisted pattern analysis (McDonald et al., 2000). The rep-PCR is useful in the identification of bacterial pathogens and disease diagnosis (Louws et al., 1999). Some fungal pathogens can also be identified by this technique (Jedryczka et al., 1999).

DIAGNOSIS OF CROP FUNGAL DISEASES

Several fungal pathogens survive in vegetative propagating materials and may serve as the primary inoculum source for the spread of the disease. Early diagnosis of the inoculum may help in eradicating the disease in the field. As many fungicides are specific only to certain pathogens or groups of pathogens, early diagnosis will be useful in the selection of the most appropriate treatment. Current detection of pathogenic fungi is based on visual inspection of characteristic symptoms. When sporulation is present, general confirmation can be made by light microscopy (Miller, 1996). The pathogen may be isolated by transferring diseased plant tissue to an agar medium in petri dishes. Sometimes a selective medium may have to be used. A bioassay is required to test the pathogenicity of isolates, particularly when closely related saprophytic species are common.

Several immunoassays have been developed to diagnose fungal diseases. Immunoassay kits are commercially available to detect species of *Rhizoctonia, Phytophthora,* and *Pythium* and *Septoria nodorum* (= *Stagonospora nodorum*), *S. tritici,* and *Pseudocercosporella herpotrichoides* (Miller, 1996). *Tilletia controversa* in wheat, *Phytophthora fragariae* in strawberry, and *Rhizoctonia* spp. in poinsettia are detected by DAS-ELISA (Miller, 1996). Indirect ELISA assays are used to detect *Spongospora subterranean* in potato (Harrison et al., 1994), *Colletotrichum acutatum* and *Phytophthora* spp. in strawberry, *Leptosphaeria maculans* in canola, and *Fusarium oxysporum* f. sp. *narcissi* in narcissus (Miller, 1996). *Verticillium dahliae* in potato is detected by immunofluorescence assays (Miller, 1996).

Gaeumannomyces graminis in wheat can be detected by Southern blot with cloned mitochondrial DNA labeled with ^{32}P or digoxigenin as a probe (Bateman et al., 1992). *Phytophthora capsici, P. cinnamomi,* and *P. palmivora* can be detected by PCR and dot-blot with an rDNA ITS (internal transcribed spacer) probe, labeled with ^{32}P (Lee, White, et al., 1993). *Phytophthora parasitica* and *P. citrophthora* can be detected by dot-blot with cloned chromosomal DNA, as a probe labeled with ^{32}P (Goodwin, English, et al., 1990; Goodwin, Kirkpatrick, et al., 1990). Southern blot with cloned chromosomal DNA as a probe, labeled with biotin, is useful in detecting *P. infestans* (Moller et al., 1993).

Fusarium culmorum can be detected by Southern blot with cloned genomic DNA as a probe, labeled with digoxigenin (Koopmann et al., 1994). *Leptosphaeria maculans* in rapeseed can be detected using RAPD-PCR with amplified chromosomal DNA as a probe, labeled with digoxigenin (Schafer and Wostemeyer, 1994). Dot-blot with cloned total DNA as a probe, labeled with ^{32}P is used to detect *Pyrenophora teres* and *Pyrenophora graminea* (Husted, 1994). Slot-blot with cloned genomic DNA as a probe, labeled with ^{32}P will be useful to detect *Pseudocercosporella herpotrichoides* in cereals (Nicholson and Rezanoor, 1994). *Pythium ultimum* can be detected by PCR and dot-blot using rDNA ITS as a probe, labeled with digoxigenin (Levesque et al., 1994).

Botrytis cinerea infection in pear stems can be detected by plating stem halves on a selective medium and by ELISA, which is a more sensitive method (Meyer et al., 2000). The PCR assay is highly sensitive and reproducible as a tool for the detection and identification of fungi when species-specific primers are carefully selected. Several fungal pathogens have been detected using this method, including *Verticillium* spp., *Fusarium* spp., *Rhizoctonia oryzae, Gaeumannomyces graminis, Magnaporthe poae, Leptosphaeria korrae,* and *Phiolophora gregata* (Fouly and Wilkinson, 2000).

Nuclear rDNA of fungi consists of small and large subunits, a 5.8S region, and an internal transcribed spacer region(s). Each subunit and region base sequence is variable among the genera and species of fungi. This variability can be used to detect fungal pathogens. For example, the ITS region of *Gaeumannomyces* is highly variable among its species and less variable among its varieties. However, the small subunit of nuclear rDNA is distinctly variable among the varieties of *G. graminis*. This small subunit of nuclear rDNA (18S rDNA) is used for the detection of *G. graminis* varieties using PCR amplification (Fouly and Wilkinson, 2000). A PCR-based diagnostic assay has been developed to detect *Rhynchosporium secalis* in barley (Lee et al., 2001). RAPD analysis is useful to characterize *P. infestans* isolates from potato and tomato (Mahuku et al., 2000). Species-specific primers were designed based on sequence data of a region consisting of the 5.8S RNA gene and internal transcribed spacers 1 and 2 of *R. secalis* (Lee et al., 2001). An oligonucleotide primer set, RS8 and RS9, was used in detecting *R. secalis*. This primer amplified a 264 bp fragment from the DNA of all *R. secalis* isolates. This is used to detect *R. secalis* in infected barley tissues (Lee et al., 2001). A species-specific PCR has been developed to detect black sigatoka and yellow sigatoka leaf-spot pathogens in banana (Johanson et al., 2000). The Rep-PCR technique is used to detect species of *Fusarium, Stagonospora, Septoria, Tilletia,* and *Leptosphaeria* (Jedryczka et al., 1999; McDonald et al., 2000).

DIAGNOSIS OF CROP BACTERIAL DISEASES

Several diagnostic techniques have been developed to diagnose fire blight of apple and pear. The available methods include isolation of the pathogen on the semiselective media of Miller and Schroth, Crosse and Goodman agar, or crystal violet-cycloheximide-thallium nitrate (CCT) agar

(Merighi et al., 2000). A minimal medium 2-copper sulfate agar was developed to specifically identify *Erwinia amylovora* (Bereswill et al., 1998). Isolation followed by pathogenicity tests will also be useful to detect the pathogen. Miller (1983) described an immunofluorescent microscopic method for the detection of *E. amylovora*. The double-antibody sandwich indirect enzyme-linked immunosorbent assay (DASI-ELISA) was developed to detect the pathogen (Gorris et al., 1996). An analysis of fatty acid methyl esters by gas chromatography (GC-FAME) is also used to detect *E. amylovora* (van der Zwet and Wells, 1993). *E. amylovora* can be detected by nested PCR, PCR dot-blot, and reverse-blot hybridization methods (McManus and Jones, 1995). PCR techniques are extensively used to detect the bacterium (Bereswill et al., 1995; Guilford et al. 1996). A PCR-ELISA was developed by Merighi et al. (2000) to detect the pathogen. A PCR-based method detected *E. amylovora* effectively in pear (Sobiczewski et al., 1999). A Rep-PCR technique is used to identify several bacterial pathogens (Louws et al., 1999).

An immunofluorescence test is used to detect *Clavibacter michiganensis* subsp. *michiganensis* in tomato, *C. michiganensis* subsp. *sepedonicus* in potato, and *Erwinia chrysanthemi* in carnation (De Boer et al., 1996). ELISA is useful in detecting *C. michiganensis* subsp. *michiganensis* in tomato, *C. michiganensis* subsp. *sepedonicus* in potato, *E. chrysanthemi* in carnation, *Pantoea stewartii* in carnation, *Xanthomonas campestris* pv. *campestris* in cabbage, and *Pseudomonas savastanoi* pv. *phaseolicola* in bean (De Boer et al., 1996). *X. translucens* pv. *undulosa* in wheat and *X. vesicatoria* in pepper and tomato are detected by dot immunoassay (De Boer et al., 1996).

Nucleic acid probes have been employed to detect some bacterial pathogens. A dot-blot assay has been employed to detect *Erwinia carotovora*, *Xanthomonas axonopodis* pv. *citri*, *X. axonopodis* pv. *phaseoli*, *Pseudomonas syringae* pv. *tomato*, and *Clavibacter michiganensis* subsp. *michiganensis*. Colony blot assay is used to detect *Agrobacterium tumefaciens*, *E. amylovora*, *Pseudomonas savastanoi* pv. *phaseolicola*, and *P. syringae* pv. *morsprunorum* (De Boer et al., 1996). PCR is used to detect *A. tumefaciens* (Dong et al., 1992), *Clavibacter michiganensis* subsp. *sepedonicus* (Schneider et al., 1993), *Ralstonia solanacearum* (Seal et al., 1992), and *X. axonopodis* pv. *citri* (Hartung et al., 1993). PCR and RFLP-based techniques are useful in detection of *A. vitis* in grape (Burr and Otten, 1999).

DIAGNOSIS OF CROP VIRAL DISEASES

A number of different serological techniques have been developed for detecting *Citrus tristeza virus* (CTV). These include ELISA (Rocha-Pena and Lee, 1991), sodium dodecyl sulfate (SDS)-immunodiffusion (Gransey et al., 1978), immunoelectron microscopy, radioimmunosorbent assay (Rocha-Pena and Lee, 1991), immunogold assay, Western blot assay (Rocha-Pena and Lee, 1991), dot immunobinding assay (Rocha-Pena et al., 1991), direct tissue blot immunoassay (Lin et al., 2000), and in situ immunofluorescence (ISIF) (Brlansky et al., 1988). Specific monoclonal antibodies produced by hybridoma technology are used for ELISA. In spite of the high serological variability of CTV, a mixture of two monoclonal antibodies (3DFI and 3CA5) is able to detect all CTV isolates tested. These antibodies were patented in 1984 and are considered an international reference for CTV diagnosis. With these commercially available monoclonal antibodies, approximately 2 million samples have been analyzed since 1990 (Terrada et al., 2000). Lin et al. (2000) developed an in situ immunobioassay that does not use fluorescent dyes. This assay is a simple and specific procedure that detects CTV in infected citrus plants in about 2 h.

Terrada et al. (2000) obtained single-chain variable fragment antibodies that bind specifically to CTV from the hybridoma cell lines 3DF1 and 3CA5. These scFv were genetically fused with dimerization domains as well as with alkaline phosphatase, and diagnostic reagents were produced by expressing these fusion proteins in *E. coli* cultures. The engineered antibodies were successfully used for CTV diagnosis in citrus plants by tissue print ELISA and DAS-ELISA. The fully recombinant ELISAs were as specific and as sensitive as conventional ELISAs performed with the parental monoclonal antibodies (Terrada et al., 2000).

ELISA and Western blots using monoclonal antibodies are recommended to identify grapevine leafroll-associated viruses. The monoclonal antibodies of these viruses are commercially available (Monis, 2000). High-quality polyclonal antisera have been raised against several members of the genus *Tospovirus* including *Tomato spotted wilt virus, Groundnut ringspot virus, Tomato chlorotic spot virus,* and *Watermelon silverleaf mottle virus,* and *Impatiens necrotic spot virus.* These antisera are now widely applied for *Tospovirus* detection with the aid of a DAS-ELISA (Griep et al., 2000). Polyclonal antisera, although widely used in routine diagnosis, are available in limited amounts, and their specificity varies from batch to batch. The antisera are being replaced by monoclonal antibodies, which can be produced indefinitely. Several monoclonal antibodies against tospoviruses have been developed. Griep et al. (2000) used recombinant antibody technology in combination with phage display technology to produce *Tomato spotted wilt virus*-specific antibodies. A panel of recombinant single-chain variable antibodies against the N protein and G1 and G2 glycoproteins of TSWV was retrieved from a human combinatorial scFv antibody library using the phage display technique. Antibodies were obtained after subcloning the encoding

DNA sequences in the expression vector pSKAP/S, which allowed the scFvs to be expressed as alkaline phosphatase fusion proteins. An antibody, N56-AP/S, at a concentration of 0.1 µg/ml, can detect as little as 1 ng of N protein of TSWV in a DAS-ELISA. The CL (mouse light-chain) ZIP (leucine zipper) fusion protein of scFv N56 was an effective coating and detecting reagent in a DAS-ELISA or detection of TSWV (Griep et al., 2000).

ELISA is commonly used to detect *Peanut stripe virus* and *Peanut mottle virus* in peanut seed and vegetative tissues. Gillaspie et al. (2000) developed an immunocapture reverse-transcription polymerase chain reaction (IC-RT-PCR) for detection of peanut virus diseases. Peanut tissue slices were extracted in a buffer and centrifuged, and a portion of the supernatant was incubated in a tube that had been coated with an antiserum to either PStV or PeMV. Following immunocapture of the virus, the tube was washed and the RT-PCR mix (with primers designed from conserved sequences within the capsid region of each virus) was placed in the same tubes. The IC-RT-PCR method was more sensitive than ELISA in the detection of peanut viruses (Gillaspie et al., 2000). *Prunus necrotic ringspot virus, Prune dwarf virus,* and *Apple mosaic virus* affect stone-fruit (plum, almond, apricot, cherry, and peach) trees. Nonisotopic molecular hybridization and multiplex reverse-transcription polymerase chain reaction methodologies have been developed to detect all these viruses simultaneously (Saade et al., 2000). For multiplex RT-PCR, a degenerate antisense primer was designed and used in conjunction with three virus-specific sense primers. The amplification efficiencies for the detection of the three viruses in the multiplex RT-PCR reaction were identical to those obtained in the single RT-PCR reactions for individual viruses (Saade et al., 2000). RT-PCR is the most sensitive test for the detection of *Banana bract mosaic virus* (BBrMV) in banana plants followed by the F(ab')$_2$ indirect DAS-ELISA and IC-PCR (Rodoni et al., 1999).

The coat protein gene of *Grapevine rupestris stem pitting-associated virus* (GRSPaV) was amplified with primers based on a completely sequenced GRSPaV isolate. The protein expressed in *Escherichia coli* was used to raise an antiserum in rabbit. This antiserum was used to detect the virus in infected grapevine extracts by dot immunobinding (by spotting on polyvinyl difluoride membranes) or by Western blot. ELISA was ineffective in detecting the virus in grapevine (Minafra et al., 2000). Hailstones et al. (2000) developed a specific seminested RT–PCR assay that detects *Citrus tatter leaf virus* in citrus trees. The sensitivity of the assay is at least 500 times greater than that of ELISA-based methods and allows detection directly from field trees.

DIAGNOSIS OF VIROID DISEASES

Bidirectional electrophoresis is used to test chrysanthemums for the chrysanthemum stunt viroid. Molecular hybridization methods are also used to detect this viroid (Dinesen and van Zaayen, 1996). Primer pairs and nucleic acid preparations were used with RT-PCR to detect peach latent mosaic viroid (PLMVd) from stone fruits (Osaki et al., 1999). PCR methods are useful in detecting hop stunt viroid in hops, apple scar skin viroid and pear rusty skin viroid in apple and pear, citrus exocortis viroid in citrus, and grape viroids in grapes (Henson and French, 1993).

DIAGNOSIS OF PHYTOPLASMA DISEASES

Phytoplasmas are detectable microscopically in phloem by means of Dienes' stain (Sinclair et al., 1996). The DAPI (4',6-diamidino-2-phenylindole-2HCL) test is more useful in the diagnosis of phytoplasmas (Seemüller, 1976). DAPI binds to DNA and causes it to fluoresce under UV. When longitudinal sections of twigs, petioles, or small roots of ash yellows-affected *Fraxinus* and *Syringa* trees are treated with DAPI and examined with a fluorescence microscope, phytoplasmal DNA appears as blue-white fluorescent specks or aggregations in sieve tubes, whereas normal sieve tubes remain dark (Sinclair et al., 1996). The DAPI test is considered nonspecific because DNA of any organism fluoresces under the test conditions.

Several DNA-based techniques are available to detect phytoplasmas. PCR is highly useful in detecting several phytoplasmas. PCR primers that are commonly used are based on sequences in the 16S ribosomal RNA gene that are common to all phytoplasmas, but do not occur in plants (Lee, Hammond, et al., 1993). A DNA segment of characteristic size is amplified from any phytoplasma. The organism is then identified by using this initially amplified DNA segment as a template for further PCRs using primers that amplify DNA from only particular phytoplasmas. These primers are based on nucleotide sequences between the positions of the first primer pair on the 16S rDNA (Lee et al., 1994).

In another method, the initial PCR product is subjected to restriction fragment length polymorphism (RFLP) analysis, in which the amplified segment is digested with certain restriction endonucleases and separated into fragments by gel electrophoresis. Phytoplasmas in different groups have different RFLP profiles (Guo et al., 2000). Ash Y phytoplasmas can be distinguished from others by RFLP analysis of 16S rDNA with the restriction enzyme *Alu*I (Lee, Hammond, et al., 1993).

Phytoplasmas can be diagnosed by means of DNA hybridizations using probes that hybridize to group-specific sequences. Immunological tests were developed to identify phytoplasmas. Monoclonal antibodies to peach eastern X-disease agent have been developed and their use in disease detection has been demonstrated (Lin and Chen, 1985). Immunocapture PCR tests have also been developed to diagnose phytoplasma diseases (Sinclair et al., 1996).

PLANT CLINICS AND DIAGNOSTIC SERVICE LABORATORIES

Several plant clinics have been set up in the United States and many other countries, and they play an important role in crop disease diagnosis. Plant clinics exist typically as part of a plant pathology department of a state's land grant university, as a part of a state's department of agriculture, or as a private or commercial service. Clinics operate throughout North America as the primary source of diagnostic information and as the primary focal point for the submission of plant-disease specimens (Barnes, 1994). Extension-university-supported clinics and state department of agriculture clinics represent the backbone of disease diagnostic services. These clinics routinely use ELISA-based diagnostic procedures and PCR technology. Many clinics routinely use highly technical diagnostic procedures. The clinics provide clients with an accurate diagnosis, and the diagnostic report is sent by mail, fax, phone, modem, or computer network.

Several diagnostic service laboratories were established in the United States since the 1970s. They charge a nominal fee for diagnosis. Plant disease clinics have an instructional component, whereas diagnostic service laboratories simply diagnose plant diseases (Barnes, 1994).

DIGITALLY ASSISTED DIAGNOSIS

Diseases are recognized mostly by visual disease symptoms. Photographic images of plant disease symptoms and signs can be useful in diagnosis. Modern telecommunications systems permit individuals to share high-resolution digital images among multiple locations within seconds (Holmes et al., 2000). These digital-imaging and digital-image transfer tools are used for crop disease diagnosis. Digitally assisted diagnosis permits long-distance consultation and accurate diagnosis of plant disease problems. In the United States, the University of Georgia introduced in 1997 the Web-based delivery system for digitally assisted diagnosis, and they named it Distance Diagnostics through Digital Imaging (DDDI). County extension offices are provided with a computer, digital camera, dissecting and compound microscopes, and have access to the DDDI system. In cases in which the disease shows unique symptoms, adequate background information is available, and a high quality digital image is obtained, accurate diagnosis can be made within minutes with the help of experienced technicians. It is now possible for the grower to send and receive images from anywhere in the world, including fields, using cellular modems (Holmes et al., 2000). Several Web sites, which describe digitally assisted diagnostics programs, are now available: <http://www.dddi.org>, <http://ddis.ifas.ufl.edu/>, <http://cf.uwex.edu/ces/ag/disdiag>, <http://www.ent.iastate.edu/rdi>, <http://www.ca.uky.edu/agcollege/plantpathology/PPAExten/digiimag.htm> (Holmes et al., 2000).

REFERENCES

Barnes, L. W. (1994). The role of plant clinics in disease diagnosis and education: A North American perspective. *Annu Rev Phytopathol,* 32:601-609.

Bateman, G. L., Ward, E., and Antoniw, J. F. (1992). Identification of *Gaeumannomyces graminis* var. *tritici* and *G. graminis* var. *avenae* using a DNA probe and non-molecular methods. *Mycological Research,* 96:737-742.

Bereswill, S., Bugert, P., Bruchmuller, I., and Geider, K. (1995). Identification of fireblight pathogen, *Erwinia amylovora,* by PCR assay with chromosomal DNA. *Appl Environ Microbiol,* 61:2636-2642.

Bereswill, S., Jock, S., Bellemann, P., and Geider, K. (1998). Identification of *Erwinia amylovora* by growth morphology on agar containing copper sulfate and by capsule staining with lectin. *Plant Dis,* 82:158-164.

Brlansky, R. H., Lee, R. F., and Gransey, S. M. (1988). In situ immunofluorescence for the detection of citrus tristeza virus inclusion bodies. *Plant Dis,* 72:1039-1041.

Burr, T. J. and Otten, L. (1999). Crown gall of grape: Biology and disease management. *Annu Rev Phytopathol,* 37:53-80.

De Boer, S. H., Cuppels, D. A., and Gitaitis, R. D. (1996). Detecting latent bacterial infections. *Adv Bot Res,* 23:27-57.

Dinesan, I. G. and van Zaayen, A. (1996). Potential of pathogen detection technology for management of diseases in glasshouse ornamental crops. *Adv Bot Res,* 23:137-170.

Dong, L. C., Sun, C. W., Thies, K. L., Luthe, D. S., and Graves, C. H. (1992). Use of polymerase chain reaction to detect pathogenic strains of *Agrobacterium. Phytopathology,* 82:434-439.

Fouly, H. M. and Wilkinson, H. T. (2000). Detection of *Gaeumannomyces graminis* varieties using polymerase chain reaction with variety-specific primers. *Plant Dis,* 84:947-951.

Fox, R. T. V. (1998). Plant disease diagnosis. In D. G. Jones (Ed.) *The Epidemiology of Plant Diseases.* Kluwer Academic Publishers, Dordrecht, pp.14-41.

Gillaspie, A. G. Jr., Pittman, R. N., Pinnow, D. L., and Cassidy, B. G. (2000). Sensitive method for testing peanut seed lots for *Peanut stripe* and *Peanut mottle viruses* by immunocapture-reverse transcription-polymerase chain reaction. *Plant Dis,* 84:559-561.

Goodwin, P. H., English, J. T., Neher, D. A., Duniway, J. M., and Kirkpatrick, B. C. (1990). Detection of *Phytophthora parasitica* from soil and host tissue with a species-specific DNA probe. *Phytopathology,* 80:277-281.

Goodwin, P. H., Kirkpatrick, B. C., and Duniway, J. M. (1990). Identification of *Phytophthora citrophthora* with cloned DNA probes. *Appl Environ Microbiol,* 56:669-674.

Gorris, M. T., Cambra, M., Llop, P., Lopez, M. M., Lecomte, P., Chartier, R., and Paulin, P. J. (1996). A sensitive and specific detection of *Erwinia amylovora* based on the ELISA-DASI enrichment method with monoclonal antibodies. *Acta Hortic,* 411:53-56.

Gransey, S. M., Gonsalves, D., and Purcifull, D. E. (1978). Rapid diagnosis of citrus tristeza virus infection by sodium dodecyl sulfate-immunodiffusion procedure. *Phytopathology,* 68:88-95.

Griep, R. A., Prins, M., van Twisk, C., Keller, H. J. H. G., Kerschbaumer, R. J., Kormelink, R., Goldbach, R. W., and Schots, A. (2000). Application of phage display in selecting *Tomato spotted wilt virus*-specific single-chain antibodies (scFvs) for sensitive diagnosis in ELISA. *Phytopathology,* 90:183-190.

Guilford, P. J., Taylor, R. K., Clark, R. G., Hale, C. N., and Forster, R. L. S. (1996). PCR-based techniques for the detection of *Erwinia amylovora. Acta Hortic,* 441:53-56.

Guo, Y. H., Cheng, Z.-M., and Walla, J. A. (2000). Characterization of X-disease phytoplasmas in chokecherry from North Dakota by PCR-RFLP and sequence analysis of the rRNA gene region. *Plant Dis,* 84:1235-1240.

Hailstones, D. L., Bryant, K. L., Broadbent, P., and Zhou, C. (2000). Detection of *Citrus tatter leaf virus* with reverse transcription-polymerase chain reaction (RT-PCR). *Australasian Plant Pathol,* 29:240-248.

Harrison, J. G., Lowe, R., Wallace, A., and Williams, N. A. (1994). Detection of *Spongospora subterranea* by ELISA using monoclonal antibodies. In A. Schots, F. M. Dewey, and R. Oliver (Eds.), *Modern Assays for Plant Pathogenic Fungi: Identification, Detection, and Quantification.* CAB International, Oxford, pp. 23-27.

Hartung, J. S., Daniel, J. F., and Pruvost, O. P. (1993). Detection of *Xanthomonas campestris* pv. *citri* by the polymerase chain reaction method. *Appl Env Microbiol,* 59:1143-1148.

Henson, J. M. and French, R. (1993). The polymerase chain reaction and plant diagnosis. *Annu Rev Phytopathol,* 31:61-109.

Holmes, G. J., Brown, E. A., and Ruhl, G. (2000). What's a picture worth? The use of modern telecommunications in diagnosing plant diseases. *Plant Dis,* 84: 1256-1265.

Husted, K. (1994). Development of species-specific probes for identification of *Pyrenophora graminea* and *P. teres* by dot-blot or RFLP. In A. Schots, F. M. Dewey, and R. Oliver (Eds.), *Modern Assays for Plant Pathogenic Fungi: Identification, Detection, and Quantification.* CAB International, Oxford, pp. 191-197.

Huttinga, H. (1996). Sensitivity of indexing procedures for viruses and viroids. *Adv Bot Res,* 23:59-71.

Jedryczka, M., Rouxel, T., and Balesdent, M. H. (1999). Rep-PCR based genomic fingerprinting of isolates of *Leptospheria maculans* from Poland. *Eur J Plant Pathol,* 105:813-823.

Johanson, A., Tushemereirwe, W. K., and Karamura, E. B. (2000). Distribution of sigatoka leaf spots in Uganda as determined by species-specific polymerase chain reaction (PCR). *Acta Hortic,* 540:319-324.

Kohler, G. and Milstein, C. (1975). Continuous cultures of fused cells secreting antibody of predetermined specificity. *Nature,* 256:495-497.

Koopmann, B., Karlovsky, P., and Wolf, G. (1994). Differentiation between *Fusarium culmorum* and *Fusarium graminearum* by RFLP and with species-specific DNA probes. In A. Schots, F. M. Dewey, and R. Oliver (Eds.), *Modern Assays for Plant Pathogenic Fungi: Identification, Detection, and Quantification.* CAB International, Oxford, pp. 37-46.

Lee, H. K., Tewari, J. P., and Turkington, T. K. (2001). A PCR-based assay to detect *Rhynchosporium secalis* in barley seed. *Plant Dis,* 85:220-225.

Lee, I.-M., Gundersen, D. E., Hammond, R. W., and Davis, R. E. (1994). Use of mycoplasmalike organism (MLO) group-specific oligonucleotide primers for nested-PCR assays to detect mixed-MLO infections in a single host plant. *Phytopathology,* 84:559-566.

Lee, I.-M., Hammond, R. W., Davis, R. E., and Gundersen, D. E. (1993). Universal amplification and analysis of pathogen 16S rDNA for classification and identification of mycoplasmalike organisms. *Phytopathology,* 83:834-842.

Lee, S. B., White, T. J., and Taylor, J. W. (1993). Detection of *Phytopthora* species by oligonucleotide hybridization to amplified ribosomal DNA spacers. *Phytopathology,* 83:177-181.

Levesque, C. A., Vrain, T. C., and De Boer, S. H. (1994). Development of a species-specific probe for *Pythium ultimum* using amplified ribosomal DNA. *Phytopathology,* 84:474-478.

Lin, C. P. and Chen, T.-A. (1985). Monoclonal antibodies against the aster yellows agent. *Science,* 227:1233-1235.

Lin, Y., Rundell, P. A., Xie, L., and Powell, C. A. (2000). In situ immunoassay for detection of *Citrus tristeza virus. Plant Dis,* 84:937-940.

Louws, F. J., Rademaker, J. L. W., and de Bruijn, F. J. (1999). The three Ds of PCR-based genomic analysis of phytobacteria: Diversity, detection and disease diagnosis. *Annu Rev Phytopathol,* 37:81-125.

Mahuku, G., Peters, R. D., Platt, H. W., and Daayf, F. (2000). Random amplified polymorphic DNA (RAPD) analysis of *Phytophthora infestans* isolates collected in Canada during 1994 to 1996. *Plant Pathol,* 49:252-260.

Maule, A. J., Hull, R., and Donson, J. (1983). The application of spot hybridization to the detection of DNA and RNA viruses in plant tissues. *J Virol Methods,* 6:183-191.

McDonald, J. G., Wong, E., and White, G. P. (2000). Differentiation of *Tilletia* species by rep-PCR genomic fingerprinting. *Plant Dis,* 84:1121-1125.

McManus, P. S. and Jones, A. L. (1995). Detection of *Erwinia amylovora* by nested PCR, PCR dot-blot, and reverse-blot hybridization. *Phytopathology,* 85:618-623.

Merighi, M., Sandrini, A., Landini, S., Ghini, S., Girotti, S., Malaguti, S., and Bazzi, C. (2000). Chemiluminescent and colorimetric detection of *Erwinia amylovora* by immunoenzymatic determination of PCR amplicons from plasmid pEA29. *Plant Dis,* 84:49-54.

Meyer, U. M., Spotts, R. A., and Dewey, F. M. (2000). Detection and quantification of *Botrytis cinerea* by ELISA in pear stems during cold storage. *Plant Dis,* 84:1099-1103.

Miller, H. J. (1983). Some factors influencing immunofluorescence microscopy as applied in diagnostic phytobacteriology with regards to *Erwinia amylovora. Phytopathol Z,* 108:235.

Miller, S. A. (1996). Detecting propagules of plant pathogenic fungi. *Adv Bot Res,* 23:73-102.

Minafra, A., Casati, P., Elcio, V., Rowhani, A., Saldarelli, P., Savino, V., and Martelli, G. P. (2000). Serological detection of a grapevine rupestris stem pitting-associated virus (GRSPaV) by a polyclonal antiserum to recombinant virus coat protein. *Vitis,* 39:115-118.

Moller, E. M., de Cock, A. W. A. M., and Prell, H. H. (1993). Mitochondrial and nuclear DNA restriction enzyme analysis of the closely related *Phytophthora* species *P. infestans, P. mirabilis,* and *P. phaseoli. J Phytopathology,* 139:309-321.

Monis, J. (2000). Development of monoclonal antibodies reactive to a new grapevine leafroll-associated closterovirus. *Plant Dis,* 84:858-862.

Nicholson, P. and Rezanoor, H. N. (1994). DNA probe for the R-type of eyespot disease of cereals *Pseudocercosporella herpotrichoides.* In A. Schots, F. M. Dewey, and R. Oliver (Eds.), *Modern Assays for Plant Pathogenic Fungi: Identification, Detection, and Quantification.* CAB International, Oxford, pp.17-22.

Osaki, H., Yamaguchi, M., Sato, Y., Tomita, Y., Kawai, Y., Miyamoto, Y., and Ohtsu, Y. (1999). Peach latent mosaic viroid isolated from stone fruits in Japan. *Ann Phytopathol Soc Japan,* 65:3-8.

Rocha-Pena, M. and Lee, R. F. (1991). Serological technique for detection of citrus tristeza virus. *J Virol Methods,* 34:311-331.

Rocha-Pena, M., Lee, R. F., and Niblett, C. L. (1991). Development of a dot-immunobinding assay for detection of citrus tristeza virus. *J Virol Methods,* 34:297-309.

Rodoni, B. C., Dale, J. L., and Harding, R. M. (1999). Characterization and expression of the coat protein-coding region of banana bract mosaic potyvirus, development of diagnostic assays and detection of the virus in banana plants from five countries in southeast Asia. *Arch Virol,* 144:1725-1737.

Saade, M., Aparicio, F., Sanchez-Navarro, J. A., Herranz, M. C., Myrta, A., Di Terlizzi, B., and Pallas, V. (2000). Simultaneous detection of the three ilarviruses affecting stone fruit trees by nonisotopic molecular hybridization and multiplex reverse-transcription polymerase chain reaction. *Phytopathology,* 90: 1330-1336.

Salinas, J. and Schots, A. (1994). Monoclonal antibodies-based immunofluorescence test for detection of conidia of *Botrytis cinerea* on cut flowers. *Phytopathology,* 84:351-356.

Sano, T., Smith, C. L., and Cantor, C. R. (1992). Immuno-PCR: Very sensitive antigen detection by means of specific antibody-DNA complexes. *Science,* 258:120-122.

Schafer, C. and Wostemeyer, J. (1994). Molecular diagnosis of rapeseed pathogen *Leptosphaeria maculans* based on RAPD-PCR. In A. Schots, F. M. Dewey, and R. Oliver (Eds.), *Modern Assays for Plant Pathogenic Fungi: Identification, Detection, and Quantification.* CAB International, Oxford, pp. 1-8.

Schneider, J., Zhao, J. L., and Orser, C. (1993). Detection of *Clavibacter michiganensis* subsp. *sepedonicus* by DNA amplification. *FEMS Microbiology Letters,* 109:207-212.

Seal, S. E., Jackson, L. A., and Daniels, M. J. (1992). Isolation of a *Pseudomonas solanacearum*-specific DNA probe by subtraction hybridization and construction of species-specific oligonucleotide primers for sensitive detection by the polymerase chain reaction. *Appl Environ Microbiol,* 58:3751-3758.

Seemüller, E. (1976). Investigations to demonstrate mycoplasmalike organisms in diseased plants by fluorescence microscopy. *Acta Hortic,* 67:109-112.

Sinclair, W. A., Griffiths, H. M., and Davis, R. E. (1996). Ash yellows and Lilac Witches'-broom: Phytoplasmal diseases of concern in forestry and horticulture. *Plant Dis,* 80:468-475.

Sobiczewski, P., Pulawska, J., Berezynski, S., and Konicka, M. (1999). Fireblight detection and control in Poland. *Acta Hortic,* 489:115-120.

Southern, E. (1975). Detection of specific sequences among DNA fragments separated by gel electrophoresis. *J Mol Biol,* 98:503-517.

Terrada, E., Kerschbaumer, R. J., Giunta, G., Galeffi, P., Himmler, G., and Camba, M. (2000). Fully "recombinant enzyme-linked immunosorbent assays" using genetically engineered single-chain antibody fusion proteins for detection of *Citrus tristeza virus. Phytopathology,* 90:1337-1344.

van der Zwet, T. and Wells, J. M. (1993). Application of fatty acid gas analysis for the detection and identification of *Erwinia amylovora. Acta Hortic,* 338:233.

CROP DISEASE EPIDEMIOLOGY

14

Assessment of Disease Progress

Epidemiology is the science that describes the progress of a disease as it becomes epidemic. Depending on the rate of disease progress, diseases are broadly classified as simple interest and compound interest diseases. Several models have been developed to describe the progression of these two kinds of diseases. The area under disease progress curve (AUDPC) provides a valid statistical description of disease progress data.

WHAT IS EPIDEMIOLOGY?

A widespread temporary increase in the incidence of an infectious disease is called an *epidemic*. A plant disease is described as epidemic when the amount of disease present increases rapidly from a low level to a high one. This is in contrast to an *endemic* disease, which is a disease established in moderate or severe form in a defined area. For an endemic disease, the disease level remains almost constant.

The study of epidemic diseases is known as *epidemiology*. Epidemiology describes the rapid multiplication of the invading pathogen and the subsequent increase in disease levels within a large population of plants that have no immunity, but varying levels of resistance, to the particular pathogen. Epidemiology comprehensively analyzes the interaction between the constituents of the *disease triangle:* the host, the pathogen, and the environment.

Epidemiology is concerned with diseases in populations, that is, with host populations interacting with pathogen (or vector) populations. Epidemiology describes the host population and pathogen population and the effect of the environment on these populations. The primary goal of epidemiology is to understand the factors that influence disease development, develop efficient disease management strategies, and reduce losses due to diseases (Van der Plank, 1963; Leonard and Fry, 1986; Manners, 1993; Campbell, 1998).

DISEASE PROGRESS CURVE

The amount of disease present on the plant varies with time, and a rapid increase in disease incidence and severity with time indicates development of an epidemic. The progress of a disease may be assessed as disease incidence (proportion of diseased plants in a plant population) or as disease severity (proportion of diseased tissue in a plant). When the results are plotted against time, the curve obtained is commonly S-shaped. This curve is known as the progress curve of the disease.

Van der Plank (1963) classified diseases into two kinds, based on the disease progress curve: *simple interest diseases* and *compound interest diseases*. In simple interest diseases, only one cycle of infection and inoculum production per growing season occur (i.e., they are *monocyclic*). These diseases are called simple interest diseases because the rate of disease increase follows simple interest pattern. Simple interest diseases are initiated from a given inoculum quantity at the beginning of the disease cycle and multiply once per season, usually to leave a legacy of a much higher quantity of inoculum in the soil or on planting material in readiness for the next season's crop. In these diseases, the amount of inoculum is constant throughout the season, even though the number of plants infected or the plant tissue area affected may rise. Examples for these diseases include wilts, some root rots, and internally seedborne smut diseases, such as loose smut of wheat.

In compound interest diseases, cycles of infection and inoculum production occur repeatedly, and thus many disease cycles occur during a growing season (i.e., they are *polycyclic*). Compound interest diseases multiply rapidly in a season with several generations of spore production. These diseases are called compound interest diseases because the disease increase follows the pattern of a bank's compound interest calculation. Examples for these diseases include rusts, powdery mildews, and other foliar diseases.

Van der Plank (1963) developed two biological models, a monomolecular model and a logistic model, for describing the progression of simple interest and compound interest diseases. The monomolecular model describes simple interest diseases, whereas the logistic model describes compound interest diseases. The monomolecular model can be written as:

$$\frac{dy}{dt} = k(1 - y)$$

where dy/dt is the absolute rate of disease increase, k is the rate parameter, and y is the proportion of increase at time (t). The maximum value that y can reach is assumed to be 1. The model assumes that the absolute rate of disease increase is proportional to the amount of healthy tissue remaining at any given time.

The logistic model can be written as

$$\frac{dy}{dt} = ky(1 - y).$$

In this case, the absolute rate of disease increase is assumed to be proportional to the amount of disease and of healthy tissue present.

Another important model developed to describe the disease progress curve is the Gompertz model. It is written as

$$\frac{dy}{dt} = rg \; y\left[\left(\ln(k)\right) - \ln(y)\right]$$

where *rg* is the rate of increase specific to the Gompertz model, *k* is the maximum level of disease, *y* is the disease at time of observation, and *t* is the time interval being considered, and the symbol In indicates natural logarithm.

These three models are often used in a linearized form. The disease progress curves are transformed to a linear form with the assumption that the maximum amount of disease, $(y_{max} = k)$, is 100 percent or proportionally 1.0. Linear regression analysis is then used to provide estimates of the initial amount of disease (y_o) and the apparent rate of increase (r). The appropriateness of a model for describing disease progress data is then judged by goodness-of-fit criteria, including the magnitude of the coefficient of determination (R^2) and mean square error (MSE) (Campbell, 1998).

Linear models are more useful than nonlinearized models (e.g., the monomolecular, logistic, and Gompertz models) because the software is able to perform linear regression analysis quite easily. However, software for nonlinear regression is now available. The fitting of linear and nonlinear regression models is based on the same idea, i.e., choosing parameter estimates that minimize the sums of squared errors. Since fewer assumptions are made with nonlinear regression analysis, this analysis is preferred for disease progress data (Campbell, 1998).

Other models employed less frequently to assess the disease progress data include the Weibull, Richards, and Gaussian models. These models are used to calculate the rate of disease increase, and in some cases meaningful and more or less constant *r* values can be obtained. However, in many cases frequent changes in *r* values occur due to large fluctuations in environmental factors. Data on environmental factors have not been incorporated in these models for analysis.

Many limitations exist in using the Weibull, Richards, and Gaussian models. For example, they may not be sufficient to describe all disease progress curves. If the weather becomes unfavorable for disease development during the course of an epidemic, a plateau may occur within the disease progress curve. When new flushes of host growth due to agronomic factors occur, the disease progress may decline. These models can be useful in regions where weather conditions are reasonably constant throughout the period under examination, and where no appreciable number of leaves or plants are added or removed.

The disease progress curves are useful to assess the development of an epidemic. In the long run, a high rate of interest is more important than a large balance in the bank today (Van der Plank, 1963). Similarly, a high rate of increase *(r)* will contribute to an epidemic. A modest decrease in *r* may prevent an epidemic.

AREA UNDER DISEASE PROGRESS CURVE (AUDPC)

The area under disease progress curve (AUDPC) is an alternative method that provides a valid statistical description of disease progress data. AUDPC is the amount of disease integrated between two times of interest, and is calculated without regard to curve shape (Campbell, 1998). It is a valid descriptor of an epidemic under the hypothesis that injury to a host plant is proportional to the amount and duration of the disease. AUDPC is calculated by using the following formula:

$$\text{AUDPC} = \sum_{i=1}^{n}\left[\left(x_{i+1} + x_i\right)/2\right]\left(t_{i+1} - t_i\right)$$

where x_i = proportion tissue affected (disease intensity) at the *i*th observation, t = time (days) after inoculation at the *i*th observation, and n = total number of observations. Σ is the sum of areas of all of the individual trapezoids or areas from i to $n-1$. i and $i+1$ represent observations from 1 to n (Shaner and Finney, 1977).

If the epidemic duration differs, AUDPC values are normalized by dividing the AUDPC by the total area of the graph (= the number of days from inoculation to the end of the observation period $\times$ 1.0). The normalized AUDPC is referred to as "relative AUDPC." Units for AUDPC are days – proportion or days – percent, depending upon the measure of disease intensity. Relative AUDPC has no units. Greater frequency of disease assessments over time result in a more accurate estimate of AUDPC than when only a few assessments are made.

APPARENT INFECTION RATE

The apparent infection rate is useful in assessing the disease in the field. It is calculated as follows (Van der Plank, 1963):

$$r = 1/t_2 - t_1 \log_e \left[x_2 \left(1 - x_1\right) / x_1 \left(1 - x_2\right) \right]$$

where x = the proportion of disease at any one time, x_1 = the amount of disease present at time t_1, x_2 = the amount of disease present at time t_2, t = time during which the infection has occurred, e = base of natural logs, and r = the average infection rate/apparent infection rate.

REFERENCES

Campbell, C. L. (1998). Disease progress in time: Modelling and data analysis. In D. G. Jones (Ed.) *The Epidemiology of Plant Diseases.* Kluwer Academic Publishers, Dordrecht, The Netherlands, pp. 181-204.

Leonard, K. J. and Fry, W. E. (1986). *Plant Disease Epidemiology, Population Dynamics and Management, Vol. 1.* Macmillan Publishing Co., New York, p. 372.

Manners, J. G. (1993). *Principles of Plant Pathology.* Cambridge University Press, Cambridge, p. 343.

Shaner, E. and Finney, R. E. (1977). The effect of nitrogen fertilization on the expression of slow-mildewing resistance in Knox wheat. *Phytopathology,* 67: 1051-1056.

Van der Plank, J. E. (1963). *Plant Diseases: Epidemics and Control.* Academic Press, New York, p. 349.

15

Factors Involved in Disease Increase

Several environmental factors are involved in the temporary widespread increase in fungal and bacterial diseases (i.e., epidemic). The life of a fungal pathogen may be divided into sporulation, dispersal (removal, transport, and deposition), germination, infection, and incubation phases. Environmental factors involved in all phases of the fungal life cycle are discussed. Environmental factors affecting bacterial spread and infection are also described.

ENVIRONMENTAL FACTORS AFFECTING SPORULATION

Sporulation of fungal pathogens in the host appears to be influenced by weather factors. The production of ascospores of *Mycosphaerella citri* on citrus leaves is greatest from March to July, and few ascospores are produced thereafter in Florida (Timmer, Roberts, et al., 2000). Moisture, temperature, humidity, and light influence sporulation. Asexual spores (pycnidiospores) of *Septoria tritici* are important in the epidemiological development cycle of the pathogen in wheat. The production of pycnidiospores is strongly dependent on moisture and temperature (Verreet et al., 2000). Sigatoka (yellow Sigatoka) and black leaf streak (black Sigatoka) pathogens (*M. musicola* and *M. fijiensis,* respectively) produce conidia abundantly in wet weather or in dew on the surface of leaf lesions (Burt et al., 1999). Sporulation of the apple sooty blotch pathogens *Peltaster fructicola* and *Leptodontium elatius* was greatest at relative humidities of 97 to 99 percent (Johnson and Sutton, 2000). The optimum temperature for sporangial production of *Phytophthora palmivora* on a citrus fruit surface was 24°C, with sporangial production decreasing rapidly at higher or lower temperatures (Timmer, Zitko, et al., 2000). A few sporangia were produced with 18 h of fruit wetness, and numbers increased as the duration of wetness increased to 72 h (Timmer, Zitko, et al., 2000). Sporulation on soybean root surfaces infected with *Fusarium solani* f. sp. *glycines* is more frequent during or immediately following high moisture in soil (Roy, 1997). Temperature and moisture influence the apothecium production of *Sclerotinia sclerotiorum,* the stem rot pathogen of soybean. High light intensity favors the development of apothecia (Sun et al., 2000). Moderate temperatures (14 to 28°C) combined with high relative humidity (>90 percent) are conducive to conidial production of *Claviceps sorghi* (Bandyopadhyay et al., 1998). Sporulation of *Pseudocercosporella herpotrichoides,* the wheat eyespot pathogen, may occur within the temperature range 1 to 20°C, with an optimum temperature of 5°C (Fitt et al., 1988). Sporulations of the peanut leaf spot pathogens *Cercospora arachidicola* and *Phaeoisariopsis personata* occur on lesions, and sporulation is favored by high relative humidity (about 100 percent RH) and temperatures ranging from 24 to 28°C. Sporulation increases with increased wet period duration (Jacobi et al., 1995). Temperature and wetness duration have the greatest effect on inoculum production of *Botrytis cinerea,* the strawberry gray mold pathogen. The optimum temperature for sporulation in dead leaf tissue is 18°C, and sporulation increases with longer wetness duration (Sosa-Alvarez et al., 1995).

ENVIRONMENTAL FACTORS AFFECTING SPORE RELEASE

The spore dispersal phase of the pathogen life cycle has three stages: removal, transport, and deposition. Rain, wind, humidity, and temperature may play important roles in spores release from the host surface. Conidia of the wheat powdery mildew pathogen *Erysiphe graminis* may be removed from infected leaves by wind gusts or by acceleration forces as leaves flap in the wind. They may also be removed by raindrops, through puff or tap mechanisms. As a raindrop spreads out on a dry leaf, it generates a puff of air that may remove dry conidia. The impact (tap) of the raindrop on the leaf may also release dry spores as kinetic energy is transferred to the leaf. Dry, airborne conidia of *E. graminis* are a common component of air spora during dry summer weather in the United Kingdom (Fitt and McCartney, 1986).

Pycnidiospores of *Septoria nodorum* are generally produced in mucilage, which prevents their removal from infected wheat plants by wind. However, during rainy weather, the first raindrops dissolve the mucilage to leave a spore suspension that is prone to splash dispersal by subsequent raindrops (Fitt and McCartney, 1986). Lighter or shorter rainfalls enhance the removal of urediniospores of the wheat leaf (brown) rust pathogen *Puccinia triticina* from wheat leaf surfaces (Sache et al., 2000). Ascospores of the sigatoka pathogen *Mycosphaerella fijiensis* are produced on older banana leaves in perithecia sunken into leaf tissues. Under wet conditions, spores are released and are dispersed by air. Rainfall, combined with a high temperature, may lead to peaks of ascospore release (Burt et al., 1999).

The release of large numbers of conidia of the citrus brownspot pathogen *Alternaria alternata* from citrus leaves was triggered by sudden drops in relative humidity or by simulated rainfall events. Vibration induced the release of low numbers of conidia (Timmer et al., 1998). The release of ascospores of *Anisogramma anomala* was monitored over a six-year period in European hazelnut orchards in the United States. The ascospores that were released were correlated with the amount of precipitation (Pinkerton et al., 1998). A suction-impaction mini-spore trap was developed to study the effect of light initiation and decreasing relative humidity (RH) on spore release of the lettuce pathogen *Bremia lactucae* in a controlled environment. Three light periods (from 0400 to 1600, 0600 to 1800, and 0800 to 2000 h, circadian time) at a constant RH of 99 to 100 percent were used for studying the effect of light initiation on spore release (Su et al., 2000). Only a few spores were released during the dark periods. Spore release increased sharply after the initiation of light periods and reached a maximum 1 to 2 h after light initiation. When relative humidity (RH) decreased from 100 to 94 percent two hours before light initiation, spore release increased within 1 h. These results suggest that both light initiation and reduction in RH can trigger spore release of *B. lactucae* (Su et al., 2000).

The release of ascospores of the grapevine powdery mildew pathogen *Uncinula necator* was monitored under natural conditions in France during a five-year period (Jailloux et al., 1999). Ascospore release always began after bud burst and generally ended before blossoming. Ascospore release was always associated with a rainfall higher than 2 mm, a wetting duration greater than 2.5 h, an average temperature generally above 11°C, and a daily mean temperature sum from November 1 to the first ascospore release above 1100°C (Jailloux et al., 1999). The release of ascospores of the sour cherry black knot pathogen *Apiosporina morbosa* was dependent on rainfall and temperature, but not on the duration of wetness (McFadden-Smith et al., 2000). Byrne et al. (2000) showed that fluctuations in relative humidity (either positive or negative) increase the release of conidia of *Oidium* sp. in poinsettia under greenhouse conditions in the United States. Watering resulted in an immediate increase (25 percent) followed by a rapid decrease in RH (32 percent) beginning 1 to 2 h later. Conidial release increased by about 89 percent within 3 h following greenhouse watering (Byrne et al., 2000). Cool, wet weather in the fall and spring provides conditions conducive to ascospore release of *Sclerotinia sclerotiorum,* the cabbage stem rot pathogen (Hudyncia et al., 2000).

ENVIRONMENTAL FACTORS
AFFECTING SPORE DISPERSAL

Wind and rain are the important factors involved in spore dispersal. Urediniospores of *Puccinia triticina* are responsible for epidemics of leaf rust. The spores are wind-borne. Local dispersal of urediniospores within a wheat canopy from an infection focus occurs most efficiently (dispersal to greater distances before becoming deposited) when the wind direction is perpendicular to the wheat rows, rather than when it is parallel to them. The resistance of wheat plants creates a turbulence that tends to lift spores into the airspace just above the canopy and keeps them suspended for a longer period of time. This may allow the spores to travel further before settling back into the canopy and being deposited on the tissues of wheat plants or the soil beneath them (Kramer and Eversmeyer, 2000). Light rain is able to wash airborne *Puccinia triticina* spores from the atmosphere and deposit them on wheat leaves. However, violent or extended rain events inhibit further spore production and removal for more than six hours. Rain can clear out the spore load in the atmosphere within 50 minutes. Susceptible trap plants exposed during violent rain events are much less infected by rain than those sheltered from rain, indicating that rain is able to wash off spores already deposited on leaves by incidental raindrops (Sache et al., 2000). Thus, rainfall patterns may decide the spread of the spores.

The urediniospores of the wheat stem rust pathogen *Puccinia graminis* f. sp. *tritici* are carried long distances by the wind. In the United States, it has been reported that the spores can travel more than 1,000 miles in about two days. Spores from Texas spread throughout the Midwest and into Canada annually (Roelfs, 1985). Rust spores are carried by wind from Australia to New Zealand. In India, urediniospores from South India are carried through cyclonic wind to North India, crossing thousands of miles; the path is called Puccinia path. Seasonal rain in India washes the spores from the atmosphere and deposits them on wheat leaves.

Venturia inaequalis, the apple scab pathogen, survives the winter in diseased apple leaves on the ground. Pseudothecia and ascospores develop in these fallen leaves. Ascospores produced on diseased leaves in the leaf litter are released into the air at ground level from ascus. The great majority of ascospores are propelled less than 1 cm into the air. Once airborne, they are simultaneously transported downwind by the horizontal motion of the wind and diffused vertically by turbulent fluctuations. A portion is also deposited on the vegetation and ground. Spore concentration decreases rapidly with height above the ground and with increasing downwind distance from a source. The rapid decrease of ascospores in the air with increasing distance and height is due to wind shear, turbulent diffusion, and losses due to wash-out by rain. The ascospores are deposited onto plant surfaces either by rain-drops containing ascospores or by impaction and sedimentation of asco-spores that are contained in the air (Aylor, 1998). Rain plays a major role in the long-distance (several km) transport of ascospores. Rain tends to reduce the effect of spore dilution in the vertical direction because raindrops sweep through the entire column of air containing spores and bring the spores to the ground (Aylor, 1998). The effect of rain on ascospore dispersal was further studied in Italy in detail (Rossi et al., 2000). Rain events were the only oc-currences that allowed ascospores to become airborne. A rain event is a period with measurable rainfall of ≥0.2 mm/h lasting one to several hours, uninterrupted or interrupted by a maximum of two dry hours (Rossi et al., 2001). In the absence of rain, dew is insufficient to allow ascospores to disperse into the air. In some cases, rain events did not cause ascospore dispersal, in-cluding when rain fell at night and was followed by heavy dew deposition that persisted some hours after sunrise (Rossi et al., 2001). The intensity of rain may determine the amount of spores dispersed. A rain density of 30 mm/h results in significantly greater rain-splash dispersal of *Colletotrichum acu-tatum* spores infecting strawberry (Ntahimpera et al., 1999).

The pycnidiospores of the glume blotch pathogen *Stagonospora nodor-um (Septoria nodorum)* are dispersed by rain splash from lower infected leaves to successively higher (younger) leaves (De Wolf and Francl, 2000). Vertical and horizontal spread of conidia of *Septoria tritici,* a wheat patho-gen, takes place through the kinetic energy of striking raindrops. Dispersal and infection takes place when total precipitation exceeds 10 mm within three days, or total precipitation exceeds 5 mm over two days, and there ex-ists a continuous three-day-long leaf wetness period (Verreet et al., 2000). The conidia of the banana sigatoka pathogen *Mycosphaerella fijiensis* are dispersed by water droplets to other leaves (Burt et al., 1999).

MEASUREMENT OF SPORE DISPERSAL

The accuracy of a spore dispersal model depends, to a large extent, on the quality of the experimental data on which it is based or with which it is tested. Spore samplers (spore traps) are used for estimating the number of airborne or splash-borne spores. Different types of spore samplers are avail-able. Spore samplers may collect spores on dry, sticky surfaces or in liquids.

They may be *passive,* collecting the spores that reach them, or *volumetric,* sampling a known volume of air. The following are important spore traps.

Dry samplers:
> Passive spore traps—Horizontal slide, vertical cylinder, volumetric
> cylinder
> Volumetric spore traps—Burkard, Cascade impactor, Rotorod, Hirst

Liquid samplers:
> Passive spore traps—Funnel
> Volumetric spore traps—Cyclone, Impinger

In the horizontal slides and volumetric cylinders, spores are collected on a surface coated with an adhesive such as Vaseline, on which spores do not germinate. In Hirst and Burkard spore traps, air is sucked into the traps at a controlled rate and impinged onto a sticky surface moved by a clockwise mechanism past the orifice. Thus, the number of spores per unit volume of air at any given time can be calculated.

The Cascade impactor is similar to the Burkard trap in its function. The Rotorod sampler consists of a U-shaped rod attached at its midpoint to the shaft of a small, battery-operated electric motor. The surface of the rod is covered with a Vaselined strip of transparent cellophane, which can be stripped off and mounted. From the area of the strip and the speed of rotation, the volume of air sampled can be calculated. In Cyclone and Impinger spore traps, spores are impacted by air suction into the samplers.

Living plants are also used as spore traps. The number of pustules developed on plants exposed for a given time (usually 24 h) and then returned to a clean air chamber for incubation are counted to assess the spore load in the air. A mobile nursery containing seedlings of the differential hosts for the pathogen is exposed in the field for a suitable time, and the plants are removed to optimum concentrations for incubation in a spore-proof glasshouse. The physiologic races to which any pustules belong are then recorded.

The choice of spore sampler depends on biological factors such as the mode of spore spread, spore size, and spore concentration in the air. When sampling above crops, it is necessary to use an inertial sampler to collect dry airborne spores. Vertical cylinders, Burkard traps, and Hirst traps collect many more spores per square centimeter than horizontal slides. Because impaction efficiencies of small spores are small, a volumetric sampler (Burkard, Hirst, Cascade impactor, Rotorod) will sample spores better than a passive vertical cylinder, particularly at low concentrations and low wind speeds.

Venturia inaequalis ascospores (measuring 14 × 7 μm) can be collected more easily using Burkard and Rotorod samplers than using sticky cylinders. However, simple cylinders are appropriate samplers for spores such as those of *Erysiphe graminis,* which are large (30 ×15 μm), present in high concentrations, and dispersed at high wind speeds. Volumetric suction samplers may underestimate concentrations of the largest pathogen spores, such as those of *Helminthosporium* and *Alternaria,* and the cylinders may be better samplers for them.

In choosing samplers for spores carried in splash droplets, the size of the droplets is more important than the size of the spores themselves. The range of particle sizes to be sampled is greater than those of dry airborne spores. Spore-carrying droplets may be 10 to 1,000 μm in diameter, and different samplers may be required for the efficient sampling of different sizes of droplets. Spores of many splash-borne pathogens are mostly carried in large ballistic droplets. Samplers appropriate for dry airborne spores are generally unsuitable for collecting large spore-carrying droplets because the spores can be washed off vertical sampling surfaces. Horizontal slides (under rain shields to prevent wash off), funnels (draining into beakers), and impingers collect more such spores. Slides probably lose some spores through runoff. Funnels collect most spores, but the liquid samples require concentration before spores can be counted. The impinger collects more spores per square centimeter than the funnel, but requires power to operate its suction pump. Small, airborne spore-carrying droplets, which are generally present in low concentrations, would not be collected efficiently by samplers with horizontal sampling surfaces near ground level. Volumetric suction samplers, such as impingers or high volume cyclone separators, are more appropriate.

The positioning of spore samplers (traps) in the field depends on the position of the source of spores, the mode of spore dispersal (especially whether spores are airborne or in ballistic splash droplets), and the type of samplers used. The number of spores of *Erysiphe graminis* collected by suction samplers, which measure the airborne spore concentration, decreased rapidly with height above a barley crop (Bainbridge and Stedman, 1979). Similarly, the numbers of spores collected on horizontal slides (mostly by impaction) increased. This increase was due to the increase in wind speed with height, and these differences between samplers illustrate that it is important to understand the sampling characteristics of a device before interpreting the results it gives (Bainbridge and Stedman, 1979). All three sampler types, i.e., suction samplers, vertical cylinders, and horizontal slides, collected fewer spores when positioned further upwind or downwind of the barley crop. Such decreases in the number of spores collected with a distance, or spore dispersal gradients, are extremely important in disease spread and in epidemiological models. The upwind gradient was steeper than the downwind gradient, which is usually the case for airborne spores. Splash-dispersed spores usually show a steep gradient, both with height above and with distance away from a source (Bainbridge and Stedman, 1979).

Therefore, for measurement of background airborne spore concentrations, samplers should be placed above and away from active sources of spores. For measurement of spore dispersal gradients, they should be close to the source, particularly when splash-dispersed spores are being studied. With ground-level sources, samplers must be near the ground. When spore dispersal within crops is of interest, they should be positioned so that there is minimum disturbance to the crop, although there must be some gap around the sampler so that infected leaves do not rub against it.

The decreases in the concentrations of spores with distance from and height above a strip of infected plants suggest the existence of spore dispersal gradients. The decreases in spore concentration with distance away from a source result in spore deposition gradients as portions of the spore clouds are deposited on the ground or crops. These gradients are important in determining the rate of spatial spread of epidemics.

DISEASE GRADIENTS

Although disease progress curves represent disease development in time, disease gradients represent disease development in space. The nature of a disease gradient in any one disease depends on the dispersal method adopted by the pathogen and on the nature and stage of development of the host. A disease gradient will occur when the primary source of infection is a focus. This may consist of what is virtually a point source, such as a single lesion, of an area, a patch of disease within a crop, or of virtually a line, such as a hedgerow. When the source is uniformly distributed over a crop, as is the case with a cloud of spores blown by the wind from a large, distant source, no gradient will occur.

Various equations have been used to describe disease gradients (Mundt, 1989). The following two formulae are widely used.

$$\text{The inverse power law: } y = as - b$$
$$\text{The negative exponential law: } y = ae - bs$$

For these formulae, y is the amount of disease at a distance (s) from the source, a is a function of the total disease amount, b is the slope of the line, which is obtained when the data are transformed (using a log – log or log –

linear transformation) to obtain an approximation to a straight line, and *e* is the basis of logarithms.

Gradients for some dry airborne spores fit the power law model better than the exponential model, while data for others fit an exponential model better. Such gradients are always steep. When multiple foci occur, the analysis of disease spread is complex (Mundt, 1989).

ENVIRONMENTAL FACTORS AFFECTING SPREAD OF BACTERIAL PATHOGENS

Xanthomonas axonopodis pv. *citri,* the citrus canker bacterial pathogen, appears in rainwater running over the surfaces of lesions and splashing onto new shoots. The pathogen enters the host tissues through stomata or wounds. Storms such as typhoons or hurricanes encourage outbreaks of citrus canker. Rainstorms favor distant dispersal of the bacteria. The pathogen can be dispersed in small raindrops also. Disease spread is independent of wind direction because rain-splash dispersal and rapid development of secondary foci of disease prevail. In contrast, windblown rain is responsible for the highly directional spread from a disease focus in the orchard. Subsequent spread from secondary loci is rapid and in many directions (Graham and Gottwald, 1991). Young citrus canker lesions rapidly exude up to 10^5-10^6 colony forming units per lesion within hours after wetting. Wind-driven rain and water-soaking of tissue appear to be essential for dissemination and ingress of bacteria and for epidemic development of citrus canker.

In the case of rice bacterial blight caused by *Xanthomonas oryzae* pv. *oryzae,* bacterial cells ooze from fresh lesions whenever dew or rain occurs. The drops dry to become minute beads that easily fall into water in the field. The bacteria are carried by irrigation water from field to field. Splash dispersal of the pathogen has been demonstrated (Mundt et al., 1999).

ENVIRONMENTAL FACTORS AFFECTING INFECTION AND DISEASE DEVELOPMENT

The severity of citrus Alternaria leaf and fruit spot disease is positively related to the amount of rainfall, duration of leaf wetness, and average daily temperatures (Timmer et al., 1998). Moisture appears to be a critical element in favoring *Tilletia indica* infection of wheat kernels. Teliospore germination requires at least 82 percent relative humidity and, preferably, free water. Several rainy days during the flowering stage favor disease incidence. The longer the period of high relative humidity and rainy weather, the greater the number of seeds infected. Moisture at the time of flowering is the most critical factor in establishment of the disease (Bonde et al., 1997).

Wetness duration, temperature, and light conditions determine the epidemic of wheat stripe rust *(Puccinia striiformis)* disease. Rain events also determine disease incidence. Light and short rainfall enhances disease incidence, whereas heavy violent and extended rain events decrease disease incidence (Vallavieille-Pope et al., 2000). Under high temperature and prolonged leaf wetness periods, infection by *Mycosphaerella fragariae* becomes severe in strawberry. The number of lesions on young leaves increases gradually from 5 to 25°C and decreases sharply from 25 to 30°C. The optimal temperature for infection is 25°C. But for most temperatures, a minimum of 12 h of leaf wetness is necessary for infection (Carisse et al., 2000).

Relative humidity and rainfall appear to contribute to the development of wheat leaf and glume blotch disease epidemics (De Wolf and Francl, 2000). Leaf wetness periods of more than 48 h and relative humidity greater than 98 percent result in epidemics of the disease caused by *Septoria tritici* in wheat (Verreet et al., 2000). The optimal temperature of pycnidiospore germination is in the range of 22 to 24°C. Infection is high at a temperature between 16 and 21°C. Around 7°C, infection is inhibited (Verreet et al., 2000). The optimal temperature for development of potato common scab disease *(Streptomyces scabies)* is about 20°C. Low soil moisture increases disease incidence. Soil characteristics can greatly affect the severity of potato scab. Scab is most severe in soils with a pH of 5.2 to 7.0 (Loria et al., 1997). Ascospores of *Monosporascus cannonballus* germinate in cucurbit rhizosphere. Optimum germination occurs at temperatures ranging from 25 to 35°C. The ascospore population within 500 μm of a root is capable of germination and subsequent penetration of cantaloupe roots. The fungus multiplies in the root tissues and brownish lesions appear (Stanghellini et al., 2000). Rainfall greater than 10 mm and a maximum temperature greater than 5°C during this favorable period can result in peach leaf curl *(Taphrina deformans)* disease incidence (Giosue et al., 2000). Daytime vapor pressure deficit and nighttime temperature have the greatest effect in strawberry flower infection by *Botrytis cinerea*. Infection is favored by low day vapor pressure deficit and high night temperature (Xu et al., 2000).

The optimum temperature for germination of conidia of *Cercospora arachidicola,* the peanut early leaf spot pathogen, is 19 to 25°C, and the conidia require free water for germination. Periods of leaf wetness exceeding 10 h and minimum temperatures during the wetness period above 21°C are the most crucial conditions for early leaf spot disease development (Jacobi et al., 1995). High relative humidity favors development of pear leaf scab caused by *Venturia pirina*. The length of the dry period reduces disease severity. No scab lesions occur if leaves are dry for more than 12 h (Villata et al., 2000b). Disease severity increases with increasing leaf wetness duration. The optimum temperature for infection is 20°C (Villata et al., 2000a). When the winter is cold and rainy, the onset of *V. pirina* epidemics occurs earlier in the season, and disease intensity reaches the highest levels. In warm and less rainy winters, the onset of epidemics is delayed and the epidemic is less extensive (Sobreiro and Mexia, 2000).

REFERENCES

Aylor, D. E. (1998). The aerobiology of apple scab. *Plant Dis,* 82:832-849.

Bainbridge, A. and Stedman, O. J. (1979). Dispersal of *Erysiphe graminis* and *Lycopodium clavatum* spores near to the source in a barley crop. *Ann Appl Biol,* 91:187-198.

Bandyopadhyay, R., Frederickson, D. E., McLaren, N. W., Odvody, G. N., and Ryley, M. J. (1998). Ergot: A new threat to sorghum in the Americas and Australia. *Plant Dis,* 52:356-367.

Bonde, M. R., Peterson, G. L., Schaad, N. W., and Smilanick, J. L. (1997). Karnal bunt of wheat. *Plant Dis,* 81:1370-1377.

Burt, P. J. A., Rosenberg, L. J., Rutter, J., Ramirez, F., and Gonzales, O. H. (1999). Forecasting the airborne spread of *Mycosphaerella fijiensis,* a cause of black Sigatoka disease of banana: Estimations of numbers of perithecia and ascospores. *Ann Appl Biol,* 135:369-377.

Byrne, J. M., Hausbeck, M. K., and Shaw, B. D. (2000). Factors affecting concentrations of airborne conidia of *Oidium* sp. among poinsettias in a greenhouse. *Plant Dis,* 84:1089-1095.

Carisse, O., Bourgeois, G., and Duthie, J. A. (2000). Influence of temperature and leaf wetness duration on infection of strawberry leaves by *Mycosphaerella fragariae. Phytopathology,* 90:1120-1125.

De Wolf, E. D. and Francl, L. J. (2000). Neural network classification of tan spot and Stagonospora blotch infection periods in a wheat field environment. *Phytopathology,* 90:108-113.

Fitt, B. D. L., Goulds, A., and Polley, R. W. (1988). Eyespot *(Pseudocercosporella herpotrichoides)* epidemiology in relation to prediction of disease severity and yield loss in winter wheat—a review. *Plant Pathol,* 37:311-328.

Fitt, B. D. L. and McCartney, H. A. (1986). Spore dispersal in relation to epidemic models. In K. J. Leonard and W. E. Fry (Eds.), *Plant Disease Epidemiology.* Vol. 1, *Population Dynamics and Management.* MacMillan Publishing Company, New York, pp. 311-345.

Giosue, S., Spada, G., Rossi, V., Carli, G., and Ponti, I. (2000). Forecasting infections of the leaf curl disease on peaches caused by *Taphrina deformans. Eur J Plant Pathol,* 106:5634-5671.

Graham, J. H. and Gottwald, T. R. (1991). Research perspectives on eradication of citrus bacterial diseases in Florida. *Plant Dis,* 75:1193-1200.

Hudyncia, J., Shew, H. D., Cody, B. R., and Cubeta, M. A. (2000). Evaluation of wounds as a factor to infection of cabbage by ascospores of *Sclerotinia sclerotiorum*. *Plant Dis*, 84:315-320.

Jacobi, J. C., Backman, P. A., Davis, D. P., and Brannen, P. M. (1995). AU-Pnuts Advisory I: Development of a rule-based system for scheduling peanut leaf spot fungicide applications. *Plant Dis*, 79:666-671.

Jailloux, F., Willocquet, L., Chapuis, L., and Froidefond, G. (1999). Effect of weather factors on the release of ascospores of *Uncinula necator*, the cause of grape powdery mildew, in the Bourdeaux region. *Can J Bot*, 77:1044-1051.

Johnson, E. M. and Sutton, T. B. (2000). Response of two fungi in the apple sooty blotch complex to temperature and relative humidity. *Phytopathology*, 90:362-367.

Kramer, C. L. and Eversmeyer, M. G. (2000). Dispersal of leaf rust urediniospores within a wheat canopy. *Acta Phytopathol et Entomol Hungarica*, 35:343-348.

Loria, R., Bukhalid, R. A., Fry, B. A., and King, R. R. (1997). Plant pathogenicity in the genus *Streptomyces*. *Plant Dis*, 81:836-846.

McFadden-Smith, W., Northover, J., and Sears, W. (2000). Dynamics of ascospore release by *Apiosporina morbosa* from sour cherry black knots. *Plant Dis*, 84:45-48.

Mundt, C. C. (1989). Modeling disease increase in host mixtures. In K. J. Leonard and W. E. Fry (Eds.), *Plant Disease Epidemiology, Vol.2, Genetics, Resistance, and Management*. McGraw-Hill, New York, pp. 150-181.

Mundt, C. C., Ahmed, H. U., Finckh, M. R., Niera, L. P., and Alfonso, R. F. (1999). Primary disease gradients of bacterial blight of rice. *Phytopathology*, 89:64-67.

Ntahimpera, N., Wilson, L. L., Ellis, M. A., and Madden, L. V. (1999). Comparison of rain effects on splash dispersal of three *Colletotrichum* species infecting strawberry. *Phytopathology*, 89:555-563.

Pinkerton, J. N., Johnson, K. B., Stone, J. K., Ivors, K. L. (1998). Maturation and seasonal discharge pattern of ascospores of *Anisogramma anomala*. *Phytopathology*, 88:1165-1173.

Roelfs, A. P. (1985). Epidemiology in North America. In W. R. Bushnell and A. P. Roelfs (Eds.), *The Cereal Rusts, Vol. 2*. Academic Press, New York, pp. 403-434.

Rossi, V., Ponti, I., Marinelli, M., Giosue, S., and Bugiani, R. (2001). Environmental factors influencing the dispersal of *Venturia inaequalis* ascospores in the orchard air. *J Phytopathol*, 149:11-19.

Roy, K. W. (1997). Sporulation of *Fusarium solani* f. sp. *glycines*, the causal agent of sudden death syndrome, on soybean plants symptomatic for the disease in the midwestern and southern United States. *Plant Dis*, 566-569.

Sache, I., Suffert, F., and Huber, L. (2000). A field evaluation of the effect of rain on wheat rust epidemics. *Acta Phytopathol et Entomol Hungarica*, 35:273-277.

Sobreiro, J. and Mexia, A. (2000). The simulation of pear scab *(Venturia pirina)* infection periods and epidemics under field conditions. *Acta Hortic*, 525:153-160.

Sosa-Alvarez, M., Madden, L.V., and Ellis, M. A. (1995). Effects of temperature and wetness duration on sporulation of *Botrytis cinerea* on strawberry leaf residues. *Plant Dis*, 79:609-615.

Stanghellini, M. E., Kim, D. H., and Waugh, M. (2000). Microbe-mediated germination of ascospores of *Monosporascus cannonballus. Phytopathology,* 90:243-247.

Su, H., van Bruggen, A. H. C., and Subbarao, K. V. (2000). Spore release of *Bremia lactucae* on lettuce is affected by timing of light initiation and decrease in relative humidity. *Phytopathology,* 90:67-71.

Sun, P., and Yang, X. B. (2000). Light, temperature, and moisture effects on apothecium production of *Sclerotinia sclerotiorum. Plant Dis,* 84:1287-1293.

Timmer, L. W., Roberts, P. D., Darhower, H. M., Bushong, P. M., Stover, E. W., Peever, T. L., and Ibanez, A. M. (2000). Epidemiology and control of citrus greasy spot in different citrus growing areas in Florida. *Plant Dis,* 84:1294-1298.

Timmer, L. W., Solel, Z., Gottwald, T. R., Ibanez, A., and Zitko, S. E. (1998). Environmental factors affecting production, release, and field populations of conidia of *Alternaria alternata,* the cause of brown spot of citrus. *Phytopathology,* 88:1218-1223.

Timmer, L. W., Zitko, S. E., Gottwald, T. R., and Graham, J. H. (2000). *Phytophthora* brown rot of citrus: Temperature and moisture effects on infection, sporangium production, and dispersal. *Plant Dis,* 84:157-153.

Vallavieille-Pope, C. De., Huber, L., Leconte, M., and Bethenod, O. (2000). Using controlled and natural conditions to assess infection efficiency of *Puccinia striiformis* and *P. triticina* on wheat. *Acta Phytopathol et Entomol Hungarica,* 35:295-298.

Verreet, J. A., Klink, H., and Hoffmann, G. M. (2000). Regional monitoring for disease prediction and optimization of plant protection measures: The IPM wheat model. *Plant Dis,* 84:815-826.

Villata, O., Washington, W. S., Rimmington, G. M., and Taylor, P. A. (2000a). Effects of temperature and leaf wetness duration of pear leaves by *Venturia pirina. Australian J Agric Res,* 51:97-106.

Villata, O., Washington, W. S., Rimmington, G. M., and Taylor, P. A. (2000b). Influence of spore dose and interrupted wet periods on the development of pear scab caused by *Venturia pirina* on pear *(Pyrus communis)* seedlings. *Australasian Plant Pathology,* 29:255-262.

Xu, X., Harris, D. C., and Berrie, A. M. (2000). Modeling infection of strawberry flowers by *Botrytis cinerea* using field data. *Phytopathology,* 90:1367-1374.

16

Forecasting Models

Several disease-forecasting models were created to help develop decision support systems for timing chemical applications to control fungal and bacterial diseases. The important forecasting models developed are described here.

A model is any representation of an object, system, or idea in some form other than that of the entity itself. Models are by necessity limited sections of reality, having some but not all of the properties of real-life situations. Models are seldom complete, and modeling is an iterative (repetitive) process in which the model provides a closer and closer approximation of reality with each successive iteration. Thus, a model is never an end point in itself (Teng, 1985).

TYPES OF MODELS

Analytic Models

Different types of epidemic models exist, including analytic, simulation, empirical, and hybrid models. Analytic models are used to analyze epidemics on a theoretical basis without taking into consideration the effects of external variables. They are mathematical models that attempt to describe epidemics as a whole by single mathematical equations. They are also called statistical models or biological models. Statistical models are mathematical formulas with parameter values that have been chosen to adequately describe disease progress for specific data sets, but lack a precise biological interpretation. Weibull, Richards, Gompertz, and Gaussian models are also mathematical models. Biological models are also mathematical models but are based on prior assumptions about the mechanism of disease increase. Monomolecular and logistic models are biological models. These models are mostly theoretical or simple regression models. They are also called "synthetic models" by Van der Plank (1984).

Simulation Models

Simulation is the process of designing a model of a real system and conducting experiments with this model for purposes of either understanding the behavior of the system or evaluating various strategies for the operation of the system (Teng, 1985). Simulation models try to mimic observed epidemics or certain aspects, such as spore dispersal, that are influenced by measured meteorological conditions (Hau, 1990). Simulation models are generated by creating conditions for the development of epidemics and carrying out experiments on different aspects of disease increase. In the simulation approach, differential equations are composed by considering the life cycle of fungal pathogens, by compartmentalizing the various components of the life cycle (e.g., sporophores, spores, dispersed spores, lesions), and by specifying the rates of change of each variable (Leonard and Fry, 1986). For example, a simulation model to analyze the epidemiology of potato late blight was developed (Bruhn and Fry, 1981). Both the number of lesions and the mean surface area of lesions in each age class were simulated in response to the environment, host factors including cultivar resistance, and fungicide application. The simulation model included the deposition and weathering rate of the fungicide chlorothalonil, the fungicide application method and dosage, interactions with plant growth and architecture, and environmental effects such as rainfall. Simulation models are concerned only with particular, or a set of particular, behaviors and hence may not be suited for the development of theories concerning the real system.

Empirical and Hybrid Models

These models are based on actual experiments and observations and not on theory or simulation. Empirical models are complex models that attempt to mimic many environmental factors and other influences.

Hybrid models that combine analytic and simulation approaches are also being developed.

SPECIFIC MODELS

Potato Late Blight Forecasting Models

Several models were developed to forecast late blight of potato *(Phytophthora infestans)*. The earliest forecasting model was developed by van Everdingen (1926). The model is based on four criteria known as "Dutch rules." The Dutch rules prescribe the following conditions to indicate the time to apply control measures against the disease: the occurrence of dew for at least four hours at night, a minimum temperature of 10°C, a mean cloudiness on the next day of 0.8 or more, and at least 0.1 mm of rain in the following 24 h. The Dutch rules were initially used in the United Kingdom. Beaumont (1947) modified these rules to state that when there are two days with a minimum temperature not less than 10°C and relative humidity not below 75 percent, the growers in the vicinity should begin routine spray programs. These two days were recognized as the Beaumont period, and it became the standard blight warning system in the United Kingdom until the mid 1970s.

"Irish rules," formulated by Bourke (1953), prescribe the following criteria to be satisfied for an outbreak of late blight: a humid period of at least 12 h with the temperature at least 10°C and relative humidity (RH) 90 percent or above (conditions favorable for sporangia formation), and free moisture on the leaves for a subsequent period of at least 4 h (conditions favorable for germination and reinfection). If there is no rainfall, the alternative requirement is for a further 4 h beyond the initial 12 h with an RH of at least 90 percent.

Smith (1956) came out with another model describing the critical two days to forecast the late blight epidemic. Temperature and humidity requirements of disease development have been deduced after several years of studies, and have been incorporated into Smith periods. A *Smith period* is defined as two consecutive days (ending at 0900 hours) when the temperature has not been less than 10°C and the relative humidity has been above 90 percent for at least 11 hours of each day. In the United Kingdom, Smith periods are used to forecast disease incidence. Notifications of Smith periods are issued by the Agriculture Development Advisory Service (ADAS) of the British government and communicated through television, bulletins, and farming press.

In the United States, a computerized forecast called BLITECAST was developed (Krause et al., 1975). It is based on temperature and relative humidity thresholds. Data required for BLITECAST include daily maximum and minimum temperatures, the number of hours when relative humidity is equal to or above 90 percent, the maximum and minimum temperature during the period when RH is 90 percent and above, and the daily rainfall figure to the nearest 1 mm. The program is available on microcomputer, which is coupled to a weather data logger. The system is available in the potato field itself. BLITECAST advises not only the date to begin spraying, but also the timing of subsequent applications.

In Denmark, Hansen et al. (1995) developed a forecasting model called NEGFRY. The model takes into account cultivar susceptibility, emergence date, and irrigation to advise on the date of first fungicide application and timing of subsequent sprays. The BLITECAST system was modified by incorporating host resistance and fungicide weathering, and this system was called SIM-CAST (Fry et al., 1983). SIM-CAST uses a decision rule similar to that of BLITECAST, i.e., accumulating disease severity values, but does not forecast the occurrence of the first spray (Grunwald et al., 2000). The SIM-CAST system was equally effective when compared to BLITECAST (Hijmans et al., 2000). Similar systems have been reported in several European countries. The BLITECAST system can also be used in tomato to forecast late blight in this crop.

Tomato Early Blight Forecasting Models

A forecaster of *Alternaria solani* on tomato (FAST) was developed to identify periods when environmental conditions are favorable for early blight development and provide an efficient fungicide application schedule (Madden et al., 1978). Daily weather data (maximum and minimum air temperature, hours of leaf wetness, maximum and minimum temperature during wet periods, hours of relative humidity more than 90 percent, and daily rainfall) are used for forecasting. FAST consists of two submodels: the dew model and the rain model. Each submodel calculates a daily rating of the severity of risk of an early blight outbreak. The dew model is based on the duration of wet periods and average air temperature during the wet periods. The hours of leaf wetness and mean air temperature during the wet period are combined to derive disease severity (S) values. The S values are rated daily from zero (conditions unfavorable for *A. solani* spore formation) to four (conditions highly favorable for spore formation). The rain model uses average air temperature for the last five days, total hours during the past five days with RH >90 percent, and total rainfall for the past seven days to calculate disease severity rating values from zero (conditions unfavorable for *A. solani* spore formation and infection) to three (conditions highly favorable for spore formation and infection) (Gleason et al., 1995). The forecasting program analyzes daily environmental data and maintains a record of the (1) total of all S values (TS) since the beginning of the growing season, (2) seven-day cumulative severity value (CS) calculated by totaling S values for the past seven days, and (3) five-day cumulative rating value (CR) calculated by totaling R values for the past five days. The first early blight spray application is recommended when TS reaches a "critical level" of 35 and the plants are in the field for at least five weeks. Subsequent fungicide applications are scheduled when CS or CR equal or exceed prespecified critical limits (Madden et al., 1978). The program was computerized for rapid and accurate analyses. The FAST system has resulted in lower fungicide application than is used in commercial fields, without sacrificing disease control and crop yield (Gleason et al., 1995). The FAST system has been modified during the past two decades because of feedback from the field and fine-tuned. A modified FAST program called TOM-CAST (Tomato Forecaster) has been developed (Gleason et al., 1995). In this system, the disease severity values (DSVs) are calculated solely on the basis of the dew model of FAST. Another departure from the FAST model is to sum DSVs, after the first fungicide spray, from the date of the last spray rather than over a five- or seven-day time window as in FAST (Gleason et al., 1995). TOMCAST can be used, not only for early blight prediction, but also to advise growers when to apply fungicides to control anthracnose and Septoria leaf spot. Use of the forecasting model TOM-CAST results in lower fungicide application and results in high quality fruit (Hong et al., 2000).

Apple Fire Blight Forecasting Models

An American predictive model, MARYBLYT, predicts the occurrence of fire blight. The predictive criteria used for blossom blight are: (1) accumulation of at least 110 degree hours $\geq 18.3^{\circ}C$ starting at full pink in apples; (2) an average daily temperature of $\geq 16.6^{\circ}C$; (3) a wetting event with 0.25 mm rain on the day before or wetness of >6 hours; and (4) flowers must be open with petals intact. All four conditions need to be fulfilled on the same day for an infection event to occur. These predictions match observed fire blight incidences (Gouk et al., 1999). This model has been used for the past 18 years in the United States and it accurately predicted the occurrence of blossom blight in orchards.

Another predictive program, FireWork, implements MARYBLYT rules using software written in Delphi 4 for Windows 95 and Windows NT. The program is driven by weather data stored in the Orchard 2000 weather database. This program is accessible to farmers for orchard management decisions about antibiotic spray timing (Gouk et al., 1999). Several other fire blight prediction models are available. Three fire blight predictive models, MARYBLYT, Cougarblight (Smith's model), and the Mean Temperature Line model, were compared to predict fire blight incidence in California. The models accurately predicted low fire blight potential during the susceptible bloom period (Holtz et al., 1999). Rain events during this period were associated with cold temperature, which reduced disease potential. Warm rain (>16° C) favors disease incidence (Holtz et al., 1999).

An automated weather and fire blight model system was developed and is currently being used by farmers in Pennsylvania. Simulated weather data available for past, present, and future weather events are interfaced with disease models. The weather information is provided as both "observed" and "forecast" for specific farm locations. In addition, the growers receive daily fire blight alerts, which include a record of the fire blight infection risk for the previous days in the month and a seven-day forecast of the fire blight risk. The growers receive simulated weather reports and daily disease risk alerts for fire blight via e-mail or fax. It was reported that the growers found the simulated weather and fire blight alerts useful (Travis et al., 1999).

Billing (1980) developed a system in Europe to predict when outbreaks of fire blight would likely be found in the field. The method is based on experimental studies of the growth rate of *E. amylovora* in vitro and historical fire blight outbreaks in southeast England. This model determines the potential doubling of bacteria (PD) from daily maximum and minimum temperatures. The PD values are used to determine the duration of an incubation period. The incubation period is completed and infection should be apparent in the field when the PD value accumulated from an infection day exceeds a certain threshold determined by temperatures and rainfall (Billing, 1980). This system was further modified and Billing's Revised System (BRS) and Billing's Integrated System (BIS95) were developed (Billing, 1999). BIS95 aims to assess risks during the growing season, including low and high incidence risks. For simplicity, degree day sums are used in this BIS95 system. Following a warm period, wetting of flowers by dew is sufficient for infection. When temperatures are high ($\geq 27°C$), blossom infection may occur without wetting. Degree day sums above a temperature of $13.0°$ have provided good guidance on times when, following infection, early signs of disease may be seen. The critical degree day sum for apple blossom blight is suggested to be ≥ 47 (Billing, 1999). BIS95 provides rules based on both weather data and field infection data. The efficacy of both MARYBLYT and BIS95 systems were compared for their relative efficacy in prediction of fire blight in Hungary and Turkey. Both the MARYBLYT and BIS95 systems were equally useful in providing guidance for obtaining daily results about the progress of different infection events, including symptom development, and for making decisions on control. The Cougarblight Fire Blight Risk Assessment Model was developed in the Northwest United States to address the daily fire blight infection risk in orchards (Smith, 1999). Host susceptibility, relative pathogen presence, and potential growth rate of bacterial colonies on stigma were all addressed prior to potential flower wetting. This model is successfully used by growers in the Pacific Northwest.

Citrus Postbloom Fruit Drop and Anthracnose Forecasting Model

The citrus postbloom fruit drop and anthracnose *(Glomerella cingulata)* forecasting system was developed in Florida (Timmer and Zitko, 1993). The most important parameters in disease prediction were the number of affected blossoms already on the tree (inoculum availability) and rainfall during the previous five days. Timmer and Zitko (1993) developed the following equation to predict the disease 3 to 4 days in advance:

$$y = -7.5 + 1.28 \, (TD)^{1/2} + 0.44 \, (R \times 100)^{1/2}$$

where y = predicted percentage of flowers affected, TD = total number of flowers affected on 20 trees, and R = rainfall total (mm) for the last five days. Sprays should be applied when more than 20 percent disease is predicted. Model-based decisions on fungicide applications result in reduced disease, large increases in fruit production, and elimination of unnecessary sprays (Timmer and Zitko, 1996).

Grapevine Downy Mildew Forecasting Models

Several forecasting models have been developed to predict downy mildew *(Plasmopara viticola)* epidemic in grapevine. The prediction systems were based on a detailed simulation of the disease cycle (Park et al., 1997), simplified models of one or more components of the disease cycle that can be used to predict a relative risk of disease development (Ellis et al., 1994), or empirical models based on long-term climatic conditions, with little direct appraisal of the disease cycle (Vercesi et al., 1994). Some of these prediction systems (such as EnviroCaster) are sold commercially as electronic predictive systems (Ellis et al., 1994). An electronic warning system based on models for the infection of leaves of the American grape, *Vitis lambrasca,* production of sporangia by *P. viticola* in lesions, and sporangial survival has been found to be useful to plan strategies for control of the disease (Madden et al., 2000).

Wheat Tan Spot Forecasting Model

De Wolf and Francl (2000) developed a model to predict wheat tan spot *(Pyrenophora tritici-repentis)* epidemic. A leaf wetness period of more than ten hours and precipitation of more than 3.4 mm are highly correlated with the tan spot epidemic. Leaf wetness duration and precipitation predicted validation cases of tan spot with about 72 percent accuracy in a sensitive analysis of single independent variable disease forecasting models (De Wolf and Francl, 2000). When the leaf wetness duration variable was withheld from the disease-forecasting model, prediction accuracy on validation cases decreased by 20 percent compared to the full model.

Wheat Septoria Forecasting Model

A Septoria forecasting model (Septoria-Timer) was developed (Verreet et al., 2000). The Septoria-Timer is placed in the crop at the beginning of the vegetation period. The equipment consists of a sensor for the detection of leaf surface wetness and a datalogger. From the subsequent information, the starting point for *Septoria tritici* infections is calculated on the basis of a leaf wetness period of more than 98 percent. The result is then shown via an easily read display. After a Septoria warning is given, the initial outbreak has to be diagnosed and quantified. If both criteria for infection are fulfilled (warning given by the Septoria-Timer and an initial infection >50 percent disease incidence), a fungicide application is recommended (Verreet et al., 2000).

Peanut Early and Late Leaf Spot Forecasting Models

A peanut early leaf spot disease *(Cercospora arachidicola)* forecasting system (Early Leaf Spot Advisory) was developed in Georgia in 1966 using the duration of relative humidity ≥95 percent and the minimum temperature during the high humidity period (Jensen and Boyle, 1966). This system was later computerized and designated as 81-ADV. This system has been used successfully in Virginia. A second advisory system (89-ADV) was developed in Virginia and has replaced the 81 ADV. A survey showed that 94 percent of peanut growers in Virginia used the 89-ADV system (Phipps, 1993). A predictive system using an EnviroCaster weather monitoring station and microprocessor was developed for late leaf spot *(Phaeoisariopsis personata).* This model is based on the accumulation of hours of leaf wetness. Another model, AU-Pnuts advisory was developed by Auburn University in Alabama in 1995 (Jacobi et al., 1995). The model was developed to schedule initial and subsequent fungicide applications for control of both early and late leaf spots of peanut. The advisory uses the number of days with precipitation greater than 2.5 mm and National Weather Service precipitation probabilities to predict periods favorable for development of early and late leaf spot.

Strawberry Gray Mold Forecasting Models

Three forecasting systems for strawberry gray mold *(Botrytis cinerea)* have been developed (Xu et al., 2000). The models were developed to relate fungal inoculum and weather conditions to the incidence of flower infection; by inoculum only, by weather variables only, and by both inoculum and weather variables. Models using weather variables only gave more accurate predictions than models using inoculum only. Models using both weather variables and inoculum gave the best predictions, but the improvement over the models based on weather variables only was small. Hence, models based on weather variables (low day vapor pressure deficit and high night temperature) will be highly useful in prediction of the disease (Xu et al., 2000).

Pear Brown Spot Forecasting Model

A forecasting model (BSPcast) was developed for the prediction of pear brown spot *(Stemphylium vesicarium)* (Llorente et al., 2000). This forecasting system is based on an empirical model. In this model, daily wetness duration *(W)* and mean air temperature during wetness periods *(T)* are used to compute a daily disease severity *(S)* according to the following equation:

$$\text{Log}_{10}(S) = -1.70962 + 0.0289T + 0.04943W + 0.00868TW - 0.002362W^2 - 0.0002238T^2W$$

Since the maximum daily disease severity predicted by the equation is 3.7942, relative daily infection risk *(R)* is calculated as:

$$R = S/3.7942$$

Therefore, *R* ranged from 0 to 1. The three-day cumulative daily infection risk (CR) was computed by totaling *R* values for the past three days and was used as an action threshold for spraying fungicides.

Mango Anthracnose Forecasting Models

Two predictive models were developed to predict mango anthracnose *(Colletotrichum gloeosporioides)*. Fitzell et al. (1984) studied the requirements of temperature and wetness duration for the production of dark appressoria from conidia applied to detached young mango leaves under laboratory conditions. They obtained the following equation:

$$Y = \ln P/q = -16.114 + 1.120T - 0.0225T^2 + 1.0862 \ln t$$

Where P = percentage of conidia forming dark appressoria, T = temperature (C), t = wet period (h), and $q = 100 - P$. The symbol ln indicates natural logarithm. This system was built into a microprocessor-based forecasting system (Fitzell et al., 1984). Infection levels were estimated from temperature and wetness duration following at least 0.2 mm of rain. If the model predicted that 10, 20, or 40 percent of conidia would form dark appressoria, it would indicate that low, medium, or high levels of anthracnose infection, respectively, would occur (Fitzell et al., 1984). A similar system was developed by Dodd et al. (1991). This system differs from the previous model, in that it was developed using combined data from fruit and leaf inoculations and includes relative humidity in addition to wetness and temperature (Arauz, 2000).

Corn Stewart's Wilt Forecasting Model

A predictive system to forecast corn Stewart's disease (bacterial wilt) was developed in the United States. Epidemic development of the disease is correlated with winter temperatures. Epidemics develop following relatively mild temperatures. A winter temperature index is calculated by adding the mean temperatures (expressed as °F) for December, January, and February. When the winter temperature index is less than 90, it is predicted that there will not be any disease incidence and there is no need for application of insecticides. If the temperature index is 90 to 100, the disease severity may be intermediate, and if the temperature index is more than 100, the disease severity will be severe. This predictive system works well in various areas of the United States (Pataky et al., 2000).

REFERENCES

Arauz, L. F. (2000). Mango anthracnose: Economic impact and current options for integrated management. *Plant Dis,* 84:600-616.

Beaumont, A. (1947). The dependence on the weather of the dates of potato blight epidemics. *Trans Brit Mycol Soc,* 31:45-53.

Billing, E. (1980). Fire blight *(Erwinia amylovora)* and weather: A comparison of warning systems. *Ann Appl Biol,* 95:365-369.

Billing, E. (1999). Fire blight risk assessment: Billing's integrated system (BIS) and its evaluation. *Acta Hortic,* 489:399-405.

Bourke, P. M. A. (1953). *Potato Blight and the Weather, a Fresh Approach.* Technical Note No. 12, Department of Industry and Commerce Meteorological Service, Dublin.

Bruhn, J. A. and Fry, W. E. (1981). Analysis of potato late blight epidemiology by simulation modeling. *Phytopathology,* 73:1054-1059.

De Wolf, E. D. and Francl, L. J. (2000). Neural network classification of tan spot and Stagnospora blotch infection periods in a wheat field environment. *Phytopathology,* 90:108-113.

Dodd, J. C., Estrada, A. B., Matcham, J., Jeffries, P., and Jeger, M. J. (1991). The effect of climatic factors on *Colletotrichum gloeosporioides,* causal agent of mango anthracnose in the Philippines. *Plant Pathol,* 40:568-575.

Ellis, M. A., Madden, L. V., and Lalancette, N. (1994). A disease forecasting program for grape downy mildew in Ohio. In D. M. Gaduory and R. C. Seem (Eds.), *Proceedings of International Workshop on Grapevine Downy Mildew Modeling, New York Agricultural Experimental Station Special Report,* 68, pp. 92-95.

Fitzell, R. D., Peak, C. M., and Darnell, R. E. (1984). A model for estimating infection levels of anthracnose disease of mango. *Ann Appl Biol,* 104:451-458.

Fry, W. E., Apple, A. E., and Bruhn, J. A. (1983). Evaluation of potato blight forecasts modified to incorporate host resistance and fungicide weathering. *Phytopathology,* 73:1054-1059.

Gleason, M. L., MacNab, A. A., Pitblado, R. E., Ricker, M. D., East, D. A., and Latin, R. X. (1995). Disease warning systems for processing tomatoes in eastern North America: Are we there yet? *Plant Dis,* 79:113-121.

Gouk, S. C., Spink, M., and Laurenson, M. R. (1999). FireWork—a Windows-based computer program for prediction of fire blight on apples. *Acta Hortic,* 489:407-412.

Grunwald, N. J., Rubio-Covarrubias, O. A., and Fry, W. E. (2000). Potato late-blight management in the Toluca Valley: Forecasts and resistant cultivars. *Plant Dis,* 84:410-416.

Hansen, J. G., Andersson, B., and Hermansen, A. (1995). NEGFRY—a system for scheduling chemical control of late blight in potatoes. In L. J. Dowley, E. Bannon, L. R. Cooke (Eds.), *Phytophthora infestans* 150. *European Association of Potato Research,* Dublin, pp. 201-208.

Hau, B. (1990). Analytical models of plant disease in a changing environment. *Annu Rev Phytopathol,* 28:221-245.

Hijmans, R. J., Forbes, G. A., and Walker, T. S. (2000). Estimating the global severity of potato late blight with GIS-linked disease forecast models. *Plant Pathol,* 49:697-705.

Holtz, B. A., Teviotdale, B., and Turini, T. (1999). Comparison of models to predict the occurrence of fire blight in the San Joaquin Valley of California. *Acta Hortic,* 489:437-443.

Hong, J.-H., Mills, D. J., Coffman, C. B., Anderson, J. D., Camp, M. J., and Gross, K. C. (2000). Tomato cultivation systems affect subsequent quality of fresh-cut fruit slices. *J Amer Soc Hort Sci,* 125:729-735.

Jacobi, J. C., Backman, P. A., Davis, D. P., and Brannen, P. M. (1995). AU-Pnuts Advisory I: Development of a rule-based system for scheduling peanut leaf spot fungicide applications. *Plant Dis,* 79:666-671.

Jensen, R. E. and Boyle, L. W. (1966). A technique for forecasting leafspot on peanut. *Plant Dis Reptr,* 50:810-814.

Krause, R. A., Massie, L. B., and Hyre, R. A. (1975). Blitecast: A computerized forecast of potato late blight. *Plant Dis Reptr,* 59:95-98.

Leonard, K. J. and Fry, W. E. (1986). *Plant Disease Epidemiology. Population Dynamics and Management, Vol.1.* Macmillan Publishing Co., New York, p. 372.

Llorente, I., Vilardell, P., Bugiani, R., Gherardi, I., and Montesinos, E. (2000). Evaluation of BSPcast disease warning system in reduced fungicide use programs for management of crown spot of pear. *Plant Dis,* 84:631-637.

Madden, L. V., Ellis, M. A., Lalancette, N., Hughes, G., and Wilson, L. L. (2000). Evaluation of a disease warning system for downy mildew of grapes. *Plant Dis,* 84:549-554.

Madden, L., Pennypacker, S. P., and MacNab, A. A. (1978). FAST, a forecast system for *Alternaria solani* on tomato. *Phytopathology,* 68:1354-1358.

Park, E. W., Seem, R. C., Gadoury, D. M., and Pearson, R. C. (1997). DMCAST: A forecasting model for grape downy mildew development. *Vitic Enol Sci,* 52:182-189.

Pataky, J. K., Toit, L. J., and Freeman, N. D. (2000). Stewart's wilt reactions of an international collection of *Zea mays* germ plasm inoculated with *Erwinia stewartii. Plant Dis,* 84:901-906.

Phipps, P. M. (1993). IPM in peanuts: Developing and delivering working IPM systems. *Plant Dis,* 77:307-309.

Smith, L. P. (1956). Potato blight forecasting by 90 percent humidity criteria. *Plant Pathology,* 5:83-87.

Smith, T. J. (1999). Report on the development and use of Cougarblight 98C—a situation-specific fire blight risk assessment model for apple and pear. *Acta Horticulturae,* 489:429-436.

Teng, P. S. (1985). A comparison of simulation approaches to epidemic modeling. *Annu Rev Phytopathol,* 23:351-379.

Timmer, L. W. and Zitko, S. E. (1993). Relationships of environmental factors and inoculum levels to the incidence of postbloom fruit drop of citrus. *Plant Dis,* 77:501-504.

Timmer, L. W. and Zitko, S. E. (1996). Evaluation of a model for prediction of postbloom fruit drop of citrus. *Plant Dis,* 80:380-383.

Travis, J. W., Felland, C. E., Hickey, K. D., Truxall, D., and Russo, J. (1999). Automated weather and fire blight model and fire blight model delivery to growers. *Acta Horticulturae,* 489:531-537.

Van der Plank, J. E. (1984). Disease Resistance in Plants, Second Edition. Academic Press, New York.

van Everdingen, E. (1926). Het verband tusschen de weersgesteldheid en de aardappelziekte *(P. infestans). Tijdschr Plantenzeikten,* 32:129-140.

Vercesi, A., Cortesi, P., Zerbetto, F., and Bisiach, M. (1994). Evaluation of the EPI model for downy mildew of grapevine in Northern Italy. In D. M. Gaduory and R. C. Seem (Eds.), *Proceedings of International Workshop on Grapevine Downy Mildew Modeling, New York Agricultural Experimental Station Special Report* 68, pp. 55-73.

Verreet, J. A., Klink, H., and Hoffmann, G. M. (2000). Regional monitoring for disease prediction and optimization of plant protection measures: The IPM wheat model. *Plant Dis,* 84:816-826.

Xu, X., Harris, D. C., and Berrie, A. M. (2000). Modeling infection of strawberry flowers by *Botrytis cinerea* using field data. *Phytopathology,* 90:1367-1374.

CROP DISEASE MANAGEMENT

17

Biological Control—Microbial Pesticides

Biological control using microbial pesticides has become important in recent years. Introduced microbial biocontrol agents (as opposed to a natural population of microbes) are called "microbial pesticides" according to the Environmental Protection Agency (EPA) (Harman, 2000). Several microbial pesticides are now commercially available. Fungal, bacterial, and viral biocontrol agents are available. They effectively control fungal, bacterial, and viral diseases. Modes of action of these biocontrol agents are described. Conditions favorable for effective action of them in controlling diseases are also discussed.

TRICHODERMA

Trichoderma species are the most important biocontrol agents. Many have been developed as commercial products. In 1999, retail sales of a product (Topshield and Rootshield) based on a single strain of *T. harzianum,* strain T-22, totaled around 3 million dollars in the United States (Harman, 2000). Commercial production of *Trichoderma* has been reported from France, New Zealand, Sweden, Poland, Denmark, Russia, Israel, Bulgaria, China, and India. There are many constraints in developing *Trichoderma* as biocontrol agents. *Trichoderma* colonizes in the spermosphere effectively, but they normally do not survive well in the rhizosphere. *Trichoderma* spp. achieve only transitory localized dominance of the rhizosphere, and these are active in only some soils and seasons (Deacon, 1994). Hence, *Trichoderma* species are likely to be effective for seed and seedling diseases, but not against diseases of a mature crop. However, crop losses will be greater when mature crops are affected, and seed and seedling diseases can be effectively controlled by seed treatment with chemicals at a very low cost. Under such conditions, the use of *Trichoderma* spp. will be limited. Another important constraint is that *Trichoderma* spores are quiescent and inactive in soil. Hence, *Trichoderma* strains cannot be added as spores. It may be easier to mass multiply fungi in the form of spores.

Several technologies were recently developed to make use of *Trichoderma* in the control of soil-borne diseases of crops at different maturity stages. *Trichoderma harzianum* strain T-22 with rhizosphere competence was developed by protoplast fusion technology (Harman, 2000). *Rhizosphere competence* is defined as the ability of a microorganism to grow and function in the developing rhizosphere. Strains that were fused were T-95 of *T. harzianum*, a rhizosphere competent mutant produced from strain T-12. T-12 was more capable of competing with spermosphere bacteria than T-95 under iron-limiting conditions; both were strong biocontrol agents. Some strains of *Trichoderma*, such as the strain T39, can induce systemic resistance, and such strains can induce resistance against diseases at any stage of the crop (De Meyer et al., 1998). Technology to apply *Trichoderma* in the form of actively growing germlings instead of spores was developed to obtain an active population of *Trichoderma* in soil to control soil-borne pathogens.

Formulations of Trichoderma

Liquid media based on molasses and molasses yeast have been used widely for the production of *Trichoderma*. The addition of complex organic materials, such as V8 juice, yeast extract, or protease peptone, increased conidial production in *T. harzianum*. The addition of osmotica such as polyethylene glycol improved conidial production of *T. harzianum* and resistance of conidia to desiccation (Whipps, 1996). *Trichoderma harzianum* has been produced in diatomaceous earth granules impregnated with 10 percent molasses. Spores, cells, or biomass are concentrated directly from liquid media by centrifugation and filtration. Biomass may be dried, milled, and incorporated into a range of dusts, alginate granules, pellets or prills, wettable powders, emulsifiable liquids or gels. Talc formulations, kaolin-based microgranules, and alginate pellet, prill, or granule formulations are available.

Conidia of *Trichoderma* are added to a bran-sand mixture, and after one to three days of incubation, this germling preparation is added to soil where colony-forming units of the antagonists continue to increase. This method provides a means of achieving an active population of antagonists in the soil. A medium supplemented with ground corn cobs was developed for applying *T. koningi* in the field (Latunde-Dada, 1993). Alternatively, a fermenter biomass of *Gliocladium* and *Trichoderma* spp. was added to a vermiculite-bran mixture moistened with 0.05 M HCl. After drying, the preparation can be remoistened with 0.05 M HCl and germlings produced as before (Lewis et al., 1991).

Seed-coating formulations have also been developed. A liquid coating formulation comprises a suspension of aqueous binder (pelgel or polyox—N-10), finely ground solid particulate matter (Agro-Lig or muck soil), and the biocontrol agent (Taylor et al., 1991). This is sprayed onto seeds in a tumbling drum. Agro-Lig has chemical and physical characteristics favorable for the growth of *Trichoderma*. This type of formulation was very effective in the control of damping-off of cucumber caused by *Pythium* (Taylor et al., 1991). Adding compounds to the seed coating that specifically enhance growth of *Trichoderma* is highly beneficial. Inclusion of specific polysaccharides and polyhydroxy alcohols improves biocontrol activity of *Trichoderma* (Nelson et al., 1988).

Method of Application of Trichoderma

Mathre et al. (1999) stated that nearly all commercialized microorganisms rely upon application of the antagonist "directly and precisely to the infection court" when and where needed. Seed treatment is the most effective method (Mathre et al., 1999). Seed-coating formulations will be useful. Seed priming is also recommended. Seed priming is the process in which hydration of the seed is controlled to a level that permits pregerminative metabolic activity to take place without emergence of the radicle. Two priming systems are available. Osmopriming utilizes aerated aqueous solutions of salts or polyethylene glycol, generating osmotic potential in the primary solution. Solid matrix priming (SMP) involves the use of moist, porous solid materials, such as powdered coal or peat, generating matric potential. Combining SMP with *Trichoderma* spp. for control of seedling diseases has been used successfully on a wide range of plants (Harman et al., 1989).

Actively growing germling populations can be applied to soil. *Trichoderma* can be applied as granules or as a drench. A single application of *T. harzianum* T-22 as RootShield granules in a greenhouse provided protection of a tomato crop against *Fusarium* crown and root rot of the mature crop. An in-furrow drench was more effective in both root colonization and disease control than a seed treatment (Harman, 2000).

In crops that are transplanted, the granules can be applied in the nursery. Tomatoes were grown in a potting mix containing the granular formulation of *T. harzianum* T-22, which permitted roots to become colonized, and then transplanted to the field. This treatment reduced *Fusarium* crown and root rot at the harvest of mature fruit (Datnoff et al., 1995). *Trichoderma* can also be applied as a spray. *Trichoderma harzianum* T-22 is effective in the control of fruit and foliar diseases when applied as a spray to these plant parts. T-22 should be applied at least once every 10 days when disease pressure is high, because it cannot extensively grow on and colonize newly formed leaf tissues. The fungus colonizes grape or strawberry flowers and immature fruits (Harman, 2000). Diseases controlled by foliar spray include powdery mildews of *Catharanthus* and pumpkins, *Botrytis cinerea* on strawberry, downy mildew of snapdragons, and turf-grass pathogens such as *Rhizoctonia solani* and *Pythium* spp. (Harman, 2000).

Bumble bees *(Bombus impatiens)* and honey bees *(Apis mellifera)* have been used to deliver *Trichoderma* to the flowers of crop plants. Bees exiting the hive pass through a device that requires them to come into contact with *Trichoderma* products containing these spores. They subsequently deliver substantial amounts of *Trichoderma harzianum* T-22 or similar fungi to the strawberry or other flowers. This method of delivery was more effective than spray applications for control of *B. cinerea* and has proven effective over several years and trials as standard chemical applications (Kovach et al., 2000; Harman, 2000).

Time of application of *Trichoderma* is also important. *Trichoderma* can be overwhelmed by heavy disease pressure. Therefore, *T. harzianum* may be used strictly as a preventative measure; it cannot cure infections. *Trichoderma* is less effective against systemic diseases than against more superficial ones. It cannot control existing diseases, and so a good systemic fungicide must be used if diseases already exist. In conditions of high or very high disease pressure, T-22 should be used as part of an integrated chemical-biological system. A combination of chemical treatment with *Trichoderma* will be highly effective in the control of diseases. A tank mix with chemical fungicides or an alternating spray with chemical fungicides is the ideal method of application of *Trichoderma* (Harman, 2000). A combination of ozone fumigation and *T. harzianum* treatment was on par with the standard methyl bromide treatment, and the combination was significantly better than either *T. harzianum* alone or ozone fumigation alone in control of strawberry root diseases (Harman, 2000).

A single strain of *Trichoderma* may not be sufficient to be effective under all conditions and against all diseases. Mathre et al. (1999) suggested that almost invariably, a different agent might be needed for each disease. Cook (1993) stated that biological control is widely recognized as being highly disease-specific. He advocated an approach to biological control that uses mixtures of numerous agents for each disease. A mixture of *Trichoderma* spp. has been developed as commercial formulations. *T. harzianum* + *T. polysporum* (Binab—T) and *T. harzianum* + *T. viride* (Trichodowels) are the important complex products (Whipps, 1996). *Trichoderma* has been combined also with other biocontrol agents. The combination of *T. harzianum* T-22 and the mycorrhizal fungus *Glomus intradices* was more effective than either organism alone (Datnoff et al., 1995). There are also reports that a single strain of *Trichoderma* may be capable of controlling diverse pathogens under diverse conditions (Chet, 1987; Harman, 2000).

Diseases Controlled by Trichoderma

Trichoderma has been reported to control *Rhizoctonia, Fusarium, Phytophthora,* and *Pythium* diseases in many crops, tomato root and crown rot, pumpkin and *Catharanthus* powdery mildews, gray mold *(Botrytis cinerea)* of strawberry, root rots of several crops caused by *Macrophomina phaseolina,* wheat take-all caused by *Gaeumannomyces graminis* var. *tritici, Sclerotinia* and *Verticillium* diseases of fruit trees, *Armillaria mellea* infection in trees, Dutch elm disease *(Ceratocystis ulmi),* Chestnut blight *(Endothia parasitica),* silver leaf disease of trees *(Chondrostereum purpureum),* and stem and root rot of pine *(Heterobasidion annosum)* (Harman et al., 1989; Nelson et al., 1988; Maplestone et al., 1991; Whipps, 1992, 1996; Datnoff et al., 1995; Nemec et al., 1996; De Meyer et al., 1998; Elad et al., 1999; Howell et al., 1999; Burns and Benson, 2000; Harman, 2000).

Mycoparasitism of Trichoderma

Trichoderma spp. may control diseases caused by various fungal pathogens by their various types of actions. Their modes of action include mycoparasitism, antibiosis, induced resistance, competition for nutrients or space, and inactivation of the pathogen's enzymes. Mycoparasitism involves tropic growth of the biocontrol agent toward the target fungi, lectin-mediated coiling of attachment of *Trichoderma* hyphae to the pathogen, and finally attack and dissolution of the target fungal cell wall by activity of enzymes, which may be associated with physical penetration of the cell wall. More than 20 separate genes may be involved in mycoparasitism. *Trichoderma* produces ten different chitinases and several β-1,3-glucanases and proteases. The importance of chitinases produced by the antagonist has been demonstrated in several ways. A 42-kDa endochitinase is induced before *T. harzianum* comes into contact with *Botrytis cinerea* (Zeilinger et al., 1999). A strain of *T. harzianum* deficient in the ability to produce endochitinase had reduced ability to control *B. cinerea* (Woo et al., 1999). Endochitinase was disrupted or overproduced in *T. virens,* and the resulting strains were found to have decreased or increased biocontrol activity, respectively (Back et al., 1999). Expression of endochitinase from *T. harzianum* in transgenic apple increases resistance to apple scab (Bolar et al., 2000). The endochitinase gene from *Trichoderma* confers resistance in many other transgenic plants

(Lorito et al., 1998). Contradicting reports state that chitinase may not be involved in an antagonistic action of *Trichoderma*. The activity of endochitinase was disrupted or overproduced in *T. harzianum,* but these changes had no effect on its biocontrol ability against *Rhizoctonia solani* or *Sclerotium rolfsii* (Carsolio et al., 1999). A strain of *T. harzianum* deficient in the ability to produce endochitinase had increased ability to control *R. solani* (Woo et al., 1999). These results suggest that other gene products may also be involved in the action of *Trichoderma.*

Antibiotic Production by Trichoderma

Forty-three antibiotic substances were reported to be produced by *Trichoderma* spp. (Sivasithamparam and Ghisalberti, 1998). Of these, alkyl pyrones, isonitriles, polyketides, peptaibols, diketopiperazines, sesquiterpenes, and steroids are important and found to be associated with biocontrol activity of *Trichoderma* spp. Mutation to eliminate production of specific antibiotics is associated, in some strains, with a loss of activity against particular pathogens (Howell, 1998).

Induced Resistance by Trichoderma

Some *Trichoderma* spp. induce systemic resistance against pathogens. *Trichoderma harzianum* induces systemic resistance against powdery mildews (Elad et al., 1999), *B. cinerea* infections (De Meyer et al., 1998), and root rot of cotton (Howell et al., 1999). *Trichoderma* products (BINABTF-WP and BINABT vector) induced systemic acquired resistance in strawberry against *Botrytis cinerea* (Ricard and Jorgensen, 2000). Different elicitors, including xylanase, have been isolated from *Trichoderma* and they trigger the synthesis of various defense compounds, including phytoalexins (Calderon et al., 1993). *Trichoderma* induces the synthesis of phytoalexins involved in disease resistance in cotton (Howell et al., 1999), modifies and strengthens plant cell walls in cucumber (Yedidia et al., 1999), and increases activities of chitinase and peroxidase in cucumber tissues (Yedidia et al., 1999). Cucumber roots treated with *T. harzianum* exhibited higher activities of chitinase, β-1,3-glucanase, and peroxidase (Yedidia et al., 2000). For more information on the role of peroxidases, β-1,3-glucanases and chitinases in conferring disease resistance, see Chapter 34.

Competition for Space or Nutrients by Trichoderma

Trichoderma may compete for space or nutrients with pathogens and suppress their development. *Botrytis cinerea* conidia require external nutrients for germination and infection. When conidia of *T. harzianum* T39 were applied to leaves, germination of conidia of the pathogen was slowed, an effect attributed in part to competition (Elad et al., 1999).

Inactivation of the Pathogen's Enzymes by Trichoderma

B. cinerea depends upon production of pectolytic, cutinolytic, and cellulolytic enzymes to infect living plants. However, conidia of two strains of *T. harzianum* (T39 and NCIM 1185), when applied to the leaves, produce a serine protease that is capable of degrading the pathogen's plant cell wall degrading enzymes and thereby reducing the ability of the pathogen to infect the plant (Elad and Kapat, 1999). The biocontrol activity of T39 could be enhanced by adding additional quantities of its protease. Several protease inhibitors reduced the biocontrol activity of T39 (Elad and Kapat, 1999).

PSEUDOMONAS SPECIES

Pseudomonas spp. are the important group of biocontrol agents that have been developed as commercial products. Fluorescent pseudomonads form the major bacterial group surviving in the rhizosphere of crop plants. They are also known as plant growth promoting rhizobacteria (PGPR) because they promote plant growth by secreting auxins, gibberellins, and cytokinins (Vidhyasekaran, 1998). These pseudomonads survive in soil, rhizoplane, phylloplane, pistils, nectarines, and fruits of the plants. *Pseudomonas fluorescens, P. putida,* and *P. cepacia (Burkholderia cepacia)* are important *Pseudomonas* species and are known as highly effective biocontrol agents. They have been reported to control soil-borne, seed-borne, and air-borne fungal, bacterial, and viral pathogens.

Diseases Controlled by Pseudomonas

The following is a list of diseases that have been reported to be controlled by saprophytic pseudomonads (Levy et al., 1998; Alstrom, 1991; Wei et al., 1991; Liu et al., 1992, 1995a,b; Wilson and Lindow, 1993; Maurhofer, Scherer, et al., 1994; Tosi and Zazzerini, 1994; Hoffland et al., 1995; Leeman et al., 1995b; Raaijmakers et al., 1995; Vidhyasekaran and Muthamilan, 1995, 1999; Benhamou et al., 1996; M'Piga et al., 1997; Vidhyasekaran, 1998, 2001; Vidhyasekaran, Rabindran, et al., 1997; Vidhyasekaran, Sethuraman, et al., 1997; Vidhyasekaran et al., 2000):

Apple—gray mold *(Botrytis cinerea)*
Bean—halo blight *(Pseudomonas savastanoi* pv. *phaseolicola)*, root
   rot *(Sclerotium rolfsii)*
Chickpea—wilt *(Fusarium oxysporum* f. sp. *ciceris)*
Cotton—bacterial blight *(Xanthomonas axonopodis* pv.
   malvacearum), Fusarium wilt *(Fusarium oxysporum*
   Schlechtend.: Fr. f. sp. *vasinfectum)*, seedling blight *(Rhizoctonia
   solani)*, and seedling rot *(Pythium ultimum)*
Cucumber—wilt *(Fusarium oxysporum* f. sp. *cucumerinum)*, damp-
   ing-off *(Pythium aphanidermatum)*, anthracnose *(Colletotrichum
   orbiculare)*, angular leaf spot *(Pseudomonas syringae* pv.
   lachrymans), and mosaic *(Cucumber mosaic virus)*
Pea—wilt *(Fusarium oxysporum* f. sp. *pisi)*, root rot *(Aphanomyces
   euteiches)*, and damping-off *(Pythium ultimum)*
Peanut—late leaf spot *(Phaeoisariopsis personata)*, stem rot (south-
   ern blight) *(Sclerotium rolfsii)*, and rust *(Puccinia arachidis)*
Pear—fire blight *(Erwinia amylovora)*
Pigeonpea—wilt *(Fusarium oxysporum* f. sp. *udum)*
Potato—bacterial wilt and brown rot *(Ralstonia solanacearum)*,
   black leg and soft rot *(Erwinia carotovora)*
Radish—wilt *(Fusarium oxysporum* f. sp. *raphani)*
Rice—blast *(Pyricularia oryzae)*, sheath blight *(Rhizoctonia solani)*,
   bacterial blight *(Xanthomonas oryzae* pv. *oryzae)*, sheath rot
   (Sarocladium oryzae), and tungro *(Rice tungro virus)* diseases
Safflower—rust *(Puccinia carthami)*
Sugar beet—leaf spot *(Phoma betae)*, damping-off *(Pythium
   ultimum)*, and root rot *(Aphanomyces euteiches)*
Tobacco—black root rot *(Thielaviopsis basicola)*, necrosis *(Tobacco
   necrosis virus)*
Tomato—wilt *(Fusarium oxysporum* f. sp. *lycopersici)*
Wheat—take-all *(Gaeumannomyces graminis* var. *tritici)*, leaf
   (brown) rust *(Puccinia triticina=Puccinia recondita)*,
   and Septoria speckled leaf blotch *(Septoria tritici)*

Induction of Systemic Resistance by Pseudomonas Spp.

Several rhizobacterial strains were shown to elevate plant resistance against several pathogens. Strains of *Pseudomonas fluorescens, P. putida,* and *P. aeruginosa* induced systemic resistance (ISR) in rice (Ohno et al., 1992; Vidhyasekaran, Rabindran, et al., 1997; Vidhyasekaran et al., 2000), cucumber (Liu et al., 1995a,b; Meyer et al., 1992; Raupach, 1996), radish (Leeman et al., 1995b), tobacco (Maurhofer et al., 1994a), tomato (Van Wees et al., 1997; M'Piga et al., 1997), bean (Meier et al., 1993), pea (Benhamou et al., 1996), carnation (Van Peer and Schippers, 1992), and *Arabidopsis* (Van Wees et al., 1997).

Rhizobacterial strains have been shown to induce different defense genes in plants. Maurhofer, Hase, et al. (1994) showed that ISR by *P. fluorescens* strain CHA0 in tobacco was associated with accumulation of various pathogenesis-related (PR) proteins (PR-1, PR-2, and PR-3 proteins). Inoculation of bean leaves with cells of *P. fluorescens* induced the accumulation of transcripts for chalcone synthase (CHS), chitinase, and lipoxygenase (Meier et al., 1993). Increase in peroxidase activity as well as an increase in the level of mRNAs encoding for phenylalanine ammonia-lyase (PAL) and CHS could be recorded in the early stages of the interaction between roots and various bacterial endophytes (Zdor and Anderson, 1992). Treatment with *P. fluorescens* causes increases in activities of peroxidase, lysozyme, and PAL in tobacco (Schneider and Ullrich, 1994). Van Peer et al. (1991) showed massive accumulation of phytoalexins in carnation roots colonized by the rhizobacterial strains after pathogen challenge. M'Piga et al. (1997) reported that *P. fluorescens* strain 63-28 induced accumulation of an electron-dense material in epidermal and outer cortical cells and coating of most intercellular spaces with similar substances in tomato. This aggregated material appeared to be mainly composed of phenolic compounds, especially phenols containing O-hydroxy groups. The deposition of β-1,3-glucans (callose) was also observed in host cell walls. Chitinases were also induced (M'Piga et al., 1997). Benhamou et al. (1996) demonstrated that pea root bacterization with *P. fluorescens* triggered a set of defense reactions that resulted in the elaboration of permeability barriers. Increases in plant lignification, phytoalexins, various lytic enzymes, and other PR proteins have been observed upon treatment of plants with different specific strains of rhizobacteria (Albert and Anderson, 1987; Frommel et al., 1991; Kloepper et al., 1993; Sayler et al., 1994; M'Piga et al., 1997; Vidhyasekaran et al., 2000).

Studies on the mode of induction of defense genes by the rhizobacterial strains reveal that different bacterial cell wall components may act as elicitors of induction of the signal transduction system. Lipopolysaccharides (LPS) of *P. fluorescens* strain WCS417r act as elicitors and induce resistance against different diseases (Duijff et al., 1997). LPS and LPS-containing cell wall preparations of *P. fluorescens* WCS417r are as effective as living WCS417r bacteria in inducing ISR in radish (Leeman et al., 1995a).

The O-antigenic side chain of the outer membrane LPS of the strain WCS417r appears to be the main determinant for induction of ISR in radish and carnation (Van Peer and Schippers, 1992; Leeman et al., 1995a; Duijff et al., 1997).

Leeman et al. (1996) demonstrated that the siderophore of WCS374r can act as an elicitor of ISR in radish. *Pseudomonas fluorescens* strain CHA0 induced systemic resistance in tobacco. A siderophore (pyoverdine)-deficient derivative of this strain no longer induced ISR (Maurhofer, Hase, et al., 1994). Leeman et al. (1996) reported that the purified siderophore, pseudobactin, from *P. fluorescens* strain WCS374 induced ISR in radish. However, a pseudobactin-deficient, *P. fluorescens* 374PSB, retained ISR-inducing activity. These results suggest that siderophore production by this strain was only partially responsible for the induction of systemic resistance in radish.

Pyochelin, a siderophore, is produced by several rhizobacteria. Salicylic acid is a precursor of pyochelin synthesis (Leeman et al., 1996). Several genera of bacteria, including fluorescent pseudomonads, are known to synthesize salicylic acid (Dowling and O'Garra, 1994). *Pseudomonas fluorescens* strain CHA0 produces salicylic acid (Meyer et al., 1992). Leeman et al. (1996) reported that *P. fluorescens* strain WCS374 produced salicylic acid in quantities that were iron dose-dependent. *Pseudomonas fluorescens* strain WCS417r has the capacity to produce salicylic acid (Leeman et al., 1996). Salicylic acid was responsible for the induction of resistance in radish (Leeman et al., 1996). Meyer et al. (1992) reported that salicylic acid itself might function as an endogenous siderophore. Leeman et al. (1996) reported that rhizobacteria-mediated ISR is affected by iron concentration. Salicylic acid production is iron (Fe^{3+}) regulated (Leeman et al., 1996). Salicylic acid production is promoted by low iron concentrations. Increasing ferric iron concentrations in vitro reduced salicylic acid production below detectable limits by bacteria (Meyer et al., 1992). When ferric iron was applied as a soil drench, ferric iron concentration increased in planta; but it significantly reduced the level of ISR observed in cucumber. This suggests that salicylic acid may not be involved in ISR; but some other siderophores mediated by iron may be involved in induction of ISR in cucumber.

Massive accumulation of phytoalexins could be detected in roots of carnation plants treated with *P. fluorescens* only after challenge inoculation with pathogen (Van Peer et al., 1991). The induction of phenolics and callose in tomato by *P. fluorescens* strain 63-28 was substantially amplified upon infection with the pathogen (M'Piga et al., 1997). These results suggest that the rhizobacterial strains may be capable of evoking transcriptional activation of plant defense genes, the expression of which may be subsequently latent until the plant perceives signals originating from contact with the pathogen. It is also possible that besides the rhizobacterial signal molecules, pathogen's signal molecules may also be involved in the induction of ISR. Induction of H_2O_2 production due to *P. fluorescens* treatment in plants has also been reported (Jakobek and Lindgren, 1993), and H_2O_2 is known as a second messenger that triggers the synthesis of defense chemicals.

Production of Antibiotics by Pseudomonas

Pseudomonas spp. produce antibiotics that have been assigned a role in disease control. *Pseudomonas fluorescens* CHA0 produces 2,4-diacetyl-phloroglucinol, pyoluteorin, and pyrrolnitrin (Voisard, 1994; Maurhofer, Sacherer, et al., 1994). *Pseudomonas fluorescens* 2-79 produces phenazine-carboxylic acid, pyoverdine, and anthranilic acid, with biocontrol activity mainly due to phenazine production (Hamdan et al., 1991). *Pseudomonas fluorescens* Pf-5 produces 2,4-diacetylphloroglucinol, pyrrolnitrin, and pyo-luteorin, with pyoluteorin being the most effective against *Pythium ultimum* (Loper et al., 1994). *Pseudomonas fluorescens* DR54 produces the antibi-otic viscosinamide (Thrane et al., 1999).

The importance of antibiotics in the biocontrol activity of *Pseudomonas* spp. was demonstrated by several studies. Biocontrol activity could be in-creased by increasing the production of antibiotics. Phenazine production could be increased by introducing extra copies of biosynthetic or activator genes. This increased the biocontrol activity of the *Pseudomonas* strain (Thomashow and Pierson, 1991). Additional evidence for a role of antibiot-ics in biocontrol by *Pseudomonas* spp. was obtained by heterologous ex-pression of complementary genes in strains that naturally do not produce the antibiotic, thus increasing their biocontrol activity (Hara et al., 1994) or by use of Tn5 mutagenesis to inactivate specific genes and affect pathways of antibiotic production (Schnider, Keel, Voisard, et al., 1995). In situ detection of antibiotics in the rhizosphere of plants treated with antibiotic-producing strains was demonstrated for phenazine-carboxylic acid (Thomashow et al., 1990), 2,4-diacetylphloroglucinol (Maurhofer et al., 1995), pyrrolnitrin (Kempf et al., 1994), and pyoluteorin (Maurhofer et al., 1995). In situ ex-pression of the genes required for the synthesis of pyoluteorin, phenazines, and Oomycin A could be demonstrated by fusing reporter genes such as those encoding β-galactosidae or ice-nucleating activity to promoters of genes encoding for antibiotics (Loper and Lindow, 1994; Kraus and Loper, 1995).

A two-component regulatory system for antibiotic production in *Pseudo-monas* spp. was reported. The system is based on two protein components, an environmental sensor and a cytoplasmic regulator or global activator (GacA) that mediates changes in response to sensor signals. A response regulator gene, *gacA,* was identified in *P. fluorescens* CHA0. Strains with a mutation of this gene lost the ability to produce pyoluteorin and 2,4-diacetyl-phloroglucinol (Laville et al., 1992). A *lemA*-like *apd* gene was found in *P. fluorescens* Pf-5. Strains with a mutation in this region failed to produce pyoluteorin and pyrrolnitrin, and lost the ability to inhibit *Rhizoctonia solani* in culture (Corbell and Loper, 1995). Antibiotic production in *Pseudomonas* spp. may be further controlled by the action of housekeeping sigma factors encoded by the *rpoS* or *rpoD* genes (Schnider, Keel, Blumer, et al., 1995).

Role of HCN, Siderophore, and β-1,3-Glucanase Produced by Pseudomonas

Pseudomonas spp. produce hydrocyanic acid (HCN), which was shown to be important in biocontrol activity (Laville et al., 1992; Voisard et al., 1994; Loper et al., 1994; Corbell and Loper, 1995). The importance of a *Pseudomonas*-produced pyoverdine siderophore in biocontrol activity was demonstrated (Hamdan et al., 1991; Maurhofer, Hase, et al., 1994; Voisard et al., 1994; Loper et al., 1994; Kraus and Loper, 1995). β-1,3-Glucanase may be involved in biocontrol activity of *Pseudomonas (Burkholderia) cepacia* (Fridlender et al., 1993).

Factors Involved in Biocontrol Efficacy of Pseudomonas

Several factors determine the efficacy of *Pseudomonas* in controlling crop diseases in the field. Effective strain selection is important (Vidhyasekaran, 1998). The antagonist inoculum dose determines the efficacy of the antagonist in controlling diseases. A threshold population density of the fluorescent pseudomonad strains of approximately 10^5 cfu per g of root is required for significant suppression of fusarium wilt of radish. When rhizosphere population densities of the strains dropped below this threshold level, the efficacy of these strains to suppress the fusarium wilt was almost lost (Raaijmakers et al., 1995). Similar results were obtained in the control of rice sheath blight by *P. fluorescens* strain Pf1 (Vidhyasekaran and Muthamilan, 1999) and *P. fluorescens* PFALR2 (Rabindran and Vidhyasekaran, 1996). The efficacy of fluorescent pseudomonads is affected drastically by increasing disease pressure. Significant wilt disease suppression in radish by fluorescent pseudomonads was observed when the disease incidence in the control (untreated) field was less than 80 percent and the antagonists did not suppress fusarium wilt when disease incidence exceeded 80 percent in the control field (Raaijmakers et al., 1995). The antagonists may be highly useful in moderately resistant varieties rather than highly susceptible varieties (Leeman et al., 1995b). It was suggested that the antagonists should be integrated with chemical fungicides, with the two being applied alternatively (Vidhyasekaran, 1998).

The method of application also determines the efficacy of pseudomonads. Seed treatment with these antagonists appears to be very effective. These bacteria establish well in the rhizosphere when introduced through seed treatment (Vidhyasekaran, Sethuraman, et al., 1997; Vidhyasekaran and Muthamilan, 1999). Seed treatment of peas with *Pseudomonas aureofaciens (P. chlororaphis)* protects pea plants against *R. solani* (Koch et al., 1998). Biopriming of treated seeds increases the population of the antagonists in seed and effectively controls diseases (Callan et al., 1990; Vidhyasekaran and Muthamilan, 1995). Soil application was shown to be effective in control of soilborne diseases (Hagedorn et al., 1993). Root dip treatment for transplanted crops can be highly useful. A root dip into suspensions of *P. aureofaciens* protects strawberry against *Phytophthora fragariae* var. *fragariae* (Koch et al., 1998). Several workers have successfully used foliar spray of these antagonists to control foliar diseases (Mew and Rosales, 1986; Vidhyasekaran, Rabindran, et al., 1997).

Pseudomonads should be used as powder formulations only. Ten-day-old bacterial cultures in a liquid nutrient medium are ineffective in controlling diseases. Several formulations with different carrier materials were developed for application in the field. Peat- and talc-based formulations are commonly used (Hagedorn et al., 1993; Hofte et al., 1991; Vidhyasekaran, Sethuraman, et al., 1997). Granular preparations of *P. aureofaciens* UKM B-111 based on clay minerals show high survival, preservation of antagonistic activity, and stability of composition during long-term storage (Kurdish et al., 1999). Vermiculite-based formulation of *P. putida* is effective in controlling *Fusarium oxysporum* f. sp. *cucurbitacearum* infection in cucumber (Amer and Utkhede, 2000).

OTHER ORGANISMS

Gliocladium

The fungus *Gliocladium catenulatum* (Primastop) was developed as a biofungicide. It was registered in the United States by Kemira Agro OY. The same product will be registered as PreStop in Europe (Niemi and Lahdenpera, 2000). This product has proven effective in the control of damping-off of vegetables, herb, and ornamental seedlings, root and stem rot diseases in vegetables and ornamentals, *Didymella* in cucumber and tomato, and gray mold in ornamentals. It is effective against various soilborne diseases caused by *Pythium*, *Phytophthora*, and *Rhizoctonia*, foliar diseases caused by *Didymella*, *Alternaria*, and *Botrytis*, and storage diseases caused by *Helminthosporium* and *Rhizoctonia* (Niemi and Lahdenpera, 2000).

Pythium Oligandrum

Pythium oligandrum was developed as a commercial product named Polygandron. The addition of *P. oligandrum* zoospores to soil reduced the ability of sclerotia of *Sclerotinia sclerotiorum* to germinate. *Pythium oligandrum* reduces the survival of *S. sclerotiorum* present naturally in soils through mycoparasitic activity (Madsen and Neergaard, 1999). The cell-wall-degrading enzymes N-acetyl-β-D-glucosaminidase, endo-chitinase, β-glucanase, β-glucosidase, cellobiohydrolase, and protease were detected in culture filtrates of *P. oligandrum* cultivated with *S. sclerotiorum* (Madsen and Neergaard, 1999). *P. oligandrum* may also act by inducing resistance in plants. A low molecular weight protein termed *oligandrin* was obtained from culture filtrates of *P. oligandrum* (Picard et al., 2000). This protein induces plant defense reactions that help restrict stem cell invasion by *Phytophthora nicotianae* var. *parasitica*. Oligandrin is similar to elicitins. Oligandrin-treated plants show reduced disease incidence in tomato caused by *P. nicotianae* var. *parasitica* (Picard et al., 2000).

Verticillium Lecanii

Verticillium lecanii is a promising biocontrol agent of rusts and powdery mildews that effectively controls cucumber powdery mildew caused by *Spherotheca fuliginea* (Askary et al., 1998). *Verticillium lecanii* grows over a wide temperature range, and a water film, or at least high humidity, is required for conidial germination. Thus, water is an important environmental factor in the control of powdery mildews on cucumber and rose by *V. lecanii* (Verhaar et al., 1996). The development of *V. lecanii* in pustules of *Puccinia striiformis* was best at 95 to 100 percent relative humidity (RH), whereas no development was observed at 80 percent RH (Mendgen, 1981). This suggests that high humidity is required for effective control of the pathogen by *V. lecanii*. *Verticillium lecanii* produces antibiotics and hydrolytic enzymes and they may be involved in the parasitism of *V. lecanii* on *Penicillium digitatum* (Benhamou and Brodeur, 2000).

Yeasts

The yeast *Torulopsis candida* (= *Candida famata*) effectively controls *Penicillium digitatum* infection on citrus fruits (Arras et al., 1999). Another yeast, *Debaryomyces hansenii*, reduces *Penicillium digitatum* decay on orange fruits. The yeast elicited production of phytoalexins, scopoletin, and scoparone, and did not produce toxic substances against the pathogens, *Penicillium digitatum* and *Botrytis cinerea*. This suggests that the yeast may reduce the fungal infection by activating host's defense mechanisms (Arras and Arru, 1999). The yeasts *Candida saitoana* and *C. oleophila* control postharvest diseases of apple and citrus fruits (El-Ghaouth et al., 2000). The yeast *(Candida guilliermondii)* suspension, when sprayed two to five times at 7 to 10 day intervals, reduces decay caused by *B. cinerea* in both table grapes (cultivars Thomson Seedless and Superior Seedless) and wine grapes (cultivar Sauvignon blane), and rots caused by *Aspergillus niger* in wine grapes (Zhavi et al., 2000). Another yeast, *Pichia membranefaciens*, controls storage rot of nectarine fruits caused by *Rhizopus stolonifer* (Fan and Tian, 2000).

Aureobasidium

A cosmopolitan yeastlike fungus, *Aureobasidium pullulans*, colonizes leaf surfaces and is a potential biocontrol agent for plant pathogens. It controls *Botrytis cinerea* on apples, *Penicillium digitatum* on grapefruits, *B. cinerea*, *Rhizopus stolonifer*, and *Aspergillus niger* on table grapes and *B. cinerea* and *R. stolonifer* on cherry tomatoes (Schena et al., 1999). Preharvest application of *Aureobasidium pullulans* isolate L47 on table grapes results in a significant reduction of postharvest rot caused by *B. cinerea* (Schena et al., 1999). *Aureobasidium pullulans* controlled apple decay caused by *B. cinerea* and *Penicillium expansum*. The yeastlike fungus induced a transient increase in β-1,3-glucanase, chitinase, and peroxidase activities in apple tissues, and all three enzymes are involved in host defense mechanisms (Ippolito et al., 2000; Castoria et al., 2001). The biocontrol agent also has capacity to out-compete pathogens for nutrients and space (Ippolito et al., 2000; Castoria et al., 2001).

Penicillium

Talaromyces flavus (anamorph *Penicillium dangeardii*) is known to control *Verticillium* wilt of potato, artichoke, and olive. *Talaromyces flavus* isolate Tf-1 suppressed *Verticillium* wilt incidence in eggplant (Fravel and

Roberts, 1991). Purified glucose oxidase from *T. flavus* significantly reduced the growth rate of *V. dahliae* in the presence, but not in the absence, of eggplant roots. This suggests that glucose from the roots is metabolized by glucose oxidase to form hydrogen peroxide, which is toxic to *V. dahliae* (Fravel and Roberts, 1991).

Coniothyrium

The fungal mycoparasite *Coniothyrium minitans* applied as a spray reduces *Sclerotinia sclerotiorum* (white mold) infection in bean, potato, carrot, and chicory (Gerlagh et al., 1999). *C. minitans*-based formulations have been developed as Coniothyrin and Contans in Russia and Germany, respectively.

Nonpathogenic Isolates of Pathogens

Nonpathogenic isolates may induce resistance against pathogenic isolates of the same pathogen and other pathogens. Nonpathogenic isolates of *Rhizoctonia* (np-R) protect seedlings against damping-off caused by virulent isolates of *Rhizoctonia* species of different anastomosis groups (Sneh, 1999). Some np-R isolates induced plant resistance against *R. solani, Pythium aphanidermatum,* and *Pseudomonas syringae* pv. *lachrymans* in cucumber (Sneh, 1999). Binucleate *Rhizoctonia* fungi are another group of biocontrol agents that effectively control diseases of potato, bean, sugar beet, cucumber, pepper, *Catharanthus,* and turf grass caused by *Rhizoctonia* and *Pythium* spp. Colonization of host tissues by nonpathogenic isolates triggers production of host defense compounds such as peroxidases, glucanases, and chitinases (Burns and Benson, 2000).

Reduced-pathogenicity isolates of *Colletotrichum gloeosporioides (Glomerella cingulata)* delayed anthracnose symptom development in avocado fruits induced by virulent isolates (Yakoby et al., 2001). Preinoculation of avocado fruit with reduced-pathogenicity isolates induced resistance that was accompanied by an increase in the levels of preformed antifungal dienes (Yakoby et al., 2001). Less aggressive strains of *Ralstonia solanacearum* induced resistance against aggressive strains of *R. solanacearum* (Trigalet et al., 1998). Similarly, nonpathogenic isolates of *Fusarium oxysporum* induced resistance against pathogenic *F. oxysporum* strains in tomato, carnation, and sweet potato (Whipps, 1996). Commercial formulations of nonpathogenic strains of *F. oxysporum* and *R. solanacearum* are available in Europe.

Ulocladium

The fungus *Ulocladium atrum* competes saprophytically with *Botrytis* spp. during the colonization of necrotic plant tissues. The inoculum potential of *B. cinerea* is reduced by antagonistic interaction with *U. atrum*, leading to slower disease epidemic (Kohl et al., 1998). In cyclamen *(Cyclamen persicum)*, naturally senesced leaves within the dense canopy play a crucial role in *Botrytis* epidemics. Since healthy leaves are normally resistant to conidial infections, *B. cinerea* depends on dead tissues for initial entry into the plant. Stimulated by this food base, the inoculum potential of the pathogen increases within the canopy of the single plant to such a level that healthy petioles and leaf blades can then be infected (Kohl et al., 2000). Biocontrol of *B. cinerea* by *U. atrum* could be achieved by competitively excluding the pathogen from colonizing necrotic leaves present within the cyclamen canopy. Repeated applications of conidial suspensions of *U. atrum* controlled the disease as effectively as the grower's standard fungicide program (Kohl et al., 1998, 2000).

Phialophora

Phialophora spp. are known to control wheat take-all caused by *Gaeumannomyces graminis* var. *tritici*. A *Phialophora* sp. (isolate I-52) was isolated from soil in a wheat field exhibiting suppression of take-all disease (Mathre et al., 1998). I-52 was grown on a variety of autoclaved organic substrates, including oat, millet, and canola seed. Each of these provided significant disease control when added to the seed furrow. Seed treatment was ineffective (Mathre et al., 1998). A *Phialophora* strain was commercialized in Australia (Wong et al., 1996).

Cryphonectria Hypoviruses

Some viruses may affect the virulence of crop pathogens. The hypovirulence caused by *Cryphonectria* hypoviruses (CHVs) on *Cryphonectria parasitica*, the causal agent of chestnut blight, is a typical example for this group (Robin et al., 2000). Infections by hypovirulent isolates result in superficial cankers on both European *(Castanea sativa)* and American *(C. dentata)* chestnut trees, whereas virus-free isolates cause deep, lethal cankers. CHVs are cytoplasmic double-stranded RNA viruses that move into conidia, but not into ascospores, and can be transmitted from an infected isolate to a virus-free isolate through hyphal anastomosis (Robin et al., 2000). In France, the Ministry of Agriculture has promoted an intensive release of hypovirulent isolates of *C. parasitica* for control of chestnut blight on 180,000 ha since 1974 (Robin et al., 2000). Mixtures of hypovirulent isolates, selected according to the vegetative compatibility types present in the populations of *C. parasitica,* were released in France. Every year in these orchards, all newly formed cankers are treated with hypovirulent isolates by introducing the mycelia into holes in the margin of the canker. This treatment shows a curative effect and contributes to the healing of the canker when the virus is successfully transmitted. The hypoviruses have a preventative effect depending on the spread of hypovirulent strains within the area where they are released. Hypoviruses are used to manage *C. parasitica* in Italy also (Bisiach et al., 1991). The major problem in exploiting hypoviruses as biocontrol agents is the existence of several vegetative compatibility (VC) types in the populations of *C. parasitica.* Most of the virus isolates are specific to the VC types. The rate of transmission of the virus is negatively correlated with the number of vegetative incompatibility genes that differ between mycelia that anastomose. In North America, the lack of spread of hypoviruses is characterized by high VC diversity. In Europe, low VC-type diversity favors the spread of hypovirulent isolates of *C. parasitica* and the recovery of European chestnut in many areas (Bissegger et al., 1997).

Bacillus Species

Different *Bacillus* spp. have been developed as microbial pesticides. *Bacillus subtilis* is available as a commercial formulation named FZB24. It is produced in a multi-stage liquid fermentation process from a stock culture that guarantees a uniform strain identity. The spores formed in this process are separated from the culture broth and then dried and formulated together with protective colloids, inert material, and other additives. The formulated end product has a storage stability of at least two years. This product was registered in Germany (Junge et al., 2000). The mode of action of this strain has been reported. The bacterium might act by competition by temporary colonization of the rhizosphere and rhizoplane. *Bacillus subtilis* has the ability to form antibiotics in vitro, but the in vivo production of antibiotics has not been demonstrated. *Bacillus subtilis* induces resistance by activation of defense genes. The bacterium also promotes plant and root growth, probably by producing cytokinins and auxins. The stronger root system ultimately leads to an uptake of water and nutrients. The growth promotion leads to the possibility of disease escape (Kilian et al., 2000). *Bacillus subtilis* strain BACT-0 formulations with vermiculite or kaolin as carrier material effectively control *Pythium aphanidermatum* infection in lettuce (Amer and Utkhede, 2000). *Bacillus cereus* applied as seed treatment was as effective as the fungicide metalaxyl in the management of seedling diseases in lucerne (Kazmar et al., 2000). *Bacillus subtilis* strain EBW-1 as a root dip effectively controls *Agrobacterium tumefaciens* crown gall in apples (Utkhede, 1999). Addition of nitrate to soil increased the efficacy of *B. subtilis* in control of soil-borne fungal diseases (Knox et al., 2000). When used as a spray, *B. polymyxa* decreases the spread of rose powdery mildew and rust on leaves of *Antirrhinum* (Saniewska et al., 1998).

Pantoea

Pantoea agglomerans (Erwinia herbicola) is an effective biocontrol agent against the apple and pear fire blight pathogen *Erwinia amylovora* (Johnson et al., 2000). *Pantoea agglomerans* was sprayed on pear and apple cultivars and the effect of environmental factors on the spread of the biocontrol agent was assessed. The introduced bacteria colonized the blossoms of the inoculated trees, and temperature was found to be the important variable affecting successful spread of this biocontrol agent from blossom to blossom. The bacterial populations were positively correlated with mean degree hours per day during bloom and negatively correlated with the proportion of days with rain (Johnson et al., 2000). *Pantoea agglomerans* is spread by honey bees *(Apis mellifera)* in apple and pear orchards in New Zealand (Vanneste et al., 1999) and the United States (Pusey, 1999). *Pantoea agglomerans* strain E25 was highly effective in the control of the fire blight pathogen in the United States (Pusey, 1999). *Pantoea agglomerans* effectively controls apple storage diseases caused by *Botrytis cinerea* and *Penicillium expansum* (Sobiczewski and Bryk, 1999).

Lyophilized, talc-based, and whey-based formulations of *P. agglomerans* were developed (Ozaktan et al., 1999). In these formulations, the bacteria survived up to 180 days of storage at 10°C and up to 60 days at 24°C. Talc-based formulations were more effective in reducing pear fruit and blossom blight caused by *E. amylovora* than the lyophilized and whey-based formulations (Ozaktan et al., 1999).

Agrobacterium, Serratia, Streptomyces, and Rhizobium

Agrobacterium radiobacter is an effective biocontrol agent that controls crown gall of various crops caused by *Agrobacterium tumefaciens*. Commercial formulations of *A. radiobacter* are available in the United States, Australia, and New Zealand (Whipps, 1996). *Agrobacterium radiobacter* produces the bacteriocin agrocin 84, which inhibits *A. tumefaciens*. *Agrobacterium radiobacter* strains, which lack agrocin 84 production, are not effective against the crown gall pathogen. Increased doses of the antagonist may reduce the disease severity only to certain extent. The amount of disease suppression per unit of antagonist dose decreased with increasing antagonist dose (Johnson and DiLeone, 1999).

Serratia marcescens controls several diseases. It controls Sclerotinia minor infection in lettuce (El-Tarabily et al., 2000). Serratia plymuthica strain RIGG4 stimulates defense reactions in cucumber seedlings inoculated with Pythium ultimum (Benhamou et al., 2000). The antagonist induced cell wall apposition in cucumber, and callose, pectin, and cellulose appeared in the wall appositions (Benhamou et al., 2000). Streptomyces griseoviridis is another important biocontrol agent. Commercial formulations of the bacterium are available and it controls Alternaria brassicicola in cauliflower and damping off of pepper (Whipps, 1996). Some Rhizobium strains also acted as biocontrol agents (Simpfendorfer et al., 1999).

COMMERCIALLY AVAILABLE MICROBIAL PESTICIDES

Commercialization of biocontrol agents as microbial pesticides requires several steps, beginning with initial discovery and then proceeding through testing of efficacy, prototyping, and then commercial production, extensive large-scale field testing, toxicology and environmental tests, registration, and marketing (Harman, 2000). Toxicological tests, such as oral, dermal, ocular, respiratory, and health hazards using test animals and fish, should show no adverse effects and the microorganism should not be a pathogen. Success in selling biocontrol products requires that potential users and distributors be educated and convinced about the value of a biological product that is probably more conceptually difficult to use than standard pesticides. It requires several years and millions of dollars to bring a single biocontrol agent to market and to become profitable. Hence, only a few biocontrol agents have been developed as commercial products. They are mainly produced by small companies in different countries. Actizyme is a commercial granulated formulation of Bacillus subtilis (Walker and Morey, 1999). Tri-D25 is a dry formulation containing Trichoderma koningii and T. harzianum (Walker and Morey, 1999). FZB24 is a product of Bayer AG, Germany, and it consists of B. subtilis (Kilian et al., 2000). The yeast Candida oleophila is commercially available under the trade name Aspire (El-Ghaouth et al., 2000). Two strains of the bacterium Pseudomonas syringae are available under the trade names Biosave-100 and Biosave-110 (El-Ghaouth et al., 2000). The following are the other commercially available microbial pesticides:

Agrobacterium radiobacter—Diegall (Fruit growers Chemical Co., New Zealand), Galltrol (AgBio Chem, Inc., California), Norbac 84-C (New Bioproducts Inc., California), NoGall (Root Nodule, Pty Ltd., Australia; Bio-Care Technology, Pty Ltd., Australia)

Ampelomyces quisqualis—AQ10 (Ecogen, Langhorne, Pennsylvania)

Bacillus subtilis—Kodiak, Kodiak (A-13), and Epic (MB 1600) (Gustafson, Inc., Texas), Bactophyt (NPO Vector, Novosibirsk, Russia), System 3 (GBO3) (Helena Chemical Co., Tennessee)

Coniothyrium minitans—Coniothyrin (Russian Govt.), Contans (Prophyta Biologischer Pflanzenschutz GmbH, Germany)

Fusarium oxysporum (nonpathogenic)—Fusaclean (Fo47) (Natural Plant Protection, Noguerres, France), Biofox-C (S.I.A.P.A., Bologna, Italy)

Gliocladium catenulatum—PrimaStop (Kemira OY, Finland), PreStop (Kemira OY, Finland)

Peniophora (Phlebia) gigantea—Pg suspension (Ecological Laboratory Ltd., U.K.), Rotstop (Kemira OY, Finland)

Pseudomonas (Burkholderia) cepacia—Intercept (Soil Technologies, Fairfield, Iowa), Blue Circle and Deny (CTT Corporation, Carlsbad, California)

Pseudomonas fluorescens—BioCoat (SandG Seeds, BV, the Netherlands), Conqueror (Mauri Foods, Australia), Dagger (no longer available)

Pythium oligandrum—Polygandron (Vyzkummy ustov rastlinnej, Slovak Republic)

Ralstonia solanacearum (nonpathogenic)—PSSOL (Natural Plant Protection, Noguerres, France)

Streptomyces griseoviridis –Mycostop (Kemira OY, Finland), Stimagrow (Kemira OY, Finland)

Trichoderma harzianum T-22—Topshield, aka 1295-22, KRL—AG2, ATC 20847 (Bioworks, Geneva, NY), T-22G and T-22B (TGT Inc., New York), RootShield (Bioworks, Geneva, NY)

Trichoderma harzianum—T-35 (Makhteshim-Agan Chemicals, Israel), Harzian 20 and Harzian 10 (Natural Plant Protection, Noguerres, France), F-stop (Eastman Kodak Co., United States TGT Inc., New York), Supraavit (Bonegaard and Reitzel, Denmark)

Trichoderma harzianum strain T39—Trichodex (Makhteshim-Agan Chemicals, Israel)

Trichoderma harzianum + *T. polysporum*—BINAB-T and W (Bio-Innovation AB, Toreboda, Sweden)

Trichoderma harzianum + *T. viride*—Trichodowels, Trichoject, Trichopel, and Trichoseal (Agrimm Technologies Ltd., New Zealand)

Trichoderma spp.—Trichodermin (Bulgarian and Russian Governments), Promot (JH Biotech, Inc., Ventura, California), Solsain, Hors-solsain,

Plantsain (Prestabiol, Montpellier, France), ANTI-FUNGUS
(Grondontsmettingen De Ceuster, Belgium), Ty (Mycontrol, Israel)
Trichoderma virens (Gliocladium virens)—GlioGard and SoilGard
(Grace-Sierra Co., Maryland)
Trichoderma viridae—Bip T (Poland)

REFERENCES

Albert, F. and Anderson, A. J. (1987). The effect of *Pseudomonas putida* colonization on root surface peroxidase. *Plant Physiol,* 85:537-541.

Alstrom, S. (1991). Induction of disease resistance in common bean susceptible to
halo blight bacterial pathogen after seed bacterization with rhizosphere pseudomonads. *J Gen Appl Microbiol,* 37:495-501.

Amer, G. A. and Utkhede, R. S. (2000). Development of formulations of biological
agents for management of root rot of lettuce and cucumber. *Can J Microbiol,*
46:809-816.

Arras, G. and Arru, S. (1999). Integrated control of postharvest citrus decay and induction of phytoalexins by *Debaryomyces hansenii. Adv Hort Sci,* 13:76-81.

Arras, G., Dessi, R., Sanna, P., and Arru, S. (1999). Inhibitory activity of yeasts
isolated from fig fruits against *Penicillium digitatum. Acta Horticulturae,* 485:
37-46.

Askary, H., Carriere, Y., Belanger, R. R., and Brodeur, J. (1998). Pathogenicity of
the fungus *Vetricillium lecanii* to aphids and powdery mildew. *Biocontrol Sci
Technol,* 8:23-32.

Back, J.-M., Howell, C. R., and Kenerley, C. M. (1999). The role of an extracellular
chitinase from *Trichoderma virens* Gv29-8 in the biocontrol of *Rhizoctonia
solani. Curr Genet,* 35:41-50.

Benhamou, N. and Brodeur, J. (2000). Evidence for antibiosis and induced host defense reactions in the interaction between *Verticillium lecanii* and *Penicillium
digitatum,* the causal agent of green mold. *Phytopathology,* 90:932-943.

Benhamou, N., Gagne, S., Le Quere, D., and Dehbi, L. (2000). Bacterial-mediated
induced resistance in cucumber: Beneficial effect of the endophytic bacterium
Serratia plymuthica on the protection against infection by *Pythium ultimum.
Phytopathology,* 90:45-56.

Benhamou, N., Kloepper, J. W., Quadt Hallmann, A., and Tuzun, S. (1996). Induction of defense-related ultrastructural modifications in pea root tissues inoculated with endophytic bacteria. *Plant Physiol,* 112:919-929.

Bisiach, M., De Martino, A., Intropido, M., and Molinari, M. (1991). Nuove
esperienze di protezione biologica contro il cancro della corteccia del castagno.
Frutticoltura, 12:55-58.

Bissegger, M., Rigling, D., and Heiniger, U. (1997). Population structure and disease development of *Cryphonectria parasitica* in European chestnut forests in
the presence of natural hypovirulence. *Phytopathology,* 87:50-59.

Bolar, J. P., Norelli, J. L., Wong, K.-W., Hayes, C. K., Harman, G. E., and Aldwinckle, H. S. (2000). Expression of endochitinase from *Trichoderma harzianum* in transgenic apple increases resistance to apple scab and reduces vigor. *Phytopathology,* 90:72-77.

Burns, J. R. and Benson, D. M. (2000). Biocontrol of damping-off of *Catharanthus roseus* caused by *Pythium ultimum* with *Trichoderma virens* and binucleate *Rhizoctonia* fungi. *Plant Dis,* 84:644-648.

Calderon, A. A., Zapata, J. M., Munoz, R., Pedreno, A. A., and Barcelo, A. R. (1993). Reveratrol production as a part of the hypersensitive-like response of grapevine cells to an elicitor from *Trichoderma viride. New Phytol,* 124:455-463.

Callan, N. W., Mathre, D. E., and Miller, J. B. (1990). Biopriming seed treatment for biological control of *Pythium ultimum* pre-emergence damping-off in Sh2 sweet corn. *Plant Dis,* 74:368-372.

Carsolio, C., Benhamou, N., Haran, S., Cortes, C., Gutierrez, A., Chet, I., and Herrera-Estrella, A. (1999). Role of the *Trichoderma harzianum* endochitinase gene, *ech42,* in mycoparasitism. *Appl Environ Microbiol,* 65:929-935.

Castoria, R., De Curtis, F., Lima, G., Caputo, L., Pacifico, S., and De Cicco, V. (2001). *Aureobasidium pullulans* (LS-30) an antagonist of postharvest pathogens of fruits: Study on its modes of action. *Postharvest Biol Technol,* 22:7-17.

Chen, C., Bauske, E. M., Musson, G., Rodriquez Kabana, R., and Kloepper, J. W. (1990). Biological control of *Fusarium* wilt in cotton by use of endophytic bacteria. *Biol Control,* 5:83-91.

Chet, I. (1987). *Trichoderma*—Application, mode of action, and potential as a biocontrol agent of soilborne plant pathogenic fungi. In I. Chet (Ed.), *Innovative Approaches to Plant Disease Control.* John Wiley and Sons, New York, pp. 137-160.

Cook, R. J. (1993). The role of biological control in the 21st century. In R. D. Lumsden and J. L. Vaughn (Eds.), *Pest Management: Biologically Based Technologies.* American Chemical Society, Washington, DC, pp. 10-20.

Corbell, N. and Loper, J. E. (1995). A global regulator of secondary metabolite production in *Pseudomonas fluorescens* Pf-5. *J Bacteriol,* 177:6230-6236.

Datnoff, L. E., Nemec, S., and Pernezny, K. (1995). Biological control of Fusarium crown and root rot of tomato in Florida using *Trichoderma harzianum* and *Glomus intraradices. Biol Control,* 5:427-431.

De Meyer, G., Bigirimana, J., Elad, Y., and Hofte, M. (1998). Induced systemic resistance in *Trichoderma harzianum* biocontrol of *Botrytis cinerea. Eur J Plant Pathol,* 104:279-286.

Deacon, J. W. (1994). Rhizosphere constraints affecting biocontrol organisms applied to seeds. In T. Martin (Ed.), *Seed treatment. Progress and Prospects.* British Crop Protection Council, Farnham, U.K., pp. 315-326.

Dowling, D. N. and O'Garra, F. (1994). Metabolites of *Pseudomonas* involved in the biocontrol of plant disease. *Trends in Biotechnology,* 12:133-141.

Duijff, B. J., Gianinazzi-Pearson, V., and Lemanceau, P. (1997). Involvement of the outer membrane lipopolysaccharides in the endophytic colonization of tomato root by biocontrol *Pseudomonas fluorescens* strain WC417r. *New Phytol,* 135: 325-334.

Elad, Y., David, D. R., Levi, T., Kapat, T., Kapat, A., Kirshner, B., Guvrin, E., and Levine, A. (1999). *Trichoderma harzianum* T39-mechanisms of biocontrol of foliar pathogens. In H. H. Lyr (Ed.), *Modern Fungicides and Antifungal Compounds II.,* Intercept Ltd., Andover, Hampshire, U.K., pp. 459-467.

Elad, Y. and Kapat, A. (1999). The role of *Trichoderma harzianum* protease in the biocontrol of *Botrytis cinerea. Eur J Plant Pathol,* 105:177-189.

El-Ghaouth, A., Smilanick, J. L., Brown, G. E., Ippolito, A., Wisniewski, M., and Wilson, C. L. (2000). Application of *Candida saitoana* and glycochitosan for the control of postharvest diseases of apple and citrus fruit under semi-commercial conditions. *Plant Dis,* 84:243-248.

El-Tarabily, K. A., Soliman, M. H., Nassar, A. H., Al-Hassani, H. A., Sivasithamparam, K., McKenna, F., Hardy, G. E. S. J. (2000). Biological control of *Sclerotinia minor* using a chitinolytic bacterium and actinomycetes. *Plant Pathol,* 49:573-583.

Fan, Q. and Tian, S. (2000). Postharvest biological control of Rhizopus rot of nectarine fruits by *Pichia membranefaciens. Plant Dis,* 84:1212-1216.

Fravel, D. R. and Roberts, D. P. (1991). In situ evidence for the role of glucose oxidase in the biocontrol of Verticillium wilt by *Talaromyces flavus. Biocontrol Sci Technol,* 1:91-99.

Fridlender, M., Inbar, J., and Chet, I. (1993). Biological control of soil-borne pathogens by a β-1,3 glucanase-producing *Pseudomonas cepacia. Soil Biol Biochem,* 25:1211-1221.

Frommel, M. I., Nowak, J., and Lazarovits, G. (1991). Growth enhancement and developmental modifications of in vitro grown potato (*Solanum tuberosum* ssp. *tuberosum*) affected by a nonfluorescent *Pseudomonas* sp. *Plant Physiol,* 96: 928-936.

Gerlagh, M., Goossen-van de Geijn, H. M., Fokkema, N. J., and Vereijken, P. F. G. (1999). Long-term biosanitation by application of *Coniothyrium minitans* on *Sclerotinia sclerotiorum*-infected crops. *Phytopathology,* 89:141-147.

Hagedorn, C., Gould, W. D., and Bardinelli, T. R. (1993). Field evaluation of bacterial inoculants to control seedling disease pathogens on cotton. *Plant Dis,* 77:278-282.

Hamdan, H., Weller, D. M., and Thomashow, L. S. (1991). Relative importance of fluorescent siderophores and other factors in biological control of *Gaeumannomyces graminis* var. *tritici* by *Pseudomonas fluorescens* 2-79 and M4-80R. *Appl Environ Microbiol,* 57:3270-3277.

Hara, H., Bangera, M., Kim, D.-S., Weller, D. M., and Thomashow, L. S. (1994). Effect of transfer and expression of antibiotic biosynthesis genes on biological control activity of fluorescent pseudomonads. In M. H. Ryder, P. M. Stephens, and G. D. Bowen (Eds), *Improving Plant Productivity with Rhizosphere Bacteria,* Commonwealth Scientific and Industrial Research Organization (CSIRO), Division of Soils, Adelaide, Australia, pp. 247-249.

Harman, G. E. (2000). Myths and dogmas of biocontrol. Changes in perceptions derived from research on *Trichoderma harzianum* T-22. *Plant Dis,* 84:377-393.

Harman, G. E., Taylor, A. G., and Stasz, T. E. (1989). Combining effective strains of *Trichoderma harzianum* and solid matrix priming to improve biological seed treatments. *Plant Dis,* 73:631-637.

Hoffland, E., Hakulinen, J., Pelt, J. A. V. (1996). Comparison of systemic resistance induced by avirulent and nonpathogenic *Pseudomonas* species. *Phytopathology,* 86:757-762.

Hoffland, E., Pieterse, C. M. J., Bik, L., and Van Pelt, J. A. (1995). Induced systemic resistance in radish is not associated with accumulation of pathogenesis-related proteins. *Physiol Mol Plant Pathol,* 46:309-320.

Hofte, M., Boelens, J., and Verstrete, W. (1991). Seed protection and promotion of seedling emergence by the plant growth beneficial *Pseudomonas* strains 7NSK2 and ANPI5. *Soil Biol Biochem,* 23:407-410.

Howell, C. R. (1998). The role of antibiosis in biocontrol. In G. E. Harman and C. P. Kubicek (Eds.), *Trichoderma and Gliocladium, Vol. 2.* Taylor and Francis, London, pp. 173-184.

Howell, C. R., Hanson, L. E., Stipanovic, R. D., and Puckhaber, L. S. (1999). Induction of terpenoid synthesis in cotton roots and control of *Rhizoctonia solani* by seed treatment with *Trichoderma virens. Phytopathology,* 90:248-252.

Ippolito, A., El-Ghaouth, A., Wilson, C. L., and Wisniewski, M. (2000). Control of postharvest decay of apple fruit by *Aureobasidium pullulans* and induction of defense responses. *Postharvest Biol Technol,* 19:265-272.

Jakobek, J. L. and Lindgren, P. B. (1993). Generalized induction of defense responses in bean is not correlated with the induction of the hypersensitive response. *Plant Cell,* 5:49-56.

Johnson, K. B. and DiLeone, J. A. (1999). Effect of antibiosis on antagonist dose-plant disease response relationships for the biological control of crown gall of tomato and cherry. *Phytopathology,* 89:974-980.

Johnson, K. B., Stockwell, V. O., Sawyer, T. L., and Sugar, D. (2000). Assessment of environmental factors influencing growth and spread of *Pantoea agglomerans* on and among blossoms of pear and apple. *Phytopathology,* 90:1285-1294.

Junge, H., Krebs, B., and Kilian, M. (2000). Strain selection, production, and formulation of the biological plant vitality enhancing agent FZB24 *Bacillus subtilis. Pflanzenschutz Nachrichten Bayer,* 53:94-104.

Kazmar, E. R., Goodman, R. M., Grau, C. R., Johnson, D. W., Nordheim, E. V., Undersander, D. J., and Handelsman, J. (2000). Regresssion analyses for evaluating the influence of *Bacillus cereus* on alfalfa yield under variable disease intensity. *Phytopathology,* 90:657-665.

Kempf, H.-J., Sinterhauf, S., Muller, M., and Paclatko, P. (1994). Production of two antibiotics by a biocontrol bacterium in the spermosphere of barley and in the rhizosphere of cotton. In M. H. Ryder, P. M. Stephens, and G. D. Bowen (Eds.), *Improving Plant Productivity with Rhizosphere Bacteria,* CSIRO, Division of Soils, Adelaide, Australia, pp.114-116.

Kilian, M., Steiner, U., Krebs, B., Junge, H., Schmiede-Knecht, G., and Hain, R. (2000). FZB24 *Bacillus subtilis*—mode of action of a microbial agent enhancing plant vitality. *Pflanzenschutz-Nachrichten Bayer,* 53:72-93.

Kloepper, J. W., Tuzun, S., Liu, L., and Wei, G. (1993). Plant-growth promoting rhizobacteria as inducers of systemic resistance. In R. D. Lumsden and J. L. Vaughn (Eds.), *Pest Management: Biologically Based Technologies.* American Chemical Society Press, Washington, DC, pp. 156-165.

Knox, O. G. G., Killham, K., and Leifert, C. (2000). Effects of increased nitrate availability on the control of plant pathogenic fungi by the soil bacterium *Bacillus subtilis. Appl Soil Ecol,* 15:227-231.

Koch, E., Kempf, H. J., and Hessenmuller, A. (1998). Characterization of the biocontrol activity and evaluation of potential plant growth-promoting properties of selected rhizobacteria. *Z Pflanzenkrankh Pflanzenschutz,* 105:567-580.

Köhl, J., Gerlagh, M., De Haas, B. H., and Krijger, M. C. (1998). Biological control of *Botrytis cinerea* in cyclamen with *Ulocladium atrum* and *Gliocladium roseum* under commercial growing conditions. *Phytopathology,* 88:568-575.

Köhl, J., Gerlagh, M., and Grit, G. (2000). Biocontrol of *Botrytis cinerea* by *Ulocladium atrum* in different production systems of cyclamen. *Plant Dis,* 84:569-573.

Kovach, J., Petzoldt, R., and Harman, G. E. (2000). Use of honey bees and bumble bees to disseminate *Trichoderma harzianum* 1295-22 to strawberries for *Botrytis* control. *Biol Control,* 18:235-242.

Kraus, J. and Loper, J. E. (1995). Characterization of a genomic region required for production of the antibiotic pyroluteorin by the biological control agent *Pseudomonas flurorescens* Pf-5. *Appl Environ Microbiol,* 61:849-854.

Kurdish, I. K., Roi, A. A., Gragulya, A. D., and Kiprianova, E. A. (1999). Survival and antagonistic activity of *Pseudomonas aureofaciens* UKM-111 stored in fine materials. *Microbiology (New York),* 68:332-336.

Latunde-Dada, A. O. (1993). Biological control of southern blight disease of tomato caused by *Sclerotium rolfsii* with simplified mycelial formulations of *Trichoderma koningii. Plant Pathol,* 42: 522-529.

Laville, J., Voisard, C., Keel, C., Maurhofer, M., Defago, G., and Haas, D. (1992). Global control in *Pseudomonas fluorescens* mediating antibiotic synthesis and suppression of black root rot of tobacco. *Proc Natl Acad Sci USA,* 89:1562-1566.

Leeman, M., Den Ouden, F. M., Van Pelt, J. A., Dirkx, F. P. M., Steijl, H., Bakker, P. A. H. M., and Schippers, B. (1996). Iron availability affects induction of systemic resistance to fusarium wilt of radish by *Pseudomonas fluorescens. Phytopathology,* 86:149-155.

Leeman, M., Van Pelt, J. A., Den Ouden, F. M., Heinsbroek, M., Bakker, P. A. H. M., and Schippers, B. (1995a). Induction of systemic resistance against fusarium wilt of radish by lipopolysaccharides of *Pseudomonas fluorescens. Phytopathology,* 85:1021-1027.

Leeman, M., Van Pelt, J. A., Den Ouden, F. M., Heinsbroek, M., Bakker, P. A. H. M., and Schippers, B. (1995b). Induction of systemic resistance by *Pseudomonas fluorescens* in radish cultivars differing in susceptibility to fusarium wilt, using a novel bioassay. *Eur J Plant Pathol,* 101:655-664.

Levy, E., Eyal, Z., and Chet, I. (1998). Suppression of *Septoria tritici* blotch and leaf rust on wheat seedlings by pseudomonads. *Plant Pathol,* 37:551-557.

Lewis, J. A., Papavizas, G. C., and Lumsden, R. D. (1991). A new formulation system for the application of biocontrol fungi to soil. *Biocontrol Sci Technol,* 1:59-69.

Liu, L., Kloepper, J. W., and Tuzun, S. (1992). Induction of systemic resistance against cucumber mosaic virus by seed inoculation with selected rhizobacterial strains. *Phytopathology,* 82:1109.

Liu, L., Kloepper, J. W., and Tuzun, S. (1995a). Induction of systemic resistance in cucumber against bacterial angular leaf spot by plant growth-promoting rhizobacteria. *Phytopathology,* 85:843-847.

Liu, L., Kloepper, J. W., and Tuzun, S. (1995b). Induction of systemic resistance in cucumber against fusarium wilt by plant growth-promoting rhizobacteria. *Phytopathology,* 85:695-698.

Loper, J. E., Corbell, N., Kraus, J., Nowak-Thompson, B., Henkels, M. D., and Carniege, S. (1994). Contributions of molecular biology towards understanding mechanisms by which rhizosphere pseudomonads effect biological control. In M. H. Ryder, P. M. Stephens, and G. D. Bowen (Eds.), *Improving Plant Productivity with Rhizosphere Bacteria.* CSIRO, Division of Soils, Adelaide, Australia, pp. 89-96.

Loper, J. E. and Lindow, S. E. (1994). A biological sensor for iron available to bacteria in their habitats on plant surfaces. *Appl Environ Microbiol,* 60:1934-1941.

Lorito, M., Woo, S. L., Garcia Fernandez, I., Colucci, G., Harman, G. E., Pintor-Toro, J. A., Filippone, E., Mucciflora, S., Lawrence, C. B., Zoina, A., Tuzun, S., and Scala, F. (1998). Genes from mycoparasitic fungi as a source for improving plant resistance to fungal pathogens. *Proc Natl Acad Sci USA,* 95:7860-7865.

Madsen, A. M. and Neergaard, E. De (1999). Interactions between the mycoparasite *Pythium oligandrum* and sclerotia of the plant pathogen *Sclerotinia sclerotiorum. Eur J Plant Pathol,* 105:761-768.

Maplestone, P. A., Whipps, J. M., and Lynch, J. M. (1991). Effect of peat-bran inoculum of *Trichoderma* species on biological control of *Rhizoctonia solani* in lettuce. *Plant Soil,* 136:257-263.

Mathre, D. E., Cook, R. J., and Callan, N. W. (1999). From discovery to use: Traversing the world of commercializing biocontrol agents for plant disease control. *Plant Dis,* 83:972-983.

Mathre, D. E., Johnston, R. H., and Grey, W. E. (1998). Biological control of take-all disease of wheat caused by *Gaeumannomyces graminis* var. *tritici* under field conditions using a *Phialophora* sp. *Biocontrol Sci Technol,* 8:449-457.

Maurhofer, M., Hase, C., Meuwly, P., Metraux, J. P., and Defago, G. (1994). Induction of systemic resistance of tobacco to tobacco necrosis virus by the root colonizing *Pseudomonas fluorescens* strain CHA0: Influence of the *gagA* gene and of pyoverdine production. *Phytopathology,* 84:139-146.

Maurhofer, M., Keel, C., Hass, D., and Defago, G. (1995). Influence of plant species on disease suppression by *Pseudomonas fluorescens* strain CHA0 on its disease suppressive capacity. *Phytopathology,* 82:190-195.

Maurhofer, M., Sacherer, P., Keel, C., Haas, D., and Defago, G. (1994). Role of some metabolites produced by *Pseudomonas fluorescens* strain CHA0 in the suppression of different plant diseases. In M. H. Ryder, P. M. Stephens, and G. D. Bowen (Eds.), *Improving Plant Productivity with Rhizosphere Bacteria.* CSIRO, Division of Soils, Adelaide, Australia, pp. 89-96.

Meena, R., Radhajeyalakshmi, R., Marimuthu, T., Vidhyasekaran, P., Sabitha, D., and Velazhahan, R. (2000). Induction of pathogenesis-related proteins, phenolics, and phenylalanine ammonia-lyase in groundnut by *Pseudomonas fluorescens. J Plant Dis Protection,* 107:514-527.

Meier, B., Shaw, N., and Slusarenko, A. J. (1993). Spatial and temporal accumulation of defense gene transcripts in bean *(Phaseolus vulgaris)* leaves in relation to bacteria-induced hypersensitive cell death. *Mol Plant-Microbe Interact,* 6:453-466.

Mendgen, K. (1981). Growth of *Verticillium lecanii* in pustules of stripe rust *(Puccinia striiformis). Phytopathol Z,* 102:301-309.

Mew, T. W. and Rosales, A. M. (1986). Bacterization of rice plants for control of sheath blight caused by *Rhizoctonia solani. Phytopathology,* 76:1260-1264.

Meyer, J. M., Azelvandre, P., and Georges, C. (1992). Iron metabolism in *Pseudomonas:* Salicylic acid, a siderophore of *Pseudomonas fluorescens* CHA0. *Biofactors,* 4:23-27.

M'Piga, P., Belanger, R. R., Paulitz, T. C., and Benhamou, N. (1997). Increased resistance to *Fusarium oxysporum* f. sp. *radicis lycopersici* in tomato plants treated with the endophytic bacterium *Pseudomonas fluorescens* strain 6328. *Physiol Mol Plant Pathol,* 50:301-320.

Nelson, E. B., Harman, G. E., and Nash, G. T. (1988). Enhancement of *Trichoderma*-induced biological control of *Pythium* seed rot and preemergence damping-off of peas. *Soil Biol Biochem,* 20:145-150.

Nemec, S., Datnoff, L. E., and Strandberg, J. (1996). Efficacy of biocontrol agents in planting mixes to colonize plant roots and control root diseases of vegetables and citrus. *Crop Prot,* 15:735-742.

Niemi, M. and Lahdenpera, M. L. (2000). *Gliocladium catennulatum* J1446—a new biofungicide for horticultural crops. *17th Danish Plant Protection Conference, Horticulture, Tjele, Denmark, DJF Rapport, Havebrug,* 12:81-88.

Ohno, Y., Okuda, S., Natsuaki, T., and Teranaka, M. (1992). Control of bacterial seedling blight of rice by fluorescent *Pseudomonas* spp. *Proc Kanto-Tosan Plant Prot Soc,* 39:9-11.

Ozaktan, H., Bora, T., Sukan, S., Sargin, S., and Sukan, F. V. (1999). Studies on determination of antagonistic potential and biopreparation of some bacteria against the fireblight pathogen. *Acta Horticulturae,* 489:663-668.

Picard, K., Ponchet, M., Blein, J. P., Rey, P., Tirilly, Y., and Benhamou, N. (2000). Oligandrin: A proteinaceous molecule produced by the mycoparasite *Pythium oligandrum* induces resistance to *Phytophthora parasitica* infection in tomato plants. *Plant Physiol,* 124:379-395.

Pusey, P. L. (1999). Laboratory and field trials with selected microorganisms as biocontrol agents for fire blight. *Acta Horticulturae,* 489:655-661.

Raaijmakers, J. M., Leeman, M., Van Oorschot, M. M. P., Van der Sluis, I., Schippers, B., and Bakker, P. A. H. M. (1995). Dose response relationship in biological control of *Fusarium* wilt of radish by *Pseudomonas* sp. *Phytopathology*, 85:1075-1081.

Rabindran, R. and Vidhyasekaran, P. (1996). Development of a formulation of *Pseudomonas fluorescens* for management of rice sheath blight. *Crop Protection*, 14:714-721.

Raupach, G. S., Liu, L., Murphy, J. F., Tuzun, S., and Kloepper, J. W. (1996). Induced systemic resistance in cucumber and tomato against cucumber mosaic cucumovirus using plant growth-promoting rhizobacteria (PGPR). *Plant Dis*, 80:1107-1108.

Ricard, T. and Jorgensen, H. (2000). BINAB's effective, economical, and environment compatible *Trichoderma* products as possible Systemic Acquired Resistance (SAR) inducers in strawberries. *17th Danish Plant Protection Conference, Horticulture, Tjele, Denmark, DJF Rapport, Havebrug*, 12:67-75.

Robin, C., Anziani, C., and Cortesi, P. (2000). Relationship between biological control, incidence of hypovirulence, and diversity of vegetative compatibility types of *Cryphonectria parasitica* in France. *Phytopathology*, 90:730-737.

Saniewska, A., Orlikowski, L. B., and Wojdyla, A. T. (1998). *Bacillus polymyxa* in the control of soil-borne and leaf pathogens. *Prog Plant Protect*, 38:198-203.

Sayler, R. J., Wei, G., Kloepper, J. W., and Tuzun, S. (1994). Induction of β-1,3-glucanases and chitinases in tobacco by seed treatment with select strains of plant growth-promoting rhizobacteria. *Phytopathology*, 84:1107-1108.

Schena, L., Ippolito, A., Zhavi, T., Cohen, L., Nigro, F., and Droby, S. (1999). Genetic diversity and biocontrol activity of *Aureobasidium pullulans* isolates against postharvest rots. *Postharvest Biol Technol*, 17:189-199.

Schneider, S. and Ullrich, W. R. (1994). Differential induction of resistance and enhanced enzyme activities in cucumber and tobacco caused by treatment with various abiotic and biotic inducers. *Physiol Mol Plant Pathol*, 45:715-721.

Schnider, U., Keel, C., Blumer, C., Troxler, J., Defago, G., and Haas, D. (1995). Amplification of the housekeeping sigma factor in *Pseudomonas fluorescens* CHA0 enhances antibiotic production and improves biocontrol abilities. *J Bacteriol*, 177:5387-5392.

Schnider, U., Keel, C., Voisard, C., Defago, G., and Haas, D. (1995). Tn5-directed cloning of *pqq* genes from *Pseudomonas fluorescens* CHA0: Mutational inactivation of the genes results in overproduction of the antibiotic pyoluteorin. *Appl Environ Microbiol*, 61:3856-3864.

Simpfendorfer, S., Harden, T. J., and Murray, G. M. (1999). The in vitro inhibition of *Phytopthora clandestina* by some rhizobia and the possible role of *Rhizobium trifolii* in biological control of Phytophthora root rot of subterranean clover. *Australian J Agric Res*, 50:1469-1473.

Sivasithamparam, K. and Ghisalberti, E. L. (1998). Secondary metabolism in *Trichoderma* and *Gliocladium*. In C. P. Kubicek and G. E. Harman (Eds.), *Trichoderma and Gliocladium, Vol. 1*. Taylor and Francis, London, pp.137-191.

Sneh, B. (1999). Biological control of *Rhizoctonia* diseases. 2. Use of non-pathogenic isolates of *Rhizoctonia* in biological control. *Summa Phytopathologica,* 25:102-106.

Sobiczewski, P. and Bryk, H. (1999). The possibilities and limitations of biological control of apples against gray mold and blue mold with bacteria *Pantoea agglomerans* and *Pseudomonas* sp. *Prog Plant Protection,* 39:139-147.

Taylor, A. G., Min., T.-G., Harman, G. E., and Jin, X. (1991). Liquid coating formulation for the application of biological seed treatments of *Trichoderma harzianum. Biol Control,* 1:16-22.

Thomashow, L. S., and Pierson, L. S., III (1991). Genetic aspects of phenazine antibiotic production by fluorescent pseudomonads that suppress take-all disease of wheat. In H. Hennecke and D. P. S. Verma (Eds.), *Advances in Molecular Genetics of Plant-Microbe Interactions.* Kluwer Academic Publishers, Dordrecht, The Netherlands, pp. 443-449.

Thomashow, L. S., Weller, D. M., Bonsall, R. F., and Pierson, L. S. III (1990). Production of the antibiotic phenazine-1-carboxylic acid by fluorescent *Pseudomonas* species in the rhizosphere of wheat. *Appl Environ Microbiol,* 56:908-912.

Thrane, C., Olsson, S., Nielsen, T. H., and Sorensen, J. (1999). Vital fluorescent stains for detection of stress in *Pythium ultimum* and *Rhizoctonia solani* challenged with viscosinamide from *Pseudomonas fluorescens* DR54. *FEMS Microbiol Ecol,* 30:11-23.

Tosi, L. and Zazzerini, A. (1994). Evaluation of some fungi and bacteria for potential control of safflower rust. *J Phytopathol,* 142:131-140.

Trigalet, A., Trigalet-Demery, D., and Feuillade, R. (1998). Aggressiveness of French isolates of *Ralstonia solanacearum* and their potential use in biocontrol. *Bulletin OEPP,* 28:101-107.

Utkhede, R. S. (1999). Treatment with *Bacillus subtilis* for control of crown gall on young apple trees. *Allelopathy J,* 6:261-266.

Van Peer, R., Niemann, G. J., and Schippers, B. (1991). Induced resistance and phytoalexin accumulation in biological control of fusarium wilt of carnation by *Pseudomonas* sp. strain WCS417r. *Phytopathology,* 81:728-734.

Van Peer, R. and Schippers, B. (1992). Lipopolysaccharides of plant growth-promoting *Pseudomonas* spp. strain WCS 417r induce resistance in carnation to *Fusarium* wilt. *Neth J Plant Pathol,* 98:129-139.

Van Wees, S. C. M., Pieterse, C. M. J., Trijssenaar, A., Westende, Y. A. M. van't, Hartog, F., and Van Loon, L. C. (1997). Differential induction of systemic resistance in *Arabidopsis* by biocontrol bacteria. *Mol Plant-Microbe Interact,* 10: 716-724.

Vanneste, J. L., Cornish, D. A., Voyle, M. D., Haine, H. M., Goodwin, R. M. (1999). Honey bees to distribute beneficial bacteria to apple and asian pear flowers. *Acta Horticulturae,* 489:615-617.

Verhaar, M. A., Hijwegen, T., and Zadoks, J. C. (1996). Glasshouse experiments on biocontrol on cucumber powdery mildew *(Sphaerotheca fuliginea)* by the mycoparasites *Verticillium lecanii* and *Sporothrix rugulosa. Biol Control,* 6:353-360.

Verhaar, M. A., Hijwegen, T., and Zadoks, J. C. (1998). Selection of *Verticillium lecani* isolates with high potential for biocontrol of cucumber powdery mildew by means of components analysis at different humidity regimes. *Biocontrol Sci Technol,* 8:465-477.

Vidhyasekaran, P. (1990). Mycorrhiza-induced resistance, a mechanism for management of crop diseases. In B.L. Jalali and H. Chand (Eds.), *Current Trends in Mycorrhizal Research.* Proceedings of National Conference on Mycorrhiza, Hissar, India, pp. 91-93.

Vidhyasekaran, P. (1998). Biological suppression of major diseases of field crops using bacterial antagonists. In S. P. Singh and S. S. Hussaini (Eds.), *Biological Suppression of Plant Diseases, Phytoparasitic Nematodes and Weeds,* Project Directorate of Biological Control, Bangalore Publication, Bangalore, India, pp. 81-95.

Vidhyasekaran, P. (2001). Induced systemic resistance for the management of rice fungal diseases. In S. Sreenivasaprasad and R. Johnson (Eds.), *Major Fungal Diseases of Rice: Recent Advances.* Kluwer Academic Publishers, the Netherlands, pp. 347-358.

Vidhyasekaran, P., Kamala, N., Ramanathan, A., Rajappan, K., Paranidharan, V., and Velazhahan, R. (2001). Induction of systemic resistance by *Pseudomonas fluorescens* Pf1 against *Xanthomonas oryzae* pv. *oryzae* in rice leaves. *Phytoparasitica,* 29:155-166.

Vidhyasekaran, P. and Muthamilan, M. (1995). Development of formulations of *Pseudomonas fluorescens* for control of chickpea wilt. *Plant Dis,* 79:782-786.

Vidhyasekaran, P. and Muthamilan, M. (1999). Evaluation of *Pseudomonas fluorescens* for controlling rice sheath blight. *Biocontrol Sci Technol,* 9:67-74.

Vidhyasekaran, P., Rabindran, R., Muthamilan, M., Kamala N., Rajappan, K., Subramanian, N., and Vasumathi, K. (1997). Development of a powder formulation of *Pseudomonas fluorescens* for management of rice blast. *Plant Pathol,* 46:291-297.

Vidhyasekaran, P., Sethuraman, K., Rajappan, K., and Vasumathi, K. (1997). Powder formulations of *Pseudomonas fluorescens* to control pigeonpea wilt. *Biological Control,* 8:166-171.

Vidhyasekaran, P., Velazhahan, R., and Balasubramanian, P. (2000). Biological control of crop diseases exploiting genes involved in systemic induced resistance. In R. K. Upadhyay, K. G. Mukherji, and B. P. Chamola (Eds.), *Biocontrol Potential and Its Exploitation in Sustainable Agriculture. Volume 1: Crop Diseases, Weeds, and Nematodes.* Kluwer Academic/Plenum Publishers, New York, pp. 1-8.

Voisard, C., Bull, C., Keel, C., Laville, J., Maurhofer, M., Schnider, U., Defago, G., and Haas, D. (1994). Biocontrol of root diseases by *Pseudomonas fluorescens* CHA0: Current concepts and experimental approaches. In F. O'Gara, D. Dowling, and B. Boesten (Eds.), *Molecular Ecology of Rhizosphere Microorganisms.* VCH Publishers, Weinheim, Germany, pp. 67-89.

Walker, G. E. and Morey, B. G. (1999). Effects of chemicals and microbial antagonists on nematodes and fungal pathogens of citrus roots. *Aust J Expt Agric,* 39:629-637.

Wei, G., Kloepper, J. W., and Tuzun, S. (1991). Induction of systemic resistance of cucumber to *Colletotrichum orbiculare* by select strains of plant growth-promoting rhizobacteria. *Phytopathology,* 81:1508-1512.

Wei, G., Kloepper, J. W., and Tuzun, S. (1996). Induced systemic resistance to cucumber diseases and increased plant growth by plant growth-promoting rhizobacteria under field conditions. *Phytopathology,* 86:221-224.

Whipps, J. M. (1992). Status of biological disease control in horticulture. *Biocontrol Sci Technol,* 2:3-24.

Whipps, J. M. (1996). Development in the biological control of soil-borne plant pathogens. *Adv Bot Res,* 26:1-134.

Wilson, M. and Lindow, S. E. (1993). Interactions between the biological control agent on *Pseudomonas fluorescens* A50b and *Erwinia amylovora* in pear blossoms. *Phytopathology,* 83:117-123.

Wong, P. T. W., Mead, J. A., and Holley, M. P. (1996). Enhanced field control of wheat take-all using cold tolerant isolates of *Gaeumannomyces graminis* var. *graminis* and *Phialophora* sp. (lobed hyphopodia). *Plant Pathol,* 45:285-293.

Woo, S. L., Donzelli, B., Scala, F., Mach, R., Harman, G. E., Kubicek, C. P., Del Sorbo, G., and Lorito, M. (1999). Disruption of the *ech42* (endochitinase-encoding) gene affects biocontrol activity in *Trichoderma harzianum* P1. *Mol Plant-Microbe Interact,* 12:419-429.

Yakoby, N., Zhou, R., Kobiler, I., Dinoor, A., and Prusky, D. (2001). Development of *Colletotrichum gloeosporioides* restriction enzyme-mediated integration mutants as biocontrol agents against anthracnose disease in avocado fruits. *Phytopathology,* 91:143-148.

Yedidia, I., Benhamou, N., and Chet, I. (1999). Induction of defense responses in cucumber plants (*Cucumis sativus* L.) by the biocontrol agent *Trichoderma harzianum*. *Appl Environ Microbiol,* 65:1061-1070.

Yedidia, I., Benhamou, N., Kapulnik, Y., and Chet, I. (2000). Induction and accumulation of PR proteins activity during early stages of root colonization by the mycoparasite *Trichoderma harzianum* strain T-203. *Plant Physiol Biochem,* 38:863-873.

Zdor, R. E. and Anderson, A. J. (1992). Influence of root-colonizing bacteria on the defense responses of bean. *Plant Soil,* 140:99-107.

Zeilinger, S., Galhaup, C., Payer, K. L., Woo, S., Mach, R., Fekete-Csaba, L., Lorito, M., and Kubicek, C. P. (1999). Chitinase gene expression during mycoparasitic interaction of *Trichoderma harzianum* with its host. *Fungal Genet Biol,* 26:131-140.

Zhang, L. and Birch, R. G. (1997). Mechanisms of biocontrol by *Pantoea dispersa* of sugarcane leaf scald disease caused by *Xanthomonas albilineans*. *J Appl Microbiol,* 82:448-454.

Zhavi, T., Cohen, L., Weiss, B., Schena, L., Daus, A., Kaplunov, T., Zutkhi, J., Ben-Arie, R., and Droby, S. (2000). Biological control of *Botrytis, Aspergillus, and Rhizopus* rots on table and wine grapes in Israel. *Postharvest Biol Technol,* 20:115-124.

Zhou, T. and Paulitz, T. C. (1994). Induced resistance in the biocontrol of *Pythium aphanidermatum* by *Pseudomonas* spp. on cucumber. *J Phytopathol,* 142:51-63.

18

Biological Control—Mycorrhiza

Mycorrhizal fungi are symbiotic organisms that live on the roots of several plants. Among them, endomycorrhizae are important because they confer resistance against pathogens. Several mycorrhizal fungi have been developed as commercial products and they can be used to reduce losses caused by diseases.

WHAT IS A MYCORRHIZA?

Mycorrhiza is a symbiotic association of a fungus with the roots of a plant. Three distinct classes of mycorrhiza are recognized: *ectomycorrhiza, endomycorrhiza,* and *ectendomycorrhiza.* In the case of ectomycorrhiza, the fungal symbionts penetrate intercellularly and partially replace the middle lamella between the cortical cells of the roots. This hyphal arrangement around such cortical cells is called the "Hartig net." The ectomycorrhizal fungal symbionts also form a dense, usually continuous, hyphal network or fungal mantle over the feeder root surface. Ectomycorrhizal association is normally seen on tree species. The majority of ectomycorrhizal fungi belong to the division Basidiomycota and the families of Amanitaceae, Tricholomataceae, Rhizopogonaceae, and Boletaceae. In contrast, the endomycorrhizal fungus penetrates the cortical cells of the feeder roots intracellularly. Such fungal symbionts form large vesicles and arbuscules in cortical tissues, and hence, they are called *vesicular-arbuscular mycorrhiza* (VAM) or *arbuscular mycorrhiza* (AM). Such fungi, however, do not form a dense fungal mantle. Instead they develop a loose, intermittent arrangement of mycelium with large spores on the root surface. Endomycorrhizal colonization takes place in many annual crops that do not form ectomycorrhiza, and the fungal symbionts belong to the division Oomycota and the family Endogonaceae.

Ectendomycorrhiza is the third class of mycorrhiza. This type of mycorrhiza is present on the roots of certain tree species under specific ecological situations. This type of mycorrhiza resembles ectomycorrhiza, in that it forms a Hartig net and a fungal mantle, and also resembles endomycorrhiza, because of the intracellular penetration of cortical tissue by these fungi. Mycorrhizal symbiosis plays a key role in nutrient cycling in the ecosystem and protects plants against environmental stress (Barea and Jeffries, 1995). Mycorrhizal fungi alter host physiology and induce biochemical changes in the host metabolism. An altered host metabolism may result in resistance against plant pathogens (Vidhyasekaran, 1990).

DISEASES CONTROLLED BY MYCORRHIZAL FUNGI

Since the 1970s, several researchers have demonstrated that endomy-corrhizal fungi, such as *Glomus mosseae, G. fasciculatum, G. intraradices,* and *G. multiculae,* could effectively control various diseases. The following are important diseases that are controlled by mycorrhizal fungi (Caron, 1989; Jalali, 1990; Linderman, 1994; Azcon-Aguilar and Barea, 1996; Cordier, 1998; Sharma and Adholeya, 2000):

> Alfalfa—root rot *(Thielaviopsis basicola),* wilt *(Verticillium dahliae)*
> Citrus—root rot *(Phytophthora parasitica),* root rot *(Thielaviopsis basicola)*
> Cotton—wilt *(Verticillium dahliae),* root rot *(Thielaviopsis basicola)*
> Cucumber—root rot *(Rhizoctonia solani),* damping-off (*Pythium* spp.)
> Onion—pink root *(Pyrenochaeta terrestris)*
> Pea—root rot *(Thielaviopsis basicola)*
> Peanut—root rot *(Sclerotium rolfsii)*
> Poinsettia—damping-off *(Pythium ultimum)*
> Soybean—root rot *(Fusarium solani),* root rot (*Phytophthora megasperma* var. *sojae*)
> Tobacco—root rot *(Thielaviopsis basicola)*
> Tomato—wilt (*Fusarium oxysporum* f. sp. *lycopersici*), crown and root rot (*F. oxysporum* f. sp. *radicis lycopersici*), root rot *(Thielaviopsis basicola),* damping-off *(Pythium aphanidermatum),* blight *(Phytophthora parasitica)*
> Wheat—take-all *(Gaeumannomyces graminis)*

MECHANISMS INVOLVED IN BIOCONTROL BY MYCORRHIZAE

Several hypotheses have been proposed to explain the mechanisms involved in the control of diseases by mycorrhizal fungi (Larsen, 2000).

Mycorrhizae may enhance nutrient uptake by plants and may strengthen the plants against pathogens. These fungi enhance root growth, and robust root development may compensate damage caused by pathogens. Exudates of roots colonized by arbuscular mycorrhizal fungi may affect the pathogen population in the mycorrhizosphere. Experiments in vitro showed that after 48 h in the presence of exudates from strawberry roots colonized by *Glomus etunicatum* and *G. monosporum,* sporulation of *Phytophthora fragariae* was reduced by about 64 and 67 percent, respectively (Norman and Hooker, 2000). A similar trend was observed in an in vivo system, with a 68 percent reduction in sporulation of *P. fragariae* in the mycorrhizosphere of colonized plants relative to sporulation in the mycorrhizosphere of uncolonized plants (Norman and Hooker, 2000).

An altered host metabolism may contribute to disease resistance induced by mycorrhizae (Benhamou et al., 1994; Cordier, 1998). Accumulation of phenolics in roots of plants due to mycorrhizal infection has been widely reported (Krishna and Bagyaraj, 1983; Cordier, 1998), and the role of phenolics in disease resistance is known. Increased accumulation of phytoalexins due to mycorrhizal infection has also been reported (Morandi et al., 1984; Harrison and Dixon, 1993). Chitinase transcripts accumulate in bean colonized by arbuscular mycorrhizal fungus *Glomus intraradices* (Blee and Anderson, 1996). β-1,3-Glucanase transcripts accumulate in and around arbuscule containing cells (Lambias and Mehdy, 1995). Chitinases and β-1,3-glucanases are important pathogenesis-related proteins that are involved in disease resistance (Vidhyasekaran, 1997, 2002). Dehne and Schoenbeck (1978) showed increased lignification of cells in the endodermis of mycorrhizal tomato and cucumber plants. Lignification is involved in cell wall thickening. Thus, mycorrhizal fungi may induce resistance against plant pathogens by several methods.

MASS PRODUCTION OF MYCORRHIZAL FUNGI

Endomycorrhizal fungi are obligate symbionts and hence they cannot be produced in a nutrient medium. They should be produced on living roots. This method is tedious and the risk of contamination with pathogens exists. AM fungi are mass multiplied on plants growing in disinfested soil. A highly susceptible trap plant is used for the multiplication of AM fungi. Stock cultures of these fungi are maintained in the form of colonized roots. These roots (with spores) are used to produce large amounts of inoculum in soil-based media. Soil in nursery beds are sterilized with methyl bromide. The AM fungal propagules are added to the soil and seeds of the trap plant (particularly monocots) are sown. The beds are watered regularly and kept free from weeds. After about three months, the spore production in infected roots is assessed and used for application in the field.

Soilless substrates are also used for multiplication of AM fungi. Shredded bark, calcined montmorillonite clay, expanded clay aggregates, pertile, soil-rite, and vermiculite have been used as inert substrates. The plants are grown in these substrates in the presence of AM fungal propagules, with frequent addition of nutrient solutions. Colonized roots and spores can be produced in hydroponics. Precolonized plants on sterile substrate are needed for this system. AM fungi can be propagated by growing precolonized plants in a defined nutrient solution that flows over the host roots. In an aeroponic system, the roots of the host are bathed in a fine mist of defined nutrient solutions suspended in air. Root-organ culture (axenic culture) is also used for multiplication of the mycorrhizal fungi (Sharma and Adholeya, 2000).

COMMERCIALIZATION OF MYCORRHIZAE

A few companies have taken up commercial production of arbuscular mycorrhizas. AGC Microbio of Cambridge, United Kingdom, produces mycorrhizal fungi in the trade name of Vaminoc. The other companies producing mycorrhizal fungi include: Bio-Enhancement Technologies, Camarillo, California; Horticultural Alliance, Sarasota, Florida; Plants Health Care, Pittsburg, Pennsylvania; Tree Pro, West Lafyette, Illinois; Biological Crop Protection, Wye, United Kingdom; Mikko-Tek Labs, Timmons, Ontario, Canada; Premier Tech, Quebec, Canada; Biorize, Dijon, France; Central Glass Co., Tokyo, Japan; and Global Horticare, Lelystad, The Netherlands.

REFERENCES

Azcon-Aguilar, C. and Barea, J. M. (1996). Arbuscular mycorrhizas and biological control of soil-borne plant pathogens—an overview of the mechanism involved. *Mycorrhiza,* 6:457-464.

Barea, J. M. and Jeffries, P. (1995). Arbuscular mycorrhizas in sustainable soil plant systems. In A. Varma and B. Hock (Eds.), *Mycorrhiza: Structure, Function, Molecular Biology and Biotechnology.* Springer-Verlag, Heidelberg, Germany, pp. 521-560.

Benhamou, N., Fortin, J. A., Hamel, C., St. Arnaud, M., and Shatilla, A. (1994). Resistance responses of mycorrhizal Ri T-DNA-transformed carrot roots to infection by *Fusarium oxysporum* f. sp. *chrysanthemi. Phytopathology,* 84:958-968.

Blee, K. A. and Anderson, A. J. (1996). Defense-related transcript accumulation in *Phaseolus vulgaris* L. colonized by arbuscular mycorrhizal fungus *Glomus intraradices* Schenk and Smith. *Plant Physiol,* 110:675-688.

Caron, M. (1989). Potential use of mycorrhizae in control of soil-borne diseases. *Can J Plant Pathol,* 11:177-179.

Cordier, C. (1998). Cell defense responses associated with localized and systemic resistance to *Phytophthora parasitica* induced in tomato by an arbuscular mycorrhizal fungus. *Mol Plant-Microbe Interact,* 11:1017-1028.

Dehne, H. W. and Schoenbeck, F. (1978). Investigation on the influence of endotrophic mycorrhiza on plant diseases. 3. Chitinase activity and ornithine cycle. *Z Pflanzenkrankh. Pflanzenschutz,* 85:666-678.

Harrison, M. J. and Dixon, R. A. (1993). Isoflavonoid accumulation and expression of defense gene transcripts during the establishment of vesicular-arbuscular mycorrhizal associations in roots of *Medicago truncata. Mol Plant-Microbe Interact,* 6:643-654.

Jalali, B. L. (1990). Mycorrhiza—a tool for biocontrol. In P. Vidhyasekaran (Ed.), *Basic Research for Crop Disease Management.* Daya Publishing House, Delhi, India, pp. 298-305.

Krishna, K. R. and Bagyaraj, D. J. (1983). Interaction between *Glomus fasciculatum* and *Sclerotium rolfsii* in peanut. *Can J Bot,* 41:2349-2351.

Lambias, M. R. and Mehdy, M. C. (1995). Differential expression of defense related genes in arbuscular mycorrhiza. *Can J Bot,* 73:533-540.

Larsen, J. (2000). Biological control of plant pathogenic fungi with arbuscular mycorrhiza. In *17th Danish Plant Protection Conference, Horticulture, Tjele, Denmark, DJF Rapport, Havebrug,* 12:43-49.

Linderman, R. G. (1994). Role of VAM fungi in biocontrol. In F. L. Pfleger and R. G. Linderman (Eds.), *Mycorrhizae and Plant Health.* APS Press, St. Paul, Minnesota, pp. 1-26.

Morandi, D., Bailey, J. A., and Gianinazzi-Pearson, V. (1984). Isoflavonoid accumulation in soybean roots infected with vesicular—arbuscular mycorrhizal fungi. *Physiol Mol Plant Pathol,* 24:357-364.

Norman, J. R. and Hooker, J. E. (2000). Sporulation of *Phytophthora fragariae* shows greater stimulation by exudates of non-mycorrhizal than by mycorrhizal strawberry roots. *Mycol Res,* 104:1069-1073.

Sharma, M. P. and Adholeya, A. (2000). Sustainable management of arbuscular mycorrhizal fungi in the biocontrol of soil-borne plant diseases. In R. K. Upadhyay, K. G. Mukherji, and B. P. Chamola (Eds.), *Biocontrol Potential and Its Exploitation in Sustainable Agriculture, Vol. I.* Kluwer Academic/Plenum Publishers, New York, pp. 117-138.

Vidhyasekaran, P. (1990). Mycorrhiza-induced resistance, a mechanism for management of crop diseases. In B.L. Jalali and H. Chand (Eds.), *Current Trends in Mycorrhizal Research.* Proceedings of National Conference on Mycorrhiza, Hissar, India, pp. 91-93.

Vidhyasekaran, P. (1997). *Fungal Pathogenesis in Plants and Crops: Molecular Biology and Host Defense Mechanisms.* Marcel Dekker, New York.
Vidhyasekaran, P. (2002). *Bacterial Disease Resistance in Plants: Molecular Biology and Biotechnological Applications.* The Haworth Press, Inc., Binghamton, NY.

19

Breeding for Disease Resistance—
Conventional Breeding

The most effective method of disease management is the use of disease-resistant plant varieties. This can be highly cost-effective, and growers need not spend anything to manage the diseases. Several genes are involved in conferring disease resistance in plants. Several types of genetic resistance have been reported. These resistances are described here. Epistasis, additive effect, and interactive effect of clusters of resistance genes have been reported. Molecular markers linked to resistance genes were developed and they are now used for selection of resistant plants from segregating populations of crosses. Various methods of breeding for disease resistance are described.

TYPES OF DISEASE RESISTANCE

Disease resistance can be classified into two major types: the host either resists the establishment of a successful parasitic relationship by restricting the infection site and the infection process, or it resists the subsequent colonization and reproduction of the parasite following successful infection. The first type of resistance affects disease onset by reducing the amount of effective initial inoculum. The other type of resistance affects the apparent infection rate and the amount of disease that finally develops in the plants (Nelson, 1978). The first type of resistance can be called "vertical resistance," "qualitative resistance," "race-specific resistance," "specific resistance," "major-gene resistance," "monogenic resistance," "complete resistance," or "true resistance." The resistance is "vertical" because the resistance is specific to races of the pathogen and is susceptible to other races (a race is a taxon of pathogens, particularly characterized by specialization to different cultivars of one host species). The resistance is "qualitative" because the disease symptom is almost completely suppressed, mostly by development of a hypersensitive pinpoint fleck reaction. It is called "race-specific/specific resistance" because the resistance is specific against races of the pathogen. It is called "major-gene resistance" because the resistance is governed by a major gene. Major genes are genes that have large, distinct phenotypic expressions showing clear Mendelian segregation. The resistance is also called "monogenic resistance" because it is governed by a single gene. It is called "complete/true resistance" because the resistance is expressed by complete suppression of disease symptom development.

The second type of resistance is called "quantitative resistance," "horizontal resistance," "race-nonspecific resistance," "nonspecific resistance," "minor gene resistance," "polygenic resistance," "general resistance," "partial resistance," or "field resistance." It is called "quantitative resistance" because the resistance is based on the amount of disease symptom development, quantified by different infection types such as necrotic flecks, necrotic and chlorotic areas with restricted sporulation, sporulation with chlorosis, abundant sporulation without chlorosis, lesion size, lesion area, etc. It is called "horizontal resistance" because this resistance is against several races of the pathogen. It is called "race-nonspecific/nonspecific resistance" because the resistance is not specific to a particular race only. It is called "minor gene resistance" because the resistance is governed by minor genes. Minor genes have small effects on the expression of the phenotype for resistance and show quantitative segregation. This resistance is also called "polygenic resistance" because it is governed by several genes. It is called "general/partial/field resistance" because this resistance is only partial and field tolerant.

Although these definitions are attractive, there are many exceptions to this broad classification. Preinfection factors may play a role in some quantitative/horizontal resistance, and sporulation is not affected in some qualitative/vertical resistance (Nelson, 1978). Separate major and minor genes may not exist. Whether major or minor, either kind of gene is inherited the same way and is capable of showing similar effects. The degree of expression is modified by environmental contribution to a large extent. Further, what works as a major gene in one genetic background could work as a minor gene in another background. Expression of resistance (major or minor) genes may be influenced or modified by other resistance genes in a different genetic background. When many major genes express together, they function exactly like minor genes.

Genetic resistance may vary depending on host age and environmental factors. Seedling resistance works when plants are young, and adult plant resistance works when the plants are mature. The wheat stem rust resistance gene *Sr25* from *Thinopyrum elongatum* is highly effective in the seedling stage of the wheat plant, but loses its effectiveness to some extent in adult wheat plants that are near maturity. The stem rust resistance gene *Sr2* is best expressed in adult wheat plants (McIntosh et al., 1995). The changes in host physiology during maturity would have contributed to this type of resistance. Seedling resistance may be based on major genes, whereas adult plant resistance is generally polygenic. However, exceptions are also common. Adult plant resistance of the durum wheat cultivar Glossy Huguenot is controlled by a single dominant gene (McIntosh, 1992). Temperatures may alter the expression of some resistance genes. The wheat stem rust resistance genes *Sr6, Sr12, Sr15, Sr17, Sr22, Sr34,* and *Sr38* are generally more effective at temperatures below 20°C. Genes *Sr13* and *Sr23* normally show more effectiveness at higher temperatures (McIntosh et al., 1995).

Two types of resistance genes have been recognized: recessive genes and dominant genes. A recessive gene is a gene that is phenotypically manifest in the homozygous state, but is masked in the presence of its dominant allele or dominant gene. Usually the dominant gene produces a functional product, whereas its recessive does not. The normal phenotype is produced if the dominant allele is present, and the mutant phenotype appears only in the absence of the dominant allele, i.e., when the recessive gene is homozygous. For example, wheat stem rust resistance genes *Sr2* and *Sr17* are recessively inherited, whereas all other *Sr* genes show dominant inheritance (McIntosh et al., 1995).

INTERACTION BETWEEN RESISTANCE GENES

Plants are endowed with several disease resistance genes. The interaction between these genes may vary. Epistasis, additive effects and interactive effects of clusters of resistance genes have been reported. Epistasis is generally observed when two or more wheat stem rust resistance *Sr* genes are present together. Only the gene conditioning the lowest infection type is expressed. An additive effect of resistance genes is observed in many plants. The wheat stem rust resistance genes *Sr7a* and *Sr12* were observed to interact, resulting in significantly higher levels of resistance than that conferred by either gene acting alone (McIntosh et al., 1995). An interactive effect of certain genes has also been reported. A gene conferring resistance to one disease may interact with a gene conferring resistance against another unrelated disease and enhance or suppress the action of the latter gene. Some suppressors that inhibit the expression of relevant resistance genes may be present on chromosomes. Suppressors on chromosomes 1D and 3D were reported to prevent the expression of stem rust resistance genes present on chromosomes 1B, 2B, and 7B in wheat. The wheat cultivar Thatcher and the backcross derivative 'Canthatch' were susceptible to several stem rust races because of a suppressor on chromosome 7DL that inhibited the expression of relevant resistance genes. Thatcher lines possessing the leaf rust resistance gene *Lr34* were more resistant to stem rust than Thatcher. In Thatcher and backcross derivatives, *Lr34* inactivates the 7DL suppressor (Kerber and Aung, 1999). Fast rusting to stem rust was reported in lines carrying leaf rust resistance genes *Lr28* and *Lr32,* suggesting that these genes might have suppressed the action of stem rust resistance genes.

MOLECULAR MARKER-ASSISTED SELECTION

The utility of molecular markers in resistance breeding is based on finding tight linkages between these markers and disease resistance genes. Such linkage permits one to infer the presence of the disease resistance gene by assaying for the marker. Incorporation of disease resistance genes into the susceptible varieties requires crosses with stocks that carry the resistance genes, followed by selection among the progeny for individuals possessing the disease resistance gene combination. Traditionally, progeny are screened for the presence of disease resistance genes by inoculation with the pathogen. However, simultaneous or even sequential screening of plants with several different pathogens can be difficult or impractical. Screening for resistance to new pathogens is also difficult because of quarantine restrictions on their shipment and use. In contrast, detecting disease resistance genes by their linkage to genetic markers makes it practical to screen for many different disease resistance genes simultaneously without the need to inoculate the population (Tanksley et al., 1989). Only the advanced generation would be required to be tested in disease nurseries. Based on host-pathogen interaction alone, it is often not possible to discriminate the presence of additional resistance (R)-genes (masking effect). With molecular marker-assisted selection (MAS), new R-gene segregation can be followed even in the presence of existing R-gene. Hence, R-genes from diverse sources can be incorporated into a single genotype for durable resistance. Pyramiding of resistance genes to fight against several races of the pathogen can be done using molecular markers.

Several types of molecular markers have been used to select resistance genes. These include restriction fragment length polymorphism, randomly amplified polymorphic DNA, simple sequence repeats (SSRs) or microsatellites, and polymerase chain reaction (PCR)-based DNA markers, such as sequence characterized amplified regions (SCAR), sequence-tagged sites (STS), and amplified fragment length polymorphisms (AFLP). RFLP is a technique in which plants may be differentiated by analysis of patterns derived from cleavage of their DNA. Genomic DNA from susceptible and resistant plants is isolated and cut with restriction endonuclease and size fractionated by gel electrophoresis. The DNA is transferred from the gel to a nitrocellulose filter and probed with a small piece of DNA (marker) in a Southern blot type experiment. The DNA fragment, when used to probe Southern blots of restricted genomic DNA from different plants, allows the visualization of variations in the size and/or number of the restriction fragments generated from the susceptible and resistant plants. RAPD and PCR-based techniques were described in Chapter 13. PCR can specifically amplify hypervariable regions within defined loci. These "minisatellite" DNA regions consist of tandemly repeated motifs of 15 to 30 nucleotides at various loci. Similarly, PCR can amplify hypervariable arrays of tandemly arranged "simple sequence" motifs of only two to ten nucleotides to detect DNA polymorphism. Another PCR-based technique is called "amplification fragment length polymorphism" (AFLP). It uses single oligonucleotide primers to amplify discrete but arbitrary stretches of DNA from a target genome. Amplification is done using a thermostable DNA polymerase.

Marker-assisted selection was used to select quantitative resistant lines against northern corn blight *(Exserohilum turcicum)* by using 87 RFLP markers and seven SSR markers (Welz and Geiger, 2000). RFLP and microsatellites or simple sequence repeats (SSRs) were used as genetic markers to select *Rsv,* a gene conferring resistance to *Soybean mosaic virus* in soybean (Yu et al., 1994). The presence of *xa13,* a recessive bacterial blight resistance gene in rice could not be identified by conventional approaches because the effect of *xa13* was masked by the dominant resistance gene *Xa21,* which was also present in those plants. The presence of *xa13* could be identified by the RFLP marker RG136 (Zhang et al., 1996). MAS is especially useful in selecting for recessive genes, such as *xa13,* where the presence of the gene in the heterozygous condition cannot be detected through traditional approaches without progeny testing. Six molecular markers have been identified to be linked to the rice bacterial blight resistance gene *Xa21.* They can be identified by RFLP and STS. These six molecular markers can be utilized in marker-assisted breeding in rice (Williams et al., 1996).

Two randomly amplified polymorphic DNA markers, $OPF6_{2700}$ and $OPH18_{2400}$, that are tightly linked to *Pi-10,* a dominant blast-resistance gene conferring complete resistance to race IB46 of the blast fungus, *Pyricularia grisea,* could be identified (Naqvi and Chattoo, 1996). Sequence characterized amplified regions (SCARs) markers were derived from $OPF6_{2700}$ and $OPH18_{2400}$, and these amplified RAPD products were cloned and sequenced. Nucleotide sequence information, obtained for each end of the two linked RAPD markers, was used to design 24-mer oligonucleotide primers for PCR amplification of the respective SCARs. Polymorphisms appearing as differences in the length of the SCARs' alternate alleles were considered for the indirect selection of *Pi-10.* Such polymorphisms converted the linked dominant RAPD loci into codominant SCAR markers. The SCAR markers will be useful in marker-assisted selection of the *Pi-10* locus (Naqvi and Chattoo, 1996).

Two SCAR markers were developed for the strawberry gene *Rpf1*, which confers resistance to red stele root rot caused by *Phytophthora fragariae* var. *fragariae*. The scar markers were developed originally from the sequence of RAPD OPO-16C$_{(438)}$ (Haymes et al., 2000). At least five RFLP marker alleles were present in all scab *(Venturia inaequalis)*-resistant selections arising from *Malus floribunda* (King et al., 1999). This was confirmed by SCAR assays and Southern-hybridization experiments. None of the marker alleles were present in the susceptible accessions (King et al., 1999). SCARs were developed based on nucleotide differences within resistance gene-like fragments isolated from a potato carrying the Ry$_{adg}$ gene, which confers extreme resistance to *Potato virus Y*. Two SCAR markers showed high accuracy for detection of the Ry$_{adg}$ gene (Kasai et al., 2000).

Fifteen AFLP markers were found linked to *Fom-2* gene, which confers resistance to races 0 and 1 of *Fusarium oxysporum* f. sp. *melonis* in melon (Wang et al., 2000). RAPD markers to the gene were also developed. Cleaved amplified polymorphic sequences (CAPS) and RFLP markers were developed by converting RAPD markers E07 and G17, respectively (Zheng et al., 1999). Both CAPS and RFLP markers were easier to score and more accurate and consistent in selecting the *Fusarium* wilt-resistant phenotype than the RAPD markers from which they were derived (Zheng et al., 1999). Fifteen AFLP markers were tightly linked to the *Vf* gene conferring scab *(Venturia inaequalis)* resistance in apple. Eleven of these AFLPs were converted into SCAR markers. Three of the converted SCAR markers were inseparable from the *Vf* gene, and these markers are useful in marker-assisted selection (Xu et al., 2001).

PCR-based markers were developed for use in marker-assisted selection. A DNA fragment (ADG2) that is 77 percent homologous to the gene *N* (resistant to *Tobacco mosaic virus*) and 53 percent homologous to *RPP5* (resistant to *Peronospora parasitica* in *Arabidopsis thaliana*) was amplified by PCR from the diploid potato genotype that carries the gene Ry$_{adg}$ located on chromosome X1 and conferring extreme resistance to *Potato virus Y* (Sorri et al., 1999). This PCR-based resistance marker, developed using a resistance analogue, will be useful in selecting resistant plants. To facilitate the transfer of the leaf rust resistance gene *Lr47* to commercial wheat varieties, the completely linked RFLP locus *Xabc465* was converted into a PCR-based marker (Helquera et al., 2000).

NEAR-ISOGENIC LINES (NIL)

Near-isogenic lines are obtained by crossing a donor plant carrying the resistance gene with a susceptible recipient parent, followed by repeated backcrosses to the recipient parent under continuous selection for the resistance trait. The donor genome is therefore progressively diluted in the introgressed lines until a short chromosomal segment is obtained around the resistance gene. When the parents and the NIL derived from them are compared with the molecular probes in NILs, only those that are linked to the resistance locus are expected to be polymorphic. NILs can be useful to select the probes for resistance genes.

COMMON BREEDING METHODS FOR DISEASE RESISTANCE

The most common method of breeding for disease resistance is pedigree selection. In this method, crosses are made between two parents (one with the resistance trait and another with good agronomic traits) and the individual plants are selected for resistance from the F2 generation. These selections are allowed to produce seed for the next generation. The selection procedure is repeated in each generation, and a higher proportion of resistant plants is obtained in each successive generation until homozygosity is obtained.

In the bulk population selection method, early segregating generations, usually F2 to F5, are bulked together without selection. In later generations, when most plants are homozygous, individual plants are selected for resistance and their progenies are evaluated for resistance as in the pedigree method.

In backcross breeding, the hybrid derived from a cross between the donor plant (with the resistance trait) and a recurrent parent (susceptible, but with good agronomic characters) is crossed back to the recurrent parent and the progeny are screened for the disease resistance trait by inoculation with the pathogen or by using molecular markers tagged to the resistance genes. The selected individuals are crossed again to the recurrent parent and the process is repeated. The selection procedure is repeated in each generation, and hence it is called "recurrent selection." After several cycles, plants are obtained that are nearly identical to the recurrent parent, with the exception that genes for resistance have been added (Tanksley et al., 1989).

The introduction of molecular markers helps to overcome major limitations of backcross breeding. If the gene(s) to be transferred are marked by tightly linked molecular markers, segregating populations of plants can be screened at the seedling stage—before the trait is expressed—for the presence of the resistance genes. Because the molecular markers can be used to mark quantitative trait loci (QTLs) as well as major genes, there are no limitations to the types of resistance that can be manipulated by the backcross selection.

Breeding for Disease Resistance Using Major Genes

Breeding strategies for major gene disease resistance involve identification of resistance genes and incorporating them into high-yielding, but susceptible varieties through pedigree/backcross breeding. Identification of disease resistance genes is carried out with the help of conventional genetic analysis employing simple inheritance studies and tests of allelism. The matching avirulence/virulence test and the gene linkages are also useful means to identify resistance genes. Several disease-resistant varieties have been developed to exploit major gene resistance. More than 40 loci for resistance to wheat stem rust have been identified and designated as *Sr* genes. Most of the designated *Sr* genes were derived from *Triticum aestivum*. However, a number of genes were derived from other *Triticum* spp., *Secale cereale,* and *Thinopyrum elongatum*. Several cultivars were developed using these genes.

Gene-for-gene hypothesis works among the major gene resistances. According to Flor (1946, 1961), a gene for resistance in the host corresponds to a gene for avirulence in the pathogen. No resistance occurs unless a resistance allele is present in the host along with a corresponding avirulence gene in the pathogen (Johnson and Knott, 1992). Several races exist in each pathogen, particularly in the case of biotrophic pathogens, and these races contain different avirulence genes. Hence, major gene resistance, which is specific against particular races, breaks down quickly whenever new races occur (Welz and Geiger, 2000). For example, 239 races of the wheat leaf rust pathogen, *Puccinia triticina,* were detected from eight wheat-growing regions of the United States from 1984 to 1999 (Long and Leonard, 2000). *Puccinia recondita* f. sp. *tritici* race 104-1,2,3,6,7,9,11, with virulence. Virulence for the wheat leaf rust resistance gene *Lr26* became widespread in Australia in 1997. This resulted in the withdrawal from cultivation of the wheat cultivar Mawson, which possesses *Lr26*. Barley varieties resistant to the powdery mildew pathogen *Blumeria graminis* f. sp. *hordei* have been developed by incorporating various resistance genes, including *Va6, Va7, Va8, Va9, Va12, Vh, Va3, Va13,* and *V(Me)*. Virulences to all these genes were detected in the population of the pathogen in Lativia (Rashal et al., 2000). Breeding for resistance that exploits major genes often results in the evolution of a matching virulence within the pathogen population, leading to an apparent breakdown of resistance. This leads to "boom-bust" cycles, in which varieties possessing effective resistance are grown on an expanding acreage (boom) until a matching virulence evolves and spreads within the pathogen population (bust) (Pink and Puddephat, 1999).

Hence, strong genes, which show resistance to many races, should be used for breeding for major gene resistance. The wheat stem rust resistance gene *Sr26,* which is derived from *Thinopyrum elongatum,* shows resistance to all stem rust cultures obtained worldwide (McIntosh et al., 1995). The wheat leaf rust resistance gene *Lr19* shows resistance to the leaf rust cultures collected worldwide, except those from Mexico.

BREEDING FOR QUANTITATIVE (PARTIAL) RESISTANCE USING QTL LINKED WITH MARKERS

Compared to major genes, partial resistance is difficult to manipulate in a breeding program. It is difficult to identify quantitative resistance. For example, quantitative resistance in spring barley cultivars with different genetic bases of resistance to scald is assessed by measuring the variables, such as the disease progress curve (AUDPC), apparent infection rate (AR), infection frequency (IF), lesion length (LL), lesion width (LW), time to disease onset at 5 percent severity (T5), final disease severity (FS), and sporulation (SP) (Xi et al., 2000). Not all of the variables may equally contribute to disease resistance. For example, sporangia production was significantly correlated to the area under disease progress curve, whereas the days to 5 percent severity threshold was negatively correlated with AUDPC in potato crosses inoculated with the late blight pathogen, *Phytophthora infestans* (Dorrance et al., 2001). It may be difficult to measure all these variables in segregating populations to select resistant plants. These variables are also affected by environmental factors.

It may be more difficult to manipulate quantitative traits in breeding programs than single gene traits. If complex traits could be resolved into their individual components, it might then be possible to deal with these traits similar to single gene traits. Molecular markers that are closely linked to genes coding for partial resistance could enable breeders to manipulate the trait by selecting for the presence of the markers associated with each of a set of minor genes.

Most of the complex quantitative traits are under the control of many genes. The genetic loci associated with complex traits are called "quantitative trait loci" (QTLs). Linkages between single genetic markers and QTLs have been reported in many cases. However, associating specific QTLs with genetic markers and locating them on a linkage group is a great task for conventional genetic analysis, because such genetic markers are not available. This limitation is remedied by the construction of complex molecular marker maps that permit systematic searches of an entire genome for QTLs influencing a trait. Molecular markers can be used to select QTLs influencing quantitative disease resistance. A cross is made between susceptible and quantitative resistant plants. The segregating progeny are obtained from the hybrid (commonly F2, backcross, or recombinant inbred lines). A number of progeny are evaluated for the QTLs and for their genotypes at the molecular marker loci at regular intervals (10 to 20 cM) throughout the genome. A search is then made for associations between the segregating molecular markers and the QTLs. If such associations are found, they should be due to linkage of the molecular marker to a gene(s) affecting the QTL (Tanksley et al., 1989). In this way, several QTLs can be identified and used for selecting quantitative resistant plants. To select for quantitative resistance, backcross breeding with a recurrent selection method using molecular markers is preferred (Bai and Shaner, 1994; Veillet et al., 1996).

Several QTLs have been detected and used in resistance breeding. The inheritance of powdery mildew and black rot resistance in grape *(Vitis vinifera)* was studied by using the cross 'Horizon' (complex interspecific hybrid) + 'Illinois 547-1' *(Vitis rupestris + V. cinerea)* (Dalbo et al., 2000). A major QTL was found in Illinois 547-1, the resistant parent. A single RAPD marker, CS25b, associated with this QTL accounted for approximately 41 percent of the phenotypic variation. Specific markers, such as Cleaved amplified polymorphic sequences (CAPS) developed for the CS25b locus, were also useful to separate resistant and susceptible individuals in the progenies of the crosses (Dalbo et al., 2000).

Breeding for Qualitative (Complete) Resistance with Additive Effects of Quantitative (Partial) Resistance Genes

A few genes in combination contribute to slow-rusting resistance in wheat to leaf rust *(Puccinia triticina)* and stripe rust *(Puccinia striiformis)*. Inheritance studies showed that individually these genes have small to intermediate effects only; combinations of three to five such genes result in a high level of resistance. Near-immunity to leaf and stripe rusts in wheat could be obtained by combining slow-rusting resistance (partial resistance) genes (Singh et al., 2000). Simple crosses were made between parents that had moderate to good levels of resistance to both rusts. The F1s were then top-crossed (three-way) with a third parent that had high yield potential and at least some resistance. From segregating generations, lines that contain high yield potential and resistance levels reaching near-immunity to both rusts could be selected. This resistance was based on additive genes (Singh et al., 2000).

Pyramiding of Genes

Combining different genes in a single plant is called "pyramiding of genes." A pyramid can be constructed with major genes and minor genes. Pyramiding different resistance genes into a genotype will facilitate in developing durable resistant varieties. Liu et al. (2000) developed powdery mildew resistant wheat varieties by molecular marker-facilitated pyramiding of different genes. Genes *(Pm)* for resistance to powdery mildew (*Erysiphe graminis* f. sp. *tritici*) were identified in wheat at more than 30 loci. Molecular markers tightly linked to about 10 of these *Pm* genes were identified. Three of these genes, *Pm2*, *Pm4a*, and *Pm21*, were used for marker-facilitated pyramiding. The following is the procedure followed by Liu et al. (2000) for pyramiding: First, near-isogenic lines (NILs), 'Yang 93-111' *(Pm4a)*, 'Yang 94-143' *(Pm2)*, and 'Yang 96 (487)' *(Pm21)* were developed by six to nine backcrosses with the local susceptible cultivar 'Yang 158' as the recurrent parent. These lines were proved to include *Pm4a*, *Pm2*, and *Pm21*, respectively, by artificial inoculation and with suitable RFLP markers. The NILs had morphological and agronomic attributes that are similar to their recurrent parent 'Yang'. To pyramid the resistance, crosses were made between the NILs, i.e., 'Yang 93-111'*(Pm4a)* × 'Yang 94-143' *(Pm2)*, 'Yang 94-143' *(Pm2)* × 'Yang 96 (487)' *(Pm21)*, 'Yang 93-111' *(Pm4a)* × 'Yang 96 (487)' *(Pm21)*. The resulting F2 populations were planted and screened at the seedling stage. A total of 75 F2 plants from cross 'Yang 93-111' × 'Yang 94-143' were obtained. RFLP with the probes BCD1871 and Whs 350 identified 53 plants carrying *Pm2*. Furthermore, RFLP analysis with probe BCD 1231, showed that 39 of these 53 plants also carried *Pm4a* (Liu et al., 2000).

Even defeated major resistance genes can be used for pyramiding for developing resistant varieties. *Xanthomonas vesicatoria* races 1, 3, 4, and 6 caused similar high levels of disease severity, followed by races 2 and 5 on susceptible bell pepper variety Early Calwonder (ECW). Race 3 caused severe disease on all isolines lacking resistance gene *Bs2*. Race 4, which defeats *Bs1* and *Bs2*, caused less disease on isoline ECW-12R (which carries *Bs1* + *Bs2*), than on isolines ECW, ECW-10R (which carries *Bs1*), and ECW-20 R (which carries *Bs2*). Race 6, which defeats all three major genes, causes less disease on isoline ECW-13R (*Bs1* + *Bs3*) and ECW-123R (*Bs1* + *Bs2* + *Bs3*) than on isolines ECW, ECW-10R, ECW-20R, and ECW-30R

(*Bs3*). Defeated major resistance genes deployed in specific gene combinations were associated with less area under the disease progress curve than when genes were deployed individually in isolines of ECW (Kousik and Ritchie, 1999). These results suggest that by pyramiding even defeated genes, improved resistance can be obtained.

Mutation Breeding

Some disease-resistant varieties have been developed by mutation breeding using physical or chemical mutagens. The wheat variety Sinvalocho gama against *Puccinia graminis* f. sp. *tritici,* the oats variety Alamo-x against *Puccinia coronata,* the rice varieties Fulgenate and RD 6 against *Pyricularia oryzae,* and the bean varieties Universal and Unima against *Colletotrichum lindemuthianum* were developed by mutation breeding (Vidhyasekaran, 1993a). Two mutants, I3-48 and I3-54, obtained by irradiation with fast neutrons from wheat cv. Hobbit 'sib' were found to be significantly more resistant than Hobbit 'sib' to *Erysiphe graminis* f. sp. *tritici*. Constitutive expression of a thaumatin-like protein gene was detected in both I3-48 and I3-54, but not in Hobbit 'sib' (Duggal et al., 2000).

Durable Resistance

Durable resistance is resistance that remains effective in a cultivar that is widely grown for a long period of time in an environment favorable to the disease. Durable resistance cannot be identified immediately after or at the time of release of a variety, as it should be grown for a long time. Durable resistance cannot be identified under greenhouse conditions or in experimental station farms because the crop should be grown in large areas. Durable resistance can probably be identified based on relative performance of other cultivars during this period. Durable resistance may be a race-nonspecific or horizontal resistance, partial resistance, field resistance, or quantitative resistance. Even a single major gene can contribute to durable disease resistance. A number of Australian wheat varieties carrying the stem rust resistance gene *Sr26* derived from *Thinopyrum elongatum* have remained resistant to stem rust since 1967, despite being grown over a large area (McIntosh et al., 1995). Sometimes, even partial resistance may break down quickly. A wheat cultivar Joss Cambier possessing partial resistance to stripe rust *(Puccinia striiformis)* released in the United Kingdom in 1969 became highly susceptible in 1971 due to a new race of *P. striiformis* (Johnson, 1993).

Several factors may contribute to durable resistance. The race-specific (major gene) resistance genes to the potato viruses A, X, and Y have shown no signs of losing effectiveness for more than 60 years in Europe. The more inoculum of a pathogen is present, the greater the chances that new races can arise. In Western Europe, potato viruses are kept at a reasonably low level due to a thoroughly controlled system of seed potato production. Due to a decreased amount of inoculum, new races may not have developed during these years (Parlevliet, 1993). When the crop is a minor crop (grown in limited, discontinuous areas), resistance may last longer. In flax in North America, the race-specific major genes against flax rust *(Melampsora lini)* lasted from 10 to more than 13 years and the flax is grown only in small areas (Parlevliet, 1993).

The parasitic ways of life of pathogens may also determine the durability of major gene resistance in the field. Race-specific resistance against biotrophs (pathogens that exploit living cells, such as rusts and powdery mildews) and hemibiotrophs (pathogens that are biotrophic at the lesion perimeter, but necrotrophic near the lesion center, such as *Phytophthora infestans* in potato and *Rhynchosporium secalis* in barley) breaks down quickly due to the development of new races. Resistance against necrotrophs (pathogens that exploit host tissue after killing it with toxins, such as *Septoria* leaf and glume blotch in wheat and *Periconia circinata* in sorghum), is generally durable.

The degree of host specialization of pathogens may also determine durability of resistance (Parlevliet, 1993). Formae speciales of *Puccinia graminis* and *Erysiphe graminis* are highly host-specific in cereals. Viruses are often less specialized and they have a wide host range. Several necrotrophs can infect multiple hosts. The means of dispersal of pathogens, such as soilborne, seed-borne, airborne, and splash-borne, is also important in the development of new races. Pathogens in which new races very often occur are biotrophic or hemibiotrophic, airborne, or splash-borne fungi. Major gene resistance against these pathogens breaks down very often.

Another group of pathogens exists, in which only a few races are known and the major gene resistance persists for longer time. Viruses such as *Tobacco mosaic virus, Bean common mosaic virus,* and *Pea seed-borne mosaic virus,* and fungi such as *Fusarium oxysporum* and *Peronospora tabacina* belong to this group. In the third group of pathogens, no races are known and the resistance used has been effective from the start (Parlevliet, 1993). Fungi such as *Cladosporium cucumerinum* in cucumber, *Pseudocercosporella herpotrichoides* in wheat, and *Helminthosporium victoriae* in oat belong to this group.

Several quantitative-resistance varieties show durable resistance. Breeding strategies, including pyramiding of major resistance genes and introgression of additive minor genes in an adapted background, may result in rapid progress in the breeding of durable disease-resistant varieties.

Multiline Mixtures

To avoid rapid breakdown of disease resistance in the field, multiline cultivars were developed (Wolfe, 1985). Two types of multilines are used. "Clean crop" multilines are those in which each variety accords a broad spectrum of resistance. "Dirty crop" multilines are those in which a susceptible variety is purposefully introduced in the multiline mixture. The susceptible variety will reduce the selection pressure on the pathogen and retard the appearance of new races. Host mixtures may restrict the spread of a disease considerably relative to the mean disease level of the components. Reduction up to 80 percent in powdery mildew infection compared with the mean disease level of the component grown as pure stands was reported in barley (Wolfe, 1985). Reduction in disease incidence may be due to the decrease in the spatial density of susceptible plants. This decrease limits the amount of susceptible tissue in a given area. It also reduces the probability of survival of spores that leave each parent lesion. Spore density declines exponentially along a gradient from the source (Vidhyasekaran, 1993b). Varietal mixture may reduce the disease incidence due to the barrier effect provided by resistant plants that fill the space between susceptible lines. Reduced disease incidence may also be due to the resistance induced by incompatible pathogenic spores. Induced resistance is always greater than induced susceptibility. Induced resistance in varietal mixtures will be great. In the varietal mixture, slower adaptation of the pathogen is common compared to single variety cultivation.

REFERENCES

Bai, G. and Shaner, G. (1994). Scab of wheat: Prospects for control. *Plant Dis,* 78:760-766.

Dalbo, M. A., Weeden, N. F., and Reisch, B. I. (2000). QTL analysis of disease resistance in interspecific hybrid grapes. *Acta Horticulturae,* 528:215-219.

Dorrance, A. E., Inglis, D. A., Helgeson, J. P., and Brown, C. R. (2001). Partial resistance to *Phytophthora infestans* in four *Solanum* crosses. *Amer J Potato Res,* 78:9-17.

Duggal, V., Jellis, G. J., Hollins, T. W., and Stratford, R. (2000). Resistance to powdery mildew in mutant lines of the susceptible wheat cultivar Hobbit 'sib.' *Plant Pathol,* 49:468-476.

Flor, H. H. (1946). Genetics of pathogenicity in *Melampsora lini. J Agric Res,* 73:335-357.

Flor, H. H. (1961). Complementary gene interaction in the rusts. *Recent Advances in Botany,* 1:574-579.

Haymes, K. M., Weg, W. E. van de, Arens, P., Maas, J. L., Vosman, B., Nijs, A. P. M. den (2000). Development of SCAR markers linked to a *Phytophthora fragariae* resistance gene and their assessment in European and North American strawberry genotypes. *J Amer Soc Hort Sci,* 125:330-339.

Helquera, M., Khan, I. A., and Dubcovsky, J. (2000). Development of PCR markers for wheat leaf rust resistance gene *Lr47. Theor Appl Genet,* 101:625-631.

Johnson, R. (1993). Durability of disease resistance in crops: Some closing remarks about the topic and the symposium. In T. Jacobs and J. E. Parlevliet (Eds.), *Durability of Disease Resistance.* Kluwer Academic Publishers, the Netherlands, pp. 283-300.

Johnson, R. and Knott, D. R. (1992). Specificity in gene-for-gene interactions between plants and pathogens. *Plant Pathol,* 41:1-4.

Kasai, K., Morikawa, Y., Sorri, V. A., Valkonen, J. P. T., Gebhardt, C., Watanabe, K. N. (2000). Development of SCAR markers to the PVY resistance gene Ry_{adg} based on a common feature of plant resistance genes. *Genome,* 43:1-8.

Kerber, E. R. and Aung, T. (1999). Leaf rust resistance gene *Lr34* associated with nonsuppression of stem rust resistance in the wheat cultivar Canthatch. *Phytopathology,* 89:518-521.

King, G. J., Tartarini, S., Brown, L., Gennari, F., and Sansavini, S. (1999). Introgression of the *Vf* source of scab resistance and distribution of linked marker alleles within the *Malus* gene pool. *Theor Appl Genet,* 99:1039-1046.

Kousik, C. S. and Ritchie, D. F. (1999). Development of bacterial spot on near-isogenic lines of bell pepper carrying gene pyramids composed of defeated major resistance genes. *Phytopathology,* 89:1066-1072.

Liu, J., Liu, D., Tao, W., Li, W., Wang, S., Chen, P., Cheng, S., and Gao, D. (2000). Molecular marker-facilitated pyramiding of different genes for powdery mildew resistance in wheat. *Plant Breeding,* 119:21-24.

Long, D. I. and Leonard, K. J. (2000). Wheat leaf rust virulence frequencies in the United States from 1984-1999. *Acta Phytopathol Entomol Hungarica,* 35:403-407.

McIntosh, R. A. (1992). Close genetic linkage of genes conferring adult-plant resistance to leaf rust and stripe rust in wheat, *Plant Pathology,* 41:523-527.

McIntosh, R. A., Wellings, C. R., and Park, R. F. (1995). *Wheat Rusts: An Atlas of Resistance Genes.* CSIRO Publications, East Melbourne, Australia.

Naqvi, N. I. and Chattoo, B. B. (1996). Development of a sequence characterized amplified region (SCAR) based indirect selection method for a dominant blast-resistance gene in rice. *Genome,* 39:26-30.

Nelson, R. R. (1978). Genetics of horizontal resistance to plant diseases. *Annu Rev Phytopathol,* 16:359-378.

Parlevliet, J. E. (1993). What is durable resistance, a general outline. In Th. Jacobs and J. E. Parlevliet (Eds.), *Durability of Disease Resistance.* Kluwer Academic Publishers, The Netherlands, pp. 23-39.

Pink, D. and Puddephat, I. (1999). Deployment of disease resistance genes by plant transformation—a "mix and match" approach. *Trends Plant Sci,* 4:71-75.

Rashal, I., Araja, I., and Kokina, I. (2000). Barley powdery mildew virulence frequencies in Lativia in 1996-1999. *Acta Phytopathol Entomol Hungarica,* 35: 397-402.

Singh, R. P., Huerta-Espino, J., and Rajaram, S. (2000). Achieving near-immunity to leaf and stripe rusts in wheat by combining slow rusting resistance genes. *Acta Phytopathol Entomol Hungarica,* 35:133-139.

Sorri, V. A., Watanabe, K. N., and Valkonen, J. P. T. (1999). Predicted kinase-3a motif of a resistance gene analogue as a unique marker for virus resistance. *Theor Appl Genet,* 99:164-170.

Tanksley, S. D., Young, N. D., Paterson, A. H., and Bonierbale, M. W. (1989). RFLP mapping in plant breeding: New tools for an old science. *Biotechnology,* 7:257-264.

Veillet, S., Filippi, M. C., and Gallais, A. (1996). Combined genetic analysis of partial blast resistance in an upland rice population and recurrent selection for line and hybrid lines. *Theor Appl Genet,* 92:644-653.

Vidhyasekaran, P. (1993a). Molecular sieve to select cells in tissue culture to develop disease resistant plants. In P. Vidhyasekaran (Ed.), *Genetic Engineering, Molecular Biology, and Tissue Culture for Crop Pest and Disease Management.* Daya Publishing House, Delhi, India, pp. 295-309.

Vidhyasekaran, P. (1993b). *Principles of Plant Pathology.* CBS Publishers, Delhi, India.

Wang, Y. H., Thomas, C. E., and Dean, R. A. (2000). Genetic mapping of a fusarium wilt resistance gene *(Fom-2)* in melon *(Cucumis melo* L.). *Mol Breeding,* 6:379-389.

Welz, H. G. and Geiger, H. H. (2000). Genes for resistance to northern corn leaf blight in diverse maize populations. *Plant Breeding,* 119:1-14.

Williams, C. E., Wang, B., Holsten, T. E., Scambray, J., Silva, F. de A. G. da, and Ronald, P. C. (1996). Markers for selection of the rice *Xa21* disease resistance gene. *Theor Appl Genet,* 93:1119-1122.

Wolfe, M. S. (1985). The current status and prospects of multiline cultivars and variety mixtures for disease resistance. *Annu Rev Phytopathol,* 23:251-273.

Xi, K., Xue, A. G., Burnett, P. A., Helm, J. H., and Turkington, T. K. (2000). Quantitative resistance of barley cultivars to *Rhynchosporium secalis. Can J Plant Pathol,* 22:217-223.

Xu, M., Huaracha, E., and Korban, S. S. (2001). Development of sequence-characterized amplified regions (SCARS) from amplified fragment length polymorphism (AFLP) markers tightly linked to the *Vf* gene in apple. *Genome,* 44:63-70.

Yu, Y. G., Saghai Maroof, M. A., Buss, G. R., Maughan, P. J., and Tolin, S. A. (1994). RFLP and microsatellite mapping of a gene for soybean mosaic virus resistance. *Phytopathology,* 84:60-64.

Zhang, G., Angeles, E. R., Abenes, M. L. P., Khush, G. S., and Huang, N. (1996). RAPD and RFLP mapping of the bacterial blight resistance gene *xa-13* in rice. *Theor Appl Genet,* 93:65-70.

Zheng, X. Y., Wolff, D. W., Baudracco-Arnas, S., and Pitrat, M. (1999). Development and utility of cleaved amplified polymorphic sequences (CAPS) and restriction fragment length polymorphisms (RFLPs) linked to the *Fom-2 Fusarium* wilt resistance gene in melon *(Cucumis melo* L.). *Theor Appl Genet,* 99:453-463.

20

Breeding for Disease Resistance—
Genetic Engineering

Genetic engineering technology has been exploited to develop disease-resistant varieties. Plant disease resistance genes and defense genes have been cloned. Transgenic plants expressing these genes have been developed, and they show resistance against fungal, bacterial, and viral diseases. Genes from fungi, bacteria, viruses, insects, and human beings have been transferred to plants, and transgenic plants expressing these genes show resistance against various diseases. The use of genetic engineering technology in the development of disease-resistant plants is discussed in this chapter.

TRANSGENIC PLANTS EXPRESSING
PLANT RESISTANCE GENES

Several fungal disease resistance genes have been cloned. *Cf9, Cf2, Cf4,* and *Cf5* genes in tomato against *Cladosporium fulvum, I2* in tomato against *Fusarium oxysporum* f. sp. *lycopersici, Pi-b* and *Pi-ta* in rice against rice blast pathogen, *LR10* against leaf rust in wheat, *HM1* against *Helminthosporium maydis, L6* and *M* genes in flax against rust, and *RPP5* against *Peronospora parasitica* in *Arabidopsis thaliana* have all been cloned. Disease resistance genes found in one plant species have also been detected in other plants. Analogs of cloned resistance genes have been detected in rice, wheat, barley, cotton, soybean, pepper, chickpea, flax, tomato, tobacco, and *Brassica napus*. Hundreds of homologues of disease resistance genes have been identified from various crops.

Resistance genes can be cloned from resistant varieties and transferred to susceptible but high-yielding varieties, making these plants resistant as well. The resistance genes can even be transferred to heterologous plants. The gene *Cf9* from tomato, which confers resistance against *Cladosporium fulvum,* has been transferred to oilseed rape. The transgenic oilseed rape shows resistance against *Leptosphaeria maculans* (Hennin et al., 2000).

Several bacterial disease resistance genes have also been cloned. These include *Pto, Pti,* and *Prf* genes in tomato against *Pseudomonas syringae* pv. *tomato, Xa21* and *Xa1* genes in rice against *Xanthomonas oryzae* pv. *oryzae,* and *RPS2* and *RPM1* in *Arabidopsis thaliana* against *Pseudomonas syringae.* Functional homologues of the resistance genes *RPM1* and *RPS2* detected in *Arabidopsis thaliana* could be detected in pea *(Pisum sativum),* bean *(Phaseolus vulgaris),* and soybean *(Glycine max).* Two disease resistance genes cloned from soybean are similar to *RPS2* in *Arabidopsis.*

The bacterial disease resistance genes cloned from resistant varieties could be transferred to susceptible but high-yielding varieties, making these plants resistant as well. The resistance gene *Xa21* has been cloned from a wild rice, *Oryza longistaminata.* The cloned gene could be transferred to different rice cultivars varying widely in their phenotypic characters, such as *O. sativa* ssp. *japonica* and *O. sativa* ssp. *indica* types (Wang et al., 1996; Tu et al., 1998). The japonica rice cultivar TP309 was transformed with the *Xa21* gene by particle-bombardment technique. The transgenic plants expressing the *Xa21* gene showed resistance to several races of the pathogen (Wang et al., 1996). The indica rice cultivar IR72 was transformed with the *Xa21* gene, and the transgenic plants showed resistance to both of the tested races (race 4 and race 6) of *X. oryzae* pv. *oryzae* (Tu et al., 1998). The transgenic *Xa21* plants showed greater resistance to various isolates of the pathogen as compared to that of the donor cultivar IRBB21. In addition, about a five to ten times stronger *Xa21*-specific hybridization was observed in the transgenic *Xa21* plants as compared to that of the donor line, indicating multiple insertions in a single locus in the transformed lines (Wang et al., 1996). Multiple copies of transgenes are often inherited as a single locus in transgenic lines generated by particle bombardment. Transgenic tomato plants expressing the cloned *Pto* gene showed high resistance to *P. syringae* pv. *tomato* (Martin et al., 1993).

A bacterial disease resistance gene from one plant species could be successfully transferred to another plant species. The cloned *Pti* gene from tomato was transferred to tobacco, and the transgenic plants showed resistance to *P. syringae* pv. *tabaci* (Zhou et al., 1995). The transgenic tobacco plants expressing the cloned tomato *Pto* gene showed resistance to *P. syringae* pv. *syringae* strains. Similar resistance was observed in *Nicotiana benthamiana* against *P. syringae* strains (Rommens et al., 1995; Thilmony et al., 1995).

So far, only three virus-disease resistance genes have been cloned. These include the *N* gene in tobacco against the *Tobacco mosaic virus* (TMV) (Whitham et al., 1996), *Sw-5* in tomato against tospoviruses (Brommonschenkel et al., 2000), and the *Rx* gene in potato against *Potato virus Y*

(PVY) (Bendahmane et al., 1999). However, several homologues of these disease resistance genes have been detected in other plants.

The resistance genes cloned from resistant varieties have been transferred to the susceptible varieties, and these transgenic plants expressed resistance. Transgenic potato plants expressing the *Rx* gene showed resistance to *Potato virus X* (PVX) (Bendahmane et al., 1999). Transgenic tomato plants carrying the cloned *N* gene of tobacco show resistance to *Tobacco mosaic virus* (Whitham et al., 1996). The potato *Rx* gene has been transferred to tobacco, and the transgenic tobacco plants show resistance to PVX (Bendahmane et al., 1995). It may be ideal to clone the resistance genes and exploit them to develop new disease-resistant varieties.

Great potential exists to exploit disease resistance genes for disease management, and more studies are needed in this aspect. However, the silencing or inactivation of the transgene in disease-resistant transgenic plants poses a problem. It is known that homologues of resistance genes exist in susceptible plants, and the interactions between an inserted gene and its DNA sequence homologues may lead to the inserted gene silencing (Jorgensen, 1990). The consequence of the interactions among the loci with DNA sequence homology may result in chromatin restructuring or DNA sequence modification by methylation of different cytosine residues or inhibition of mRNA processing, transport, export, or translation. The expression of the transgene in the transgenic plant may be finally cosuppressed and *trans*-activated. However, inactivation of the transgene may not occur in all plants. Hence, a large number of progeny should be obtained for selecting a homozygous line with a consistently high level of resistance to the pathogens.

TRANSGENIC PLANTS EXPRESSING PLANT DEFENSE GENES

Molecular biological studies have revealed that none of the resistance genes code for any toxic chemicals, which contribute to disease resistance. This job is done by another group of genes called defense genes. Resistance genes are only regulatory genes, while defense genes are the functional genes. Defense genes are quiescent in healthy plants. They are only sleeping genes, requiring specific signals and signal transduction systems for activation. When defense genes are activated, defense chemicals are synthesized, and development of the pathogen is suppressed. In susceptible varieties, the defense genes are not activated. It is now possible to clone defense genes and make them express constitutively by placing the gene under a suitable promoter. Thus, we can make the sleeping defense genes active constitutively and make the susceptible plants resistant even in the absence of functional resistance genes.

Pathogenesis-related genes are the most important defense genes, which encode for pathogenesis-related (PR) proteins. Numerous PR proteins have been detected in various plants, and most of them are highly inhibitory to fungal (Vidhyasekaran, 1997) and bacterial (Vidhyasekaran, 2002) pathogens. Several PR genes have been cloned, and transgenic plants have been developed. These transgenic plants show enhanced disease resistance. Transgenic tomato plants constitutively expressing a PR-5 gene showed resistance to *Botrytis cinerea, Oidium lycopersicum, Leveillula taurica,* and *Phytophthora infestans,* and the resistance was confirmed up to the T3 generation (Veronese et al., 1999). Rice plants overexpressing rice chitinase (PR-3) gene show increased resistance against *Rhizoctonia solani* (Lin et al., 1995). Datta and colleagues (1999, 2000) have exploited rice class I chitinase (PR-3) and PR-5 genes to develop disease-resistant rice cultivars. A rice chitinase gene *(RC7)* isolated from *Rhizoctonia solani*-infected rice plants was introduced into indica rice cultivars IR72, IR64, IR68899B, MH63, and Chinsurah Boro II, and the transformants showed increased tolerance to *R. solani* (Datta et al., 2001). Transgenic tomato plants expressing chitinase (Logemann et al., 1994), and β-1,3-glucanase (PR-2) genes (Jongedijk et al., 1995), potato plants expressing PR-10a (Matton et al., 1993) and PR-5 (Zhu et al., 1996) genes, alfalfa plants expressing β-1,3-glucanase (Masoud et al., 1996), rose plants expressing chitinase (Marchant et al., 1998), and tobacco plants expressing PR-1 (Alexander et al., 1993), PR-2 (Yoshikawa et al., 1993), PR-3 (Linthorst et al., 1990; Broglie et al., 1991), PR-5 (Melchers et al., 1993), and PR-14 (lipid transfer protein LTP2) (Molina and Garcia-Olmedo, 1997) have been developed to manage various fungal diseases.

A PR gene from one plant could be transferred to an unrelated plant, making the transgenic plant resistant. Transgenic wheat plants expressing rice PR-5 gene show resistance to scab disease caused by *Fusarium graminearum* (Chen et al., 1999). Transgenic oilseed rape and tomato plants expressing the defensin (PR-12) gene *rs-afp2* from radish showed enhanced resistance against fungal pathogens (Parashina et al., 1999, 2000). Transgenic chrysanthemum plants expressing rice chitinase showed resistance to *Botrytis cinerea* (Takatsu et al., 1999). Transgenic tomato plants expressing an acidic endochitinase (pcht28) isolated from *Lycopersicon chilense* showed resistance to *Verticillium dahliae* race 2 (Tabaeizadeh et al., 1999). Tomato plants expressing the tobacco PR-9 (peroxidase) gene (Lagrimini et al., 1993), potato plants expressing the tobacco PR-5 gene (Liu et al., 1994), cucumber plants expressing the rice chitinase gene (Tabei et al., 1998), soybean plants expressing the tobacco PR-2 gene (Yoshikawa et al., 1993), and tobacco plants expressing rice chitinase (Zhu et al., 1993), barley chitinase (Jach et al., 1995), sugarbeet chitinase (Nielsen et al., 1993), bean chitinase (Roby et al., 1990), peanut chitinase (Kellmann et al., 1996), barley PR-2 (Jach et al., 1995), radish PR-12 (Terras et al., 1995), barley PR-13 (thionin) (Florack et al., 1994), barley ribosome-inactivating protein (Logemann et al., 1992), and maize ribosome-inactivating protein (Maddaloni et al., 1997) genes have been developed, and most of them show enhanced resistance to pathogens.

Some pathogenesis-related proteins have been shown to have antibacterial activity. These PR genes do not express constitutively. The PR genes have been cloned, and transgenic plants constitutively expressing the PR genes have been developed. Some transgenic plants constitutively expressing PR proteins show enhanced resistance to bacterial pathogens. Tobacco plants expressing barley thionin genes show resistance to *P. syringae* pv. *tabaci* (Carmona et al., 1993). Transgenic tobacco plants expressing barley lipid-transfer proteins 2 (LTP2) show high resistance to *P. syringae* pv. *tabaci* (Molina and Garcia-Olmedo, 1997). Similarly, transgenic *Arabidopsis* plants expressing barley LTP2 showed enhanced resistance to *P. syringae* pv. *tomato* (Molina and Garcia-Olmedo, 1997).

Pathogenesis-related proteins also contribute to virus-disease resistance. Transgenic plants expressing genes responsible for activation of synthesis of PR proteins show increased disease resistance. Transgenic tobacco plants expressing the *hrmA* gene from *P. syringae* pv. *syringae* showed elevated synthesis of PR proteins. These transgenic plants showed resistance to the *Tobacco mottling virus* and the *Tobacco etch virus* (TEV) (Shen et al., 2000). An antiviral protein has been isolated from pokeweed, *Phytolacca americana*. Transgenic tobacco plants expressing pokeweed antiviral protein (PAP) or a variant (PAP-v) showed resistance to a broad spectrum of plant viruses including *Tobacco mosaic virus* and *Potato virus X*. Expression of PAP-v in transgenic plants induces synthesis of pathogenesis-related proteins. The enzymatic activity of PAP may be responsible for generating a signal that may activate defense mechanisms against viruses (Smironv et al., 1997). A mutant of pokeweed antiviral protein, PAPn, was nontoxic to viruses when expressed in transgenic tobacco plants. However, the tobacco transgenic plants expressing PAPn induced basic PR proteins, proteinase inhibitor II (PR-6 protein), and protein kinase and showed resistance to viruses (Zoubenko et al., 2000).

Although several transgenic plants expressing PR genes have been developed, field performance of them has not yet been demonstrated. The increased resistance observed in the transgenic plants was also not high. Probably, a single PR protein may not be sufficient to confer resistance. Synergism between PR proteins has been reported. Coordinated induction of the different families of PR proteins in the same genome may indeed confer a level of resistance considerably broader than any protein itself. Most of the phyto-pathogenic fungi that have been tested in vitro are resistant to β-1,3-glucanases and chitinases, but their growth is highly inhibited by a combi-nation of these PR proteins. Synergism of LTPs with thionins has been reported. Hence, transgenic plants expressing more than one PR protein should be developed. Transfer of more genes with suitable promoters may be difficult to achieve; scientists prefer to transfer one or two genes rather than multiple genes. Veronese and colleagues (1999) constructed cassettes containing three target genes, PR-1, PR-3, and PR-5 genes, used for trans-formation of tomato plants. One good method to transfer multiple genes has been suggested by Zhu and colleagues (1994). They developed transgenic tobacco plants expressing a gene encoding rice basic chitinase and trans-genic tobacco plants expressing alfalfa acidic glucanase. Hybrid plants were generated by crossing transgenic parental lines. Homozygous selfed progeny that expressed the two transgenes could be selected from the hy-brids. Logemann and colleagues (1994) could develop transgenic tomato plants simultaneously expressing class I chitinase and β-1,3-glucanase. This type of approach will be useful in developing disease-resistant plants.

Phytoalexins are the second group of defense chemicals in plants. Over-production and early production of phytoalexins may confer resistance in susceptible varieties. Genes encoding phytoalexin-biosynthetic enzymes have been cloned. Isoflavone-O-methyl transferase is an essential enzyme in biosynthesis of the phytoalexin medicarpin in legumes. Transgenic alfalfa plants overexpressing the enzyme show increased induction of phytoalexin pathway transcripts after infection with *Phoma medicaginis*. These trans-genic plants showed resistance to *P. medicaginis* (He and Dixon, 2000). Resveratrol synthase is the key enzyme in biosynthesis of the phytoalexin resveratrol in grapevine. The gene encoding this enzyme has been cloned and transferred to tobacco. The transgenic tobacco plants showed resistance to *Botrytis cinerea* (Hain et al., 1993). Transgenic rice plants expressing grapevine stilbene synthase show resistance to *Magnaporthe grisea* (Stark-Lorenzen et al., 1997). However, the transgenic tomato plants expressing grapevine resveratrol synthase did not show much enhanced resistance to *Phytophthora infestans*. This type of specificity of phytoalexins to patho-gens is well known (Vidhyasekaran, 1997). Alfalfa was transformed with resveratrol synthase gene from grapevine. Transgenic plants accumulated a new phytoalexin, reveratrol-3-O-β-D-glucopyranoside and showed resis-tance to *Phoma medicaginis* (Hipskind and Paiva, 2000). Transgenic wheat plants expressing a grapevine stilbene synthase gene have also been devel-oped (Fettig and Hess, 1999).

Phenolics are another important group of toxic chemicals inhibitory to fungal and bacterial pathogens. Overproduction of phenolics may contribute to disease resistance. Anthocyanins are one important group of phenolics. Transgenic rice plants expressing maize anthocyanin genes showed increased blast resistance (Madhuri et al., 2001). Lignins are another important group of phenolics, and peroxidase is the important enzyme in biosynthesis of lignins. Transgenic tobacco plants expressing the peroxidase gene *Shpx2* from *Stylosanthes humilis* show resistance to *Phytophthora nicotianae* pv. *nicotianae* and *Cercospora nicotianae* (Way et al., 2000). *Rir1b* is one of a set of putative defense genes, and its transcripts accumulate in incompatible interactions in rice. Transgenic rice plants overexpressing this gene confer resistance to *Magnaporthe grisea,* the blast pathogen (Schffrath et al., 2000).

TRANSGENIC PLANTS EXPRESSING GENES INVOLVED IN A SIGNAL TRANSDUCTION SYSTEM

It is known that defense genes are abundant even in the susceptible varieties, but they are also quiescent in them. Specific signals are needed to activate the defense genes. Once the defense genes are activated, the susceptible plants will become resistant. Several genes involved in the activation of a signal transduction system have been cloned from various bacteria, fungi, and viruses and transferred to plants. Elicitins are the elicitors produced by *Phytophthora* spp. Transgenic tobacco plants expressing a *Phytophthora cryptogea* gene encoding the highly active elicitor cryptogein were developed. The transgenic plants showed enhanced resistance to *Thielaviopsis basicola, Erysiphe cichoracearum,* and *Botrytis cinerea* (Keller et al., 1999). Harpin is a protein produced by several *Erwinia* species. It elicits disease resistance in incompatible hosts. Transgenic *Arabidopsis thaliana* plants expressing a harpin gene *(hrpN)* showed resistance to *Peronospora parasitica* (Bauer et al., 1999). Transgenic potato plants expressing a harpin protein gene from the apple blight pathogen *Erwinia amylovora* showed resistance to *Phytophthora infestans* (Li and Fan, 1999). The *hrmA* gene is a component of the *hrp* genes in *Pseudomonas syringae* pv. *syringae.* Transgenic tobacco plants expressing the *hrmA* gene from *P. syringae* pv. *syringae* showed elevated synthesis of PR proteins. These transgenic plants showed resistance to *Phytophthora nicotianae* var. *parasitica* and *Pseudomonas syringae* pv. *tabaci* (Shen et al., 2000). Pectic enzymes can form unsaturated oligogalacturonides from plant cell-wall pectin, and these galacturonates activate genes encoding phytoalexins, PR proteins, phenolics, and cell-wall proteins. Transgenic potato plants expressing a pectate lyase gene,

PL3, of *Erwinia carotovora* ssp. *atroseptica* show enhanced resistance to *E. carotovora* (Wegener et al., 1996). The *PL3* gene induced the defense gene encoding phenylalanine ammonia-lyase (Wegener et al., 1996). Polygalactunose inhibitor proteins have been detected in plant cells, and they inhibit polygalacturonases produced by pathogens resulting in production of the endogenous elicitors, unsaturated oligogalacturonates. Transgenic tomato plants expressing the pear fruit polygalacturonase inhibitor protein show resistance to *Botrytis cinerea* (Powell et al., 2000). An antiviral protein has been isolated from pokeweed, *Phytolacca americana.* Transgenic tobacco plants expressing a variant of pokeweed antiviral protein showed increased synthesis of PR-6 pathogenesis-related protein and protein kinase and showed resistance to fungal pathogens (Zoubenko et al., 2000).

Salicylic acid has been shown to be an important systemic signal in inducing resistance against pathogens. Several genes have been shown to induce this signal. These genes have been used to develop transgenic plants with overproduction of salicylic acid. Enhanced synthesis of salicylic acid enhances disease resistance. Salicylic acid may be synthesized from chorismate, the general precursor of aromatic compounds. This substrate is converted by isochromistate synthase (ICS) to isochromismate, which is subsequently cleaved by pyruvate lyase (IPL) to yield salicylic acid. Transgenic tobacco expressing both the ICS gene from *Escherichia coli* and the IPL gene from *Pseudomonas fluorescens* showed constitutive accumulation of salicylic acid. The transgenic plants showed a 500- to 1,000-fold accumulation of salicylic acid and salicylic acid glucoside compared to control plants. These transgenic tobacco plants constitutively expressed defense genes, particularly PR genes, and showed resistance to a fungal pathogen *(Oidium lycopersicon)* (Verberne et al., 2000). Transgenic potato plants expressing a *Halobacterium halobium* gene encoding a light-driven bacterio-opsin proton pump showed higher systemic levels of salicylic acid and showed resistance to *Phytophthora infestans* (Abad et al., 1997). Transgenic tobacco plants expressing the *H. halobium* gene encoding a light-driven bacterio-opsin proton pump showed higher systemic levels of salicylic acid. These plants showed enhanced resistance to the wild-fire pathogen *P. syringae* pv. *tabaci* (Mittler et al., 1995). However, the same gene, when expressed in potato, did not offer protection against tuber infection of *E. carotovora* (Abad et al., 1997). Probably, this gene would not have expressed in tubers, although it would have expressed in its leaves (the leaves of transgenic potato plants showed resistance to the fungal pathogen *Phytophthora infestans*). Abad and colleagues (1997) have used the figwort mosaic virus promoter in the gene construction. Wegener and colleagues (1996) reported that when they used cauliflower mosaic virus (CaMV) 35S promoter for expression of the pectate lyase gene *PL3,* the expression was stronger in potato leaves and much less intense in tubers. However, when patatin B33 was used as a promoter, the gene expression was limited to tubers only. Different types of promoters should probably be tested to achieve maximum expression of the desired gene.

Transgenic tobacco plants expressing a gene encoding the A1 subunit of cholera toxin showed increased resistance to *P. syringae* pv. *tabaci* (Beffa et al., 1995). The cholera toxin is known to activate the signal transduction system dependent on G proteins in animals. The transgenic plants showed enhanced accumulation of salicylic acid (Beffa et al., 1995). H_2O_2 is an important component in the signal transduction system. A glucose oxidase gene from the fungus *Aspergillus niger* was transferred to potato. The transgenic potato plants expressing the glucose oxidase gene showed an increased level of resistance to *E. carotovora* (Wu et al., 1995). Although glucose oxidase was produced constitutively and extracellularly in the transgenic plants, a significant increase in H_2O_2 was detected only following bacterial infection. Intracellular glucose is probably released only after bacterial infection, and the released glucose in the apoplast may serve as a substrate for the extracellular glucose oxidase. The induced H_2O_2 may induce salicylic acid and induce resistance.

H_2O_2 is degraded by catalase, and catalase may suppress the signal transduction. When potato roots were infected with *E. carotovora* or *Corynebacterium sepedonicum,* a catalase gene *(Cat2St)* was induced. *Cat2St* was found to be systemically induced after compatible interaction. Hence, suppression of a catalase gene may result in H_2O_2 signal transduction and induce resistance. Transgenic tobacco plants with severely depressed levels of catalase have been developed with antisense suppression (Takahashi et al., 1997). These plants showed both enhanced accumulation of salicylic acid and enhanced disease resistance (Takahashi et al., 1997).

The cuticle forms an important barrier on plant surfaces against the ingress of pathogens into the hosts (Vidhyasekaran, 1997). The cuticle consists of cutin. Cutin monomers could serve as a signal to the plant during a pathogen attack, triggering defense responses. Cutin monomers elicit H_2O_2 and enhance the activity of other H_2O_2 elicitors (Fauth et al., 1998). Cutin monomers are derived from fatty-acid precursors. Hence, it is possible that by releasing cutin monomers, diseases can be managed. Tomato plants were transformed with the yeast Δ-9 desaturase gene. In transgenic plants, large increases in 9-hexadecenoic acid (cutin monomer) were observed, and these plants showed increased resistance to *Erysiphe polygoni*. Cutin monomers from the transgenic plants inhibited spore germination of the pathogen (Wang, Chin, and Gianfagna, 2000).

TRANSGENIC PLANTS EXPRESSING PLANT PROTEINASE-INHIBITOR GENES

Viral cysteine proteinases are involved in the processing and replication of potyviruses in plants. The potyvirus genome encodes a polyprotein that is processed by proteolysis into individual gene products. Three proteinases have been shown to be involved in the processing mechanism, and at least one of them seems to be of the cysteine type. An essential step for the replication of potyviruses is the activity of proteases identified as cysteine proteinases, which process the 346 kDa polyprotein. Cystatins are powerful and specific inhibitors of cysteine proteinases. Hence, cystatins can be used to develop virus-disease-resistant plants. Tobacco plants transformed with a rice cysteine proteinase inhibitor (oryzacystatin) gene showed resistance to the *Tobacco etch virus* and *Potato virus Y* (Gutierrez-Campos et al., 1999).

TRANSGENIC PLANTS EXPRESSING VIRAL GENES

Most successful transgenic plants with viral-disease resistance have been developed by using viral genes themselves. This phenomenon is called pathogen-derived resistance. The two most common viral genes used are those coding for the coat protein and the RNA-dependent RNA polymerase (also called replicase). The polymerase copies the viral genome. The coat protein gene tends to confer resistance to a broader range of related viruses, but resistance is often incomplete, with the plant showing delayed and milder symptoms than the untransformed plant. Resistance can be overcome by extremely high levels of inoculum or inoculation with naked viral RNA. In contrast, polymerase (replicase) gene-mediated resistance can confer complete immunity, but only to virus strains with a very high sequence homology to the one from which the transgene was derived. Besides using these two groups of viral genes, viral movement protein genes, untranslatable sense and antisense RNAs, proteases, defective-interfering RNAs, satellites, and ribozymes have been used to develop virus-resistant transgenic plants.

Transgenic plants expressing a coat protein (CP) gene have been reported by several researchers. Transgenic zucchini squash plants expressing the coat protein genes of the *Squash mosaic virus* (SqMV) showed resistance to SqMV (Jan, Pang, and Gonsalves, 2000; Jan, Pang, Tricoli, and Gonsalves, 2000). Transgenic *Nicotiana benthamiana* plants expressing the *Arabis mosaic nepovirus* (ArMV) coat protein gene showed resistance to the virus (Spielmann et al., 2000). The resistance was expressed as a delay in infection and a reduction of the percentage of infected plants (Spielmann et al., 2000). Transgenic tobacco plants expressing *Andean potato mottle virus* (APMoV) coat protein genes (*CP22* and *CP42*) show resistance to APMoV (Neves-Borges et al., 2001). *Nicotiana benthamiana* plants were transformed with the coat protein coding sequence and the 3' nontranslated region of the severe strain of *Sweet potato feathery mottle virus* (SPFMV-S), and these transgenic plants showed resistance to SPFMV-S (Sonoda et al., 1999). Tamarillos *(Cyphomandra betacea)* expressing a *Tamarillo mosaic potyvirus* (TaMV) coat protein showed resistance to TaMV (Cohen et al., 2000). Transgenic *N. benthamiana* plants expressing *Grapevine virus A* (GVA) coat protein sequences showed resistance to the virus (Radian-Sade et al., 2000). Transgenic tobacco plants expressing the *Tobacco vein mottling virus* (TVMV) coat protein gene expressed resistance against various potyviruses and transgenic tobacco plants expressing the *Alfalfa mosaic virus* (AMV) coat protein gene expressed resistance against AMV (Xu et al., 1999). Transgenic cucumber plants expressing the *Watermelon mosaic virus* (WMV-2) coat protein gene showed enhanced resistance to WMV-2 (Wang, Zhao, and Zhou, 2000). *Nicotiana benthamiana* and potato plants expressing the coat protein gene from the *Potato mop-top virus* (PMTV) showed resistance to PMTV (McGeachy and Barker, 2000). Transgenic tobacco plants expressing a *Potato virus Y* coat protein gene showed resistance to PVY (Han et al., 1999). Three tobacco cultivars expressing the nucleocapsid *(N)* gene of a dahlia isolate of the *Tomato spotted wilt virus* (TSWV) showed resistance to TSWV in the field (Herrero et al., 2000). Peanut lines transgenic for the antisense nucleocapsid *(N)* gene of a TSWV strain isolated from peanut showed resistance to TSWV under field conditions (Magbanua et al., 2000). Transgenic *Nicotiana tabacum* and *N. benthamiana* plants expressing the *N* gene of TSWV showed resistance against TSWV (Accotto et al., 2000). Both an RNA-mediated and a protein-mediated resistance may be acting in the transformed plants expressing nucleocapsid proteins (Accotto et al., 2000).

Viral replicase genes have been exploited in developing virus disease resistant plants. Russet Burbank potatoes have been genetically modified with the full-length replicase gene of the *Potato leafroll virus* (PLRV), and these plants showed resistance against PLRV. This transgenic plant has been patented and marketed under the name NewLeaf Plus (Lawson et al., 2001). Wheat plants expressing the replicase gene *NIb* of the *Wheat streak mosaic virus* (WSMV) showed resistance to WSMV (Sivamani et al., 2000). Some transgenic clones of potato expressing *Potato virus Y^N* (PVY^N) replicase showed resistance to PVY^N (Flis and Zimnoch-Guzowska, 2000). Transgenic rice plants expressing *Rice yellow mottle sobemovirus* (RYMV) RNA polymerase were resistant to RYMV strains from different African locations. In these plants, virus multiplication was suppressed. The resistance was stable over at least three generations. The viral replicase gene-mediated resistance was also associated with posttranscriptional gene silencing (PTGS) (Pinto et al., 1999). *Nicotiana benthamiana* plants transformed with a *Pepper mild mottle virus* (PMMoV) 54 kDa fragment of viral replicase showed complete resistance to PMMoV (Tenllado and Diaz-Ruiz, 1999).

Nicotiana bethamiana plants were transformed with the cytoplasmic inclusion protein (CI) gene of the *Plum pox virus* (PPV). The CI protein is an RNA helicase that contains a conserved nucleotide binding motif (NTBM), and it plays an important role in viral replication. Transgenic lines containing the gene mutated in the NTBM region remained completely symptomless after *Plum pox virus* infection (Wittner et al., 1998). Viruses move from cell to cell in the plant through plasmodesmata, which connects the cytoplasm between neighboring cells. Viral movement protein (MP) helps viral movement in plants. The movement proteins of plant viruses are localized to plasmodesmata in infected plant cells. They induce a significant increase in the plasmodesmatal exclusion limit and bind single-stranded RNA. This allows the viral genome to migrate from cell to cell. The MPs expressed in transgenic plants can complement the cell-to-cell movement of MP-defective viral mutants. Transgenic plants expressing a viral movement protein gene show increased susceptibility to viruses (Yoshikawa et al., 2000). The transgenic *Nicotiana occidentalis* plants expressing a movement protein of the *Apple chlorotic leaf spot virus* (ACLSV) showed increased susceptibility to a homologous virus (ACLSV). However, the transgenic plants showed strong resistance to the heterologous virus *Grapevine berry inner necrosis virus* (GINV) (Yoshikawa et al., 2000). Although tobacco plants expressing the *Cucumber mosaic virus* MP gene did not show resistance to CMV infection, transgenic plants expressing a deletion mutant of the CMV MP gene showed high resistance to CMV infection (Zhang et al., 1999). Transgenic potato lines carrying a mutant movement protein gene of the *Potato leafroll virus* showed resistance to potato viruses (Rohde et al., 2000). Transgenic tobacco plants expressing a defective MP of the *Tobacco mosaic virus,* are resistant to TMV and also to viruses of other taxa, i.e., *Alfalfa mosaic virus, Cucumber mosaic virus, Tobacco rattle virus,* and *Peanut streak virus* (Cooper et al., 1995). Thus, any altered nonfunctional MP gene confers resistance against a wide range of viruses. This phenomenon is called MP-derived resistance.

The viral gene-induced resistance may be mostly due to posttranscriptional gene silencing. The gene silencing may be due to host-induced, sequence-specific RNA degradation. An incoming virus containing the same sequences would also be targeted for degradation. Several studies have proven this hypothesis. Resistance in the transgenic tobacco plants expressing the APMoP coat protein gene is mediated by transcriptional gene silencing mechanism, as there was an inverse correlation between resistance and the accumulation of CP transgene mRNAs. The gene silencing was due to methylation (Neves-Borges et al., 2001). When *N. benthamiana* plants were transformed with the coat protein coding sequence and the 3' nontranslated region of the severe strain of the *Sweet potato feathery mottle virus,* seven out of 19 transgenic plants showed resistance to *Potato virus X* (Sonoda et al., 1999). In most of the resistant lines, relatively low steady-state accumulation of the coat protein gene mRNA and few or no protein products were observed. The regions of the transgene corresponding to the RNA target in the resistant lines were differentially methylated compared to the transgene sequence in a susceptible line. These results suggest that the resistance in the transgenic plants is due to a posttranscriptional gene-silencing mechanism. More convincing evidence to prove the hypothesis was presented by Jan, Pang, Tricoli, and Gonsalves (2000). Transgenic *N. benthamiana* plants expressing 218 and 110 base pair (bp) *N* gene segments of the *Tomato spotted wilt virus* did not show any resistance to TSWV. But, when the *N* gene segments were fused to the nontarget green fluorescent protein *(GFP)* gene and *N. benthamiana* plants were transformed with the chimeric genes, the transgenic plants showed resistance to TSWV. The *GFP* gene is known to induce PTGS of other genes. The GFP DNA-induced PTGS would have targeted *N* gene segments and the incoming homologous TSWV for degradation, resulting in a resistant phenotype. In another experiment, it was shown that the full-length coat protein gene of the *Turnip mosaic virus* (TuMV) was linked to 218 *N* gene segments and transformed into *N. benthamiana* (Jan, Pang, Tricoli, and Gonsalves, 2000). The transgenic lines expressing the *N* gene of the TSWV and the coat protein gene of the TuMV showed resistance to both viruses. The TuMV CP gene is known to induce PTGS and would have helped the *N* gene of the TSWV to express PTGS.

Gene silencing appears to be a common phenomenon of virus-disease resistance. RNA-mediated cross-protection has been reported in plants infected with nepoviruses, caulimoviruses, tobraviruses, and potexviruses. The upper leaves of plants infected with these viruses are symptom free and contained reduced levels of the virus. These leaves are said to be recovered. Recovery is associated with RNA-mediated cross-protection against secondary virus infection (Ratcliff et al., 1999). A transient gene expression assay showed that RNA-mediated cross-protection was functionally equivalent to posttranscriptional gene silencing (Ratcliff et al., 1999). To uncover factors that may play roles in gene silencing, sequences in the 3' part of the transcribed region of the *Potato virus Y* coat protein gene were transcribed in vitro, and the RNA fragments were incubated with cell extracts from transgenic tobacco plants expressing the coat protein gene of PVY. Ribonuclease activity was detected that appeared to be specific for this transcript in the PVY-resistant transgenic plants (Han et al., 1999).

Viral genes may also code for elicitors of defense mechanisms. The disease resistance gene *N* has been cloned in tobacco and induces resistance against TMV. The TMV-encoded elicitor of this gene has been identified as TMV replicase. The TMV replicase proteins interact with the *N*-gene product to activate defense genes of the host (Padgett et al., 1997). The elicitor of the disease resistance gene *Rx* in potato has been identified as a capsid protein of PVX (Bendahmane et al., 1995). The resistance induced by the capsid protein was shown to be due to the elicitation of a response that was not specific to the viral target sequence (Bendahmane et al., 1999).

Several plant viruses in the family *Tombusviridae* are known to generate defective interfering (DI) RNAs during their replication. In many cases, the viral symptoms are ameliorated by the DIs. Rubio and colleagues (1999) designed a DNA cassette to transcribe defective interfering RNAs of the *Tomato bushy stunt virus* (TBSV), in which DI RNA sequences were flanked by ribozymes (RzDI). When RzDI RNAs transcribed in vitro were mixed with parental TBSV transcripts and inoculated into plants, they became amplified, reduced the accumulation of the parental RNA, and mediated attenuation of the lethal syndrome characteristic of TBSV infections. Transgenic *N. benthamiana* plants expressing defective interfering RNAs from the TBSV showed protection against a wide spectrum of tombusviruses but remained susceptible to a distantly related tombus-like virus and to unrelated viruses (Rubio et al., 1999).

TRANSGENIC PLANTS EXPRESSING FUNGAL
AND BACTERIAL GENES

Transgenic Plants Expressing Fungal Genes

Chitinases are lytic enzymes which degrade chitin, a common fungal cell-wall component. Degraded products of chitin are also elicitors of defense mechanisms in plants. Several fungi, particularly saprophytes, are known to produce chitinases, and the genes encoding chitinases have been cloned. Transgenic apple plants expressing genes encoding endochitinase cloned from *Trichoderma harzianum* showed increased resistance to apple scab (Bolar et al., 2000) and powdery mildew *(Podosphaera leucotricha)* (Hanke et al., 1999). The gene *ech-42* encoding an endochitinase from *T. harzianum* was introduced alone or in combination with the osmotin gene from *N. tabacum* into *Petunia hybrida*. The transgenic plants showed resistance to *Botrytis cinerea* (Esposito et al., 2000). *ThEn-42,* an endochitinase gene cloned from strain P1 of *T. harzianum,* was introduced into the japonica rice varieties Taibei 309 and Nonghu 6. These transgenic plants expressed high resistance to the sheath blight pathogen *Rhizoctonia solani* (Qin et al., 2000). Phenolics appear to play an important role in inducing resistance against virus diseases. Plants overexpressing genes coding for phenolic synthesis have shown virus-disease resistance. Tobacco plants expressing a yeast-derived invertase (β-fructofuranosidase) gene showed increased synthesis of phenolics, particularly a caffeic acid amide (*N*-caffeoylputrescine). These transgenic plants showed resistance to *Potato virus Y* (Baumert et al., 2001).

Transgenic Plants Expressing Bacterial Genes

Non-haem chloroperoxidases of *Pseudomonas pyrrocinia* catalyzes the oxidation of alkyl acids to peracids by hydrogen peroxide. Alkyl peracids possess potent antifungal activity. Tobacco plants were transformed with a gene for chloroperoxidase from *P. pyrrocinia.* Transgenic tobacco plants expressing this gene showed enhanced resistance to pathogens. The leaf extracts from the transgenic plants inhibited growth of *Aspergillus flavus* by up to 100 percent (Jacks et al., 2000). Rajasekaran and colleagues (2000) reported that transgenic tobacco plants producing chloroperoxidase encoded by *Pseudomonas pyrrocinia* showed increased resistance against *Colletotrichum destructivum,* the causal agent of tobacco anthracnose.

Transgenic Plants Expressing Genes from Fungal Viruses

Some double-stranded RNA viruses infect fungi. They secrete killing proteins (KP). The *Ustilago maydis*-infecting virus secretes an antifungal protein called KP4. The cDNA encoding the antifungal protein KP4 was inserted behind the ubiquitin promoter of corn and genetically transferred to wheat varieties susceptible to stinking smut *(Tilletia tritici)* disease. The transgenic plants showed increased resistance to *T. tritici* (Clausen et al., 2000). The transgenic lines showed an antifungal activity against *U. maydis* and the antifungal activity correlated with the presence of the KP4 transgene (Clausen et al., 2000).

Transgenic Plants Expressing Genes from Bacteriophages

Lysozymes are known to lyse bacterial cells. Several genes coding for lysozymes have been isolated from bacteriophages. Bacteriophages lyse the bacterial cell walls producing lysozymes. A lysozyme has been purified from bacteriophage T4 and was active against both gram-negative and gram-positive bacteria. The phage T4 shows lytic action against *E. caroto-vora,* and a receptor for phage T4 has been reported in the bacterium. Attempts were made to induce disease resistance by transferring a bacterio-phage T4 lysozyme gene into plants and causing it to secrete to the inter-cellular spaces in which the bacterial pathogens multiply. In the developed transgenic tobacco plants, the T4 lysozyme was found to be secreted into the intercellular spaces. The same chimeric gene cassette was transferred to potato through *Agrobacterium tumefaciens*-mediated transformation. The T4 lysozyme could be detected in intercellular washing fluids of transgenic potato plants (During et al., 1993). The transgenic potato plants expressing the bacteriophage T4 lysozyme showed high resistance to *E. carotovora* ssp. *atroseptica* as shown by reduced tissue maceration in transgenic plants. No black leg symptoms developed in transgenic potato plants until harvest, while no control plants survived under greenhouse conditions (During et al., 1993). Transgenic potato plants expressing the T4 lysozyme gene also showed resistance to *E. carotovora* ssp. *carotovora* (Ahrenholtz et al., 2000). Transgenic apple plants expressing the bacteriophage T4 gene have been developed and show resistance to *E. amylovora* (Ko et al., 1999).

TRANSGENIC PLANTS
EXPRESSING INSECT GENES

Cationic antimicrobial proteins have been detected in insects. These pro-teins also have been exploited to develop disease-resistant plants. Cecropins are a family of homologous antibacterial peptides derived from the giant silk moth, *Hyalophora cecropia* (Mills and Hammerschlag, 1993). The cecropins have no effect on eukaryotic cells but affect prokaryotic cells through the formation of voltage-dependent ion channels. The lytic peptides form pores in bacterial membranes. A chimeric cecropin B gene was intro-duced into *Pelargonium hortorum* plants. Transgenic plants showed resis-tance to *Xanthomonas* (Renou et al., 2000). Cecropins are unstable in plants and are degraded by plant proteases. Several stable analogs of cecropin B, including SB-37, Shiva-1, and MB39, have been synthesized. Transgenic tobacco plants expressing Shiva-1 showed enhanced resistance to *Ralstonia solanacearum* (Jaynes, 1993). Transgenic apple (Norelli et al., 1999) and pear (Reynoird, Mourgues, Chevreau, et al., 1999) plants expressing SB-37 show resistance to *Erwinia amylovora.*

The stable analog of cecropin, MB39, was used to develop transgenic to-bacco plants (Huang et al., 1997). A chimeric gene fusion cassette, consist-ing of a secretory sequence from barley α-amylase joined to the MB39 coding sequence and placed under control of the promoter and terminator from the potato proteinase inhibitor (PiII) gene, was introduced into tobacco by *Agrobacterium*-mediated transformation. No disease symptom development was observed in leaf tissues infiltrated with *P. syringae* pv. *tabaci* in MB39 transgenic plants, while in control plants disease developed rapidly. The attacins are another class of lytic peptides isolated from the giant silk moth. The *attacin E (att E)* gene has been introduced into apple plants. The transgenic apple showed resistance to fire blight *(E. amylovora)* (Norelli et al., 1994; Ko et al., 2000). Transgenic pear expressing *att E* (Reynoird, Mourgues, Norelli, et al., 1999), SB-37, and Shiva-1 (Chevreau et al., 1999) showed resistance to *E. amylovora.*

Transgenic cauliflower plants expressing the Shiva protein showed resistance to *Xanthomonas campestris* pv. *campestris* (Braun et al., 2000). A synthetic gene encoding a N-terminus-modified cecropin-melittin cationic peptide (isolated from bees) chimera was introduced into two potato cultivars. The transgenic plants showed high resistance to *Erwinia carotovora* (Osusky et al., 2000). Antimicrobial peptides, sarcotoxins, isolated from the insect *Sarcophaga peregrina* show severalfold higher antibacterial activity and broader spectra than cecropin B. Transgenic tobacco plants expressing sarcotoxin IA showed resistance to *E. carotovora* and *Pseudomonas syringae* pv. *tabaci* (Ohshima et al., 1999; Mitsuhara et al., 2000).

Cecropins isolated from the giant silk moth, *H. cecropia,* are a family of homologous peptides of 35 to 37 residues with amphipathic N termini and hydrophobic C termini. Cecropins inhibited only bacterial pathogens and not fungal pathogens. Another 26-amino acid antibacterial peptide, melittin, is produced by bees. It has a predominantly hydrophobic N terminus with an amphipathic C terminus. A chimeric peptide, Cecropin-melittin antibiotic peptide (CEMA), was generated, and it contains eight amino acid residues from cecropin A and a modified melittin sequence at the C terminus. CEMA was toxic to plants. Toxicity was reduced by modifying the N-terminus while retaining the original a-helical character. A synthetic gene *(msrA1)* encoding a N-terminus-modified cecropin-melittin cationic peptide chimera was constructed and introduced into two potato cultivars. The transgenic plants showed high resistance to *Phytophthora cactorum* and *Fusarium solani* (Osusky et al., 2000). Antimicrobial peptides, sarcotoxins, isolated from the insect *Sarcophaga peregrina* show severalfold higher antimicrobial activity than cecropin B. Transgenic tobacco plants expressing sarcotoxin IA showed resistance to *Rhizoctonia solani* and *Pythium aphanidermatum* (Ohshima et al., 1999; Mitsuhara et al., 2000).

TRANSGENIC PLANTS
EXPRESSING ANIMAL GENES

Transgenic Plants Expressing Hen Genes

A hen-egg-lysozyme gene has been cloned, and transgenic tobacco plants expressing this gene have been developed (Trudel et al., 1995). Extracts from transgenic tobacco plants producing hen-egg lysozyme inhibited the growth of several species of bacteria (Trudel et al., 1995). The *chly* gene, encoding lysozyme from chicken, was introduced into potato plants. The transgenic plants showed increased resistance to the black leg disease caused by *E. carotovora* ssp. *atroseptica* (Serrano et al., 2000). A gene encoding a lytic peptide (tachyplesin) has been isolated from the horseshoe crab. Transgenic potato plants expressing this gene were developed, and these plants showed reduction in tuber rot caused by *E. carotovora* (Allefs et al., 1996).

Transgenic Plants Expressing Frog Genes

Antibacterial lytic peptides have also been isolated from frogs. The gene encoding the magainin II peptide has been cloned from the African clawed frog. The gene was modified for plant codon usage. The gene construct included a 35S promoter, the 5' leader sequence from alfalfa mosaic virus, the signal peptide from the tobacco PR-S gene to target protein export to the intracellular space, and a NOS terminator. In addition, the vector contained a *NPT II* (NOS-NPT II-NOS) gene for selection of transgenic cells. Transgenic cauliflower plants were developed using *Agrobacterium*-mediated transformation. The transgenic plants expressing frog genes showed increased resistance to *X. campestris* pv. *campestris* (Braun et al., 2000).

Transgenic Plants Expressing Human Genes

Human lysozyme is a powerful lytic enzyme degrading fungal and bacterial cell walls. A human lysozyme gene was placed under control of the constitutive CaMV 35S promoter, and the resulting expression plasmid was introduced into carrot. The transgenic plants showed strong resistance to two fungal pathogens, *Erysiphe heraclei* and *Alternaria dauci* (Takaichi and Oeda, 2000). Tobacco plants were transformed with a human lysozyme gene (Nakajima et al., 1997). These transgenic plants showed reduced disease symptoms induced by *P. syringae* pv. *tabaci* in tobacco. Lactoferrin is an iron-binding glycoprotein known to have antibacterial properties. The expression of a human lactoferrin gene in tobacco delayed the onset of symptoms caused by *Ralstonia solanacearum* from 5 to 25 days (Zhang et al., 1996). Transgenic pear plants expressing a lactoferrin gene show resistance to the fire blight pathogen, *E. amylovora* (Malnoy et al., 2000).

TRANSGENIC PLANTS EXPRESSING DETOXIFICATION GENES

Pathogens produce toxins, which have been shown to be responsible for disease symptom development. Mutants, which do not produce toxins, have reduced virulence. Toxin-insensitive plants show enhanced resistance. The resistance gene cloned from maize has been shown to be involved in detoxification of the toxin produced by *Helminthosporium maydis* (Johal and Briggs, 1992). *Eutypa lata,* the pathogen of grapevine dieback, produces a toxin called eutypine. The toxin plays a prominent role in the expression of eutypa dieback symptoms. A protein from *Vigna radiata* exhibits eutypine reductase, which degrades the toxin into a nontoxic compound. The gene encoding the enzyme has been cloned. The *VR-ERE* (eutypine reducing enzyme) cDNA encodes an NADPH-dependent reductase [alcohol dehydrogenase (NADP$^+$)] of 36 kDa exhibiting a high affinity toward eutypine. Expression of the *VR-ERE* cDNA in transformed grapevine cells conferred resistance to the toxin (Roustan et al., 2000).

Bacterial pathogens also produce toxins. Hence, it is possible to develop bacterial disease-resistant plants by inactivating toxins produced by bacterial pathogens. *Pseudomonas savastanoi* pv. *phaseolicola,* the bean halo blight pathogen, produces a toxin which induces typical symptoms of the disease in bean. The pathogen is tolerant to its own toxin, and it has been shown to be due to the bacterial enzyme ornithine carbomyl transferase. The corresponding gene *(argK)* has been cloned. Transgenic bean plants expressing *argK* showed high resistance to *P. savastanoi* pv. *phaseolicola* (Herrera-Estrella and Simpson, 1995). *P. syringae* pv. *tabaci,* the wild fire pathogen of tobacco, produces tabtoxin which is responsible for symptom development. The gene *ttr,* which encodes an enzyme that inactivates the tabtoxin, has been cloned from *P. syringae* pv. *tabaci.* The transgenic tobacco plants expressing the *ttr* gene show high resistance to the pathogen (Anzai et al., 1989).

Albicidins are the toxins produced by *Xanthomonas albilineans,* the sugarcane leaf scald pathogen. The toxins have been shown to be important in the disease development. The bacterium *Pantoea dispersa* (= *Erwinia herbicola*) was found to detoxify albicidin. The *P. dispersa* gene *(albD)* for enzymatic detoxification of albicidin has been cloned (Zhang and Birch, 1997). Transgenic sugarcane plants expressing the *albD* gene showed high resistance to the leaf scald pathogen (Birch et al., 2000).

TRANSGENIC PLANTS EXPRESSING GENES ENCODING ANTIBODIES

Monoclonal antibodies raised against a virus react with various related virus species. Antibodies can inactivate viruses. Monoclonal antibody 3-17, which was raised against the *Potyvirus Johnsongrass mosaic virus,* was shown to react strongly with 14 potyvirus species (Xiao et al., 2000). Transgenic tobacco plants expressing a single-chain variable region antibody derived from the monoclonal antibody 3-17 showed resistance to *Potato virus Y* and *Clover yellow vein virus* (ClYVV). It suggests that one single-chain construct can be used to protect plants from distinct potyviruses (Xiao et al., 2000). The *Tobacco mosaic virus* neutralizing single-chain Fv antibody fragment (scFv24) was targeted to the endoplasmic reticulum and integrated into the plasma membrane of tobacco cells. The transgenic plants showed resistance to TMV infection (Schillberg et al., 2000). The results suggest that virus diseases can be controlled by inducing the plants to produce antibodies to viruses.

PRACTICAL UTILITY OF TRANSGENIC PLANTS IN CROP-DISEASE MANAGEMENT

Several successful reports of development of transgenic plants with disease resistance have been published recently. The important criteria for exploiting this technology are the durability of expression of these foreign genes in the transgenic plants, absence of toxicity, and minimal environmental impact. Field evaluation of all these technologies is important. One major objection in adoption of transgenic plants is the presence of marker genes in the products from these plants. Several methods have been developed to limit the expression of the marker genes to the stages at which selection is made and eliminate the marker genes well before harvest. In the near future, we can expect that this type of genetic engineering strategy will result in development of a practical strategy for efficient management of diseases in crop plants. Globally, the area of cultivation of transgenic plants is increasing year by year. It was 1.7 million hectares in 1996, and it increased to 11.0 million hectares in 1997, and 27.8 million hectares in 1998 (James,

1998). Already transgenic plants expressing virus resistance are commercially available in potato, tomato, soybean, and corn.

REFERENCES

Abad, M. S., Hakimi, S. M., Kaniewski, W. K., Rommens, C. M. T., Shulaev, V., Lam, E., and Shah, D. M. (1997). Characterization of acquired resistance in lesion-mimic transgenic potato expressing bacterio-opsin. *Mol Plant-Microbe Interact,* 10:635-645.

Accotto, G. P., Allavena, A., Vaira, A. M., and Nervo, G. (2000). Inserting the nucleoprotein gene of tomato spotted wilt virus in different plant species, and screening for virus resistance. In A. D. Arencibia (Ed.), *Plant Genetic Engineering Towards the Third Millennium.* Elsevier Science Publishers, Amsterdam, The Netherlands, pp. 48-153.

Ahrenholtz, I., Harms, K., De Vries, J., and Wackernagel, W. (2000). Increased killing of *Bacillus subtilis* on the hair roots of transgenic T4 lysozyme-producing potatoes. *Appl Environ Microbiol,* 66:1862-1865.

Alexander, D., Goodman, R. M., Gut-Rella, M., Glascock, C., Weymann, K., Friedrich, L., Maddox, D., Ahl Goy, P., Luntz, T., Ward, E., and Ryals, J. (1993). Increased tolerance to two oomycete pathogens in transgenic tobacco expressing pathogenesis-related protein 1a. *Proc Natl Acad Sci USA,* 90:7327-7331.

Allefs, S. J. H. M., de Jong, E. R., Florack, D. E. A., Hoogendoom, C., and Stiekema, W. J. (1996). *Erwinia* soft rot resistance of potato cultivars expressing antimicrobial peptide tachyplesin I. *Mol Breed,* 2:97-105.

Anzai, H., Yoneyama, K., and Yamaguchi, I. (1989). Transgenic tobacco resistant to a bacterial disease by the detoxification of a pathogenic toxin. *Mol Gen Genet,* 219:492-494.

Bauer, D. W., Garr, E. R., Beer, S. V., Norelli, J. L., and Aldwinckle, H. S. (1999). New approaches to the development of transgenic plants resistant to fire blight. *Acta Horticulturae,* 489:301-304.

Baumert, A., Mock, H. P., Schmidt, J., Herbers, K., Sonnewald, U., and Strack, D. (2001). Patterns of phenylpropanoids in non-inoculated and potato virus Y-inoculated leaves of transgenic yeast-derived invertase. *Phytochemistry,* 56:535-541.

Beffa, R., Szell, M., Meuwly, P., Pay, A., Vogeli-Lange, R., Metraux, J. P., Neuhaus, G., Meins, F., and Ferenc, N. (1995). Cholera toxin elevates pathogen resistance and induces pathogenesis-related gene expression in tobacco. *EMBO J,* 14:5753-5761.

Bendahmane, A., Kanyuka, K., and Baulcombe, D. C. (1999). The *Rx* gene from potato controls separate virus resistance and cell death responses. *Plant Cell,* 11: 781-791.

Bendahmane, A., Kohm, B. A., Dedi, C., and Baulcombe, D. C. (1995). The coat protein of potato virus X is a strain specific elicitor of *Rx-1*-mediated virus resistance in potato. *Plant J,* 8: 933-941.

Birch, R. G., Bower, R., Elliott, A., Hansom, S., Basnayake, S., and Zhang, L. (2000). Regulation of transgene expression: Progress towards practical development in sugarcane, and implications for other plant species. In A. D. Arencibia (Ed.), *Plant Genetic Engineering: Towards the Third Milennium.* Elsevier Science Publishers, Amsterdam, The Netherlands, pp. 118-125.

Bolar, J. P., Norelli, J. L., Wong, K. W., Hayes, C. K., Harman, G. E., and Aldwinckle, H. S. (2000). Expression of endochitinase from *Trichoderma harzianum* in transgenic apple increases resistance to apple scab and reduces vigor. *Phytopathology,* 90:72-77.

Braun, R. H., Reader, J. K., and Christey, M. C. (2000). Evaluation of cauliflower transgenic for resistance to *Xanthomonas campestris* pv. *campestris. Acta Horticulturae,* 539:137-143.

Broglie, K., Chet, I., Holliday, M., Cressman, R., Biddle, P., Knowlton, S., Mauvais, J. C., and Broglie, R. (1991). Transgenic plants with enhanced resistance to the fungal pathogen *Rhizoctonia solani. Science,* 254:1194-1197.

Brommonschenkel, S. H., Frary, A., Frary, A., and Tanksley, S. D. (2000). The broad-spectrum tospovirus resistance gene *Sw-5* of tomato is a homolog of the root-knot nematode resistance gene *Mi. Mol Plant-Microbe Interact,* 13:1130-1138.

Carmona, M. J., Molina, A., Fernandez, J. A., Lopez-Fando, J., and Garcia-Olmedo, F. (1993). Expression of the alpha-thionin gene from barley in tobacco confers enhanced resistance to bacterial pathogens. *Plant J,* 3:457-462.

Chen, W. P., Chen, P. D., Liu, D. J., Kynast, R., Friebe, B., Velazhahan, R., Muthukrishnan, S., and Gill, B. S. (1999). Development of wheat scab symptoms is delayed in transgenic wheat plants that constitutively express a rice thaumatin-like protein gene. *Theor Appl Genet,* 99:755-760.

Chevreau, E., Mourgues, F., Reynoird, J. P., and Brisset, M. N. (1999). Gene transfer for fire blight resistance in pear. *Acta Horticulturae,* 489:297-300.

Clausen, M., Krautter, R., Schachermayr, G., Potrykus, I., and Sautter, C. (2000). Antifungal activity of a virally encoded gene in transgenic wheat. *Nature Biotechnology,* 18:446-449.

Cohen, D., van den Brink, R. C., MacDiarmid, R. M., Beck, D. L., and Forster, R. L. S. (2000). Resistance to tamarillo mosaic virus in transgenic tamarillos and expression of the transgenes in F1 progeny. *Acta Horticulturae,* 521:43-49.

Cooper, B., Lapidot, M., Heick, J. S., Dodds, J. A., and Beachey, R. N. (1995). A defective movement protein of TMV in transgenic plants confers resistance to multiple viruses whereas the functional analog increases susceptibility. *Virology,* 206:307-313.

Datta, K., Koukolikova-Nicola, Z., Baisakh, N., Oliva, N., and Datta, S. K. (2000). *Agrobacterium*-mediated engineering for sheath blight resistance of indica rice cultivars from different ecosystems. *Theor Appl Genet,* 100:832-839.

Datta, K., Tu, J., Oliva, N., Ona, I., Velazhahan, R., Mew, T. W., Muthukrishnan, S., and Datta, S. K. (2001). Enhanced resistance to sheath blight by constitutive expression of infection-related rice chitinase in transgenic elite indica rice cultivars. *Plant Sci,* 160:405-414.

Datta, K., Velazhahan, R., Oliva, N., Ona, I., Mew, T., Khush, G., Muthukrishnan, S., and Datta, S. K. (1999). Overexpression of the cloned rice thaumatin-like protein (PR-5) gene in transgenic rice plants enhances environmental friendly resistance to *Rhizoctonia solani* causing sheath blight disease. *Theor Appl Genet,* 98:1138-1145.

During, K., Porsch, P., Fladung, M., and Lorz, H. (1993). Transgenic potato plants resistant to the phytopathogenic bacterium *Erwinia carotovor. Plant J,* 3:587-598.

Esposito, S., Colucci, M. G., Frusciante, L., Filippone, E., Lorito, M., and Bressan, R. A. (2000). Antifungal transgenes expression in *Petunia hybrida. Acta Horticulturae,* 508:157-161.

Fauth, M., Schweizer, P., Buchala, A., Markstadter, C., Reiederer, M., Tadahiro, K., and Kauss, H. (1998). Cutin monomers and surface wax constituents elicit H_2O_2 in conditioned cucumber hypocotyl segments and enhance the activity of other H_2O_2 elicitors. *Plant Physiol,* 117:1373-1380.

Fettig, S. and Hess, D. (1999). Expression of a chimeric stilbene synthase gene in transgenic wheat lines. *Transgenic Research,* 8:179-189.

Flis, B. and Zimnoch-Guzowska, E. (2000). Field performance of transgenic clones obtained from potato cv. Irga. *J Appl Genet,* 41:81-90.

Florack, D. E. A., Dirkse, W. G., Visser, B., Heidekamp, F., and Stiekema, W. J. (1994). Expression of biologically active hordothionins in tobacco. Effects of pre- and pro-sequences at the amino and carboxyl termini of the hordothionin precursor on mature protein expression and sorting. *Plant Mol Biol,* 24:83-96.

Gutierrez-Campos, R., Torres-Acosta, J. A., Saucedo-Arias, L. J., and Gomez-Lim, M. A. (1999). The use of cysteine proteinase inhibitors to engineer resistance against potyviruses in transgenic tobacco plants. *Nature Biotechnology,* 17: 1223-1227.

Hain, R., Reif, H. J., Krause, E., Langebartels, R., Kindl, H., Vornam, B., Weise, W., Schmelzer, E., and Schreier, P. H. (1993). Disease resistance results from foreign phytoalexin expression in a novel plant. *Nature,* 361:153-156.

Han, S. J., Cho, H. S., You, J. S., Nam, Y. W., Park, E. K., Shin, J. S., Park, Y. I., Park, W. M., and Paek, K. H. (1999). Gene silencing-mediated resistance in transgenic tobacco plants carrying the potato virus Y coat protein gene. *Molecules and Cells,* 9:371-383.

Hanke, V., Hiller, I., Klotzsche, G., Winkler, K., Egerer, J., Richter, K., Norelli, J. L., and Aldwincke, H. S. (1999). Transformation in apple for increased disease resistance. *Acta Horticulturae,* 538:611-616.

He, X. Z. and Dixon, R. A. (2000). Genetic manipulation of isoflavone 7-*O*-methyltransferase enhances biosynthesis of 4'-*O*-methylated isoflavonoid phytoalexins and disease resistance in alfalfa. *Plant Cell,* 12:1689-1702.

Hennin, C., Hofte, M., and Diederichsen, E. (2000). Induction of an artificial hypersensitive response by application of the CF9/AVR9 system in oilseed rape and its effect on *Leptosphaeria maculans.* In *Fourteenth Forum for Applied Biotechnology, Brugge, Belgium, 2000, Proceedings Part II. Mededelingen-Faculteit Landbouwkundige en Toegepaste Biologische Wetenschappen, Universiteit Gent,* 65(3b):409-416.

Herrera-Estrella, L. and Simpson, J. (1995). Genetically engineered resistance to bacterial and fungal pathogens. *World J Microbiol Biotechnol,* 11:383-392.

Herrero, S., Culbreath, A. K., Csinos, A. S., Pappu, H. R., Rufty, R. C., and Daub, M. E. (2000). Nucleocapsid gene-mediated transgenic provides protection against tomato spotted wilt virus epidemics in the field. *Phytopathology,* 90:139-147.

Hipskind, J. D. and Paiva, N. L. (2000). Constitutive accumulation of a resveratrol-glucoside in transgenic alfalfa increases resistance to *Phoma medicaginis. Mol Plant-Microbe Interact,* 13:551-562.

Huang, Y., Nordeen, R. O., Owens, L. D., and Mc Beath, J. H. (1997). Expression of an engineered cecropin gene cassette in transgenic tobacco plants confers disease resistance to *Pseudomonas syringae* pv. *tabaci. Phytopathology,* 87:494-499.

Jach, G., Gornhardt, B., Mundy, J., Logemann, J., Pinsdorf, E., Leah, R., Schell, J., and Maas, C. (1995). Enhanced quantiative resistance against fungal disease by combinatorial expression of different barley antifungal proteins in transgenic tobacco. *Plant J,* 8:97-109.

Jacks, T. J., De Lucca, A. J., Rajasekaran, K., Stromberg, K., and van Pee, K. H. (2000). Antifungal and peroxidative activities of nonheme chloroperoxidase in relation to transgenic plant protection. *J Agric Food Chem,* 48:4561-4564.

James, C. (1998). *Global review of commercialized transgenic crops.* International Service for the Acquisition of Agri-biotech Applications, Ithaca, New York.

Jan, F. J., Pang, S. Z., and Gonsalves, D. (2000). A single chimeric transgene derived from two distinct viruses confers multi-virus resistance in transgenic plants through homology-dependent gene silencing. *J Gen Virol,* 81:2103-2109.

Jan, F. J., Pang, S. Z., Tricoli, D. M., and Gonsalves, D. (2000). Evidence that resistance in squash mosaic comovirus coat protein-transgenic plants is affected by plant developmental stage and enhanced by combination of transgenes from different lines. *J Gen Virol,* 81:2299-2396.

Jaynes, J. M., Nagpala, P., Destefano-Beltran, L., Huang, J. H., Kim, J. H., Denny, T., and Cetiner, S. (1993). Expression of a cecropin-B lytic peptide analog in transgenic tobacco confers enhanced resistance to bacterial wilt caused by *Pseudomonas solanacearum. Plant Sci,* 89:43-53.

Johal, G. S. and Briggs, S. P. (1992). Reductase activity encoded by the HM1 disease resistance gene in maize. *Science,* 258:985-987.

Jongedijk, E., Tigelaar, H., Van Roekel, J. S. C., Bres-Vloemans, S. A., Dekker, L., Vanden Elzen, P. J. M., Cornelissen, B. J. C., and Melchers, L. S. (1995). Synergistic activity of chitinases and β-1,3-glucanases enhances fungal resistance in transgenic tomato plants. *Euphytica,* 85:173-175.

Jorgensen, R. (1990). Altered gene expression in plants due to *trans* interactions between homologous genes. *Trends Biotechnol,* 8:340-344.

Keller, H., Pamboukdjian, N., Ponchet, M., Poupet, A., Delon, R., Verrier, J. L., Roby, D., and Ricci, P. (1999). Pathogen-induced elicitin production in transgenic tobacco generates a hypersensitive response and nonspecific disease resistance. *Plant Cell,* 11:223-235.

Kellmann, J. W., Kleinow, T., Engelhardt, K., and Philipp, C. (1996). Characterization of two class II chitinase genes from peanut and expression studies in transgenic tobacco plants. *Plant Mol Biol,* 30:351-358.

Ko, K., Brown, S. K., Norelli, J. L., Borejsza-Wysocka, E. E., Aldwinckle, H. S., and During, K. (1999). Effect of multiple transgenes on resistance to fire blight of 'Galaxy' apple. *Acta Horticulturae,* 489:257.

Ko, K., Norelli, J. L., Reynoird, J. P., Boresiza-Wysocka, E., Brown, S. K., and Aldwinckle, H. S. (2000). Effect of untranslated leader sequence of AMV RNA 4 and signal peptide of pathogenesis-related protein 1b on attacin gene expression, and resistance to fire blight in transgenic apple. *Biotechnology Letters,* 22: 373-381.

Lagrimini, L. M., Vaughan, J., Erb, W. A., and Miller, S. A. (1993). Peroxidase overproduction in tomato: wound-induced polyphenol deposition and disease resistance. *Hort Sci,* 28:218-221.

Lawson, E. C., Weiss, J. D., Thomas, P. E., and Kaniewski, W. K. (2001). NewLeaf Plus Russet Burbank potatoes: Replicase-mediated resistance to potato leafroll virus. *Molecular Breeding,* 7:1-12.

Li, R. G. and Fan, Y. L. (1999). Reduction of lesion growth rate of late blight of plant disease in transgenic potato expressing harpin protein. *Science in China, Series C—Life Sciences,* 42:96-101.

Lin, W., Anuratha, C. S., Datta, K., Potrykus, I., Muthukrishnan, S., and Datta, S. K. (1995). Genetic engineering of rice for resistance to sheath blight. *Biotechnology,* 13:686-691.

Linthorst, H. J. M., Van Loon, L. C., Van Rossum, C. M. A., Mayer, A., Bol, J. F., Van Roekel, C., Meulenhoff, E. J. S., and Cornelissen, B. J. C. (1990). Analysis of acidic and basic chitinases from tobacco and *Petunia* and their constitutive expression in transgenic tobacco. *Mol Plant-Microbe Interact,* 3:252-258.

Liu, D., Raghothama, K. M., Hasegawa, P. M., and Bressan, R. A. (1994). Osmotin overexpression in potato delays development of disease symptoms. *Proc Natl Acad Sci USA,* 91:1888-1892.

Logemann, J., Jach, G., Tommerup, H., Mundy, J., and Schell, J. (1992). Expression of a barley ribosome-inactivating protein leads to increased fungal protection in transgenic tobacco plants. *Biotechnology,* 10:305-308.

Logemann, J., Melchers, L. S., Trigelaar, H., Sela-Buurlage, M. B., Ponstein, A. S., van Roekel, J. S. C., Bres-Vloemans, S. A., Dekker, I., Cornelissen, B. J. C., van den Elzen, P. J. M., and Jongedijk, E. (1994). Synergistic activity of chitinase and β-1,3-glucanase enhances *Fusarium* resistance in transgenic tomato plants. *J Cell Biochem,* 18A:88.

Maddaloni, M., Forlani, F., Balmas, V., Donini, G., Stasse, L., Corazza, L., and Motto, M. (1997). Tolerance to the fungal pathogen *Rhizoctonia solani* AG4 of transgenic tobacco expressing the maize ribosome inactivating protein B-32. *Transgenic Res,* 6:393-402.

Madhuri, G., De Kochko, A., Chen, L. L., Nagabhushana, I., Fauquet, C., and Reddy, A. R. (2001). Development of transgenic rice plants expressing maize anthocyanin genes and increased blast resistance. *Molecular Breeding,* 7:73-83.

Magbanua, Z. V., Wilde, H. D., Roberts, J. K., Chowdhury, K., Abad, J., Moyer, J. W., Wetzstein, H. Y., and Parbott, W. A. (2000). Field resistance to tomato spotted wilt virus in transgenic peanut (*Arachis hypogaea* L.) expressing an antisense nucleocapsid gene sequence. *Molecular Breeding,* 6:227-236.

Malnoy, M., Chevreau, E., and Reynoird, J. P. (2000). Preliminary evaluation of new gene transfer strategies for resistance to fire blight in pear. *Acta Horticulurae,* 538:635-638.

Marchant, R., Davey, M. R., Lucas, J. A., Lamb, C. J., Dixon, R. A., and Power, J. B. (1998). Expression of a chitinase transgene in rose (*Rosa hybrida* L.) reduces development of blackspot disease (*Diplocarpon rosae* Wolf). *Molecular Breeding,* 4:187-194.

Martin, G. B., de Vicente, M. C., and Tanksley, S. D. (1993). High-resolution linkage analysis and physical characterization of the *Pto* bacterial resistance locus in tomato. *Mol Plant-Microbe Interact,* 6:26-34.

Masoud, S. A., Zhu, Q., Lamb, C., and Dixon, R. A. (1996). Constitutive expression of an inducible β-1,3-glucanase in alfalfa reduced disease severity caused by the oomycete pathogen *Phytophthora megasperma* f. sp. *medicaginis,* but does not reduce severity of chitin-containing fungi. *Transgenic Res,* 5:313-323.

Matton, D. P., Prescott, G., Bertrand, C., Camirand, A., and Brisson, N. (1993). Identification of *cis*-acting elements involved in the regulation of the pathogenesis-related gene STH-2 in potato. *Plant Mol Biol,* 22:279-291.

McGeachy, K. D. and Barker, H. (2000). Potato mop-top virus RNA can move long distance in the absence of coat-protein: Evidence from resistant, transgenic plants. *Mol Plant-Microbe Interact,* 13:125-128.

Melchers, L. S., Sela-Buurlage, M. B., Vloemans, S. A., Woloshuk, C. P., Van Roekel, J. S. C., Pen, J., van den Elzen, P. J. M., and Cornelissen, B. J. C. (1993). Extracellular targeting of the vacuolar tobacco proteins AP 24, chitinase and β-1,3-glucanase in transgenic plants. *Plant Mol Biol,* 21:583-593.

Mills, D. and Hammerschlag, F. A. (1993). Effect of cecropin B on peach pathogens, protoplasts and cells. *Plant Sci,* 93:143-150.

Mitsuhara, I., Matsufuru, H., Ohshima, M., Kaku, H., Nakajima, Y., Murai, N., Natori, S., and Ohashi, Y. (2000). Induced expression of sarcotoxin IA enhanced host resistance against both bacterial and fungal pathogens in transgenic tobacco. *Mol Plant-Microbe Interact,* 13:860-868.

Mittler, R., Shulaev, V., and Lam, E. (1995). Co-ordinated activation of programmed cell death and defense mechanisms in transgenic tobacco plants expressing a bacterial proton pump. *Plant Cell,* 7:29-42.

Molina, A. and Garcia-Olmedo, F. (1997). Enhanced tolerance to bacterial pathogens caused by the transgenic expression of barley lipid transfer protein LTP2. *Plant J,* 12:669-675.

Nakajima, H., Muranaka, T., Ishige, F., Akutsu, K., and Oeda, K. (1997). Fungal and bacterial disease resistance in transgenic plants expressing human lysozyme. *Plant Cell Rep,* 16:674-679.

Neves-Borges, A. C., Collares, W. M., Pontes, J. A., Breyne, P., Farinelli, L., and De Oliveira, D. E. (2001). Coat protein RNAs-mediated protection against Andean potato mottle virus in transgenic tobacco. *Plant Sci,* 160:699-712.

Nielsen, K. K., Mikkelsen, J. D., Kragh, K. M., and Bojsen, K. (1993). An acidic class III chitinase in sugar beet induction by transgenic tobacco plants. *Mol Plant-Microbe Interact,* 6:495-506.

Norelli, J. L., Aldwinckle, H. S., Destefano-Beltran, L., and Jaynes, J. M. (1994). Transgenic 'Malling 26' apple expressing the attacin E gene has increased resistance to *Erwinia amylovora*. *Euphytica,* 77:123-128.

Norelli, J. L., Mills, J. A. Z., Momol, M. T., and Aldwinckle, H. (1999). Effect of cecropin-like transgenes on fire blight resistance of apple. *Acta Horticulturae,* 489:273-278.

Ohshima, M., Mitsuhara, I., Okamoto, M., Sawano, S., Nishiyama, K., Kaku, H., Natori, S., and Ohashi, Y. (1999). Enhanced resistance to bacterial disease of transgenic tobacco plants overexpressing sarcotoxin IA, a bactericidal peptide of insect. *J Biochem,* 125:431-435.

Osusky, M., Zhou, G. Q., Osuska, L., Hancock, R. E., Kay, W. W., and Misra, S. (2000). Transgenic plants expressing cationic peptide chimeras exhibit broad-spectrum resistance to phytopathogens. *Nature Biotechnology,* 18:1162-1166.

Padgett, H. S., Watanabe, Y., and Beachy, R. N. (1997). Identification of the TMV replicase sequence that activates the *N* gene-mediated hypersensitive response. *Mol Plant-Microbe Interact,* 10:709-715.

Parashina, E. V., Serdobinskii, L. A., Kalle, E. G., Lavrova, N. V., Avetisov, V. A., Lunin, V. G., and Naroditskii, B. S. (2000). Genetic engineering of oilseed rape and tomato plants expressing a radish defensin gene. *Russian J Plant Physiol,* 47:471-478.

Parashinà, E. V., Shadenkov, A. A., Labrova, N. V., and Avetisov, A. A. (1999). The use of the defensive peptide (defensin) gene from radish seeds in order to improve the resistance of tomatoes towards diseases caused by fungi. *Biotekhonologiya,* 15(6):35-41.

Pinto, Y. M., Kok, R. A., and Baulcombe, D. C. (1999). Resistance to rice yellow mottle virus (RYMV) in cultivated African rice varieties containing RYMV transgenes. *Nature Biotechnology,* 17:702-707.

Powell, A. L.T., van Kan, J., Have, A. T., Visser, J., Greve, L. C., Bennett, A. B., and Labavitch, J. M. (2000). Transgenic expression of pear PGIP in tomato limits fungal colonization. *Mol Plant-Microbe Interact,* 13:942-950.

Qin, H. T., Xiao, H., Sun, Z. X., and Xu, T. (2000). Enhancing disease resistance of transgenic rice plants with an endochitinase gene *ThEn-42* from *Trichoderma harzianum. J Yunnan Agricultural University,* 15:212-215.

Radian-Sade, S., Perl, A., Edelbaum, O., Kuznetsova, L., Gafny, R., Sela, I., and Tanne, E. (2000). Transgenic *Nicotiana benthamiana* and grapevine plants transformed with grapevine virus A (GVA) sequences. *Phytoparasitica,* 28: 79-86.

Rajasekaran, K., Cary, J. W., Jacks, T. J., Stromberg, K. D., and Cleveland, T. E. (2000). Inhibition of fungal growth *in planta* and in vitro by transgenic tobacco expressing a bacterial nonheme chloroperoxidase gene. *Plant Cell Rep,* 19:333-338.

Ratcliff, F. G., MacFarlane, S. A., and Baulcombe, D. C. (1999). Gene silencing without DNA: RNA-mediated cross-protection between viruses. *Plant Cell,* 11: 1207-1215.

Renou, J. P., Mary, I., Hanteville, S., Narcy, J. P., Diolez, A., and Florack, D. (2000). Evaluation of the protection against *Xanthomonas* in transgenic *Pelargonium* containing a chimeric cecropin gene. *Acta Horticulturae*, 508:323-325.

Reynoird, J. P., Mourgues, F., Chevreau, E., Brisset, M. N., and Aldwinckle, H. S. (1999). Expression of SB-37 gene in transgenic pears enhanced resistance to fire blight. *Acta Horticulturae*, 489:243-244.

Reynoird, J. P., Mourgues, F., Norelli, J., Aldwinckle, H. S., Brisset, M. N., and Chevreau, E. (1999). First evidence for improved resistance to fire blight in transgenic pear expressing the attacin E gene from *Hyalophora cecropia*. *Plant Sci (Limerick)*, 149:23-31.

Roby, D., Broglie, K., Cressman, R., Biddle, P., Chet, I., and Broglie, R. (1990). Activation of bean chitinase promoter in transgenic tobacco plants by phytopathogenic fungi. *Plant Cell*, 2:999-1007.

Rohde, W., Jaag, C., Paap, B., Tacke, E., Schmitz, J., Kierdorf, M., Ashoub, A., Gunther, S., van Bel, A., and Prufer, D. (2000). Genetic engineering of potato for tolerance to biotic and abiotic stress. In A. D. Arencibia (Ed.), *Plant Genetic Engineering: Towards the Third Milennium*. Elsevier Science Publishers, Amsterdam, The Netherlands, pp. 177-181.

Rommens, C. M., Salmeron, J. M., Oldroyd, G. E., and Staskawicz, B. J. (1995). Inter generic transfer and functional expression of the tomato disease resistance gene *Pto*. *Plant Cell*, 7:1537-1544.

Roustan, J. P., Colrat, S., Dalmayrac, S., Guillen, P., Guis, M., Martinez-Reina, G., Deswarte, C., Bouzayen, M., Fallot, J., Pech, J. C., and Latche, A. (2000). Eutypa dieback of grapevine: Expression of an NADPH-dependent aldehyde reductase, which detoxifies eutypine, a toxin from *Eutypa lata*. *Acta Horticulturae*, 528:329-336.

Rubio, T., Borja, M., Scholthof, H. B., Feldstein, P. A., Morris, T. J., and Jackson, A. O. (1999). Broad-spectrum protection against tombusviruses elicited by defective interfering RNAs in transgenic plants. *J Gen Virol*, 73:5070-5078.

Schffrath, U., Mauch, F., Freydl, E., Schweizer, P., and Dudler, R. (2000). Constitutive expression of the defense-related *Rir1b* gene in transgenic rice plants confers enhanced resistance to the rice blast fungus *Magnaporthe grisea*. *Plant Mol Biol*, 43:59-66.

Schillberg, S., Zimmermann, S., Findlay, K., and Fischer, R. (2000). Plasma membrane display of anti-viral single chain Fv fragments confers resistance to tobacco mosaic virus. *Molecular Breeding*, 6:317-326.

Serrano, C., Arce-Johnson, P., Torres, H., Gebauer, M., Gutierrez, M., Moreno, M., Jordana, X., Venegas, A., Kalazich, J., and Holuigue, L. (2000). Expression of the chicken lysozyme gene in potato enhances resistance to infection by *Erwinia carotovora* subsp. *atroseptica*. *Amer J Potato Res*, 77:191-199.

Shen, S. H., Li, Q. S., He, S. Y., Barker, K. R., Li, D. B., and Hunt, A. G. (2000). Conversion of compatible plant pathogen interactions into incompatible interactions by expression of the *Pseudomonas syringae* pv. *syringae* 61 *hrmA* gene in transgenic tobacco plants. *Plant J*, 23:205-213.

Sivamani, E., Brey, C. W., Dyer, W. E., Talbert, L. E., and Qu, R. (2000). Resistance to wheat streak mosaic virus in transgenic wheat expressing the viral replicase *(NIb)* gene. *Molecular Breeding,* 6:469-477.

Smironv, S., Shulaev, V., and Tumer, N. E. (1997). Expression of pokeweed antiviral protein in transgenic plants induced virus resistance in grafted wild-type plants independently of salicylic acid accumulation and pathogenesis-related protein synthesis. *Plant Physiol,* 114:1113-1121.

Sonoda, S., Mori, M., and Nishiguchi, M. (1999). Homology-dependent virus resistance in transgenic plants with the coat protein gene of the sweet potato feathery mottle potyvirus: Target specificity and transgene methylation. *Phytopathology,* 89:385-391.

Spielmann, A., Krastanova, S., Douet-Orhant, V., and Gugerli, P. (2000). Analysis of transgenic grapevine *(Vitis rupestris)* and *Nicotiana benthamiana* plants expressing an *Arabis mosaic virus* coat protein gene. *Plant Sci (Limerick),* 156: 235-244.

Stark-Lorenzen, P., Nelke, B., Hanssler, G., Muhlbach, H. P., and Thomzik, J. E. (1997). Transfer of a grapevine stilbene gene of rice *(Oryza sativa* L.). *Plant Cell Rep,* 16:668-673.

Tabaeizadeh, Z., Agharbaoui, Z., Harrak, H., and Poysa, V. (1999). Transgenic tomato plants expressing a *Lycopersicon chilense* chitinase gene demonstrate improved resistance to *Verticillium dahliae* race 2. *Plant Cell Rep,* 19:197-202.

Tabei, Y., Kitade, S., Nishizawa, Y., Kikuchi, N., Kayano, T., Hibi, T., and Akutsu, K. (1998). Transgenic cucumber plants harboring a rice chitinase gene exhibit enhanced resistance to gray mold *(Botrytis cinerea). Plant Cell Rep,* 17:159-164.

Takahashi, H., Chen, Z., Du, H., Liu, Y., and Klessig, D. (1997). Development of necrosis and activation of disease resistance in transgenic tobacco plants with severely reduced catalase levels. *Plant J,* 11:993-1005.

Takaichi, M. and Oeda, K. (2000). Transgenic carrots with enhanced resistance against two major pathogens, *Erysiphe heraclei* and *Alternaria dauci. Plant Sci (Limerick),,* 153:135-144.

Takatsu, Y., Nishizawa, Y., Hibi, T., and Akutsu, K. (1999). Transgenic chrysanthemum *(Dendrathema grandiflorum* (Ramat.) Kitamura) expressing a rice chitinase gene shows enhanced resistance to gray mold *(Botrytis cinerea). Scientia Horticulturae,* 82:113-123.

Tenllado, F. and Diaz-Ruiz, J. R. (1999). Complete resistance to pepper mild mottle tobamovirus mediated by viral replicase sequences partially depends on transgene homozygosity and is based on a gene silencing mechanism. *Transgenic Research,* 8:83-93.

Terras, F. R. G., Eggermont, K., Kovaleva, V., Raikhel, N. V., Osborn, R. W., Kester, A., Rees, S. B., Torrekens, S., Leuven, F. V., Vanderleyden, J., Commue, B. P. A., and Broekaert, W. F. (1995). Small cysteine-rich antifungal proteins from radish: Their role in host defense. *Plant Cell,* 7:573-578.

Thilmony, L., Chen, Z., Bressan, R. A., and Martin, G. B. (1995). Expression of the tomato *Pto* gene in tobacco enhances resistance to *Pseudomonas syringae* pv. *tabaci* expressing *avrPto. Plant Cell,* 7:1529-1536.

Trudel, J., Porvin, C., and Asselin, A. (1995). Secreted hen lysozyme in transgenic tobacco: Recovery of bound enzyme and in vitro growth inhibition of plant pathogens. *Plant Sci,* 106:55-62.

Tu, J., Ona, I., Zhang, Q., Mew, T. W., Khush, G. S., and Datta, S. K. (1998). Transgenic rice variety 'IR72' with *Xa21* is resistant to bacterial blight. *Theor Appl Genet,* 97:31-36.

Verberne, M. C., Verpoorte, R., Bol, J. F., Mercado-Blanco, J., and Linthorst, H. J. M. (2000). Overproduction of salicylic acid in plants by bacterial transgenes enhances pathogen resistance. *Nature Biotechnology,* 18:779-783.

Veronese, P., Crino, P., Tucci, M., Colucci, F., Yun, D. J., Hasegawa, M. P., Bressan, R. A., and Saccardo, F. (1999). Pathogenesis-related proteins for the control of fungal diseases of tomato. In G. T. Scarascia Mugnozza, E. Porceddu, and M. A. Pagnotta (Eds.), *Genetics and Breeding for Quality and Resistance.* Kluwer Academic Publishers, Dordrecht, Germany, pp. 15-24.

Vidhyasekaran, P. (1997). *Fungal Pathogenesis in Plants and Crops: Molecular Biology and Host Defense Mechanisms.* Marcel Dekker, New York.

Wang, C. L., Chin, C. K., and Gianfagna, T. (2000). Relationship between cutin monomers and tomato resistance to powdery mildew infection. *Physiol Mol Plant Pathol,* 57:55-61.

Wang, G.-L., Song, W.-Y., Runan, D.-L., Sideris, S., and Ronald, P. C. (1996). The cloned, *Xa 21,* confers resistance to multiple *Xanthomonas oryzae* pv. *oryzae* isolates in transgenic plants. *Mol Plant-Microbe Interact,* 9:850-855.

Wang, H. Z., Zhao, P. J., and Zhou, X. Y. (2000). Regeneration of transgenic *Cucumis melo* and its resistance to virus diseases. *Acta Phytophylactica Sinica,* 27:126-130.

Way, H. M., Kazan, K., Goulter, K. C., Birch, R. G., and Manners, J. M. (2000). Expression of the *Shpx2* peroxidase gene of *Stylosanthes humulis* in transgenic tobacco leads to enhanced resistance to *Phytophthora parasitica* pv. *nicotianae. Molecular Plant Pathology,* 1:223-232.

Wegener, C., Bartling, S., Olsen, O., Weber, J., and Von Wettstein, D. (1996). Pectate lyase in transgenic potatoes confers preactivation of defence against *Erwinia carotovora. Physiol Mol Plant Pathol,* 49:359-376.

Whitham, S., McCormick, S., and Baker, B. (1996). The *N* of tobacco confers resistance to tobacco mosaic virus in transgenic tomato. *Proc Natl Acad Sci USA,* 93:8776-8781.

Wittner, A., Palkovics, L., and Balazs, E. (1998). *Nicotiana benthamiana* plants transformed with the plum pox virus helicase genes are resistant to virus infection. *Virus Research,* 53:97-103.

Wu, G., Shortt, B. J., Lawrence, E. B., Levine, E. B., Fitzsimmons, K. C., and Shah, D. M. (1995). Disease resistance conferred by expression of a gene encoding H_2O_2-generating glucose oxidase in transgenic potato plants. *Plant Cell,* 7: 1357-1368.

Xiao, X. W., Chu, P. W. G., Frenkel, M. J., Tabe, L. M., Shukla, D. D., Hanna, P. J., Higgins, T. J. V., Muller, W. J., and Ward, C. W. (2000). Antibody-mediated improved resistance to ClYVV and PVY infections in transgenic tobacco plants expressing a single-chain variable region antibody. *Molecular Breeding,* 6:421-431.

Xu, D. M., Collins, G. B., Hunt, A. G., and Nielsen, M. T. (1999). Agronomic performance of transgenic barley tobaccos expressing TVMV or AMV coat protein genes with and without virus challenges. *Crop Sci*, 39:1195-1202.

Yoshikawa, M., Tsuda, M., and Takeuchi, Y. (1993). Resistance to fungal diseases in transgenic tobacco plants expressing the phytoalexin elicitor-releasing factor, β-1,3-endoglucanase, from soybean. *Naturwissenchaften*, 80:417-420.

Yoshikawa, N., Gotoh, S., Umezawa, M., Satoh, N., Satoh, H., Takahashi, T., Ito, T., and Yoshida, K. (2000). Transgenic *Nicotiana occidentalis* plants expressing the 50-kDa protein of apple chlorotic leaf spot virus display increased susceptibility to homologous virus, but strong resistance to grapevine berry inner necrosis virus. *Phytopathology*, 90:311-316.

Zhang, L. and Birch, R. G. (1997). Mechanisms of biocontrol by *Pantoea dispersa* of sugarcane leaf scald disease caused by *Xanthomonas albilineans*. *J Appl Microbiol*, 82:448-454.

Zhang, Z. C., Li, D.W., Zhang, L., Yu, J. L., and Liu, Y. (1999). Virus movement protein gene mediated resistance against cucumber mosaic virus infection. *Acta Botanica Sinica*, 41:585-590.

Zhang, Z. Y., Coyne, D. P., and Mitra, A. (1996). Gene transfer for enhancing plant disease resistance to bacterial pathogens. *Annu Report Bean Improvement Cooperative*, 39:52-53.

Zhou, J., Loh, Y. T., Bressen, R. A., and Martin, G. B. (1995). The tomato gene *Pti* encodes a serine/threonine kinase that is phosphorylated by *Pto* and is involved in the hypersensitive response. *Cell*, 83:925-935.

Zhu, B., Chen, T. H. H., and Li, P. H. (1996). Analysis of late-blight disease resistance and freezing tolerance in transgenic potato plants expressing sense and antisense genes for an osmotin-like protein. *Planta*, 198:70-77.

Zhu, Q., Doerner, P. W., and Lamb, C. J. (1993). Stress induction and developmental regulation of a rice chitinase promoter in transgenic tobacco. *Plant J*, 3:203-212.

Zhu, Q., Maher, E. A., Masoud, S., Dixon, R. A., and Lamb, C. J. (1994). Enhanced protection against fungal attack by constitutive co-expression of chitinase and glucanase genes in transgenic tobacco. *Biotechnology*, 12:807-812.

Zoubenko, O., Hudak, K., and Tumer, N. E. (2000). Non-toxic pokeweed antiviral protein mutant inhibits pathogen infection via a novel salicylic acid-independent pathway. *Plant Mol Biol*, 44:219-229.

21

Breeding for Disease Resistance— In Vitro Selection

Plant cells can be cultured in vitro and plants can be regenerated from the cultured cells. These regenerants show variability, and some of the variable characters are heritable. These heritable characters also include disease resistance. In vitro selections from calluses, somaclones, gametoclones, organ cultures, and somatic hybrids have resulted in developing disease-resistant plants. However, in vitro selection requires proper selective agents. Furthermore, gene silencing appears to be common among somaclones. Most of the somaclonal variations are due to epigenetic changes, and DNA modifications in the somaclones are not very stable. However, some studies report stability of somaclonal characteristics for several generations. In vitro selection can be one alternative technology for breeders to develop high-yielding and disease-resistant varieties, particularly when useful genes are not available in the germplasm for improvement of the cultivars.

WHAT IS IN VITRO SELECTION?

Plant cells, tissues, and organs can be cultured in vitro, and these cultures can be developed into whole plants. Cells and protoplasts (cells without walls) can be obtained by developing calluses by plating excised plant tissues from root, shoot, leaf, meristem, seed, embryo, and pollen in a specific nutrient medium. The cultured cells and protoplasts are totipotent, developing into whole plants under defined conditions. Some organs, such as cotyledons and young leaves, can be developed through organogenesis into whole plants without callus formation.

Populations of cultured plant cells generally contain more genetic variability than do whole plants. Some of these variations may occur due to pre-existing mutations in cells of the explant material. However, most of the variations are due to mutations induced in the cell culture. Plant-growth regulators such as 2,4-D and kinetin used in tissue-culture media may act as mutagens. Plants regenerated from cell, tissue, and organ cultures are called *somaclones.* Vast variations are seen among somaclones as they are developed from the mutated cultured cells. Many of the variations have been found to be stable and heritable. The stable and heritable variation displayed among somaclones is called *somaclonal variation.* Somaclonal variation occurs more frequently than the spontaneous mutations occurring in field-grown plants. Variation has been detected in the progeny of up to 15 percent of regenerated plants in contrast to spontaneous mutation rates of one in one million. Somaclonal variations have been exploited to develop disease-resistant varieties.

TYPES OF IN VITRO SELECTION

In Vitro Selection Among the Regenerants

In this technique, thousands of calluses are developed from different explants (whole seed, embryo, root, stem, and leaf) or cultured plant cells or protoplasts. All the developed calluses are transferred to a regeneration medium. Some organ cultures are also regenerated, and all the regenerated plants are tested for their reaction to pathogens. By this method, somaclones showing resistance to various fungal, bacterial, and viral pathogens could be selected in various crops. The following somaclones have been selected: in rice, somaclones resistant to the blast pathogen *Pyricularia oryzae* (Araujo et al., 1999) and the brown spot pathogen *Helminthosporium oryzae* (Ling et al., 1985); in wheat, somaclones resistant to Karnal bunt and powdery mildew (Sharma et al., 1998); in apple, somaclones resistant to fire blight (Chevreau et al., 1998); in potato, somaclones resistant to late and early blights (Matern et al., 1976) and soft rot (*Erwinia carotovora* ssp. *carotovora*) (Polgar et al., 1999); in peppermint, somaclones resistant to verticillium wilt (Sink and Grey, 1999); and in tomato, somaclones showing resistance to wilt caused by *Fusarium oxysporum* f. sp. *lycopersici* (Shahin and Spivey, 1986), *Cucumber mosaic virus* (Hanus-Fajerska et al., 2000), and bacterial wilt caused by *Ralstonia solanacearum* (Mandal, 1999).

Resistant plants have been selected by exploiting somaclonal variation in many other crops. Rust *(Puccinia arachidis)* and late and early leaf spots (*Mycosphaerella* spp.) are the most important diseases in groundnut, and somaclones showing resistance to these diseases have been identified (Eapen et al., 1997). The most serious disease worldwide in banana is Sigatoka leaf spot caused by *Mycosphaerella musicola.* Two groups of researchers could select resistant banana plants exploiting somaclonal variation (Trujillo et al., 1999; Vidal et al., 2000). Somaclone resistant to *Fusarium* wilt of banana has also been obtained (ChingYan et al., 2000).

In Vitro Selection at the Callus Stage

Although somaclonal variation can be identified among regenerants, it is a difficult task to achieve. Each regenerant must be individually tested for agronomic traits under greenhouse or, preferably, under field conditions. Similarly, each regenerant must be tested for its reaction to diseases by inoculating spores of individual pathogens and providing a suitable environment (temperature, humidity, dew, etc.) for disease development. In the case of viral diseases, transmission of viruses by suitable vectors should be tried. There will be many "escapes" in the plant population if the pathogen inoculations were not done scrupulously. Great skill is needed to score individual plants. Furthermore, most of the somaclones show only reduced disease intensity (reduction in lesion size, lesion number, intensity of sporulation, etc., otherwise called quantitative resistance rather than qualitative resistance). Scoring individual plants for quantitative resistance is very difficult. To overcome this difficulty, selection of somaclones can be done at the callus stage itself. Specific selective agents can be used to discriminate the calluses, which may yield somaclones with good agronomic traits and disease resistance.

Calluses from susceptible plants are profusely colonized by pathogens, while colonization in the calluses from resistant plants is significantly decreased in many host-pathogen interactions. Hence, this characteristic can be used to select pathogen-resistant calluses. Oilseed rape callus tissues were inoculated with *Phoma lingam* spores and embryogenic calluses, which were free of attack, were selected. The regenerants showed moderate resistance to the pathogen (Sacristan, 1982). Sun and colleagues (1986) inoculated 365 calluses from a susceptible rice variety with the bacterial blight pathogen *Xanthomonas oryzae* pv. *oryzae*. Most of the calluses died as the result of bacterial infection. However, 63 calluses showed sectional proliferation. From those calluses, 45 plants could be regenerated. All but one were resistant to the disease.

Many fungal and bacterial pathogens produce toxins, and these toxins have been shown to be involved in disease development. Toxin-insensitive plants show less susceptibility or even high resistance. Toxin-deficient mutants are less virulent or are avirulent. When calluses that show resistance to toxins are regenerated, the regenerants show increased resistance to pathogens, which produce those toxins. Vidhyasekaran and colleagues selected rice somaclones showing resistance to *Helminthosporium oryzae* based on their insensitivity to the host-specific toxin produced by the pathogen (Vidhyasekaran et al., 1986; Ling et al., 1985; Vidhyasekaran et al., 1990). Toxin-resistant calluses were selected, and regenerants from those calluses were tested for their susceptibility to the pathogens. Some of the regenerants showed stable and heritable resistance to diseases. By this method, the following disease-resistant plants have been developed: wheat plants resistant to *Helminthosporium sativum;* potato plants resistant to *Fusarium oxysporum* and *Phytophthora infestans;* tomato plants resistant to *Alternaria solani, Phytophthora infestans,* and *Fusarium oxysporum* f. sp. *lycopersici;* maize plants resistant to *Helminthosporium maydis* and *Phyllosticta maydis;* sugarcane plants resistant to *Helminthosporium sacchari;* alfalfa plants resistant to *F. oxysporum* f. sp. *medicaginis;* oat plants resistant to *Helminthosporium victoriae;* barley plants resistant to *H. sativum;* tobacco plants resistant to *Pseudomonas syringae* pv. *tabaci* and *Alternaria alternata;* and peach plants resistant to *Xanthomonas campestris* pv. *pruni* (Vidhyasekaran, 1990, 1993). The selection of disease-resistant plants using fungal and bacterial toxins was not significantly different from the selection without any toxin. Both of these methods select only the mutants generated in tissue-culture medium.

In Vitro Selection from Organ Cultures

Several somaclonal variations have been reported among somaclones obtained from organ cultures without callus formation. Somaclonal variants of groundnut having field tolerance to rust and leaf spot diseases were obtained from shoot-tip regenerants (Eapen et al., 1997). Four somaclonal variants regenerated from adventitious buds of the apple variety Greensleeves (susceptible to fire blight) showed resistance to the fire blight caused by *Erwinia amylovora* (Chevreau et al., 1998). Some of the tomato somaclones, obtained through adventitious organogenesis initiated on leaf explants of cultivated tomato, showed increased tolerance to the cucumber mosaic virus (Hanus-Fajerska et al., 2000).

In Vitro Selection from Somatic Hybrids

Protoplast fusion among otherwise sexually incompatible species facilitates transfer of genes from widely divergent species. This technique helps to produce *somatic hybrid* plants. Somaclonal variations have also been observed among somatic hybrids. *Sinapis alba* is resistant to *Alternaria brassicae,* which causes a serious disease in cabbage *(Brassica oleracea).* Somatic hybrids between *S. alba* and *B. oleracea* were generated for transferring resistance against *A. brassicae* to *B. oleracea* (Hansen and Earle, 1997). Twenty-seven plants were regenerated from protoplast fusion. Some of the plants obtained from cuttings from the somatic hybrids showed a resistance to *A. brassicae* that was similar to that found in *S. alba* (Hansen and Earle, 1997).

Three somatic hybrid lines between potato *(Solanum tuberosum)* and *Solanum brevidens* were obtained (Polgar et al., 1999). The lines originated from the same callus showed different reactions to the soft rot pathogen *Erwinia carotovora* ssp. *carotovora,* and some of them showed less susceptibility to the pathogen (Polgar et al., 1999). The variations might have occurred in the early callus stage of development.

In Vitro Selection Exploiting Gametoclonal Variation

Anther culture-derived plants are called *gametoclones* and are haploid. Gametoclonal variations have also been observed in many plants. Blast disease-resistant gametoclones have been identified in rice (Chawla and Wenzel, 1987). Some wheat gametoclones have shown resistance to powdery mildew (Vidhyasekaran, 1993).

STABILITY AND HERITABILITY OF RESISTANCE EVOLVED THROUGH IN VITRO SELECTION

Somaclonal variations have been shown to be unstable after two to three generations with few exceptions. In sugarcane, 73 percent of the *Helminthosporium sacchari*-resistant regenerants showed resistance even up to five generations, while others reverted to being susceptible (Larkin and Scowcroft, 1983). *Helminthosporium oryzae*-resistant rice regenerants showed resistance up to the R_3 generation (Vidhyasekaran et al., 1990). Somaclonal variant CIEN BTA-03 resistant to yellow Sigatoka *(Mycospherella musicola)* was obtained from a susceptible banana clone. This somaclone exhibited the disease resistance in the field for five consecutive years of asexual reproduction (Vidal et al., 2000).

The inheritance pattern of resistance obtained through somaculture has been studied in tomato (Shahin and Spivey, 1986). *Fusarium* wilt-resistant plants were regenerated from protoplast-derived calluses. Analysis of R_1 progenies obtained from self-fertilization of selected R_0 individuals showed a typical 3:1 (resistance:susceptible) ratio of segregation. Analysis of R_2 progenies showed that resistant plants were either homozygous or heterozygous dominant for the gene conferring resistance. Out of 23 R_1 plants, six were homozygous mutants, six were homozygous normal (susceptible), and 11 were heterozygous, suggesting a theoretical ratio of 1:2:1 and confirming that the original regenerated plant was heterozygous for a dominant resistance allele. Dominant-gene mutation has been shown in tissue-culture-derived rice plants to *Helminthosporium oryzae* (Ling et al., 1985). The wilt-disease resistance observed in tissue-culture derived alfalfa plants also has been reported to be due to mutation at a single dominant gene (McCoy, 1987). The blast-disease resistance observed in three rice somaclones appears to be controlled by a single gene (Araujo et al., 1999). DNA polymorphisms were detected in somaclones derived from somatic hybrids (*Solanum tuberosum* and *Solanum brevidens*). Loss of certain DNA segments on chromosomes 5, 6, 9, and 11 in the somaclones has been observed (Polgar et al., 1999).

Studies conducted so far indicate that somaclonal variation may be due to cytological abnormalities, frequent qualitative and quantitative phenotypic mutation, sequence change, and gene activation and silencing (Kaeppler et al., 2000). Activation of quiescent transposable elements and retrotransposons has been observed in somaclones. It indicates that epigenetic changes may occur through the culture process. Epigenetic activation of DNA elements further suggests that epigenetic changes may also be involved in cytogenetic instability through modification of heterochromatin, and as a basis of phenotypic variation through the modulation of gene function. DNA methylation patterns are highly variable among regenerated plants and their progeny. It suggests that DNA modifications are less stable in somaclones than in seed-grown plants (Kaeppler et al., 2000).

REFERENCES

Araujo, L. G., Prabhu, A. S., and Filippi, M. C. (1999). Inheritance of resistance to leaf blast in somaclones of rice cultivar "Araguaia." *Fitopatologia Brasileira,* 24:182-184.

Chawla, H. S. and Wenzel, G. (1987). In vitro selection of barley and wheat for resistance against *Helminthosporium sativum. Theor Appl Genet,* 74:841-845.

Chevreau, E., Brisset, M. N., Paulin, J. P., and James, D. J. (1998). Fire blight resistance and genetic trueness-to-type of four somaclonal variants from the apple cultivar Greensleeves. *Euphytica,* 104:199-205.

ChingYan, T., ChengChiang, L., ChiHsieh, T., TsanHsiu, T., and ShinChuan, H. (2000). Improvement of the horticultural traits of Cavendish banana (*Musa* spp., AAA group). Selection and evaluation of a semi-dwarf clone resistant to *Fusarium* wilt. *J Chinese Soc Hort Sci,* 46:173-182.

Eapen, S., Kale, D. M., Murty, G. S. S., and Leela, G. (1997). Somaclonal variation in peanut. In P. B. K. Kishor (Ed.), *Plant Tissue Culture and Biotechnology: Emerging Trends.* Universities Press (India) Ltd., Hyderabad, India, pp. 240-243.

Hansen, L. N. and Earle, E. D. (1997). Somatic hybrids between *Brassica oleracea* L. and *Sinapis alba* L. with resistance to *Alternaria brassicae* (Berk.) Sacc. *Theor Appl Genet,* 94:1078-1085.

Hanus-Fajerska, E., Lech, M., Pindel, A., and Miczynski, K. (2000). Selection for virus resistance in tomato exposed to tissue culture procedures. *Acta Physiologiae Plantarum,* 22:317-324.

Kaeppler, S. M., Kaeppler, H. F., and Young, R. (2000). Epigenetic aspects of somaclonal variation in plants. *Plant Mol Biol,* 43:179-188.

Larkin, P. J. and Scowcroft, W. R. (1983). Somaclonal variation and eyespot toxin tolerance in sugarcane. *Plant Cell Tissue Organ Culture,* 2:111-121.

Ling, D. H., Vidhyasekaran, P., Borromeo, E. S., Zapata, F. J., and Mew, T. (1985). In vitro screening of rice germplasm for resistance to brown spot disease using phytotoxin. *Theor Appl Genet,* 71:133-135.

Mandal, A. B. (1999). Efficient somaculture system and exploitation of somaclonal variation for bacterial wilt resistance in tomato. *Indian J Hort,* 56:321-327.

Matern, U., Strobel, G., and Shepard, J. (1976). Reaction of phytotoxin in a potato population derived from mesophyll protoplast. *Proc Natl Acad Sci USA,* 75: 4935-4939.

McCoy, T. J. (1987). Tissue culture selection and evaluation of *Fusarium* resistance in alfalfa. *In Vitro,* 23:60A.

Polgar, Z., Wielgus, S. M., Horvath, S., and Helgeson, J. P. (1999). DNA analysis of potato×*Solanum brevidens* somatic hybrid lines. *Euphytica,* 105:103-107.

Sacristan, M. D. (1982). Resistance response to *Phoma lingam* of plant regenerated from selected cells and embryogenic cultures of haploid *Brassica napus. Theor Appl Genet,* 61:193-212.

Shahin, E. A. and Spivey, R. (1986). A single dominant gene for *Fusarium* wilt resistance in protoplast-derived tomato plants. *Theor Appl Genet,* 61:193-206.

Sharma, K. D., Singh, B. M., and Chauhan, R. S. (1998). Variation in Karnal bunt and powdery mildew resistance among somaclones and doubled haploids of bread wheat cv. Sonalika. *Indian Phytopathol,* 51:324-328.

Sink, K. C. and Grey, W. E. (1999). A root-injection method to assess verticillium wilt resistance of peppermint (*Mentha X piperita* L.) and its use in identifying resistant somaclones of cv. Black Mitcham. *Euphytica,* 106:223-230.

Sun, L. H., She, J. M., and Lu, X. F. (1986). In vitro selection of *Xanthomonas oryzae*-resistant mutants in rice. I. Induction of resistant callus and screening regenerated plants. *Acta Genet Sinica,* 13:188-193.

Trujillo, I., Garcia, E., and Berroteran, J. L. (1999). Evaluation of in vitro-derived banana plants. *Anales de Botanica Agricola,* 6:29-35.

Vidal, M., Del, C., and De Garcia, E. (2000). Analysis of a *Musa* spp. variant resistant to yellow Sigatoka. *Plant Mol Biol Reptr,* 18:23-31.

Vidhyasekaran, P. (1990). In vitro screening for disease resistance. In P. Vidhyasekaran (Ed.), *Basic Research for Crop Disease Management.* Daya Publishing House, New Delhi, India, pp. 27-36.

Vidhyasekaran, P. (1993). Molecular sieve to select cells in tissue culture to develop disease resistant plants. In P. Vidhyasekaran (Ed.), *Genetic Engineering, Molecular Biology and Tissue Culture for Crop Pest and Disease Management.* New Delhi, Daya Publishing House, New Delhi, India, pp. 295-309.

Vidhyasekaran, P., Borromeo, E. S., and Mew, T. W. (1986). Host specific toxin production by *Helminthosporium oryzae. Phytopathology,* 76:261-266.

Vidhyasekaran, P., Ling, D. H., Borromeo, E. S., Zapata, F. J., and Mew, T. W. (1990). Selection of brown spot-resistant rice plants from *Helminthosporium oryzae* toxin resistant calluses. *Ann Appl Biol,* 117:515-523.

– 22 –

Chemical Control—Bacterial Diseases

Many fungicides have been developed, and several chemical industries have invested several millions of dollars in developing fungicides. More than 100 active fungicide ingredients have been registered, and several hundred fungicide formulations are available in the market. In contrast, very few bactericides are available to control bacterial diseases (Research Information, 1998). In this chapter, the bactericides available in the market are listed and their uses are given.

COMMON AND TRADE NAMES OF THE BACTERICIDES AND THEIR USES

The following is the list of common name, trade name, and uses of bactericides:

Ammoniacal copper sulfate—Copac E, controls bacterial diseases in pears, stone fruit, vegetables, and ornamental plants.

Bordeaux mixture—Comac, Bordeaux 13 percent, Bordeaux Mixture 13 percent, Bordeaux Mixture 20 percent, Bordeaux Mixture 27 percent, Bouillie Bordelaise RSR, mixture of copper sulfate and lime (calcium hydroxide). Copper sulfate (10 lb) is added to 50 gallons of water, and then this solution is added to another 50 gallons of water containing 10 lb of calcium hydroxide. This mixture must be prepared freshly before application in the field. Ready-formulated Bourdeaux mixture is also available; however, these formulations are inferior to freshly prepared Bordeaux mixture. Bordeaux mixture controls fire blight of pear. It is recommended to apply as a spray at a concentration of 1-1-100 (copper sulfate lb-calcium hydroxide lb-water gallon), and the spray should be applied at five- to seven-day intervals. It is applied at 4-4-50 concentration to control angular leaf spot (*Pseudomonas syringae* pv. *lachrymans*) of cucurbits. It controls citrus canker, bacterial wilt of pepper, bacterial gummosis and canker *(P. syringae),* wild fire of tobacco, and bacterial leaf spots of tomato.

333

Bordeaux paste—paste prepared by dissolving 1 lb copper sulfate in 3 qt water, then mixing with 1.5 lb calcium hydroxide in 3 qt water. Bordeaux paste is used to control crown gall of apples, peaches, almonds, apricots, avocados, cherries, nectarines, and pecans. Diseased areas are cut away and the cut ends are covered with the paste.

Bronopol—Bronotak, antibacterial seed treatment for control of black arm and bacterial blight of cotton.

Chloropicrin—Chlor-O-Pic, preplant soil fumigant for control of soilborne bacterial pathogens.

Copper hydroxide—KOH-Hydroxide, KOP Oxy, Kocide, controls bacterial diseases of citrus, apples, pears, vegetables, and ornamentals; controls bean bacterial blight (*Xanthomonas axonopodis* pv. *phaseoli*) and halo blight (*Pseudomonas savastanoi* pv. *phaseolicola*), cabbage black rot (*X. campestris* pv. *campestris*), cucumber angular leaf spot (*P. syringae* pv. *lachrymans*), fire blight of pears *(Erwinia amylovora),* walnut blight (*X. arboricola* pv. *juglandis*), tomato bacterial speck (*P. syringae* pv. *tomato*), bacterial spot of peppers *(X. vesicatoria),* and bacterial spot of peaches and nectarines (*X. arboricola* pv. *pruni*).

Copper oxychloride—Cobox, Rokkol 400 SC, controls bacterial diseases in fruit trees and vegetables; controls fire blight of apples and pears *(E. amylovora),* walnut blight (*X. arboricola* pv. *juglandis*), tomato bacterial spot *(X. vesicatoria),* and angular leaf spot *(P. syringae* pv. *lachrymans*) and bacterial wilt *(Erwinia tracheiphila)* of cucumbers.

Copper oxychloride sulfate—C-O-C-S, a mixture of basic copper sulfate and basic copper chloride, controls fire blight of apples and pears *(E. amylovora),* walnut blight (*X. arboricola* pv. *juglandis*), tomato bacterial spot *(X. vesicatoria),* and angular leaf spot *(P. syringae* pv. *lachrymans*) and bacterial wilt *(Erwinia tracheiphila)* of cucumbers.

Copper sulfate tribasic—KOP 300, controls bacterial diseases in fruit trees, nuts, and vegetables; controls bean bacterial blight, cucurbit angular leaf spot, pepper bacterial spot, tomato bacterial spot, and celery bacterial spot (*P. syringae* pv. *apii*).

Dodine—Syllit, controls bacterial spot of peaches (*X. arboricola* pv. *pruni*).

Flumequine—effectively controls fire blight of apples and pears.

Fosetyl-Al—Aliette, controls *Erwinia amylovora* on pears.

Galltrol (Agrobacterium radiobacter)—biocontrol product, controls crown gall in apple, pear, peach, almond, apricot, plum, cherry, blueberry, kiwi, prune, bushberry, walnut, pecan, and nectarine.

Gentamicin—Agri-gent, controls bacterial diseases caused by *Agrobacterium tumefaciens* and *Xanthomonas campestris*.

Hypochlorite—calcium chlorohypochlorite (Perchloran, HTH) and sodium hypochlorite (Chlorox, Purex) formulations are available, used as
surface disinfectants of plant propagative material such as grape and
rose cuttings; the cuttings for rootings and rose stems are treated 20
minutes in 0.5 percent hypochlorite solution to eliminate *Agrobacterium
tumefaciens* that may be carried on the surface (Schroth and McCain,
1991).

Kasugamycin—Kasumin, controls bean halo blight (*Pseudomonas
savastanoi* pv. *phaseolicola*) and potato soft rot *(Erwinia carotovora);*
also controls *Xanthomonas* and *Clavibacter* diseases.

Kasugamycin + copper oxychloride—Kasuran, controls citrus canker.

Octhilinone—Pancil-T, RH-893, controls bacterial diseases of top fruit
and citrus bacterial canker.

Oxine-copper—Quinondo, controls various bacterial diseases.

Oxytetracycline—Terramycin, controls fire blight of pear, bacterial spot
of peach, a tree injection formulation is available in the control of eastern X disease of peach, cherry, and nectarine. Trees are treated by gravity infusion or by pressure treatment with oxytetracycline (1,320 ppm)
using 950 ml per tree after harvest but before leaf fall (Schroth and
McCain, 1991).

Probenazole—Probenazol, Oryzemate, controls rice bacterial blight.

Quaternary ammonium compounds—Bionol, Conide, Consan, Culsan,
Dichloran, Germital, Hyamine, Onyxide, Physan, Risosan, Roccal,
Shield, Sterosept, cuttings and rootings are treated with this compound
to eliminate bacterial pathogens.

Quintozene—Brassicol, Tubergran, Terraclor, Folosan, Tritisan, Saniclor,
Avicol, Kobutol, Pentagen, Botrilex, Kobu, Earthcide, Turfcide,
RTUPCNB, PCNB, Quintozene, Tilcarex, TriPCNB, controls potato
scab caused by *Streptomyces scabies.*

Salicylic acid—several formulations, induces systemic resistance and reduces disease incidence; effective against *Xanthomonas oryzae* pv.
oryzae.

Streptomycin—Agrimycin 17, Agri-strep, Plantomycin, controls soft rot
and black leg of potato, fire blight of apple and pear, wild fire of tobacco, bacterial wilt of chrysanthemum, bacterial blight of celery, bacterial spot of tomato, black arm of cotton, rice bacterial blight, and citrus canker; controls *Xanthomonas, Erwinia, Ralstonia, Pseudomonas,*
and *Agrobacterium* diseases.

Sulfur—That Big 8, Top Cop, reduces bacterial diseases.

Sulfur flowable—That Flowable Sulfur, has antibacterial action; reduces
bacterial diseases.

Tecloftalam—Shirahagen-S, systemic bactericide; controls rice bacterial blight.

Tetracycline—controls bacterial diseases.

Trichlamide—Hataclean, controls potato scab *(Streptomyces scabies)*.

2,4-Xylenol + meta-cresol—Gallex, controls crown gall and olive knot *(Pseudomonas savastanoi)* on apple, pear, peach, almond, apricot, plum, cherry, blueberry, kiwi, prune, bushberry, walnut, pecan, and nectarine.

EFFECTIVENESS OF BACTERICIDES

Although a few bactericides are available, their uses in management of bacterial diseases are limited. Their efficacy in the field varies widely. Most of the bacterial pathogens are systemic, but most of the bactericides are surface protectants. Bacteria also develop resistance to antibiotics very quickly. Persistence of antibiotics in plants is very low, requiring that antibiotics be applied once every four to five days. This is impractical and will be uneconomical. The current bactericides cannot reach sites where bacteria overwinter, such as blighted wood, cankers, and lesions in the case of woody plants. Only partial control of bacterial diseases is achievable with the available bactericides.

REFERENCES

Research Information, Ltd. (1998). *International Pesticide Directory,* Sixteenth Edition. Research Information Ltd., Herts, U. K.

Schroth, M. N. and McCain, H. D. (1991). The nature, mode of action, and toxicity of bactericides. In D. Pimentel (Ed.), *CRC Handbook of Pest Management in Agriculture,* Volume II. CRC Press, Boca Raton, Florida, pp. 497-505.

23

Chemical Control—Fungal Diseases

Fungicides are the most important components in the management of fungal diseases. More than 100 fungicides have been developed, and several hundreds of fungicide formulations are available. They have been formulated as dustable powders, wettable powders, emulsifiable concentrates, flowables, liquids, granules, and gases. Special formulations have also been developed specifically for seed treatment and soil application. Formulations consist of several adjuvants in addition to the active ingredients. Formulations of mixtures of fungicides are also available. The mode of action of different classes of fungicides varies widely, and each class of fungicides controls different groups of diseases. Specificity of fungicides appears to be restricted to the fungal genera level, not the species level. Various types of application equipment are available to apply fungicides to crops. Fungi develop resistance to certain fungicides, and several techniques have also been developed to manage fungicide resistance in pathogens. All of these issues are discussed in this chapter.

HISTORY OF CHEMICAL CONTROL

Among the various methods of disease management, chemical methods appear to be the most important. Many farmers believe that control of crop diseases is possible only by using fungicides, as they reduce diseases dramatically. Chemical control of diseases has been attempted since the occurrence of diseases became known. There is a mention of the use of a fungicide in the Indo-Aryan Vedas, which date from perhaps around 1500 B.C. Around 1000 B.C., the Greeks were applying practical chemical control against diseases. The Roman Caius Plinius Secundus (23 to 79 A.D.), in his writings *Historia Naturalis,* described treatment of cereal seeds with wine or a concoction of cypress leaf extract to control mildew. In his text *Pelzbuch,* Gottfried von Franken, a German who lived around 1230 A.D., gave advice on treatment of cankers of cherry and some other fruit diseases.

William Forsyth (1737-1804), from Scotland, published a book in 1802 in which he recommended urine and lime-water treatment to control mildew of fruit trees. He also stated that decoction of tobacco, sulfur, unslaked lime, and elder buds *(Sambuscus)* in boiling water controls powdery mildew. Benedict Prevost (1755-1819), in a monograph published in 1807, recommended the use of copper sulfate as a seed treatment to control bunt disease of wheat. He demonstrated that at a concentration of 1:10,000 copper sulfate killed the fungus, without damaging the seeds. Thomas Andrew Knight (1759-1838) in 1817 reported that sulfur controls scab *(Venturia pirina)* in pear trees. John Robertson (1824) from Ireland showed that mildew *(Sphaerotheca pannosa* var. *persicae)* is controlled by a mixture of sulfur in soapsuds, the soap acting as a spreader and sticker. In 1833, William Kenrick (1789-1872) from the United States published *The New American Orchardist,* in which he advocated the use of sulfur and quick lime, mixed in boiling water and diluted for use with cold water, to control diseases. Knight (1842) also showed that application of lime and sulfur controlled peach leaf curl *(Taphrina deformans).*

Several fungicides have been developed out of necessity. In 1848, the vine industry in France was severely affected because of outbreak of powdery mildew *(Uncinula necator)* in grapevine. In 1855, a fine form of sulfur was produced to control grape powdery mildew (McCallan, 1967). Later, downy mildew became severe in grapevine. The most important discovery in the field of chemical control is that of Bordeaux mixture in 1885. Vine growers in Bordeaux, France, were having problems with people stealing grapes from different parts of the vineyard, which bordered a road. They sprayed a mixture of copper sulfate and hydrated lime over the vines to deter people from stealing. This spray gave the grapes a blue color and reduced the amount of theft. Pierre Millardet, a professor working at the University of Bordeaux, observed that vines treated with this mixture of chemicals were almost free from downy mildew. In 1885, Millardet demonstrated the efficacy of this mixture, called Bordeaux mixture, in the control of grapevine downy mildew in a field experiment. Since then, Bordeaux mixture has become the most used fungicide, and it is still in use today, even after more than 100 years, for the control of fungal diseases on a wide range of crops.

Organomercurial compounds were developed for seed treatment in the early 1900s. The first organomercurial seed treatment for control of bunt of wheat was introduced by E. Riehm (1914). The Bayer company from Germany introduced the first commercial organomercurial fungicide, Uspulam. Subsequently, Imperial Chemical Industries (ICI) (U.K.) developed the organomercurial fungicides Ceresan in 1929 and Agrosan G in 1933. Mercurial fungicides were popular for more than five decades, and now many countries have banned their use because of environmental pollution problems. In 1934, the DuPont Company discovered the first organic fungicides, the dithiocarbamates. Several dithiocarbamates such as zineb, maneb, mancozeb, and thiram were introduced within a span of another ten years. All of these sulfur, copper, mercury, and dithiocarbamate fungicides are protectants or surface fungicides. They do not penetrate plants and cannot cure disease. They are also subject to weathering.

In the 1960s, a new group of fungicides appeared which could enter plants and be translocated to different parts of plants. Called *systemic fungicides,* they also had curative action. Their discovery revolutionized the chemical control approach, and a new era had begun. Several chemical companies in the United States, United Kingdom, Germany, Switzerland, France, and Japan introduced numerous fungicides during the next 30 years. Several ready-formulated fungicide mixtures have also been released.

DEVELOPMENT OF COMMERCIALLY VIABLE FUNGICIDES

Several commercial companies are involved in the development of fungicides. Numerous compounds are screened to select compounds with fungicidal action. The selected compounds are tested against a wide range of pathogens in the laboratory, particularly against economically important pathogens affecting crops of high commercial value and causing heavy losses in those crops cultivated in vast areas. When thousands of chemicals are screened, a large number of chemicals with some fungitoxicity can be identified. These compounds undergo stringent secondary tests which identify the concentration at which the compound loses its efficacy, compared to a suitable standard fungicide. The compound should be active at a low concentration (normally at 10 to 25 ppm) to merit elevation to the next stage. Further tests include laboratory and greenhouse studies to assess the efficacy of the selected compounds against test diseases in the appropriate hosts. Normally, these tests will be initially conducted with a single target pathogen (most important pathogen), and then the selected compound will be tested against other taxonomically related pathogens.

Many more tests are required to develop a compound as a commercial fungicide. The selected compounds should have a broad disease-control spectrum and an extended control period. Their mobility after application in the plant should be assessed. The curative properties of the compounds should also be explored. The absence of curative activity is a disadvantage unless systemicity or the potential to redistribute in the crop is demonstrated. Immobile protectant activity alone may limit the use of the candidate chemical. The product should be compatible with other products, and the formulations should be easy to use. The selected chemical should be safe to the crops (without phytotoxicity). Several fungi are known to develop resistance to fungicides. Antiresistance activity of the selected chemicals should be tested. The selected chemical should be cost effective, the dosage required to control diseases should be low, and it should require fewer treatments per season.

The candidate chemical should be safe to users and consumers of the treated product and be environmentally acceptable (Waxman, 1998). The following toxicological data are required for the registration of a fungicide: acute oral toxicity, acute dermal toxicity, acute inhalation toxicity, primary eye irritation, dermal sensitization, acute delayed neurotoxicity, subchronic oral toxicity, subchronic dermal toxicity, subchronic inhalation toxicity, subchronic neurotoxicity, chronic toxicity, oncogenicity, reproduction, teratogenecity, mutagenecity, metabolism, and pharmacology.

All of these tests are very expensive. It has been estimated that one out of about 120,000 compounds may satisfy all these requirements and be developed as a commercial product. The average cost of developing one new synthetic compound is approximately 200 million dollars, and it takes about eight years to launch a product. Two-thirds of the total cost is attributed to biological efficacy trials, and toxicological and environmental safety tests alone may account for 60 percent of this cost. An estimated 570 million dollars is spent annually on research and development by the leading 15 agrochemical companies (Knight et al., 1997).

Several chemical industries in different parts of the world are involved in the development and marketing of fungicides. Novartis (Ciba-Geigy and Sandoz), Bayer, Rhone-Poulenc, BASF, DuPont, AgrEvo, Rohm and Haas, Ishihara, and Zeneca (ICI) are the major suppliers of fungicides in the world. Their sales of fungicides exceed $4 billion annually. Mancozeb, copper, chlorothalonil, sulfur, propiconazole, tebuconazole, metalaxyl, epoxiconazole, benomyl, iprodione, cyproconazole, maneb, carbendazim, tricyclazole, prochloraz, oxadixyl, vinclozolin, flusilazole, fosetyl-Al, cyproconazole, thiophanate-methyl, edifenphos, thiabendazole, fenpropimorph, procymidone, bayleton, penconazole, triflumizole, triphenyl tin, and fenbuconazole are the major active ingredients sold in the market (Hewitt, 1998). Their sales alone exceed $3.5 billion annually.

Common, Chemical, and Trade Names of Fungicides

The common names, chemical names, and trade names of fungicides available in the world market are given in the following list. The common names are followed by chemical names (in parentheses) and trade names of each fungicide.

anilazine (2,4-dichloro-6-(2-chloroanilino)-1,3,5-triazine)—B622,
Botrysan, Direz, Dyrene, Kemate, Triasyn, Direx, Zinochlor, Triazine

azoxystrobin (β-methoxyacrylate)—Amistar, ICIA5504, Quadris,
Abound, Bankit, Heritage, Ortiva

benalaxyl (methyl N-phenylacetyl-N-2,6-xylyl-DL-alaninate)—Galben,
M9834, Tairel, Trecatol

benodanil (2-iodobenzanilide)—Calirus, BAS3170F

benomyl (methyl 1-(butylcarbamoyl) benzimidazol-2 yl carbamate)—
Benlate, Benex, Fundazol, Fibenzol, Benosan, Hockley Benomyl

binapacryl (2-*sec*-butyl-4,6-dinitrophenyl 3-methylcrotonate)—Acricid,
Endosan, Morocide, Ambox, Dapacryl, Morocid, Hoe2784

biphenyl (1,1'-biphenyl)—Lemonene, PhenadorX, Phenylbenzene

bitertanol (1-(biphenyl-4 yloxy)-3,3-dimethyl-1-(1*H*-1,2,4-triazol-1 yl)
butan-2-ol)—Baycor, Baymat, Sibutol, Biloxazol, Bay KWG 0599

blasticidin S (antibiotic)—Bla-S

Bordeaux mixture (mixture of copper sulfate and calcium hydroxide)—
Comac, Bordeaux 13 percent, Bordeaux Mixture 13 percent,
Bordeaux Mixture 20 percent, Bordeaux Mixture 27 percent,
Bouillie Bordelaise RSR

bronopol (2-bromo-2-nitro-1,3-propanediol)—Bronocot, Bronotak

bupirimate (5-butyl-2-ethylamino-6-methylpyrimidin-4-yl
dimethylsulfamate)—Nimrod, Nimrod T, PP588

buthiobate (butyl 4-*tert*-butylbenzyl *N*-(3-pyridyl) dithiocarbonimidate)—
Denmert, S1358

captafol (*cis N*-[(1,1,2,2-tetrachloroethyl)thio] 4-cyclohexene-1,2-
dicarboximide)—Difolatan, Haipen, Ortho5865, Merpafol, Pillartan,
Sanspor

captan (N-[(trichloromethyl)thio]-4-cyclohexene-1,2-dicarboximide)—
Orthocide, Pillarcap, Merpan, Vondcaptan, Captan 50, Captanex,
Captaf, Drexel

carbendazim (methyl benzimidazol-2-ylcarbamate)—Bavistin, BAS 346F,
Derosal, Battal, Focal, Hoe 17411, Delsene, Stempor, Derroprene,
Equitdazin, Kemdazin, Virolex, Lignasan, Pillarstin, Custos, Triticol

carboxin (5,6-dihydro-2-methyl-1,4-oxathiine-3-carboxanilide)—Vitavax,
Kemikar, D 735, Kisvax, Vitavax 34, Enhance, ProGro, Vitavax 30C

carpropamid (1RS,3SR)-2,2-dichloro-N-[1-(4-chlorophenyl)ethyl]-1-
ethyl-3-ethylcyclopropane carboxamide)—Win, KTU 3616

chloranil (tetrachloro-*p*-benzoquinone)—Spergon

chloroneb (1,4-dichloro-2,5-dimethoxybenzene)—Terraneb, Demosan

chloropicrin (trichloronitromethane)—Chlor-O-Pic, TriChlor,
Dojyopicrin, Dolochlor, Acquinite, PicClor

chlorothalonil (2,4,5,6-tetrachloroisophthalonitrile)—Bravo 500, Bombardier, Daconil 2787, Exotherm Termil, Clortocaffaro, Notar, Faber, Repulse, Tuffcide, Clortosip, Kavach

chlozolinate (ethyl 3-(3,5-dichlorophenyl)-5-methyl-2,4-dioxo-oxazolidine-5-carboxylate)—Serinal, Manderol, M 8164

copper hydroxide (dicopper trihydroxide)—Kocide 101

copper oxychloride (copper chloride oxide hydrate)—Blitox, Coprantol, Coptox, Cupravit, Cuprocaffaro, Cuprokylt, Cobox, Fytolan, Kauritil, Recop, Vitigran, Cuprosan, Viricuivre, Cekuper

copper sulfate (copper sulfate)—Comac Bordeaux Plus, Triangle, Blue Viking, Vencedor, Sulfacop, Phyton27

cufraneb (dithiocarbamate complex containing copper, manganese, iron, and zinc)—Cufram Z

cuprous oxide (dicopper oxide)—CopperSandoz, Cobre Sandoz, Kupfer Sandoz, Copper Nordox, Perenox, Yellow Cuprocide, Caocobre, FungiRhap

cycloheximide (4(2 R)-2[(1S, 3S, 5S)-(3,5-dimethyl-2-oxocyclohexyl)]-2-hydroxyethyl) piperidine-2,6-dione)—Actidione, Actispray, Hizarocin

cymoxanil (1-(2-cyano-2-methoxyiminoacetyl)-3-ethylurea)—Curzate, DPX 3217

cyproconazole (1,2,4-triazole)—Alto, Atemi, Sentinel, Baycor 25

cyprodanil (anilinopyrimidine)—Unix

cyprofuram (α-[*N*-(3-chlorophenyl)cyclopropanecarboxamido]-γ-butyrolactone)—Vinicur, SN 78314, Stanza

diazoben (sodium *p*-(dimethylamino)-benzene diazosulfonate)—Dexon

dichlobutrazol ((2RS,3RS)-1-(2,4-dichlorophenyl)-4,4-dimethyl-2-(1*H*-1,2,4-triazol-1-yl) pentan-3-ol)—Vigil

dichlofluanid (*N*-dichlorofluoromethylthio-*N'*,*N'*-dimethyl-*N*-phenylsulphamide)—Elvaron, Euparen, Euparene, Bay 47531

dichlone (2,3-dichloro-1,4-naphthoquinone)—Quintar, Phygon

dichlorophen (4,4'-dichloro-2,2'-methylenediphenol)—Super Mosstox, Mosstox, Bio Moss Killer, Panacide

diclomezine (6-(3,5-dichloro-4-methyl-phenyl)-3(2H)pyridazinone)—Monguard

dicloran (2,6-dichloro-4-nitroaniline)—Botran, Allisan, Resisan, Kiwi Lustr, Marisan, Scleran, Sclerosan, DCNA, Ditranil

difulmetorim (*RS*-5-chloro-N[1-(4-difluoromethoxy phenyl)propyl]-6-methyl-primidin-4-ylamine)—Pyricut, UBF002EC

dimethirimol (5-butyl-2-dimethylamino-6-methylpyrimidin-4-ol)—Milcurb

dimethomorph ((*E*,*Z*)-4-[3-(4-chlorophenyl)-3-(3,4-dimethoxyphenyl)acrylolyl] morpholine)—Acrobat, Forum diniconazole ((*E*)-(*RS*)-1-(2,4-dichlorophenyl)-4,4-dimethyl-2-(1*H*-1,2,4-triazol-1-yl)pent-1-en-3-ol)—Sumi8, Sumieight, Spotless, S3308L, Ortho Spotless, XE779L

dinobuton (2-*sec*-butyl-4,6-dinitrophenyl isopropyl carbonate)—Acrex, Dessin, Drawinol, Dinofen, Talan

dinocap (2 (or 4)-(1-methylheptyl)-4,6(or 2,6)-dinitrophenyl crotonate)—Karathane, Ezenosan, Crotothane, Caprane, Cekucap, Fenocap

ditalimfos (*O,O*-diethyl phthalimidophosphonothioate)—Plondrel, Laptran, Millie, Farmil, Dowco 199, Frutogard, Leucon

dithianon (5,10-dihydro-5,10-dioxonaphtho(2,3-*b*)-1,4-dithiin-2,3-dicarbonitrile)—Delan, DelanCol

DNOC (4,6-dinitro-*o*-cresol)—ExtarA, ExtarLin, Sandoline, Sinox, Selinon, Trifocide, Trifina, Cresofin, Chemsect, Antinonnin, Elgetol, Nitrador

dodemorph acetate (4-cyclododecyl-2,6-dimethylmorpholinium acetate)—Meltatox, BASF-Mehltaumittel, Milban

dodine (1-dodecylguanidinium acetate)—Cyprex, Melprex, Venturol, Carpene, Syllit, Curitan, Vondodine, Efuzin

drazoxolon (4-(2-chlorophenyl hydrazone)-3-methyl-5-isoxazolone)—Ganocide, MilCol, SAIsan, PP 781

edifenphos (*O*-ethyl S,S-diphenyl phosphorodithioate)—Hinosan, Bay 78418

etaconazole (1-[2-(2,4-dichlorophenyl)-4-ethyl-1,3-dioxolan-2-ylmethyl]-1*H*-1,2,4-triazole)—Benit, Sonax, Vangard, CGA 64251

ethirimol (5-butyl-2-ethylamino-6-methylpyrimidin-ol)—Milgo, Milstem, Milcurb Super, PP149

etridiazole (5-ethoxy-3-trichloromethyl-1,2,4-thiadiazole)—AAterra, Dwell, Koban, Pansoil, Terrazole, Truban

famoxadone (3-anilino-5-methyl-5-(4-phenoxyphenyl)-1,3-oxazolidine-2,4-dione)—Famoxate, DPX-JE874

fenaminosulf (sodium 4-dimethylaminobenzenediazosulphonate)—Lesan, Bay 5072, Bay 22555

fenarimol (2,4'-dichloro-α-(pyrimidin-5-yl)benzhydyl alcohol)—Bloc, EL222, Rubigan, Rimidin, Fenal

fenfuram (2-methyl-3-furanilide)—Panoram, Pano-ram

fenhexamid (*N*-(2,3-dichloro-4-hydroxyphenyl)-1-methylcyclohexane-carboxamide)—KBR 2738

fenpropidin ((RS)-1-[3-(4-*tert*-butylphenyl)-2-methylpropyl]piperidine)—Patrol, Ro 123049, Tern

fenpropimorph (*cis*-4-[3-(4-*tert*-butylphenyl)-2-methylpropyl]-2,6-dimethylmorpholine)—Corbel, Mistral fentin acetate (triphenyltin acetate)—Brestan, Chimate, Batasan, Fenilene, Fentol, Hoe 2824, LiroTin, Pandar, PhenostanA, Stanex, Suzu, Tinestan, Ucetin

fentin chloride (triphenyltin chloride)—Aquatin, Phenostat C, Tinmate

fentin hydroxide (triphenyltin hydroxide)—Brestan Flow, DuTer, Duter, Flo Tin, Haitin, SuzuH, Tubotin, Farmatin, SuperTin, Phenostat H, Triple Tin

ferbam (ferric dimethyldithiocarbamate)—AAfertis, Carbamate, Ferbam, Knockmate, Trifungol, Ferberk, Hexaferb

fluazinam (3-chloro-N-(3-chloro-5-trifluoromethyl-2-pyridyl)-α,α,α,-trifluoro-2,6-dinitro-p-toluidine)—Shirlam Flow, Shirlan

fluoromide (2,3-dichloro-N-4-fluorophenylmaleimide)—Spartcide, MK-23

fluotrimazole (1-(3-trifluoromethyltrityl)-1H-1,2,4-triazole)—Bay BUE 0620, Persulon

flusilazol (1-[·bis(4-fluorophenyl)methylsilyl·methyl]-1H-1,2,4-triazole)—Punch, Nustar, Olymp, DPX H6573

flutolanil (α,α,α-trifluoro-3'-isopropoxy-o-toluanilide)—Moncut, NNF-136, Prostar 4E, Monstar

flutriafol ((RS)-2,4'-difluoro-α-(1H-1,2,4-triazol-1-ylmethyl)benzhydryl alcohol)—Impact, PP 450, Vincit

folpet (N-(trichloromethylthio) phthalimide)—Phaltan, Fungitrol, Folpan

fosetyl-aluminium (aluminium tris ethyl phosphonate)—Aliette, LS 74783

fthalide (4,5,6,7-tetrachlorophthalide)—Rabcide, KF-32

fuberidazole (2-(2'-furyl)benzimidazole)—Voronit, Bay 33172, Fuberidazole

furalaxyl (methyl N-(2-furoyl)-N-(2,6-xylyl)-DL-alaninate)—Fonganil, Fongarid, CGA38140

furmecyclox (methyl N-cyclohexyl-2,5-dimethylfuran-3-carbohydroxamate)—Campogran, Xyligen B, BAS 389F, Epic

guazatine (acetates of a mixture of the reaction products from polyamines [comprising mainly octamethylenediamine, iminodi(octamethylene) diamine, and octamethylenebis (iminooctamethylene)diamine] and carbamonitrile)—Panoctine, Panolil, Radam, Kenopel, Rappor

hexachlorobenzene (hexachlorobenzene)—Anticarie, No Bunt, Ceku C.B.

hexaconazole (RS-2[2,4-dichlorophenyl]-1-[1H-1,2,4-triazol-1-yl] hexan-2-ol)—Anvil, PP523, Planete Aster

hydroxyquinoline sulphate (bis(8-hydroxyquinolinium)sulfate)—Chinosol, Albisal, Cryptonol

imazalil (1-(β-allyloxy-2,4-dichlorophenylethyl)imidazole)—Bromazil, Deccozail, Fecundal, Fungaflor, Fungazil, Freshgard imibenconazole (4-chlorobenzyl-*N*-(2,4-dichlorophenyl)-2-(1*H*-1,2,4-triazol-1-yl) thioacetimidate)—Manage

iminoctadine tris (1,1'-iminiodi-(octamethylene)diguanidinium)—Bellkute

iprobenfos (*S*-benzyl *O,O*-di-isopropyl phosphorothioate)—Kitazin

iprodione (3-(3,5-dichlorophenyl)-*N*-isopropyl-2,4-dioxoimidazolidine-1-carboxamide)—Rovral Flo, Rovral WP, 26019 RP, Chipco 26019, LFA 2043, Kidan, Verisan

iprovalicarb (strobilurin compound)—SZX 0722

isoprothiolane (di-isopropyl 1,3-dithiolan-2-ylidenemalonate)—Fuji-one, Isoran, Fudiolan, NNF109, SS 11946

kasugamycin (5-amino-2-methyl-6(2,3,4,5,6-pentahydroxycyclo-hexyloxy)tetrahydropyran-3-yl)amino-α-iminoacetic acid)—Kasumin

kresoxim methyl (methyl methoxyiminoacetate)—Stroby WG, Candit, BASF 490F

lime sulfur (calcium polysulfide)—Orthorix, Security Lime Sulfur

mancozeb (manganese-zinc ethylenebis dithiocarbamate)—Dithane 945, Penncozeb, Dithane M45, Manzate 200, Nemispor, Vondozeb Plus, Sandozebe, Mancozin, Manzin, Policar MZ, Manzeb, Azko Chemie Mancozeb, Policar S, Ziram-Dithane

maneb (manganese ethylenebis dithiocarbamate)—Dithane M22, Maneb, Lonacol M, Tersan, Trimangol, Nespor, Manesan, Farmaneb, Plantineb Roussel Uclaf, Polyram M, Mangavis, Manebgan, Remasan, Rhodianebe, Kypman 80, Akzo Chemie Maneb, Maneb Spritzpulver, Manex 80, Maneba, Manex, Manzate, Manzate D, MDiphar, Chloroble M, Sopranebe, Tubothane, Unicrop Maneb

mepronil (3'-isopropoxy-*o*-toluanilide)—Basitac

mercuric oxide (mercury oxide)—Santar, Kankerdood

mercurous chloride (mercury chloride)—Cyclosan

metalaxyl (methyl *N*-(2-methoxyacetyl)-*N*-(2,6-xylyl)-DL-alaninate)—Apron, Ridomil, Subdue, CGA 48988

metconazole ((1*RS*, 5*RS*; 1*RS*, 5*SR*)-5-(4-chlorobenzyl)-2,2-dimethyl-1-(1*H*-1,2,4-triazol-1-ylmethyl)cyclopentanol)—Caramba

metham-sodium (sodium methyldithiocarbamate)—Vapam, VPM, Arapam, Busan, Solasan, Trimaton, Maposol, Sistan, Sometam, Monam, Nemasol

methasulfocarb (S-(4-methylsulfonyloxy-phenyl)-*N*-methylthiocarbamate)—Kayabest

methoxyethylmercury acetate (2-methoxyethylmercury acetate)—Panogen-Metox, Panogen M, Cekusil Universal A

methoxyethylmercury chloride (2-methoxyethylmercury chloride)—Emisan 6 percent methyl bromide (bromomethane)—Meth-O-Gas, Terr-O-Gas 100, Haltox

methyl isothiocyanate (methyl isothiocyanate)—Trapex

metiram (zinc ammoniate ethylenebis(dithocarbamate)-
polyethylenethiuram disulfide)—Polyram, PolyramCombi, Carbatene

metomeclan (1-(3,5-dichlorophenyl)-3-methoxy-methyl-2,5-
pyrrolidindione)—Drawifol

metominostrobin ((E)-2-methoxyimino-*N*-methyl-2-(2-
phenoxyphenyl)acetamide)—Oribright

myclobutanil (α-butyl-α-(chlorophenyl)-1*H*-1,2,4-triazole-1-
propanenitrile)—Systhane, RH3866, Eagle, Nova, Rally

nabam (disodium ethylenebis (dithiocarbamate))—Dithane A-40,
Nabasan, Chem Bam, DSE, Parzate, Spring Bak

nitrothal-isopropyl (di-isopropyl 5-nitroisophthalate)—used as mixed formulations with other fungicides

nuarimol (2-chloro-4'-fluoro-α-(pyrimidin-5-yl) benzhydryl alcohol)—
Triminol, Trimidal, EL228, Gauntlet, Murox

octhilinone (2-octylisothiazol-3(2*H*)-one)—Pancil-T, RH-893

ofurace (2-chloro-*N*-(2,6-dimethylphenyl)-*N*-(tetrahydro-2-oxo-3-
furanyl)acetamide)—Ortho 206 15

oxadixyl (2-methoxy-*N*-(2-oxo-1,3-oxazolidin-3-yl)acet-2',6'-xylidide)—
Sandofan, SAN 371F, Recoil, Ripost, Wakil, Pulsan

oxine copper (cupric 8-quinolinoxide)—Quinolate, Cunilate, Dokirin,
Fruitdo, Quinondo, Cellu-Quin, Bioquin, Milmer

oxycarboxin (5,6-dihydro-2-methyl-*N*-phenyl-1,4-oxathiin-3-
carboxamide-4,4-dioxide)—Plantvax, F461, Oxykisvax, Carbexin

pefurazoate (pent-e-enyl *N*-furfuryl-*N*-imidazol-1-ylcarbonyl-DL-
homoalaninate)—Healthied

penconazole (1-(2,4-dichloro-β propylphenethyl)-1*H*-1,2,4-triazole)—
Topas, Topaz, Topazo, Onmex, Award, CGA 71818

pencycuron (1-(4-chlorobenzyl)-1-cyclopentyl-3-phenylurea)—
Monceren, Bay NTN 19701

pentachlorophenol (pentachlorophenol)—Penchloral, Pentacon, Penwar,
Sinituho

phenylmercury acetate (phenylmercury acetate)—Ceresol, Agrosan,
Cekusil, Celmer, Pamisan, Seedtox, PMAS, Unisan, Hong Nien, Phix,
Mersolite

phenylphenol (biphenyl-2-ol)—Dowicide 1, Nectryl

phosdiphen (bis(2,4-dichlorophenyl)ethyl phosphate)—MTO460

pimaricin ((8*E*, 14*E*, 16*E*, 18*E*, 20*E*)-(1*S*, 3*R*, 5*S*, 7*S*, 12*R*, 24*R*, 25*S*,
26*R*)-22-(3-amino-3,6-dideoxy-β-D-mannopyranosyloxy)-1,3,26-

trihydroxy-12-methyl-10-oxo-6,11,28-trioxatricyclo
(22.3.1.0^{5,7})octacosa-8,14,16,18,20-pentaene-25-carboxylic acid)]—
Delvolan

polyoxins [5-(2-amino-5-O-(aminocarbonyl)-2-deoxy-L-xylonyl)amino]-
1,5-dideoxy-1-[3,4-dihydro-5-(hydroxymethyl)-2,4-dioxo-1(2H)-
pyrimidinyl]-β-D-allofuranuronic acid (polyoxin B); 1-5-[[2-amino-5-
O-(aminocarbonyl)-2-deoxy-L-xylonyl]amino]-5-deoxy-β-D-
allofuranuronosyl]-1,2,3,4-tetrahydro-2,4-dioxo-5-pyrimidinacid
(polyoxin D)—Polyoxin AL-polyoxin B, Piomycin-polyoxin B,
Polyoxin Z-polyoxin D zinc salt, Piomy, Pio

probenazole (3-allyloxy-1,2- benzoisothiazole 1,1-dioxide)—Oryzemate,
Probenazol

prochloraz (1-N-propyl-N-[2-(2,4,6-
trichlorophenoxy)ethyl]carbamoylimidazole)—Sportak, Sporgon

procymidone (N-(3,5-dichlorophenyl)-1,2-dimethylcyclopropane-1,2-
dicarboximide)—Sumilex, Sumisclex, S-7131

propamocarb hydrochloride (propyl 3-(dimethylamino)propylcarbamate
hydrochloride)—Filex, Previcur N, SN 66752, Prevex, Banol

propiconazole (1-[2-(2,4-dichlorophenyl)-4-propyl-1,3-dioxolan-2-
ylmethyl]-1H-1,2,4-triazole)—Radar, Tilt, Desmel, Banner, CGA
64250, Bumper, Orbit, Alamo

propineb (polymeric zinc propylenebis(dithiocarbamate))—Antracol, Bay
46131, LH 30/Z, Airone, Taifen

propionamide (N-(1-cyano-1,2-dimethyl propyl)-2-(2,4-dichlorophenoxy)
propionamide—AC382042

pyrazophos (O,O-diethyl-O- (5-methyl-6-ethoxy-carbonyl-pyrazolol
(1,5a)-pyrimid-2-yl)thionophosphate)—Afugan, Curamil, Missile, Hoe
02873, Pokon Mildew Spray

pyrifenox (2',4'-dichloro-2-(3-pyridyl)acetophenone O-methyloxime)—
Dorado, ACR3651 A, Ro 151297

pyroquilon (1,2,5,6-tetrahydropyrrolo(3,2,1-ij)quinolin-4-one)—
Fongoren, Fongorene, CGA 49104

quinoxyfen (4-phenoxyquinoline)—DE795

quintozene (pentachloronitrobenzene)—Brassicol, Tubergran, Terraclor,
Folosan, Tritisan, Saniclor, Avicol, Kobutol, Pentagen, Botrilex, Kobu,
Earthcide, Turfcide, RTU-PCNB, PCNB, Quintozene, Tilcarex,
TriPCNB

spiroxamine (spiroketal)—Impulse, KWG 4168

sulfur (sulfur)—Kumulus, Solfa, Thiovit, Cosan, Thiolux, Thion, Sulfex,
Elosal, Suffa, Super Six, Sulfur Alfa, Microthiol Special

tecnazene (1,2,4,5-tetrachloro-3-nitrobenzene)—Fusarex, Bygran, Hytec, Nebulin, Hickstor, Hystore, Arena, Easytec

thiabendazole (2-(thiazol-4-yl)benzimidazole)—Tecto, Mertect, Arbotect, Thibenzole, Comfuval, Apl-Luster, Storite, Mycozol, TBZ

thiophanate (diethyl 4,4'-(*o*-phenylene)bis(3-thioallophanate))—Topsin E

thiophanate-methyl (dimethyl 4,4'-(*o*-phenylene)bis(3-thioallophanate))—Cecrobin, Topsin M, Seal&Heal, Mildothane, Enovit, Frumidor, Sipcaplant, Pelt 44, Thiophan, NF44

thiram (tetramethylthiuram disulfide)—Hexathir, Thiotox, Tripomol, Fernide, Vancide TM, Polyram Ultra, Cunitex, HyVic, Tetrapom, Spotrete, Thiramad, TMTD, AAtack, Aules, Chipco Thiram, Fermide 850, Fernasan, Mercuram, Nomersam, Pomarsol forte, SpotreteF, Tetrapom, Thimer, Thioknock, Thiotex, Thirasan, Thiuramin, Tirampa, TMTDS, Trametan, Tripomol, Tuads

tolclofos-methyl (*O*-2,6-dichloro-*p*-tolyl *O*,*O*-dimethyl phosphorothioate)—Rizolex, S3349

tolylfluanid (*N*-dichlorofluoromethylthio-*N'*,*N'*-dimethyl-*N*-*p*-tolysulphamide)—Euparen M, Bay 49854

triadimefon (1-(4-chlorophenoxy)-3,3-dimethyl-1(1*H*-1,2,4-triazol-1-yl)butanone)—Bayleton, Amiral, Bay MEB 6447, Acizol, Strike

triadimenol (1-(4-chlorophenoxy)-3,3-dimethyl-1(1*H*-1,2,4-triazol-yl)butan-2-ol)—Bayfidan, Baytan, Summit

trichlamide ((RS)-*N*(1-butoxy-2,2,2-trichloethyl)salicylamide)—Hataclean

tricyclazole (5-methyl-1,2,4-triazolol(3,4-b)benzothiozole)—Beam, Bim

tridemorph (reaction products of C_{11}-C_{14} 4-alkyl-2,6-dimethylmorpholine homologues containing 60 to 70 percent of 4-tridecyl isomers)—Bardew, Calixin, Ringer

triflumizole ((E)-4-chloro-α,α,α-trifluoro-*N*-(1-imidazol-1-yl-2-propoxyethylidene)-*o*-toluidine)—Trifmine, NF114

triforine (1,1'-piperazine-1,4-diyldi-[*N*-(2,2,2-trichloroethyl)formamide])—Saprol, Cela W524, Funginex, Denarin

validamycin A ([1S-(1α, 4α, 5β, 6α)]-(1,5,6-trideoxy-3-*O*-β-D-glucopyranosyl-5-(hydroxymethyl)-1-[[4,5,6-trihydroxy-3-(hydroxymethyl)-2-cyclohexen-1-yl]amino]-D-*chiro*-inositol])—Validacin, Valimon

vinclozolin (3-(3,5-dichlorophenyl)-5-ethenyl-5-methyl-2,4-oxazolidinedione)—Ronilan FL, Curalan, Ornalin, Vorlan, Vinclosin

zineb (zinc ethylene bisdithiocarbamate)—Dithane Z 78, Lonacol, Zinosan, Ditozin, PolyramZ, Hexathane, Zineb, Hortag, Tritoftorol

ziram (zinc dimethyldithiocarbamate)—Cuman, Corozate, Mezene,
Pomarsol Z, Carbazinc, Fungostop, Zirex, Ziram, Drupina 90, Hexazir,
Mezene, Prodaram, Tricarbamix Z, Triscabol, Zerlate, Vancide MZ96,
Zincmate, Ziram Technical, Ziram F4, Ziram W76, Ziramvis, Zirasan
90, Zirberk, Ziride, Zitox

Ready-Formulated Mixtures of Fungicides

Many chemical companies produce a mixture of fungicides, mixing different active ingredients (Research Information Ltd., 1998). These ready-formulated mixtures increase the potential spectrum of activity. The mixtures may be of site-specific chemicals with different modes of action, or involve multisite components. This mixture will reduce the risk of the development of fungicide-resistant strains, as the pathogen must mutate to overcome two modes of action simultaneously. The important commercially available mixed formulations of fungicides are listed as follows. Some mixtures contain insecticides (lindane, gamma-HCH, diazinon, furathiocarb) and/or micronutrients (molybdenum, manganese, zinc, iron, copper) in addition to fungicides. The following list provides the names of the active ingredients in the mixtures and their trade names (in parentheses).

benalaxyl+mancozeb (Galben M)
benalaxyl+mancozeb+fosetyl (Carlit)
benomyl+mancozeb (Manzate)
benomyl+thiram (Benlate T)
bitertanol+fuberidazole (Sibutol)
bromophos+captan+thiabendazole (Bromotex T)
bromuconazole+iprodione (But)
bupirimate+chlorothalonil (Dovetail)
bupirimate+hexaconazole (Oscar)
bupirimate+triforine (NimrodT)
captan/pyrifenox+dithianon or mancozeb (Rondo)
captan+DCNA (Captan DCNA 60200)
captan+dichloran (Botec)
captan+fosetyl aluminium+thiabendazole (Aliette Extra)
captan+quintozene+thiabendazole (Rival)
carbendazim+chlorothalonil (Bravocarb, Greenshield)
carbendazim+chlorothalonil+maneb (Victor)
carbendazim+fenpropimorph (Corbel-Duo, Corbel TX)
carbendazim+flutriafol (Early Impact)
carbendazim+flutriafol+chlorothalonil+pyrazophos (Impact)

carbendazim+hexaconazole (Planete R)
carbendazim+iprodione (Vitesse)
carbendazim+mancozeb (Kombat WDG)
carbendazim+mancozeb+sulfur (Kombat S)
carbendazim+maneb (MultiW FL, Delsene M Flowable, Delsene
    MX Septal, Bavistin M, Bavistin M72, Headland Dual, Legion)
carbendazim+maneb+sulfur (Bolda FL)
carbendazim+maneb+tridemorph (Cosmic FL)
carbendazim+prochloraz (Sportak Alpha)
carbendazim+propiconazole (Hispor 45 WP)
carbendazim+tecnazene (Hortag Tecnacarb Dust, Arena Plus)
carbendazim+thiram+gamma HCH (Gammalex Liquid)
carbendazim+vinclozolin (Konker)
carboxin+captan (Enhance)
carboxin+diazinon+lindane (Germate Plus)
carboxin+imazalil+thiabendazole (Cerevax Extra, Vitavax Extra)
carboxin+maneb+lindane (Enhance Plus)
carboxin+metalaxyl+quintozene (Prevail)
carboxin+PCNB (VitavaxPCNB)
carboxin+phenylmercury acetate (Murganic RPB)
carboxin+thiabendazole (Cerevax)
carboxin+thiram (Vitavax200, Vitavax 75W, RTUVitavax Thiram,
    Vitavax Flo)
carboxin+thiram+gamma HCH (Vitavax RS Flowable)
carboxin+thiram+lindane (Gustafson Vitavax-Thiram-Lindane)
chlorothalonil+carbendazim (Bravocarb)
chlorothalonil+copper oxychloride (Clortocaffaro Ramato)
chlorothalonil+fenpropimorph (Corbel CL, Corbel-Star, Corbel-
    Fort)
chlorothalonil+flutriafol (Impact Excel, Halo)
chlorothalonil+metalaxyl (Folio 575 FW, Ridomil plus Bravo W,
    Ridomil CT)
chlorothalonil+propamocarb (Tattoo C)
copper oxychloride+copper carbonate+copper sulfate+mancozeb
    (Tri-Miltox)
copper oxychloride+copper sulfate+copper carbonate+folpet
    (Trimifol)
copper oxychloride+maneb+sulfur (Ashlade SMC)
copper oxychloride+sulfur (Wacker 83 and Wacker 83v)
copper oxychloride+zineb (Miltox)
copper sulfate+cufraneb (Comac Macuprax)
copper sulfate+sulfur (Stoller, TopCop)

cymoxanil+mancozeb (Fytospore, Ripost M)
cymoxanil+propineb (Diametan, Milraz)
dimethomorph+mancozeb (Acrobat WG)
epoxiconazole+fenpropimorph (Opus Team, Opus Top)
epoxiconazole+thiophanate-methyl (Rex)
epoxiconazole+tridemorph (Tango, Opus Forte)
ethirimol+flutriafol+thiabendazole (Ferrax)
fenarimol+dodine (Rubigan Plus)
fenpropimorph+chlorothalonil+carbendazim (Corbel Triple)
fenpropimorph+fenpropidin (Boscor)
fenpropimorph+fenpropidin+prochloraz (Towrnoi)
fenpropimorph+iprodione (Sirocco)
fenpropimorph+prochloraz (Sprint, Magic)
fenpropimorph+thiram+gamma HCH (LindexPlus FS)
fenpropimorph+thiram+lindane (Lindex-Plus FS Seed Treatment)
fenpropimorph+tridemorph (Gemini, Rockett Flowable, Rockett
    Ultra)
fentin acetate+maneb (Brestan 60, Trimastan)
fentin hydroxide+maneb+zinc (Chiltern Tinman)
fentin hydroxide+metoxuron (Endspray)
fentin hydroxide+sulfur (Pennsuc S)
ferbam+maneb+zineb (Trimanzone, Tricarbamix)
flusilazole+chlorothalonil (Triumph)
flusilazole+tridemorph (Meld, Sherif/Gal)
fosetyl+captan+carbendazim (Aliette III WG)
fosetyl+folpet (Mikal)
fosetyl+folpet+cymoxanil (Valiant)
fosetyl+mancozeb+cymoxanil (R6 Triplo)
fuberidazole+imazalil+triadimenol (Baytan IM)
guazatine+imazalil (Rappor Plus)
hexaconazole+fenpropidin (Jupiter)
iodophor+tecnazene (Bygran F)
iodophor+thiabendazole (Byatran, Tubazole)
iprodione+thiophanate-methyl (Compass)
kresoxim methyl+epoxiconazole (Juwel 250, Allegro)
kresoxim methyl+mancozeb (Kenbyomix)
mancozeb+benalaxyl (Galben M)
mancozeb+cymoxanil (Fytospore)
mancozeb+fosetyl (Rhodax)
mancozeb+metalaxyl (Fubol 58 WP, Ridomil MZ)
mancozeb+ofurace (Patafol)
mancozeb+oxadixyl (Recoil, Sandofen M 8)

mancozeb+pyrifenox (Furado)
maneb+zinc oxide (Mazin)
maneb+zineb (Luxan Zinnanaat, Triziman, Vondozeb)
metalaxyl+copper (Ridomil plus)
metalaxyl+copper+folpet (Acylon)
metalaxyl+folpet (Ridomil combi)
metalaxyl+metiram (Arcerid)
metalaxyl+PCNB (ApronTerraclor, Ridomil PC)
metalaxyl+thiabendazole+furathiocarb (Rapcol TZ)
metalaxyl+thiabendazole+thiram (Apron Combi 453 FS)
methyl bromide+chloropicrin (Methyl Bromide-Chloropicrin
    Mixtures)
methylisothiocyanate+dichloropropene (Vorlex)
metiram+cymoxanil (Aviso DF, Aviso G, Aviso C, Aviso S, Aviso)
metiram+cymoxanil+copper oxychloride (Aviso Cup)
metiram+metiram zinc (Polycarbacin)
metiram+nitrothal-isopropyl (Pallinal, Pallitop)
metiram+ofurace (Aviso Combi)
nuarimol+chlorothalonil (Guardaton)
nuarimol+imazalil (Elanco Seed Treatment)
nuarimol+mancozeb (Tridal M)
oxadixyl+cymoxanil+dithiocarbamate+phtalimide (Ripost, Pulsan)
oxadixyl+propineb (Vitiril, Fruvit)
oxine-copper+copper hydroxide (Kinset WP)
penconazole+captan (Topas C)
penconazole+mancozeb (Topas MZ)
phenylmercury acetate+gamma HCH (Mergamma 30)
prochloraz+carbendazim (Sportak Alpha, Sportak Alpha/PF)
prochloraz+carbendazim+tebuconazole (Troika)
prochloraz+carboxin (Abavit)
prochloraz+cyproconazole (Sportak Delta 460, Sportak
    Delta/Tiptor, Tiptor)
prochloraz+fenpropidin (Sponsor)
prochloraz+fenpropimorph (Sprint, Sprint HF)
prochloraz+manganese chloride complex (Octave)
propamocarb+mancozeb (Tatoo)
propiconazole+carbendazim (Tilt CB)
propiconazole+chlorothalonil (Tilt CT)
propiconazole+fenpropidin (Zenit)
propiconazole+fenpropimorph (Archer,Tilt Top)
propiconazole+imazalil+thiabendazole (Benit Universal)
propiconazole+tridemorph (Tilt Turbo)

propineb+oxadixyl (Fruvit)
sulfur+nitrothal-isopropyl (Kumulan)
tebuconazole+imazalil (Premis Delta)
tecnazene+organo iodine (Bygran F)
tecnazene+thiabendazole (Hytec Super, Storite SS)
thiabendazole+imazalil (Extratect Flowable)
thiabendazole+thiram (HYTL, Ascot 480 FS, HyVic, RTU Flowable
 Soybean Fungicide)
thiabendazole+thiram+gamma HCH (New Hysede FL)
thiram+gamma-HCH (Hydraquard, Hysede FL)
triadimenol+thiram (RTU-Baytan-Thiram)
triadimenol+tridemorph (Dorin)
triticonazole+anthraquinone (Real)
triticonazole+iprodione (Raxil Complex)
vinclozolin+chlorothalonil (Ronilan Spezial)
vinclozolin+maneb (Ronilan M)
vinclozolin+thiram (Ronilan T-Combi, Silbos DF)

INGREDIENTS IN FUNGICIDE FORMULATIONS

Laboratory-tested active ingredients alone are not normally useful for field application and are rarely applied in their pure form. These active ingredients are formulated to make them suitable for large-scale field use. Formulations are vehicles that enable the active ingredient to be applied under a variety of conditions without loss in performance. These formulations should be safe to the crop, easy to handle, and compatible with other major products. The ideal formulation should deliver the fungicide (active ingredient) in a manner that maintains its intrinsic activity or enhances its performance through enhanced redistribution or mobility. The formulation should reduce losses through volatile action and increase the persistence of the product for a long time. Formulation strategies may need to be designed for each type of active ingredient. If the selected active material acts by redistribution in the crop through the vapor phase, the formulation should not inhibit volatile action of the active ingredient. Similarly, if the active ingredient acts as a surface-acting protectant, the formulation, which prevents wash-off in rain by increasing penetration of the fungicide into the plant, will lose its activity.

Active ingredients are normally mixed with inert (inactive) ingredients so that the mixture can be handled more conveniently and safely and can be applied more easily and efficiently. The formulations also consist of several adjuvants.

Adjuvants are materials added to improve some chemical or physical property of a plant protectant. Adjuvants increase efficiency of the products and reduce drift. The adjuvants used in the fungicide formulations are wetting agents, dispersing agents, spreaders, stickers, UV filters, surfactants, emulsifiers, foam suppressors, penetrants, and drift control agents.

Emulsifiers are chemicals that assist the formation of suspension of small droplets (often of colloidal dimensions) of one liquid in another in which the former is insoluble. They help in the formation of an emulsion comprising small spheres of organic solvent/fungicide in the sprayer.

Foam suppressors are surface-active substances that form a fast-draining foam to provide maximum contact of the spray to the plant surface.

Penetrants are wetting agents, oils, or oil concentrates that enhance the absorption of a systemic fungicide by the plant. Agri-Dex, Induce, and Penetrator are the commercially available penetrants.

Stickers facilitate persistence of the applied fungicide on foliage even during rain.

Surfactants are surface-active agents capable of dissolving in both water and organic solvents, and therefore, they can help in mixing water-insoluble fungicides in water. Surfactants reduce surface tension through adsorption at air-water or water-solid interfaces. The active ingredients are generally insoluble in water but soluble in lipophilic, organic solvents such as cyclohexane and xylene. The solvents are used in fungicide formulations. The fungicide dissolved in organic solvents would not be miscible with water in the spray tank and therefore would not be delivered during the application process. Surfactants help in forming suspension of fungicides in water. Surfactants have several functions. These include retention of the spray, wetting of the leaf surface, and enhancement of uptake of the active ingredient. Surfactants in fungicide formulations increase penetration of the fungicide into the plant subsurface. Surfactant sorption by plant cuticles has been demonstrated. Water-soluble surfactants increase spray droplet size compared to water alone (Ellis and Tuck, 2000). Polyethoxy derivatives (Triton X surfactants) and Tween 85, Tween 21, Tween 80, Tween 40, and Tween 20 are the important surfactants.

UV filters increase photolytic stability of the fungicide on foliage.

Wetting agents and **dispersing agents** are added to assist in particle suspension during application.

The manufactured fungicide formulations themselves often contain all the necessary adjuvants in appropriate amounts. However, addition of some adjuvants to the spray fluid may increase efficacy of the fungicide several-fold (Kwon et al., 1998). Several adjuvants are screened for their effect in increasing efficacy of various active ingredients. For example, alcohol ethoxylate surfactants increase the efficacy of dimethomorph (Grayson et al., 1996) and prochloraz (Stock, 1996). Addition of ethylene oxide-propylene oxide block copolymers and ethylene oxide-propylene oxide-propylene oxide ethylenediamines to the spray fluid forms larger spray deposits with better coverage on foliage (Richards et al., 2000). Some adjuvants such as Sandovit and SBR 0110 act as rainfasteners and improve rainfastness of fungicides.

Several proprietary wetting agents, surfactants, emulsifiers, spreading agents, and stickers are available in the market. They can be mixed in the spray tank to increase the efficacy of the fungicides. The commercially available adjuvants are listed in the following section.

Available Adjuvants

(Names of the suppliers of the adjuvants are given in parentheses)

Acer—penetrating acidifying surfactant (Tripart Farm Chemicals)

Activator 90—a low-foaming, spreader adjuvant that provides quick wetting, more uniform distribution, and increased retention of spray by reducing surface tension of the spray droplet (Loveland)

Add-It-To-Oil—a natural oil emulsifier (Drexel)

Agral—nonionic wetting and spreading agent (Zeneca)

Agrol Plus S—adjuvant to increase efficiency of benomyl, mancozeb, and zineb (Chemol)

Agura—wetting agent giving improved adhesion of spray fluid to crops (Agro-Kanesho)

Bionex—a powerful spreading, wetting, and penetrating agent that reduces the surface tension in sprays; it permits a uniform film to be formed on sprayed surfaces, ensuring the total coverage of treated foliage (Grupo Bioquimico Mexicano)

Bond—a sticker that acts as a deposition agent by binding fungicide sprays to the leaf surface; it prevents loss of fungicidal activity due to heat, wind, rain, and sprinkler irrigation (Loveland)

Carrier—a vegetable oil surfactant (Stoller)

Citowett—nonionic spreader-sticker and wetting agent (BASF)

Croptex Z.I.P.—a spreader and buffering agent; functions as wetting agent; lowers pH of spray solutions thereby improving effectiveness of many sprayed products (Hortichem)

Deposit—a drift-retardant additive that hydrates fungicides (Loveland)

Eyear—foamless wetting agent which enhances and improves emulsifiability and suspensibility of spray preparations (Agro-Kanesho)

Haiten—spreader-sticker of high efficacy (Hokko)

Headland Guard—sticking and spreading agent (Headland)

Headland Intake—wetting and penetrating agent (Headland)

KK Sticker—Sticking and wetting agent which provides enhanced retention of fungicides on plant surfaces and reduces weathering (Agro-Kanesho)

Lentus A—a synthetic latex surfactant, useful as a sticker and extender (spreader) improving rainfastness (Tripart Farm Chemicals)

Li 700—penetrating surfactant, improves penetration of the fungicide into plant tissue (Loveland)

Lo-Drift—spray additive to reduce drift (Rhone-Poulenc)

Minax—nonionic wetter for use with fungicides to improve spreading penetration (Tripart Farm Chemicals)

Neptune—a spreader-wetter (Loveland)

Nonit—wetting agent to increase the wettability of fungicide sprays (Chemol)

PBI Spreader—nonionic spreading and wetting agent (Pan Britannica)

Petan V Sticker—wetting agent with high adherence and persistence and reduced weathering (Agro-Kanesho)

Pinene II—a nonionic spreader sticker adjuvant (Drexel)

Plyac—a spreader-sticker which dries quickly to produce a water-resistant, elastic film, which increases spray deposits and reduces weathering (Loveland)

Sandovit—surface-active adjuvant (Sandoz Agro)

Sil-Fact—silicone surfactant with fast wetting properties and stomatal flooding (Drexel)

Spreader—nonionic spreading and wetting agent (Pan Britannica)

SP Super—wetting agent and sticker (Productos OSA)

Surf-AC 820—nonionic surfactant (Drexel)

Surf-AC 910—nonionic surfactant (Drexel)

Wettol—emulsifier and wetting and dispersing agent for formulating fungicides (BASF)

Wetz—a nonionic surfactant in an antifoam system for use with fungicides (Drexel)

TYPES OF FUNGICIDE FORMULATIONS

Fungicides are formulated in several ways, depending on their physical characteristics and application methods. Fungicides used for seed treatment should be formulated differently from those used as foliar spray. Fungicides that are soluble in water should be formulated differently from those that are insoluble in water. Fungicides are formulated as wettable powders, dustable powders, driftless dusts, emulsifiable concentrates, water-soluble powders, soluble concentrates, suspension concentrates, granules, flowable formulations, water-soluble bags, oil-miscible liquids, oil-dispersible powders, water-miscible concentrates, liquid formulations, aqueous suspensions, ULV formulations, oil formulations, coating agents, fumigants, dry seed treatment, liquid seed treatment, flowable seed treatment, and soil treatment formulations. These formulations consist of various types of adjuvants.

Wettable Powders

Wettable powders are solid formulations suitable for fungicides that have low solubility in water. They are produced by crushing the solid active ingredient and an inorganic diluent, such as clay, talc, or silica, in a ball mill to a particle size of about 25 μM. Surface-active agents (surfactants) are mixed with the powder. This will help the mixture to become wettable in water; otherwise, the water insoluble fungicide will be only in suspended form in water. Wetting agents and dispersion agents are added to the mixture to assist in particle suspension during application. Other adjuvants such as stickers and UV filters are also added to improve persistence. Wettable powders form a suspension rather than a true solution when added to water. Many fungicides are formulated as wettable powders. Chlozolinate, cufraneb, cymoxanil, cyprofuram, dichlone, fenaminosulf, ferbam, fosetyl-aluminium, fluoromide, pimaricin, triflumizole, and tolyfluanid are formulated only as wettable powders. Carbendazim, carboxin, chlorothalonil, copper oxychloride, copper sulfate, cuprous oxide, dichlofluanid, dicloran, diniconazole, ditalimfos, dithianon, dodine, etaconazole, etridiazole, fenaminosulf, fenarimol, fentin acetate, fentin hydroxide, folpet, flutolanil, fluotrimazole, fthalide, furalaxyl, isoprothiolane, kasugamycin, mancozeb, maneb, mepronil, metalaxyl, myclobutanil, oxine-copper, oxycarboxin, penconazole, pency-curon, pentachlorophenol, pimaricin, prochloraz, procymidone, pyrifenox, pyroquilon, thiabendazole, thiophanate, thiophanate-methyl, thiram, tolclofos-methyl, tolyfluanid, triadimenol, triforine, zineb, and ziram are mostly for-mulated as wettable powders. Several wettable powder trade products are available. Trade names and suppliers include Adagio (chlorothalonil+mancozeb, PBI Agrochemicals), Carbendazim (carbendazim, High Kite), Clortocaffaro Ramato (chlorothalonil+copper oxychloride, Caffaro), Cosan (wettable sulfur, Hoechst), Dithane M-45 (mancozeb, Rohm and Haas), Fungitex (thiram, Protex), Gilmore Benomyl (benomyl, Gilmore), Gilmore Enidod (dodine, Gilmore), Hymush (thiabendazole, Agrichem), KOP-Hy-droxide 50 (copper hydroxide, Drexel), Microcop 50 (copper oxychloride, Probe), M.S.S. Patafol (mancozeb+ofurace, Mirfields), Neoram (copper oxychloride, Caffaro), Quintozene wettable powder (quintozene, Rhone-Poulenc Environmental), Rovral (iprodione, Rhone-Poulenc), SF-101 50 SF (ditalimfos, Intermed), Suffa 90 (micronized wettable powder sulfur, Drexel), Sulfex (wettable sulfur, Excel), Topsin M (thiophanate-methyl, Elf Atochem), Triadimefon (triadimefon, High Kite), Trimanex (mancozeb, Protex), Vinclozolin (vinclozolin, High Kite), Virolex (carbendazim, Protex), and Vitigran (copper oxychloride, Hoechst).

Dustable Powder (Dust)

Dustable powder formulations are similar to wettable powders, as both are manufactured by grinding the fungicide, together with a solid diluent, in a ball mill. The particle size of the dust formulations is about 10 μM. Parti-cle size is very important. If the diameter is too small, the particles will co-agulate, and if the particle size is too large, the activity of it will be signifi-cantly reduced. Mercurous chloride is available only as a dust formulation. Dust formulations of copper oxychloride, dichlofluanid, dicloran, ditalimfos, dodine, edifenphos, etridiazole, folpet, flutolanil, fthalide, hymexazol, ipro-benfos, isoprothiolane, kasugamycin, mancozeb, mepronil, pencycuron, procymidone, tecnazene, tolclofos-methyl, sulfur, and zineb are available. Dust formulation is liable to drift, and it may drift long distances from the treated area, even when wind velocities are low. Trade names of some dustable dusts are Fluidosoufre (powdered sulfur, Elf Atochem Agri.), Arena 6 (tecnazene, Tripart Farm Chemicals), Arena Plus (tecnazene+ carbendazim, Tripart Farm Chemicals), Atlas Tecnazene 6 percent dust (tecnazene, At-las), and Maneb 80 (maneb, Pennwalt).

Driftless (DL) Dust

Dustable dust is subject to drift. To avoid this, DL dusts have been developed. The particle size of DL dust (20 to 30 µm) is bigger than conventional dust (10 µM). The bigger particles are less likely to drift than finer particles. Furthermore, the secondary aggregation of the particles is achieved by addition of paraffin, alkyl, or alkylphenyl ether phosphate as an aggregating agent, thus producing driftless formulations. DL formulations are used in the control of rice diseases in Japan.

Emulsifiable Concentrate

In emulsifiable concentrate formulations, active ingredients are dissolved in organic solvents such as xylene and cyclohexane. These solvents are not soluble in water, so emulsifiers are added to the formulation. Emulsifers, which are partly hydrophilic and partly lipophilic, enable the formation of a homogeneous and stable dispersion of small globules of the solvent in water. Wetting agents, surfactants, spreaders, and stickers are also included in emulsifiable concentrate formulations. Dinobuton, dodemorph, fenpropimorph, methyl isothiocyanate, phosdiphen, and tridemorph are formulated only as emulsifiable concentrates. Emulsifiable concentrate formulations of cuprous oxide, dichlorophen, diniconazole, ditalimfos, edifenphos, ethirimol, etridiazole, fenarimol, flusilazol, fluotrimazole, hymexazol, iprobenfos, isoprothiolane, metalaxyl, myclobutanil, oxycarboxin, penconazole, prochloraz, propiconazole, pyrifenox, tolclofos-methyl, triadimenol, and triforine are available. Trade names of some emulsifiable concentrates are Nustar (flusilazole, DuPont), Point Imazalil (imazalil, Point), Tridemorph (tridemorph, 750 g/liter emulsifiable concentrate, High Kite), and Mirage 45 EC (prochloraz, Protex).

Water-Soluble Powder

In water-soluble powder formulation, the active ingredient is finely ground. The formulation may contain a small amount of wetting agent to assist solution of the ground active ingredient in water. When the formulation is added to water in the spray tank, it dissolves immediately. Copper sulfate and 8-hydroxyquinoline sulfate are the important water-soluble powder formulations. Terraneb SP Turf Fungicide (chloroneb, Kincaid) is an important water-soluble commercial product.

Soluble Concentrate

The active ingredient in a soluble concentrate formulation is soluble in water and is formulated either with water or with a solvent such as alcohol, which mixes readily with water. This formulation forms a true solution in water in the spray tank. Soluble concentrate formulations of dichlorophen, dimethirimol, guazatine, 8-hydroxyquinoline sulphate, nabam, propamocarb, and propiconazole are available.

Suspension Concentrate

A suspension concentrate formulation is similar to a soluble concentrate, but the active ingredient is only partially soluble in water. The active ingredient is ground to a fine powder (about 5 μM), suspended in either water or in an organic liquid, and then blended with a solid inert material and suitable adjuvants. The efficacy of suspension concentrates depends on the type of adjuvants added to them. Suspension concentrate formulations form suspension in water in the spray tank. Flutriafol is available only in this form. Carboxin, dithianon, drazolon, ethirimol, fenarimol, fentin acetate, folpet, fthalide, hexaconazole, mepronil, oxine-copper, procymidone, thiophanate, and tolclofos-methyl are available as suspension concentrates. Rovral Flo (iprodione, Rhone-Poulenc) is the trade name of a suspension concentrate.

Water Miscible Concentrate

This type of concentrate formulation is miscible with water. Banol (propamocarb, AgrEvo) is the commercially available water miscible concentrate.

Flowable Formulations

Flowable formulations are also called water-dispersible suspensions. The active ingredients in these formulations may not be soluble in water or in commonly available organic solvents. The active ingredient is impregnated on a diluent such as clay and milled to an extremely fine powder. The powder is then suspended in a small amount of liquid so that the resulting formulation is thick, like paste or cream. The flowables are mixed with water to form suspensions in the spray tank. Flowables are similar to wettable powders but are in suspension. They combine the benefits of both wettable powders and emulsifiable concentrates. Flowable formulations are available as *flowable concentrate* (carboxin, iprodione), *flowable suspension* (mancozeb, maneb, copper sulfate), and *flowable liquid* (mancozeb, fenbuconazole). *Dry flowables* (maneb, carbendazim) are also available. Flowables in general are very finely divided powders.

Trade names of some flowable formulations of fungicides are Agrichem Maneb Dry Flowable (maneb, Agrichem), Anchor FL (oxadixyl, Gustafson), Captan 4L Flowable (a flowable formulation containing four pounds of captan per gallon, Drexel), Captan 4 Flowable Seed Protectant (a flowable fungicide used for seed treatment, Drexel), Captan Plus Molybdenum Flowable Liquid Seed Treatment for Soybeans (captan+molybdenum, Drexel), Carbate Flowable (carbendazim, Pan Britannica), Chloroneb 300 Fungicide Flowable (chloroneb, Kincaid), Clortocaffaro Flow (chlorothalonil, Caffaro), Cuproxat Flowable (copper sulfate tribasic, Agrolinz), Delsene (a dry flowable carbendazim, DuPont), Dithane Dry Flowable (mancozeb, Pan Britannica), Enable (fenbuconazole, liquid flowable, Rohm and Haas), Extratect Flowable (thiabendazole+imazalil, MSD Agvet), Gilmore Carbendazim 750g DF (dry flowable carbendazim, Gilmore), Gustafson Baytan 30 (a flowable systemic seed treatment triadimenol fungicide, Gustafson), Imber (flowable sulfur, Tripart Farm Chemicals), KOP-300 (copper sulfate tribasic, Drexel), Legion (carbendazim+maneb, Tripart Farm Chemicals), Manzate 200 DF (mancozeb dry flowable formulation, DuPont), Nustar (flusilazole dry flowable formulation, DuPont), Rizolex Flowable (tolclofos-methyl, AgrEvo), Pasta Caffaro (copper oxychloride, 25 percent copper metal in flowable formulation, Industrie Chimiche Caffaro), Rovral (iprodione, oily and aqueous flowable concentrates, Rhone-Poulenc), RTU-Baytan-Thiram (triadimenol+thiram, Gustafson), RTU Flowable Soybean Fungicide (thiram+thiabendazole, Gustafson), RTU-PCNB (quintozene, Gustafson), RTU-Vitavax-Thiram (carboxin+ thiram, Gustafson), Suffa (flowable 6 pound elemental sulfur, Drexel), Suffa 8 (flowable 8 pound elemental sulfur, Drexel), Super-Tin 4L (flowable fentin hydroxide, Chiltren), Tecto Flowable Turf Fungicide (thiabendazole, Vitax), That Flowable Sulfur (sulfur, Stoller), Thiovit (dry flowable micronised sulfur, Sandoz), Topsin M (thiophanate-methyl, Flowable, Elf Atochem), Victor (carbendazim+chlorathalonil+maneb, Tripart Farm Chemicals), Vitavax-200 (carboxin+thiram, Gustafson), Vitavax-30 C (carboxin, Gustafson), Vitavax-34 (carboxin, Gustafson), Vitavax-Extra (carboxin+imazalil+thiabendazole, Gustafson), Vitavax- PCNB (carboxin+quintozene, Gustafson), Wacker-Kupferkalk (copper oxychloride, Wacker), Flowable Sulphur (sulfur, CMI), Ziram Flowable (ziram, Elf Atochem), and Ziram Dry Flowable (ziram, Elf Atochem).

Water-Soluble Bag

Some wettable and soluble powders are marketed in water-soluble bags. When formulations in soluble bags are placed directly in the spray tank, the bags dissolve. This type of packaging eliminates the need to measure the fungicide for addition to the spray tank. The trade names of fungicide-containing water-soluble bags are Contact 75 (chlorothalonil, ISK Biotech) and Topsin M (thiophanate-methyl, water-soluble bag, Elf Atochem).

Water-Soluble Granule

Granules are very much like dusts except that the inert materials are much larger. Size of granule formulations ranges from 300 to 1,700 μM. Liquid formulations of the active ingredient are added to particles of clay or other porous materials such as walnut shells and corn cobs. Granular formulations are used exclusively for soil treatment. Granular formulations have the merits of little scatter and small loss. They are soluble in water. When applied to soil, the active ingredients are released gradually from the inert material. Granular formulations of flusilazol, furalaxyl, hexaconazole, iprobenfos, isoprothiolane, kasugamycin, metalaxyl, pyroquilon, and tecnazene are available in the market. Trade names of some granular formulations are Arena Granules (tecnazene, Tripart Farm Chemicals), Atlas Tecgran 100 (tecnazene, Atlas), Bygran F (tecnazene+organo iodine, Wheatley), Bygran S (tecnazene, Wheatley), Hystore 10 (tecnazene, Agrichem), Moncut (flutolanil, Nihon Nohyaku), and Thiovit (micronized wettable granular sulfur fungicide, Pan Britannica).

Water-Dispersible Granule

Water-dispersible granule formulations are similar to water-soluble granules. However, the active ingredient is not easily soluble in water. Water-dispersible formulations of copper oxychloride, cuprous oxide, and pentachlorophenol have been developed. Trade names of some water-dispersible granules are Dithane DF (mancozeb, Rohm and Haas), Kumulus DF (sulfur, BASF), Microthiol Special (micronized granular sulfur, Elf Atochem North America), Penncozeb WDG (mancozeb, Cyanamid), Polyram DF (metiram complex, BASF), Recop (copper oxychloride, Sandoz), and Topsin M (thiophanate-methyl, water dispersible granule, Elf Atochem).

Microgranule

Microgranular formulations of sulfur, fthalide, and iprobenfos have been developed. The particle size of microgranules is 180 to 710 μM. This formulation prevents drift of the active ingredient and gives uniform distribution. Trade names of some microgranule fungicides are Solfa (a free flowing microgranule of sulfur, Atlas, U.K.), and Copper Sandoz (a water dispersible microgranular cuprous oxide fungicide, Sandoz Agro). Fthalide and iprobenfos are also available as microgranule formulations.

Liquid Fungicide

Some fungicides have been developed as liquid fungicides. Liquid formulations of dodine, kasugamycin, lime sulfur, and ziram are available. Trade names of some liquid fungicides are Bencarb (carbendazim, Productos OSA), Impact (flutriafol+carbendazim+chlorothalonil+pyrazophos, Zeneca), Hy-TL (thiram+thiabendazole, Agrichem), Jupiter (hexaconazole+fenpropidin, Zeneca), Mascot Contact Turf Fungicide (vinclozolin, Rigby Taylor), Mascot Systemic Turf Fungicide (carbendazim, Rigby Taylor), Milcurb-Super (ethirimol, Zeneca), Mildothane Liquid (thiophanate-methyl, Hortichem), Mildothane Turf liquid (thiophanate-methyl, Rhone-Poulenc Environmental), Milgo E (ethirimol, Zeneca), and Pennsuc S (fentin hydroxide+sulfur, Elf Atochem Agri).

Aqueous Suspension

In this formulation, the active ingredient is formulated as aqueous suspension.Trade names of some aqueous formulations are Cuperhidro (copper oxychloride, Grupo Bioquimico Mexicano), Cupertron (copper oxychloride, Grupo Bioquimico Mexicano), and Metacid 400 TS (thiram, Grupo Bioquimico Mexicano).

Aqueous Solution

Dimethirimol and propamocarb have been developed as aqueous solution formulations.

Oil Dispersible Powder

Some fungicides have been formulated as oil dispersible powders. Trade names of oil dispersible powders are Benlate OD (benomyl, DuPont) and Gilmore Benomyl (benomyl, Gilmore).

Oil-Miscible Liquid

Oil miscible formulations of hexaconazole and pentachlorophenol are available.

ULV Formulation

An ultralow-volume (ULV) formulation is sprayed without any dilution, at the rate of one-half gallon per acre. The formulation may contain only the active ingredient or the active ingredient in a small amount of solvent. Special application equipment is needed to spray ULV formulations. Several acres of the crop can be sprayed with a small amount of liquid with this formulation. Turbair Rovral (iprodione, Turbair) and Kasugamycin ULV liquid are the commercially available ULV fungicide formulations.

Oily Formulation

The commercially available oily formulation is Vondozeb (maneb+zineb, Elf Atochem). It is mainly used in horticultural crops.

Paste Formulation

Some fungicides are developed as paste formulations. They are mainly used for treating cuttings and pruning injuries. Copper oxychloride, oxine-copper, and ziram pastes are available.

Dry or Slurry Seed Treatment

Several fungicides have been formulated specifically for seed treatment. Seed treatment wettable powder formulations of carboxin, etaconazole, etridiazole, fenfuram, fuberidazole, furmecyclox, guazatine, hexachlorobenzene, hymexazol, mepronil, metalaxyl, metiram, oxine-copper, pencycuron, phenylmercury acetate, tolclofos-methyl, and triadimenol are available. Trade names of some dry/slurry seed treatment formulations available in the market are Abavit (prochloraz+carboxin, Hoechst), Aliette III WG (fosetyl+captan+carbendazim, Rhone-Poulenc), Apron Dry (metalaxyl, Gustafson), Apron-FL (metalaxyl, Gustafson), Baytan (triadimenol, Bayer), Benit Universal (propiconazole+imazalil+thiabendazole, Novartis), Benlate T (benomyl+thiram, DuPont), Beret (fenpiclonil, Novartis), Captan 30-DD (fungicidal seed protectant [captan] with colorant, Gustafson), Captan 300 (fungicidal seed protectant without colorant, Gustafson), Carboxin (carboxin, Melchemie), Celest (fludioxonil, Novartis), Cupromox (cuprous oxide, Universal Crop), Dividend (difenoconazole, Novartis), Elanco Seed Treatment (nuarimol+imazalil, DowElanco), D-264 Captan Seed Treatment (captan+diazinon, Drexel), Emisan 6 (2-methoxyethyl mercury chloride, Excel), Enhance (carboxin+captan, Gustafson), Enhance Plus (carboxin +maneb+lindane, Gustafson), Ferrax (flutriafol+ethirimol+thiabendazole, Zeneca), Fongorene (pyroquilon, Novartis), Gambit (fenpiclonil, Novartis), Gustafson Captan 400 D (captan with red dye, Gustafson), Gustafson Captan 400 (captan without red dye, Gustafson), Kemikar (carboxin, Kemira Agro OY), Lindex-Plus F5 Seed Treatment (lindane+thiram+fenpropimorph, DowElanco), Magnate (imazalil, Makhteshim-Agan), Nordox SD-45 (cuprous oxide, NORDOX), Panoctine (guazatine, Rhone-Poulenc), Pentaclor 600TS (quintozene, Grupo Bioquimico Mexicano), PMA Technical (phenyl mercury acetate, Excel), Point Carboxin (carboxin, Point), Pomarsol (thiram, Bayer), Prelude (prochloraz, Hoechst), Prelude SP (prochloraz+manganese chloride+carbendazim, Hoechst), Prevail (carboxin+ metalaxyl+quintozene, Gustafson), Quinolate (oxine-copper, Novartis), Quinondo WP (oxine-copper, Agro-Kanesho), Rapcol TZ (furathiocarb+ metalaxyl+thiabendazole, Novartis), Raxil (tebuconazole, Bayer), Real (anthraquinone+triticonazole, Rhone-Poulenc), Seed Tox (phenyl mercury acetate, AIMCO), Sibutol (bitertanol+fuberidazole, Bayer), Spectro (difenoconazole, Novartis), Spotless (diniconazole, Valent), 42S Thiram Fungicide (thiram, Gustafson), Systhane (myclobutanil, Rohm and Haas), Thiram 50 WP (thiram, Gustafson), Tops-5 (thiophanate-methyl, Gustafson), Trimanoc (mancozeb, Elf Atochem), Trimangol (maneb, Elf Atochem), Trimisem (nuarimol, DowElanco), Trimisem Total (nuarimol+maneb+ lindane+anthraquinone, DowElanco), Triple-Noctin L (thiram+molybdenum, Gustafson), Tripomol (thiram, Elf Atochem), and Vincit (flutriafol, Zeneca).

Liquid Seed Treatment

Liquid formulations of seed treatment fungicides have also been developed. Etaconazole, ethirimol, fenfuram, guazatine, methoxyethylmercury acetate, phenylmercury acetate, prochloraz, and triforine are available as liquid seed treatment formulations. Trade names of some liquid formulations are Agrichem Flowable Thiram (thiram, Agrichem), Captan Plus Molybdenum Flowable Liquid Seed Treatment for Soybeans (captan+molybdenum, Drexel), Hydraguard (thiram+gamma-HCH, Agrichem), Hysede FL (gamma-HCH+thiram, Agrichem), Hy-Vic (thiram+thiabendazole, Agrichem), Rappor (guazatine, DowElanco), and Rappor Plus (guazatine+ imazalil, DowElanco).

Emulsifiable Seed Treatment

Emulsifiable formulations for seed treatment are available. Gustafson Terracoat LT-2N (quintozene, Gustafson) is the commercially available emulsifiable seed treatment formulation.

Flowable Seed Treatment

Several flowable formulations have been developed for seed treatment. Trade names of some of them are Gustafson Vitavax-Thiram-Lindane (carboxin+thiram+lindane, Gustafson), Gustafson Baytan 30 (a flowable systemic seed treatment triadimenol fungicide, Gustafson), Chloroneb 300 Fungicide Flowable (chloroneb, Kincaid), Captan 4 Flowable Seed Protectant (a flowable fungicide used for seed treatment, Drexel), Rival (captan+ quintozene+thiabendazole, Gustafson), RTU Flowable Soybean Fungicide (thiram+thiabendazole, Gustafson), RTU-PCNB (quintozene, a flowable seed treatment, Gustafson), RTU-Vitavax-Thiram (carboxin+thiram, Gustafson), Vitavax-30C (carboxin, Gustafson), Vitavax-34 (carboxin, Gustafson), Vitavax-200 (carboxin+thiram, Gustafson), Vitavax-Extra (carboxin+ imazalil+thiabendazole, Gustafson), and Vitavax-PCNB (carboxin+PCNB, Gustafson).

Soil Application Fungicide—Wettable Powder

Some formulations have been developed specifically for application to soil. The wettable powder formulations for soil application are Fongarid (furalaxyl, Novartis), Gilmore PCNB (quintozene, Gilmore), Botran (dicloran, Hoechst), Ipam (metham-ammonium, soil sterilant, Chemol), Pentaclor 600F (quintozene, Grupo Bioquimico Mexicano), and Terraclor 20D (quintozene, Hortichem).

Soil Application Fungicide—Fumigant

Fumigants are applied to soil to control soilborne pathogens. They are either in liquid or gaseous form. Trade names of some fumigants are Basamid-Granular (dazomet, BASF), Brom-O-Gas (methyl bromide+chloropicrin, Great Lakes), Chlor-O-Pic (chloropicrin, liquid, Great Lakes), Di-Trapex N (methyl isothiocyanate+dichloropropene, liquid fumigant, Hoechst), Methyl Bromide (methyl bromide, gas, Melchemie and Marman), Sistan (metham-sodium, liquid fumigant, Universal Crop), Unifume (metham-sodium, liquid fumigant, Universal Crop), Vapam (metham-sodium, Zeneca), and Vorlex (methyl isothiocyanate+dichloropropane, Hoechst).

Coating Agent

Octhilinone and 2-phenylphenol are used as coating agents. They can be used for control of postharvest diseases of fruits by impregnation of fruit wrappers and crates or by application as a wax directly to the fruit. They can also be used to coat pruning cuts. Mercuric oxide is used as a wound protectant.

CLASSIFICATION OF FUNGICIDES

More than 113 active ingredients have been registered as fungicides worldwide. Grouping of fungicides is needed to help the actual user (farmer) make quick decisions to select a suitable fungicide to control a particular disease. Three important criteria used to classify fungicides are their chemical strucures, their type of mobility action, and their biochemical modes of action. Fungicides are broadly divided into two groups based on their mobility in host tissues: systemic and nonsystemic.

Systemic Fungicides

Systemic fungicides penetrate into plant tissue and show apoplastic mobility (movement within cell walls and xylem tissues), symplastic mobility (movement through plasmadesmata from cell to cell through phloem tissues), or both apoplastic and symplastic mobility in plants. However, apoplastic movement alone is common. Most systemic fungicides exhibit acropetal movement (redistribution toward the plant apex or leaf margins); basipetal movement (from plant apex to bottom) is rare. Systemic fungicides mostly move in the transpiration stream. When these fungicides are applied to the roots of plants, the chemicals may be translocated over a long distance to protect young top leaves against pathogens. Some systemic fungicides possess translaminar activity. When these compounds are applied to the upper surface of a leaf, they can control disease on the lower side of the leaf, and vice versa. Very few systemic fungicides move in both directions. Fosetyl-aluminium has been shown to move through both xylem and phloem. Some systemic fungicides such as triadimefon may produce vapors with fungicidal activity and act at large spreading areas in the foliage. Systemic fungicides may have protectant (protection against sporulation, spore germination, appressorium formation, and prevention of penetration), curative (suppression of postinfection and presymptomatic phase of infection and visible effects of symptoms), and eradicant (suppressive action on the post-symptomatic stage by killing the pathogen in the host) action.

Nonsystemic Fungicides

Nonsystemic fungicides are also called surface active, surface protectant, or contact fungicides. They do not penetrate the plant surface, and they are not translocated within the plant. On application, they remain on the surface of the plant, and redistribution on the foliage of the crop may occur through the action of dew or rainfall. Nonsystemic fungicides can also be called protectant fungicides, as they are applied prophylactically to the crop. They do not have curative or eradicant action and should be applied before infection of the pathogen occurs. These fungicides may prevent fungal spore germination, appressorium formation, and penetration of host tissues. Nevertheless, once infection is established, these fungicides may not have any function. However, they cannot be called protectant fungicides, as several systemic fungicides also have protectant action. Some systemic fungicides such as quinoxyfen inhibit appressorium formation on leaf surfaces. Another systemic fungicide, tricyclozole, inhibits melanization of appressoria and the subsequent formation of the infection apparatus of the fungi. Some systemic fungicides such as probenazole and fosetyl-aluminium may act through induction of the host's defense mechanisms and act as protectants. Only a few systemic fungicides have protectant action, but all nonsystemic fungicides have protectant action.

Fungicide Families

Fungicides are further subdivided into families based on their chemical structure. The important chemical families are given in the following list. Names of the chemical families are followed by the common names of fungicides included in those families.

> Acetamide—cymoxanil, ofurace, metominostrobin
> Acylalanine—benalaxyl, furalaxyl, metalaxyl
> Amino acid amide carbamate—iprovalicarb
> Anilinopyrimidine—pyrimethanil, cyprodinil
> Antibiotic—kaugamycin, Blasticin-S, pimaricin, polyoxins, validamycin A, cycloheximide
> Benzenediazosulphonate—fenaminosulf

Benzimidazole, MBC—benomyl, carbendazim, fuberidazole,
thiabendazole, thiophanate, thiophanate-methyl
Biphenyl—2-phenylphenol
Butaylamine—*sec*-butylamine
Carbamate—propamocarb, propineb, thiophanate, thiophanate-
methyl, diethofencarb, iprovalicarb, methasulfocarb
Carboxamide—carboxin, oxycarboxin, fenfuram, benodanil,
mepronil, flutolanil, furmecyclox, fenhexamid, ethaboxam,
carpropamid
Chlorinated hydrocarbon—chloroneb
Chlorinated nitroaniline—dicloran
Chlorophenyl—chlorothalonil
Dicaboximide—iprodione, procymidone, vinclozolin
Dithiocarbamate—cufraneb, ferbam, mancozeb, maneb, metiram,
metham-sodium, zineb, ziram, thiram
Dithiolane—isoprothiolane
Guanidine—dodine, guazatine
Imidazole—imazalil, prochloraz, triflumizole, perfurazoate
Inorganic—Bordeaux mixture (tribasic copper sulfate), copper
oxychloride, copper oxide, copper hydroxide, sulfur, lime sulfur,
mercurous chloride, mercuric oxide
Isothiazoline—octhilinone
Isothiocyanate—methyl isothiocyanate
Isoxazole—drazoxolon, hymexazol
Morpholine—dodemorph, fenpropimorph, tridemorph
Nitro compound—binapacryl, bronopol, chloropicrin, dinobuton,
dinocap, DNOC, nitrothal-isopropyl
Organoaluminium—fosetyl-aluminium
Organochlorine—hexachlorobenzene, pentachlorophenol,
quintozene, tecnazene
Organomercury—methoxyethylmercury acetate, phenylmercury
acetate
Organophosphorous—ditalimfos, edifenphos, iprobenfos,
phosdiphen, pyrazophos, tolclofos-methyl
Organotin—fentin acetate, fentin hydroxide
Oxathiin—oxycarboxin, carboxin
Oxazolidine—chlozolinate, oxadixyl
Oxazolidinone—famoxadone
Phenoxyquinoline—quinoxyfen
Phenylpyrrole—fenpiclonil, fludioxonil
Phthalic acid—fthalide

Phthalimide—captan, captafol, folpet
Piperazine—triforine
Piperidine—fenpropidin
Pyridine—buthiobate, pyrifenox
Pyrimidine—bupirimate, nuarimol, dimethirimol, ethirimol, fenarimol
Pyrrole—fluoromide
Pyrroloquinoline—pyroquilon
Quinoline—8-hydroxyquinoline sulphate, oxine-copper
Quinone—dichlone, dithianon, chloranil
Quinoxaline—chinomethionat (quinomethionate)
Strobilurines—kresoxim-methyl, azoxystrobin
Sulphamide—dichlofluanid, tolyfluanid
Thiadiazole—etridiazole
Triazine—anilazine
Triazole—bitertanol, diclobutrazol, triadimefon, diniconazole, etaconazole, flusilazol, flutriafol, fluotrimazole, hexaconazole, myclobutanil, penconazole, propiconazole, triadimenol, tebuconazole, fluquinconazole, difenoconazole, epoxiconazole, triticonazole, bromuconazole, cyproconazole, metconazole, imibenconazole
Urea—cymoxanil, pencycuron

Surface protectant fungicides include inorganics, dithiocarbamates, thiadiazoles, quinones, triazines, sulphamides, quinoxalines, nitro compounds, phthalimides, guanidines, organochlorines, organomercurials, organophosphorous compounds (tolclofos-methyl, edifenphos), organotin compounds, organozinc compounds, chlorophenyls, chlorinated nitroanilines, urea fungicides, benzenediazosulphonates, furans, pyrroles, fthalides, isoxazole (drazoxolon), acetamide (cymoxanil), imidazole (prochloraz), dicarboximide (iprodione), isothiazoline, quinolines, biphenyl, and carbamate (propineb).

Systemic fungicides include acetamides (ofurace, metominostrobin), anilides, anilopyrimidines, benzanilides, benzimidazoles, carbamates, dicarboximide (procymidone), oxazolidine, pyrimidines, pyridines, triazoles, morpholines, organoaluminium, acylalanines, quinolines, isoxazole (hymexazol), imidazole (imazalil, triflumizole), organophosphorous (iprobenfos, pyrazophos), dithiolanes, carboxylates, piperazines, pyrolloquinoline, oxathiins, strobilurins, oxazolidinediones, and carboxamides.

(Some chemical families contain both systemic and surface protectant fungicides. In such families, the names of such fungicides are given within parentheses.)

MODE OF ACTION OF FUNGICIDES

Disruption of Fungal Cell Functions

Several chemical families of fungicides act by disrupting fungal cell functions (Ragsdale and Sisler, 1991). These include inorganics, dithiocarbamates, phthalimides, triazines, sulphamides, chlorophenyls, quinones, and quinoxalines. They are all nonsystemic (surface protectant) fungicides.

Sulfur

Sulfur belongs to the inorganic chemical family. Several formulations of sulfur (dustable dust, wettable powder, and flowable formulations) are available. Sulfur in combination with lime, as lime sulfur, is still an important fungicide. Many theories have been evolved to explain the mode of action of sulfur. The oxidized sulfur theory attributes toxicity of sulfur to oxidation products such as sulfur dioxide, sulfur trioxide, sulfuric acid, thiosulfuric acid, or pentathionic acid. The reduced sulfur theory ascribes toxicity of sulfur to hydrogen sulfide generated by fungal action on sulfur. The direct action theory, which is now well accepted, is based on the toxic action of sulfur as such. Sulfur may be involved in biological redox reactions (electron transport system) in the fungal metabolism. Sulfur is reduced in the electron transport system at a site between cytochrome b and c. Electrons are utilized to reduce sulfur instead of oxygen, and sites of energy conservation (ATP [adenosine triphosphate] generation) are bypassed. Suppression of the ATP generation system results in interference with the energy supply system, leading to the death of fungal pathogens. Toxicity could also result from cross-linking of proteins or other cellular components by free radicals of sulfur or from extensive oxidation of thiol groups, leading to loss of function or structural intregrity of proteins. Sulfur acts against several biochemical sites in fungal cells. It inhibits respiration, disrupts proteins, and forms chelates with heavy metals within the fungal cells.

Copper Compounds

The inorganic copper fungicides include copper oxychloride, copper sulfate, cuprous oxide, and copper hydroxide formulations. Bordeaux mixture consists of copper sulfate and lime. Copper ion is the active component in copper salts. Copper ions in the presence of CO_2 from the air and the organic acids excreted from the plant and/or fungal spores interact to produce the resultant activity. Copper ions are capable of penetrating fungal spores and inhibiting their enzyme reactions. They may chelate and remove other essential metals. They may also block or interact with sulfhydryl groups of the fungal enzymes. Copper ions cause membrane damage resulting in leakage of metabolites from fungal cells.

Mercury Compounds

Organic and inorganic mercurial fungicides are used exclusively for seed and soil treatment. All mercurial fungicides have lost their registration in the United States because of their potential environmental damage and high toxicity. However, a few mercurial fungicides are still allowed in some countries. Mercurial ion reacts with enzymes having a reactive sulfhydryl group.

Tin Compounds

Another group of inorganic fungicides consists of tin compounds. These compounds inhibit fungal growth by inhibiting oxidative phosphorylation (suppressing energy production). They may inhibit mitochondrial ATPase of fungal cells and may bind to protein membrane components and cause extensive and nonspecific alterations in membrane functions, including permeability. These fungicides bind to several proteins, including those containing sulfhydryl groups of cysteine residues. Thus, tin fungicides may act at multiple sites in fungal cells.

Dithiocarbamates

Dithiocarbamates are still one of the most important groups of organic surface-protectant fungicides. The dithiocarbamates are divided into two groups based on their modes of action: monoalkyldithiocarbamates and dialkyldithiocarbamates. Monoalkyldithiocarbamates include zineb, maneb, mancozeb, nabam, metiram, and metham, while dialkyldithiocarbamates include thiram, ziram, and ferbam. The monoalkyldithiocarbamates possess a reactive hydrogen on the nitrogen atom which permits the formation of a highly toxic isothiocyanate. The toxicity of these fungicides appears to be based on isothiocyanate generation. These fungicides alter permeability of fungal cells. They inactivate thiol groups of essential enzymes.

Dialkyldithiocarbamates are strong chelating agents. They deprive fungal cells of essential heavy metals. The metals in a chelate complex would permeate cell membranes more readily than the free metal ions. This will lead to transport of excess heavy metals into fungal cells. The internally released heavy metals would inhibit enzyme activity. These fungicides may form a complex with Cu^{2+} and bind enzymes or other cellular components of fungi. The role of Cu^{2+} in coupling dialkyldithiocarbamates to proteins has been well demonstrated. This coupling will lead to inhibition of fungal enzyme activity or of other cellular functions. These dithiocarbamates catalyze rapid incorporation of Cu^{2+} into the diethyl ester of mesoporphyrin. The toxicity of dithiocarbamates may result from incorporation of Cu^{2+} into a porphyrin precursor of an essential heme-type pigment. Dialkyldithiocarbamates may form mixed disulfides with protein thiol groups and inhibit enzyme activity. These fungicides inhibit copper-containing enzymes. Thus, dialkyldithiocarbamates may act at multiple sites in fungal cells and may be multisite biochemical inhibitors.

Phthalimides

Captan, captafol, and folpet are the important surface-protectant phthalimide fungicides. Captafol and folpet are not marketed in the United States. These fungicides may oxidize cellular thiol groups and inactivate thiol enzymes. They may reduce the level of glutathione in fungal cells. Respiration is strongly inhibited by these fungicides. Several enzymes such as glyceraldehyde-3-phosphate dehydrogenase, carboxylase, hexokinase, and aldolase in fungal cells are inhibited by these fungicides. They also inactivate coenzyme-A. They have multiple sites of action in fungal metabolic systems.

Chlorophenyls

Chlorothalonil is a chlorophenyl fungicide and an important surface protectant. It markedly depresses the level of soluble and cell-bound thiol groups in fungal cells. The fungicide inactivates the thiol enzymes alcohol dehydrogenase, glyceraldehyde-3-phosphate dehydrogenase, and malate dehydrogenase. It does not inactivate nonthiol enzymes. Its fungicidal mechanism appears to be based on thiol inactivation.

Quinones

Dichlone and dithianon are the important fungicides in the quinone group. Dichlone, which is not marketed in the United States, interferes with the electron transport system in fungal cells. It uncouples oxidation from phosphorylation in the normal electron transport pathway or diverts electrons through other pathways to oxygen not associated with high-energy phosphate generation. It also inactivates coenzyme-A and inhibits certain dehydrogenases and carboxylases of fungal cells. It binds thiol and amino groups in fungal cells. Dithianon inactivates thiol-dependent enzymes.

Triazines

Anilazine belongs to the triazine group. It reacts with free amino acid and thiol groups in fungal cells. Its mode of action seems to be similar to that of phthalimides, quinones, and chlorophenyls.

Sulphamides

Dichlofluanid and tolyfluanid belong to the sulphamide group. They are also surface-protectant fungicides. Their mode of action is similar to that of phthalimide fungicides.

Quinoxalines

Chinomethionate (quinomethionate) belongs to the quinoxaline group. It acts by binding with amino and mercapto groups resulting in the inhibition of sulfhydryl-containing enzymes in fungal cells.

Thiadiazoles

Etridiazole belongs to the thiadiazole group. It inhibits respiration of fungal cells.

Butylamine

sec-Butylamine interferes with pyruvate dehydrogenase complex and inhibits pyruvate oxidation in fungal cells.

Guanidines

Guanidine fungicides disrupt permeability of fungal cells. Alteration of membrane permeability may result from disruption of normal anionic-cationic interactions in the membrane because of the displacement of membrane cationic groups by firmly bound cationic surfactant molecules. Actions that affect membrane structure can result in enzyme inhibition.

Organochlorines

Quintozene is the important fungicide in the organochloride group. It probably binds to hydrophobic regions in the fungal cell and interferes with various processes such as membrane function, meiosis, or cell division processes. It may also affect nuclear processes.

Nitro Compounds

Nitro compounds uncouple oxidative phosphorylation and prevent the incorporation of inorganic phosphate into ATP without affecting electron transport in fungal cells. Cell death occurs due to the lack of energy for cellular metabolism.

Chlorinated Nitroanilines

Dicloran belongs to the chlorinated nitroaniline group. Inhibition of protein synthesis has been suggested as the toxic mechanism of dicloran. It affects nuclear stability in fungi and disorganizes cell growth and division.

Inhibition of Sterol Biosynthesis

Triazoles, Imidazoles, Pyridines, Pyrimidines, Piperazines, Morpholines, Piperidines, Spiroxamines

Several systemic fungicides act by inhibiting sterol synthesis in fungi. Sterols are commonly found in fungi, and ergosterol is the major fungal sterol. Ergosterol is involved in the maintenance of membrane integrity, and reduction in ergosterol synthesis results in membrane disruption and electrolyte leakage in fungal cells. The biosynthetic pathway of sterols involves several steps. Sterols are formed from acetyl-CoA. Sterol biosynthesis inhibitors can be divided into two major groups based on their primary site of action in the sterol biosynthetic pathway: (1) sterol C-14 demethylation inhibitors (DMI) and (2) sterol $\Delta^{8,7}$ isomerase and/or sterol Δ^{14} reductase inhibitors. DMIs contain the most commercially valuable group of fungicides. The chemical families include triazoles, imidazoles, pyridines, pyrimidines, and piperazines. Several triazole fungicides are available on the market. Imazalil, prochloraz, and triflumizole are the important imidazole DMIs. Nuarimol and fenarimol are the important pyrimidine DMIs. Buthiobate and pyrifenox are the important pyridines. DMIs inhibit all groups of fungi except those belonging to chromista and oomycota, as the ergosterol synthetic pathway is absent in fungi belonging to oomycota. The latter group of fungi obtains their sterol requirements directly from their hosts through mycelial uptake. DMIs inhibit sterol C-14 demethylation in fungal cells through interference with a cytochrome P-450 monooxygenase enzyme. The N atom of the fungicide binds at the position of the P-450 heme prosthetic group where O_2 would normally bind. The lipophilic substituent of the DMI fungicides binds to the region of the apoenzyme that is structurally associated with a C-14 methyl sterol.

Sterol $\Delta^{8,7}$ isomerase and/or sterol Δ^{14} reductase inhibitors include morpholines, piperidines, and spiroxamines. Dodemorph, fenpropimorph, tridemorph, fenpropidin, and spiroxamine are the important sterol $\Delta^{8,7}$ isomerase and/or sterol Δ^{14} reductase inhibitors. The mode of action of these fungicides is mediated by the interaction of the negatively charged enzyme site and the positively charged nitrogen atom in the fungicide molecule.

Inhibition of Phosphatidylcholine Biosynthesis

Edifenphos, Iprofenphos, Isoprothiolane

These three fungicides appear to have a common mode of action. Phosphatidyl choline is essential for the function of fungal cell membranes, providing a permeability barrier to the movement of ions and macromolecules and a fluid matrix for the activity of membrane-associated proteins. It is essential for fungal growth. The inhibition of phosphatidyl choline synthesis reduces its incorporation into membranes and decreases chitin synthase activity. It leads to decreased hyphal growth. Hyphal swelling is common in fungi treated with these fungicides.

Inhibition of Phosphatidylinositol Biosynthesis

Validamycin

Phosphatidylinositol is required for fungal growth. Inositol-deficient fungal mutants require a supply of exogenous inositol for normal growth. These mutants show reduced hyphal growth, increased hyphal branching, and cell lysis. Its deficiency affects carbohydrate, protein, and nucleic acid metabolism. It also affects wall-bound biosynthetic enzymes. Validamycin inhibits phosphatidylinositol biosynthesis in fungal cells, and it is the only fungicide that shows this mode of action.

Inhibition of Tubulin Synthesis

Benzimidazoles

Benzimidazoles act primarily on cell and nuclear division. Nuclear division is inhibited by the binding of benzimidazoles to the microtubular proteins involved in the synthesis of the mitotic spindle apparatus. Microtubules are formed by the assembly of tubulin, a heterodimeric protein with subunits as alternating helices of α- and β-tubulins, which forms an essential part of the fungal cytoskeleton. They are active in spindle formation and in segregation of chromosomes in cell division. Benzimidazoles show high affinity for tubulin proteins in fungi. They disrupt mitosis during cell division at metaphase in fungi. The mitotic spindle is distorted, and daughter nuclei fail to separate. These events lead to fungal cell death. The target site of benzimidazoles is β-tubulin. The modification of a single amino acid resulting from a mutational change in β-tubulin confers resistance to benzimidazoles in fungi. Benzimidazoles are highly selective in their action against fungi. They do not inhibit fungi belonging to oomycota and *Alternaria* spp., although these fungi also contain β-tubulin. This suggests that binding sites on tubulin may vary among pathogens, and binding at the specific sites alone may lead to the inhibitory action of a specific fungicide.

Inhibition of Nucleic Acid Metabolism

Acylalanines

Metalaxyl, metalaxyl M, and benalaxyl belong to the acylanine group. They act by inhibiting the synthesis of ribosomal RNA via the RNA polymerase I-template complex. Ribosomal RNA synthesis is selectively inhibited, while transfer RNA and messenger RNA syntheses are not affected. Protein, DNA, and lipid biosyntheses are affected later to a lesser degree, but respiration is not affected. This group of fungicides specifically inhibits fungi belonging to oomycota. The fungicides do not affect sporangia and zoospore germination, penetration, and formation of primary haustoria but do affect further hyphal development. The hyphal cells become thickened, leading to eventual death. The RNA-polymerase I complex is inhibited which results in accumulation of rRNA precursors, the nucleotide triphosphates that promote the activity of fungal β-1,3-glucan synthetase and the synthesis of cell wall constituents.

Pyrimidines

Ethirimol, dimethirimol, and bupirimate are the important pyrimidine fungicides. They act through the inhibition of adenosine deaminase, an enzyme in the purine salvage pathway. Adenosine deaminase is not present in plants but is found in a wide range of fungi. However, the fungicides affect the enzyme of powdery mildew fungi only.

Isoxazoles

Hymexazol is the important isoxazole fungicide. It inhibits DNA synthesis in fungal cells. RNA and protein synthesis is less affected, and DNA synthesis is regarded as the primary site of action of this fungicide. The other isoxazole fungicide, drazoxolon, inhibits oxidative phosphorylation in fungi.

Inhibition of Respiration

Carboxamides

Carboxin and oxycarboxin are the important fungicides in the carboxamide group. These fungicides are inhibitory to fungi belonging to basidiomycota. They inhibit succinate dehydrogenase complex in the respiratory chain in fungi. Carboxin blocks membrane-bound succinate-ubiquinone oxidoreductase activity in the mitochondrial electron transport chain. The iron-sulfur cluster S_3 complexed with small ubiquinone binding polypeptide(s) in a phospholipid environment appears to be the carboxin receptor in the succinic dehydrogenase complex. It also inhibits glucose and acetate oxidative metabolism. RNA and DNA syntheses are also inhibited, which may be due to a lack of cellular energy resulting from the inhibition of respiration.

Benodanil, fenfuram, mepronil, fluotanil, and furmecyclox also inhibit respiration in fungi, and their mode of action is similar to carboxin.

Strobilurins

Kresoxim-methyl and azoxystrobin are the important fungicides in the strobilurin group. They are active in the inhibition of electron transfer in complex III of the mitochondrial electron transport chain.

Famoxadone

Famoxadone inhibits the function of the ubiquinol cytochrome c oxido-reductase enzyme at complex III.

Inhibition of Chitin Biosynthesis

Polyoxins

Chitin is a component of several fungi; however, it is absent in fungi belonging to oomycota. Chitin synthase (chitin-UDP-*N*-acetylglucosaminyl transferase) transfers *N*-acetylglucosamine from UDP-*N*-acetylglucosamine to the chitin polymer. Polyoxins inhibit chitin synthesis. Polyoxins are similar to UDP-*N*-acetylglucosamine and are competitive inhibitors of chitin synthase. Polyoxins induce swelling and bursting of spore germ tubes and hyphal tips, reactions that kill the fungus.

Inhibition of Melanin Biosynthesis

Tricyclazole and Pyroquilon

The pigment melanin is important in fungal pathogenicity. Melanization of fungal appressorial walls is essential for the development of infection hyphae and penetration of the host epidermis. Tricyclazole and pyroquilon inhibit melanin synthesis and suppress fungal infection.

Carpropamid

Carpropamid inhibits scytalone dehydratase (an enzyme involved in fungal melanin biosynthesis), effectively preventing the pathogen from penetrating the host plant. The fungicide has a dual mode of action. In addition to inhibiting melanization, it also acts as an inducer of disease resistance. It induces lignification in rice plants. It also induces increased accumulation of the phytoalexins momilactone A and sakuranetin (Thieron et al., 1998).

Inhibition of Protein Synthesis

Kasugamycin and Blasticidin S

Both antibiotics are inhibitors of protein synthesis in fungal cells. Kasugamycin binds to the 30S ribosomal subunit and inhibits protein synthesis. Blasticidin S interacts with the larger ribosomal subunit and blocks the binding site for incoming aminoacyl-tRNA molecules.

Inhibition of Biosynthesis of Lytic Fungal Enzymes

Anilinopyrimidines

Cyprodanil and pyrimethanil are the important anilinopyrimidine fungicides. They inhibit production and secretion of fungal enzymes involved in plant cell wall penetration. These enzymes include pectic enzymes, cellulases, and proteinases.

Inhibition of General (Undefined) Metabolism of Fungi

Organophosphorous Compounds

Triamiphos, ditalimfos, and pyrazophos inhibit fungal spore germination and appressorium formation.

Hexachlorobenzene, Pentachlorophenol, Tecnazene, Dicloran, Chloroneb

They are organochlorine and related compounds. They induce swelling of hyphal tips and germ tubes, leading to lysis. They do not inhibit spore germination. They affect membrane function and induce mitochondrial aberrations.

Phenylpyrroles

Fenpiclonil and fludioxonil are the important fungicides in the phenylpyrrole group. They cause several changes in fungal metabolism. They inhibit glucose phosphorylation in fungal cells.

Propamocarb

Propamocarb affects the fungal cell membrane, inducing leakage of phosphates, carbohydrates, and proteins. It inhibits only oomycota fungi.

Cymoxanil

Cymoxanil affects fungal metabolic processes. It inhibits nucleic acid and protein biosynthesis in fungal cells.

Quinoxyfen

Quinoxyfen specifically inhibits powdery mildew fungi. It inhibits appressorial formation.

Dimethomorph

Dimethomorph suppresses cell wall formation in fungi belonging to oomycota. However, *Pythium* spp. are not affected by this fungicide.

Fosetyl-Aluminum

Fosetyl-aluminum shows specific activity against fungi belonging to oomycota. This seems to be the only fungicide that moves through phloem. It inhibits sporangial formation and zoospore release. It alters fungal cell wall morphology. It may modify host defense mechanisms.

Dicarboximides

Dicarboximide and related compounds include iprodione, procymidone, vinclozolin, and chlozolinate. They inhibit conidial germination and mycelial growth. They induce swelling and bursting of hyphal cells. They alter lipid metabolism and DNA synthesis in fungal cells. They may cause mitotic instability. Dicarboximides are known to inhibit certain flavin enzymes such as NADPH-cytochrome c reductase. This action may result in an abnormal electron flow producing active oxygen species such as H_2O_2 and superoxide anion. It may lead to membrane destruction and various nonspecific toxic effects.

Aminoacid Amide Carbamates

Iprovalicarb belongs to the aminoacid amide carbamate group. It inhibits the growth of germ tubes of zoospores and sporangia, mycelial growth, and sporulation of oomycota fungi.

SPECTRUM OF DISEASES CONTROLLED BY INDIVIDUAL FUNGICIDES

Numerous fungicides have been commercially developed to manage fungal diseases. The efficacy of each fungicide in controlling crop diseases is given in the following list. The common name of the fungicide alone is given. The specificity of action of fungicides is normally restricted to the level of genera of the fungi and not to the species level. For example, a fungicide that is effective against the powdery mildew fungus *Erysiphe graminis* will be effective against all other species of *Erysiphe*, including *E. polygoni, E. cichoracearum*, and *E. pisi*. Similarly, a fungicide that is effective against the rust fungus *Puccinia graminis* will be effective against all other *Puccinia* species. Hence, the diseases are indicated only by the genera of the causal fungal pathogens. Furthermore, the specificity of a fungicide appears to be against the groups of fungal pathogens that cause similar symptoms. For example, a fungicide that is effective against one powdery mildew fungus, *E. graminis,* will be effective not only against other species of *Erysiphe,* but also against other genera of fungi causing powdery mildews, such as *Sphaerotheca, Uncinula, Leveillula, Oidium, Microsphaera, Podosphaera,* and *Phyllactinia.* Names of some specific diseases are also given in some places; this is mainly to indicate that the particular fungicide is highly useful in the control of the particular disease. The common names of the fungicides are given not according to importance but in alphabetical sequence.

Anilazine—controls early and late blight of potato and tomato; glume blotch of wheat; anthracnose in cucurbits; *Pyrenophora, Botrytis, Septoria, Colletotrichum, Cercospora, Helminthosporium,* and *Alternaria* diseases of various crops

Azoxystrobin—a broad spectrum fungicide; controls powdery mildews, rusts, and downy mildews; *Mycosphaerella, Phytophthora, Phomopsis, Pyrenophora, Septoria, Rhizoctonia,* and *Alternaria* diseases; also controls *Claviceps purpurea* and *Monographella nivalis* diseases; *Aphanomyces* root rot of peas; and radish and club root of cabbage

Benalaxyl—controls diseases caused by oomycete fungal pathogens such as *Phytophthora, Plasmopara, Sclerospora, Pseudoperonospora,* and *Pythium*

Benodanil—effectively controls rust diseases; *Rhizoctonia* diseases; *Typhula* in cereals

Benomyl—controls *Botrytis* diseases; powdery mildews; *Fusarium, Verticillium, Sclerotinia, Cercospora, Septoria, Gloeosporium, Colletotrichum, Venturia, Rhizoctonia, Macrophomina, Ascochyta, Cephalosporium, Claviceps, Diaporthe, Gibberella, Phomopsis,* and *Pyricularia* diseases; also controls *Plasmodiophora brassicae*

Benzothiadiazole—controls downy mildews; *Plasmopara halstedi*

Binapacryl—effective in control of powdery mildews

Bitertanol—controls scab and *Monilinia* diseases of apple and pear; apricot leaf scorch *(Gnomonia erythrostoma);* black spot of roses; Sigatoka of banana, powdery mildews and rusts of various crops; also controls *Fusarium, Myrothecium, Phaeoisariopsis,* and *Monographella* diseases

Bordeaux mixture—controls downy mildews; *Phytophthora, Alternaria, Cercospora, Colletotrichum, Pythium,* and *Venturia* diseases

Bromuconazole—controls *Microsphaera alphitoides;* used in cereals against *Pseudocercosporella herpotrichoides* and *Septoria tritici*

Bronopol—controls bacterial diseases caused by *Xanthomonas* and *Erwinia;* black arm of cotton; also controls *Botrytis* and *Fusarium* diseases

Bupirimate—controls powdery mildews; *Pleospora betae*

Buthiobate—controls powdery mildews

Butylamine—controls *Fusarium solani, Helminthosporium solani, Phoma exigua,* and *Polyscytalum pustulans* diseases

Captafol—controls downy mildew and black rot of vines; early and late blight of potato and tomato; controls scab of apple and pear; *Botrytis, Colletotrichum, Corticium, Fusarium, Rhizoctonia,* and *Sclerotinia* diseases; used as a seed treatment to control *Pythium* and *Phoma* spp.

Captan—controls scab of apple and pear; *Alternaria, Phytophthora, Pythium, Botrytis, Gloeosporium, Fusarium, Botryodiplodia, Diaporthe, Gibberella, Glomerella, Macrophomina, Phoma, Phomopsis, Sclerotinia,* and *Colletotrichum* diseases; brown rots of cherries, apricots, peaches, plums, and citrus fruits; mostly used as a seed treatment for the control of seedborne diseases

Carbendazim—controls *Cercospora, Septoria, Colletotrichum, Fusarium, Rhizoctonia, Macrophomina, Rhynchosporium, Pseudocercosporella, Botryodiplodia, Botrytis, Ceratocystis, Corticium, Diaporthe, Erysiphe, Gibberella, Glomerella, Mycosphaerella,* and *Pyricularia* diseases

Carboxin—seed treatment for control of smuts and bunts of wheat, barley, oats, and sorghum; controls *Rhizoctonia solani, Corticium rolfsii, Stagonospora,* and *Leptosphaerulina* diseases and a wide range of seed and seedling diseases

Carpropamid—systemic fungicide used as seed treatment and foliar spray for control of *Pyricularia oryzae* in rice

Chloralformamide—controls *Leptosphaeria* and *Ustilago* diseases

Chloroneb—mostly applied as a soil drench or as a seed treatment to control seedling diseases caused by *Rhizoctonia, Sclerotium,* and *Pythium;* snow mold of turf grass

Chloropicrin—used as a soil disinfectant for control of soil fungal pathogens

Chlorothalonil—a broad spectrum protectant fungicide; controls *Cercospora, Septoria, Alternaria, Fusarium, Phytophthora, Colletotrichum, Glomerella, Pseudocercosporella, Rhizoctonia, Rhynchosporium, Taphrina, Mycosphaerella, Botryodiplodia, Sclerotium, Stemphylium,* and *Botrytis* diseases; rusts and powdery mildews

Chlozolinate—controls *Botrytis, Sclerotinia,* and *Monilinia* diseases

Copper hydroxide—controls bacterial and fungal diseases; citrus canker; control of many diseases caused by *Phytophthora, Alternaria, Cercospora, Colletotrichum, Fusarium, Rhizoctonia,* and *Septoria;* downy mildews; apple scab; coffee rust; provides long residual control, will not readily wash off, and is highly economical

Copper oxychloride—controls many diseases caused by *Phytophthora, Alternaria, Cercospora, Colletotrichum, Fusarium, Rhizoctonia, Taphrina,* and *Septoria;* downy mildews; apple scab; coffee rust; asparagus rust; controls bacterial diseases; *Pseudomonas syringae* pv. *tabaci* in tobacco

Copper sulfate—controls Dutch elm disease by tree injection; phytotoxic when used alone, but used as a general fungicide when mixed with lime to form Bordeaux mixture

Cufraneb—control of downy mildews, early and late blights of potato, apple scab, and black arm of cotton

Cuprous oxide—control of downy mildews, rusts, early and late blights of potato and tomato, and leaf spots caused by *Alternaria, Cercospora, Colletotrichum,* and *Septoria*

Cycloheximide—controls powdery mildews in ornamentals; rust and leaf spot diseases of lawn grass

Cymoxanil—Specifically effective against fungi belonging to Peronosporales such as *Phytophthora, Plasmopara,* and *Peronospora;* controls *Phoma, Erysiphe, Alternaria,* and *Fusarium* diseases

Cyproconazole—controls rusts and powdery mildews; *Cercospora, Cochliobolus, Colletotrichum, Fusarium, Gibberella, Mycosphaerella, Pseudocercosoporella, Rhynchosporium,* and *Septoria* diseases; controls coffee rust

Cyprodinil—controls apple scab; *Botrytis, Sclerotinia, Leptosphaeria,* and *Monilinia* diseases; *Pyrenophora teres, Pseudocercosporella herpotrichoides, Rhynchosporium secalis,* and *Mycosphaerella graminicola* diseases of cereals

Cyprofuram—specifically effective against fungi belonging to Peronosporales such as *Pythium, Phytophthora, Plasmopara,* and *Peronospora*

Dazomet—soil sterilant; controls *Pythium, Rhizoctonia,* and *Sclerotinia* diseases

Dichlofluanid—controls scab of apple and pears; *Botrytis* diseases on grapes, tomato, and strawberry; downy mildews; *Alternaria* and *Phytophthora* diseases; reduces powdery mildews

Dichlone—controls scab on apples and pears, brown rot of stone fruit, and *Botrytis cinerea* diseases

Dichlorophen—controls *Fusarium* patch and dollar spot in turf

Diclobutrazol—controls powdery mildews in cereals; apple, cucurbits, and vines; rusts in cereals

Diclomezine—controls *Rhizoctonia* and *Sclerotium* diseases in rice, wheat, peas, and beans

Dicloran—highly effective against *Botrytis*; controls *Monilinia, Sclerotinia, Sclerotium,* and *Rhizopus* diseases

Diethofencarb—controls *Botrytis cinerea*

Difenoconazole—controls rusts, powdery mildews, and bunts; controls black as well as yellow Sigatoka and other leaf diseases in banana; diseases caused by *Septoria, Stagonospora, Leptosphaeria, Botrytis, Cercospora, Alternaria, Cochliobolus, Diaporthe, Colletotrichum, Mycosphaerella, Pseudocercosporella, Setosphaeria, Claviceps, Diaporthe, Venturia,* and *Fusarium*

Difulmetorim—effective against rose powdery mildew *(Sphaerotheca pannosa)* and chrysanthemum white rust *(Puccinia horiana)*

Dimethirimol—systemic fungicide, absorbed by the roots, with translocation in the xylem; soil application controls powdery mildews in cucurbits, tobacco, capsicum, and some ornamentals

Dimethomorph—controls *Phytophthora, Peronospora,* and *Plasmopara* diseases

Diniconazole—very effectively controls late leaf spot, early leaf spot, rust, and white mold on peanuts; gives excellent control of powdery mildews on apples and grapes; controls rusts, smuts, bunts, apple scab, and *Septoria* diseases of wheat; also controls *Rhizoctonia* diseases

Dinobuton—controls powdery mildews

Dinocap—controls powdery mildews in cucurbits, vines, tobacco, pome fruits, and stone fruits

Diphenylamine—controls apple scald

Disulfoton—controls coffee rust

Ditalimfos—controls powdery mildews in pome fruits, cucurbits, and ornamental plants; scab on apples and pears

Dithianon—broad spectrum protectant fungicide; controls scab on apple, pear, and cherries; downy mildews; rusts; leaf spot diseases caused by *Mycosphaerella, Ascochyta, Glomerella, Diplocarpon,* and *Didymella;* peach leaf curl; *Marssonia* leaf spot on poplars; *Monilia* diseases

DNOC—controls *Phomopsis viticola* on vines

Dodemorph acetate—controls powdery mildews on roses and other ornamental plants

Dodine—protective and curative fungicide; controls scab on apple and pear; leaf spot diseases of celery, cherries, and strawberries

Drazoxolon—controls powdery mildews on roses and blackcurrants; controls coffee rust, tea blister blight, and *Ganodema* disease on rubber; used as seed treatment for control of *Pythium* and *Fusarium* diseases

Epoxiconazole—controls rusts and powdery mildews, *Claviceps purpurea* on rye; *Pyrenophora, Colletotrichum, Glomerella, Fusarium, Leptosphaeria, Septoria, Rhizoctonia solani,* and *Monographella nivalis* diseases

Ethoxyguin—controls apple scald

Etridiazole—controls *Pythium* and *Rhizoctonia* diseases in cotton, vegetables, ornamentals, and turf

Fenamiphos—controls *Fusarium* and *Phytophthora* diseases

Fenarimol—very effectively controls powdery mildews; also controls apple and pear scab and *Pleospora betae*

Fenbuconazole—a protectant wettable powder fungicide for control of brown rot of plums and peaches, cherry leaf spot, and rusts

Fenhexamid—controls *Botrytis cinerea* and *Monilinia fructigena;* effective against brown rot and gray mold in sweet cherries

Fenpiclonil—controls *Fusarium, Phoma, Helminthosporium,* and *Pyrenophora* diseases; also controls *Monographella nivalis* and *Polyscytalum pustulans* diseases; effective against a bacterial disease caused by *Streptomyces;* as seed treatment controls seedborne diseases including silver scurf and stem canker in potatoes

Fenpropidin—effective against powdery mildews, rusts, and *Rhynchosporium secalis* of cereals

Fenpropimorph—controls powdery mildews and rusts; controls *Colletotrichum, Septoria, Rhynchosporium, Glomerella, Mycosphaerella, Pleospora, Pseudocercosporella,* and *Leptosphaeria* diseases

Fentin acetate—controls early and late blights of potatoes; *Cercospora, Septoria, Phytophthora, Monilia, Rhizoctonia, Cochliobolus, Helminthosporium, Pyricularia,* and *Venturia* diseases

Fentin hydroxide—controls *Cercospora, Alternaria,* and *Phaeoisariopsis griseola* diseases

Fluazinam—controls late blight of potato, *Rhizoctonia, Corticium, Sclerotinia, Leptosphaeria, Phaeoisariopsis, Plasmodiophora brassicae,* and *Rosellinia necatrix* diseases; controls the bacterial disease caused by *Streptomyces*

Fludioxonil—controls *Botrytis, Phytophthora, Fusarium, Sclerotinia,* and *Monographella nivalis* diseases; controls dollar spot, brown patch, pink and snow mold in turf; *Rhizoctonia* in ornamentals

Fluoromide—controls apple and pear scab, scab of citrus, powdery mildews, *Monilia* diseases, pink disease of rubber, *Alternaria* leaf spots, and anthracnose diseases

Fluotrimazole—controls powdery mildews of barley, cucumbers, grapes, and peaches

Fluquinconazole—controls many cereal diseases; effective in control of *Gaeumannomyces graminis* and *Mycosphaerella graminicola* in wheat

Flurprimidol—controls *Rhizoctonia solani*

Flusilazole—controls powdery mildews, rusts and scab of apples and pears; *Colletotrichum, Glomerella, Monilinia, Leptosphaeria, Mycosphaerella, Phaeoisariopsis,* and *Taphrina deformans* diseases

Flusulfamide—has bactericidal activity; controls the bacterial pathogens *Clavibacter michiganensis* ssp. *sepedonicus* and *Streptomyces scabies;* also controls the fungal pathogen *Plasmodiophora brassicae*

Flutolanil—controls rice sheath blight, potato black scurf *(Rhizoctonia solani),* cereal *Typhula* (snow blight) diseases, *Corticium* diseases, and turf diseases

Flutriafol—controls leaf and ear diseases in cereals; as a seed treatment, controls seedborne diseases of cereals; controls powdery mildews, rusts, bunts, and scab; *Botrytis, Alternaria, Cercospora, Cochliobolus, Colletotrichum, Glomerella, Leptosphaeria, Mycosphaerella,* and *Pleospora betae* diseases

Folpet—protectant fungicide; controls downy mildews, powdery mildews, and scab diseases; *Botrytis, Alternaria, Pythium, Rhizoctonia, Mycosphaerella,* and *Pestalotiopsis* diseases

Fosetyl-aluminium—controls diseases caused by fungi belonging to Peronosporales such as *Phytophthora, Pythium, Plasmopora,* and *Bremia; Fusarium oxysporum, Fusarium solani, Oidium mangiferae, Peronosclerospora sorghi,* and *Pseudoperonospora cubensis* diseases

Fthalide—effectively controls rice blast

Fuberidazole—controls *Fusarium culmorum* and *Monographella nivalis* diseases of cereals

Furalaxyl—controls soilborne diseases caused by *Pythium* and *Phytophthora*

Furmecyclox—useful seed treatment for control of *Ustilago* spp., *Helminthosporium* spp., *Septoria* spp., *Fusarium* spp., and *Rhizoctonia* spp. in cereals, potato, and cotton

Gramicidin S—controls *Sphaerotheca fuliginea*

Guazatine acetate—useful mainly as a seed treatment for the control of *Septoria nodorum, Tilletia caries, Fusarium* spp., and *Helminthosporium* spp. in cereals; controls *Cercospora* in groundnut and pine apple disease in sugarcane; as a postharvest treatment controls

Penicillium digitatum, P. italicum, and *Geotrichum candidum* in citrus and postharvest decay of pineapple

Hexachlorobenzene—effective seed treatment for control of common bunt and dwarf bunt of wheat

Hexaconazole—controls rusts, powdery mildews, and scab diseases in apples, pears, vines, vegetables, and other crops

Hexanal—controls *Botrytis cinerea* and *Penicillium expansum*

8-Hydroxyquinoline sulfate—controls *Botrytis cinerea;* also used as soil sterilant for control of soilborne diseases such as damping-off in seed beds of vegetables

Hymexazol—controls *Fusarium, Pythium, Corticium,* and *Aphanomyces* diseases

Imazalil—controls powdery mildews, *Fusarium culmorum, Fusarium solani, Gibberella fujikuroi, Helminthosporium solani, Cochliobolus sativus, Phoma exigua,* and *Pleospora betae* diseases; storage diseases of apple, pear, citrus, and banana; also used as a seed dressing, for control of diseases of cereals and cotton; used for control of silver scurf, gangrene, skin spot, and *Fusarium* dry rot on ware and seed potatoes

Imibenconazole—broad spectrum fungicide effective against diseases in fruits, vegetables, and ornamentals; controls apple scab, rust, powdery mildew, fly speck, sooty blotch, and *Alternaria* blotch; pear scab and rust; peach scab; apricot scab; grape anthracnose; citrus scab; peanut late leaf spot; tobacco powdery mildew; rose black spot, powdery mildew; chrysanthemum rust; and watermelon powdery mildew

Iminoctadine—controls *Botrytis cinerea* and *Sphaerotheca fuliginea* diseases; *Fusarium nivale* and *Fusarium roseum* diseases in wheat; rice blast and brown spot; apple scab, blotch *(Alternaria mali),* ring rot *(Botryosphaeria berengeriana),* and black spot *(Diplocarpon mali);* pear black spot, scab, and canker diseases; brown rot and scab of peach; cucumber powdery mildew; potato early blight; watermelon anthracnose, powdery mildew, and cottony rot *(Sclerotinia sclerotiorum)*

Ipconazole—controls rice fungal diseases

Iprobenfos—controls rice blast, sheath blight, and stem rot

Iprodione—contact fungicide, controls *Botrytis, Alternaria, Monilinia, Corticium, Fusarium, Helminthosporium, Phoma, Rhizoctonia, Sclerotinia, Septoria, Cochliobolus, Diaporthe, Phomopsis, Gibberella, Glomerella, Leptosphaeria, Pleospora, Uromyces, Pseudocercosporella, Pseudoperonospora,* and *Typhula* diseases; controls postharvest decay of peaches, sweet potatoes, and grapes; used as a seed treatment for control of seedborne fungi

Iprovalicarb—shows specific activity against oomycetes; controls *Plasmopara, Phytophthora,* and *Pythium* diseases

Isoprothiolane—controls blast and stem rot of rice

Kasugamycin—controls fungal and bacterial diseases; useful for the control of rice blast, tomato leaf mold, apple scab, sugar beet leaf spot *(Cercospora beticola),* and the bacterial diseases bean halo blight *(Pseudomonas phaseolicola)* and potato soft rot *(Erwinia carotovora);* also controls *Xanthomonas* and *Clavibacter* diseases

Kresoxim—controls *Sphaerotheca pannosa* in rose, *Venturia inaequalis* in apple, *Aphanomyces* root rot of peas and radish, and club root of cabbage

Kresoxim-methyl—controls powdery mildews and apple scab; *Mycosphaerella graminicola, Pyrenophora teres, Stagonospora nodorum,* and *Monographella nivalis* diseases in cereals; *Fusarium, Rhizoctonia, Septoria,* and *Myrothecium* diseases

Lime sulfur—controls powdery mildews, anthracnose, and scab

Mancozeb—a very efficient protectant fungicide, controls early and late blights of potato and tomato, downy mildews of grapevine, hops, cucurbits, lettuce, tobacco, and leeks, powdery mildews and rusts, scab of apple and pear, anthracnose of beans and cucurbits, and *Cercospora* leaf spots of banana, peanut, sugar beet, berries, and currants; controls *Botrytis, Cochliobolus, Colletotrichum, Corticium, Diaporthe, Drechslera, Fusarium, Gibberella, Glomerella, Magnaporthe, Mycosphaerella, Pestalotiopsis, Phoma, Phytophthora, Alternaria, Helminthosporium, Rhizoctonia, Pythium,* and *Sclerotinia* diseases

Maneb—controls early and late blights of potato and tomato, rusts on cereals, roses, asparagus, beans, carnations, plums, and currants; downy mildews of vines, onion, tobacco, hops, and ornamentals; scab of apples and pears, leaf spot diseases of beet, celery, and currants; anthracnose diseases of beans and tulips; needle cast in forestry

Mepanipyrim—controls *Botrytis cinerea*

Mepronil—controls rice sheath blight, black scurf of potato, damping-off diseases, *Typhula* snow blight of wheat, barley, and oats, and rust diseases

Mercuric oxide—used as a wound protectant for pruning cuts and other bark injuries on fruit trees and ornamental trees

Mercurous chloride—as a soil application it controls club root of brassicas and white rot in onion; controls dollar spot and *Fusarium* patch on turf

Metalaxyl—controls fungi belonging to Peronosporales; *Phytophthora, Plasmopara, Peronospora, Pseudoperonospora, Pythium,* and *Sclerospora* diseases; used as a seed treatment in many crops

Metalaxyl-M—one of the two enantiomers of metalaxyl; controls *Pythium* and *Phytophthora* diseases

Metam—controls *Rhizoctonia solani, Armillaria mellea,* and *Sclerotinia sclerotiorum*

Metconazole—controls *Rhizoctonia solani, Mycosphaerella graminicola, Fusarium culmorum, Septoria tritici, Rhynchosporium secalis,* and *Leptosphaeria nodorum* in cereals; also controls powdery mildews and rusts

Metham-sodium—used as a soil fumigant; controls soil fungal pathogens such as *Verticillium, Fusarium,* and *Rhizoctonia*

Methasulfocarb—applied as a dust into nursery soil before sowing; controls rice seedling blight caused by *Pythium, Fusarium, Rhizoctonia, Rhizopus,* and *Mucor*

Methoxyethylmercury acetate—useful as a seed dressing for controlling seed-borne diseases of wheat, barley, oats, and rye

2-Methoxyethylmercury chloride—controls *Ustilago scitaminea, Rhizoctonia solani,* and *Fusarium*

Methyl bromide—a fumigant; soil fumigation controls soilborne diseases; *Fusarium, Armillaria, Rhizoctonia, Sclerotinia,* and *Pythium* diseases

Methyl isothiocyanate—soil fumigant for control of soil fungal pathogens

Methyl jasmonate—induces systemic resistance against pathogens; controls *Botrytis cinerea*

Metiram—controls scab of apple and pear, rust on asparagus, currants, and plums, late blight of potato and tomato, downy mildews on vines, hops, tobacco, and lettuce; also controls *Alternaria, Botrytis, Rhizoctonia,* and *Leptosphaeria* diseases; as a seed treatment controls pre- and postemergence diseases of vegetables

Metominostrobin—controls rice blast

Monox—controls *Pythium aphanidermatum*

Myclobutanil—controls rusts and powdery mildews; *Diplocarpon rosae* and *Sphaerotheca pannosa* in rose; *Pythium, Rhizoctonia, Monilinia,* and *Thielaviopsis basicola* diseases; used as a seed treatment for control of seed- and soilborne diseases in wheat, rice, barley, corn, and cotton; controls postharvest diseases

Nabam—fungicide with protective action; controls soilborne diseases in cotton, capsicums, and onion when applied to soil

Nimbicidine—controls *Erysiphe polygoni*

Nitrothal-isopropyl—used in combination with other fungicides for control of powdery mildews and scab of pome fruits

Norspermidine—controls *Erysiphe graminis*

Nuarimol—controls powdery mildews in pome fruits, stone fruits, vines, hops, cucurbits, and cereals; apple scab, smuts; *Cercosporella* and

Septoria diseases of cereals; used both as a foliar spray and as a seed treatment

Octhilinone—has both fungicidal and bactericidal action; controls apple and pear canker *(Nectria galligena), Phytophthora,* and *Ceratocystis* diseases of citrus; citrus bacterial canker

Ofurace—commonly used in combination with other fungicides such as mancozeb, folpet, captafol, zineb, and maneb; controls downy mildews and *Phytophthora* diseases; dead arm *(Phomopsis viticola)* and black rot *(Guignardia bidwellii)* of grapevine

Oxadixyl—in combination with contact fungicides it controls downy mildews; controls *Alternaria, Fusarium, Gibberella, Phoma,* and *Phytophthora* diseases

Oxine-copper—as a seed treatment it controls bunt, glume blotch, and snow mold of wheat; controls *Cercospora, Alternaria, Phoma, Pythium, Botrytis,* and *Sclerotinia* diseases; also controls scab and canker diseases in apple and pears

Oxycarboxin—controls rust diseases, fairy rings in turf, and coffee rust

Paclobutrazol—controls *Diplodia* and *Rhizoctonia* diseases

Penconazole—highly effective against powdery mildews *(Erysiphe, Oidium, Pseudoperonospora)*

Pencycuron—nonsystemic fungicide with specific action against *Rhizoctonia solani* diseases such as rice sheath blight and black scurf of potato

Perfurazoate—as seed treatment controls blast, brown spot, and *Fusarium moniliforme*

Phenylmercury acetate—used as dry seed dressing; controls *Gibberella fujikuroi*

2-Phenylphenol—useful for the control of storage diseases of apples, plums, peaches, citrus fruits, tomato, and capsicums by impregnation of fruit wrappers, crates, etc., or by application as a wax directly to the fruit; used for disinfection of seed boxes; controls apple canker by application during the dormant period

Phosdiphen—controls rice blast

Pimaricin—controls diseases of bulbs, especially basal rot of daffodils

Polyoxins—antibiotics; control vegetable diseases

Primisulfuron—controls *Gibberella zeae*

Prochloraz—broad spectrum fungicide; controls rusts and powdery mildews; *Fusarium, Gibberella, Botrytis, Helminthosporium, Leptosphaeria, Mycosphaerella, Phoma, Pyrenophora, Colletotrichum, Alternaria,* and *Pleospora betae* diseases; also controls *Pyrenophora tritici, Pseudocercosporella herpotrichoide,* and *Erysiphe graminis* in wheat; peach leaf curl and rice blast, bakanae, and *Helminthosporium* diseases; as a liquid seed treatment controls seedborne diseases of linseed and flax

Procymidone—controls *Botrytis, Sclerotinia, Fusarium, Alternaria, Diaporthe, Glomerella, Phomopsis, Monilinia, Cladosporium cucumerinum,* and *Stemphylium* diseases

Propamocarb—systemic fungicide effective against oomycete fungi such as *Pythium, Phytophthora,* and downy mildew fungi

Propiconazole—controls brown rot, blossom blight, and fruit rot in stone fruits; powdery mildews, bunts, and rusts; also controls *Armillaria, Cercospora, Cochliobolus, Colletotrichum, Glomerella, Gaeumannomyces, Helminthosporium, Monilinia, Mycosphaerella, Phaeoisariopsis, Phomopsis, Pleospora, Pseudocercosporella, Rhizoctonia, Rhynchosporium, Septoria, Setosphaeria, Stagonospora,* and *Pyrenophora* diseases

Propineb—controls downy mildews, early and late blights of potato and tomato, scab on apple and pear; *Glomerella, Cercospora,* and *Rhizoctonia* diseases

Pyrazophos—controls powdery mildews, black Sigatoka of bananas, and leaf spot diseases in cereals

Pyrifenox—systemic fungicide with protective and curative activity against powdery mildews, scab of apple and pears, *Monilia* in top fruit, Sigatoka in banana

Pyrimethanil—controls *Botrytis cinerea, Venturia inaequalis,* and *Sclerotinia fuckeliana* diseases

Pyroquilon—controls rice blast

Quinomethionate—controls powdery mildews of cucurbits, citrus, tobacco, and pome fruits

Quinosol—controls *Phytophthora* diseases

Quinoxyfen—effective against powdery mildews

Quintozene—a contact fungicide; controls dollar spot, red thread, and snow mold in turf; controls *Pythium, Rhizoctonia, Sclerotium, Sclerotinia, Botrytis, Diaporthe, Phomopsis, Corticium,* and *Botryodiplodia* diseases; controls club root of cabbage, bunts of wheat, potato black scurf, smut and white rot of onions, and potato bacterial disease caused by *Streptomyces scabies*

Salicylic acid—induces resistance and reduces disease incidence; effective against fungal, bacterial, and viral diseases; controls fungal pathogens *Magnaporthe grisea, Fusarium oxysporum,* and *Diplodia pinea,* bacterial pathogen *Xanthomonas oryzae* pv. *oryzae,* and the virus *cucumber mosaic virus (cucumovirus)*

Spiroxamine—controls powdery mildews and rusts; *Pyrenophora teres, Rhynchosporium secalis, Septoria tritici, Fusarium culmorum,* and *Leptosphaeria nodorum* diseases in cereals

Strobilurin—controls *Leptosphaeria nodorum, Mycosphaerella graminicola, Podosphaeria, Septoria,* and *Venturia* diseases

Sulfur—controls powdery mildews; scab on apples, pears, and peaches; controls *Colletotrichum, Corticium, Elsinoe, Glomerella,* and *Taphrina* diseases

Tebuconazole—effectively controls powdery mildews, bunts, scabs, and rusts; diseases caused by *Alternaria, Ascochyta, Botrytis, Cercospora, Colletotrichum, Diplocarpon, Fusarium, Rhizoctonia, Monilinia, Mycosphaerella, Leptosphaeria, Gibberella, Nectria, Monographella, Monilinia, Phaeoisariopsis, Phomopsis, Phytophthora, Rhynchosporium, Septoria, Sclerotinia,* and *Setosphaeria*

Tecnazene—controls dry rot in seed potatoes and ware potatoes; inhibits sprouting; smoke formulations for control of *Botrytis* on tomato, chrysanthemum, and lettuce

Tetraconazole—controls rusts; *Cercospora* and *Cochliobolus* diseases

Thiabendazole—controls *Fusarium, Gibberella, Mycosphaerella, Phoma, Phomopsis, Aspergillus, Botrytis, Colletotrichum, Rhizoctonia, Sclerotinia, Septoria, Thielaviopsis, Verticillium,* and *Diaporthe* diseases; controls dollar spot and red thread in turf

Thiophanate—controls scab of apple and pear, powdery mildews, *Cercospora* leaf spots, *Verticillium* and *Fusarium* wilts; *Botrytis, Pyricularia, Rhizoctonia,* and *Sclerotinia* diseases

Thiophanate-methyl—a systemic broad spectrum fungicide; controls apple and pear scab and powdery mildews; effective against *Rhizoctonia* and *Fusarium* diseases; also controls *Colletotrichum, Diplocarpan, Gibberella, Glomerella, Phoma, Macrophomina, Leptosphaeria,* and *Nectria* diseases

Thiram—used mostly as a seed treatment to control damping-off and emergence diseases in various crops; controls scab and *Gloeosporium* rot in apples and pears, peach leaf curl, and rusts on ornamentals; also controls *Botrytis, Monilia, Cochliobolus, Corticium, Diaporthe, Fusarium, Gibberella, Glomerella, Leptosphaeria, Macrophomina, Mycosphaerella, Phoma, Phomopsis, Rhizoctonia, Sclerotinia, Stagonospora,* and *Thielaviopsis* diseases

Tolclofos-methyl—specifically useful in control of black scurf and stem canker in potatoes; used as a seed, bulb, or tuber treatment, soil drench, foliar spray, or by soil incorporation; controls soilborne diseases caused by *Fusarium, Rhizoctonia, Sclerotium,* and *Typhula;* also controls *Phoma, Helminthosporium, Corticium,* and *Gibberella* diseases

Tolyfluanid—controls scab of apples and pears, *Botrytis* on strawberries, raspberries, blackberries, currants, and grapes; also controls *Diaporthe, Gibberella,* and *Phomopsis* diseases; reduces powdery mildew incidence

Triadimefon—controls powdery mildews in cereals, vines, pome fruit, stone fruit, hops, and cucurbits; rusts in cereals, pines, coffee, and turf grasses; *Monilinia* spp. in stone fruit, black rot of grapes, snow mold in cereals

Triadimenol—controls rusts, powdery mildews, and *Rhynchosporium* in cereals; as a seed treatment controls smuts, bunts, *Pyrenophora, Cochliobolus, Mycosphaerella,* and *Typhula* in cereals; also controls *Diaporthe, Fusarium, Phomopsis, Rhizoctonia, Pseudocercosporella,* and *Thielaviopsis* diseases

Trichlamide—controls *Aphanomyces* root rot of peas and radishes; club root of cabbage; potato scab *(Streptomyces scabies)*

Tricyclazole—controls rice blast

Tridemorph—controls powdery mildews and rusts; Sigatoka disease of banana, blister blight of tea, pink disease of coffee and tea, and *Cercospora* leaf spot of coffee; also controls *Corticium, Cerotelium, Colletotrichum, Glomerella, Mycosphaerella,* and *Pestalotiopsis* diseases

Triflumizole—controls *Diplocarpon rosae* and *Sphaerotheca pannosa* var. *rosae* in rose; controls *Corticium, Fusarium, Gibberella, Helminthosporium, Phragmidium,* and *Rhizoctonia* diseases

Triforine—controls powdery mildews and rusts; controls turf diseases and scab in apples and pears

Triticonazole—controls rusts, powdery mildews, bunts, *Rhynchosporium secalis, Pyrenophora teres,* and *Cochliobolus sativus* in cereals

Validamycin A—controls rice sheath blight, black scurf of potato, and web blight of forestry seedlings caused by *Rhizoctonia solani*

Vinclozolin—controls *Botrytis* diseases in various crops; also controls *Diplocarpon, Sphaerotheca, Diaporthe, Alternaria, Corticium, Phomopsis, Leptosphaeria, Sclerotinia, Sclerotium,* and *Phragmidium* diseases

Zinc oxide—controls *Alternaria* and *Phytophthora* diseases in potato

Zineb—controls downy mildews, early and late blights in potato and tomato, anthracnose on beans, grapevine, and citrus fruits, scab on apples and pears, and *Cercospora* leaf spots on bananas and many other crops; also controls *Botrytis* diseases

Ziram—controls *Venturia, Monilia, Septoria, Cercospora, Colletotrichum, Alternaria,* and *Taphrina* diseases

FUNGICIDE APPLICATION

Fungicide Application Based on Computer-Based Decision Support Systems

Several fungicides and ready-formulated mixtures of fungicides are available. Even to control a single pathogen, several groups of fungicides are available. Some fungicides should be applied through soil, others should be applied through seed, and others should be applied on foliage to control diseases. For example, metominostrobin has been developed as controlled-release granules. The granules are applied to seedling boxes, and after two days, the rice seedlings are transplanted in the field. The concentration of metominostrobin in plants reaches its maximum within 24 hours. The activity of the fungicide persists even after 90 days and protects the plants against the blast pathogen *Magnaporthe grisea* (almost throughout the crop period). There is almost no need to apply more fungicide after a one-time application of the granule (Tashima et al., 1999). Some fungicides, such as Ferrax (flutriafol+ethirimol+thiabendazole) and Ridomil MZ (metalaxyl+ mancozeb), when applied as seed treatment protect the seedlings against airborne infection by pathogens for at least a few weeks. Recently, some long-term systemic protectants such as quinoxyfen and Bion have been developed. These long-lasting-protection fungicides are few in number and they have premium prices. Very few fungicides offer protection for more than two to three weeks. Hence, farmers must apply fungicides several times to manage diseases. Application of several rounds of fungicides as routine prophylactic sprays, even in the absence of disease, makes crop cultivation uneconomical. It is advisable to apply fungicides only when they are needed. Unnecessary sprays should be avoided, and by this, the farmer can save money.

A number of computer-based decision systems are available worldwide for use by farmers to decide the precise application date and the fungicide to be sprayed in the most effective way (Gleason et al., 1995). In Europe, the computer-based decision system EPIPRE (EPIdemiology, PREdiction, and PREvention) developed in the Netherlands is used to manage diseases of wheat crops (Zadoks, 1981). In Germany, the decision support system PRO_PLANT is widely used. It predicts diseases in wheat, barley, rye, triticale, sugarbeet, and potatoes and predicts application date and the fungicide to be sprayed (Frahm and Volk, 1994). In the United States, the system MoreCrop (Managerial Option of Reasonable Economic Control of Rusts and Other Pathogens) is used to manage wheat diseases (Cu and Line, 1994). Another system, WDCA (Wheat Disease Control Advisory), is used in Israel. An American predictive model, MARYBLYT, is used to predict the fireblight disease outbreak in apples (Lightner and Steiner, 1990). Firework, a fire blight predictive program, is used in New Zealand for management decisions in apple crops (Gouk et al., 1999). Other available computer-based decision support systems include DESSAC (Decision Support System for Arable Crops) and Counsellor to manage diseases of various crops, AU-Pnuts (Auburn University Peanuts) Advisory I and AUPNUT systems to manage peanut diseases, FAST (Forecasting *Alternaria solani* on Tomatoes), CU-FAST (Cornell University-FAST), and TOM-CAST (Tomato Forcaster) systems to manage tomato diseases, and the BLITE-CAST (Blight Forecaster) system to manage potato late blight. (For further details, see Chapter 16, "Forecasting Models".)

Although many decision support systems are available, they are still not very popular among farmers, which may be due to the high cost involved in purchasing such a system. Furthermore, they involve additional work for the farmers. The field must be regularly monitored for disease incidence, probably more than six times during the crop period. In some systems, farmers are asked to assess disease incidence in 40 to 100 tillers or more than 100 leaves selected at random across the field by counting lesions or pustules in each sample. In other systems, weather factors must be monitored regularly; even leaf surface wetness period must be recorded. The extra workload makes many farmers hesitant to follow a computer-based decision support system.

Instead of a personal computer system, messages can also be communicated through a postal system to interact with growers. This system works slowly, so a network system is often introduced instead. Interactive advisory systems, in which observations by farmers are sent in, the risks analyzed centrally, and recommendations are sent out, may help farmers to make decisions. Government institutions may provide a major advice service to the farmers. For example, Kansas State University conducts a plant disease survey regularly and gives the names of the fields surveyed and the disease severity in those fields. These reports are prepared every week. These data are transmitted through a computer-based electronic communication system. The data can be used by growers in making fungicide application decisions. In the United Kingdom, Agricultural Development and Advisory Service (ADAS) issues leaflets on the use of fungicides to help farmers. These data may also assist extension specialists in conveying the message to growers to make the decision to spray fungicides (Sim et al., 1988). Spray warnings by mail, radio, or answering service will also help farmers to make decisions to spray fungicides.

Fungicide Application Based on Disease Threshold

Need-based application of fungicides would be ideal, although it is difficult to practice. In this method, fungicides are applied only when the disease threshold level is reached in the field. Studies on determining the disease threshold level of each pathogen are important. Threshold values establish those population densities which, when exceeded, lead to commercial (economic) loss. This threshold defines the optimal time for applying fungicide. The effectiveness of fungicide treatments depends on the flexibility and adaptability of pathogen-specific threshold values, as well as on cultural and environmental influences. Specific epidemiological control thresholds for the most important pathogens of wheat (*Stagonospora nodorum, Septoria tritici, Puccinia striiformis, Erysiphe graminis* f. sp. *tritici, Pseudocercosporella herpotrichoides,* and *Drechslera tritici-repentis*) have been developed through various field trials (Verreet et al., 2000). For example, the disease threshold level of *Erysiphe graminis* was calculated as follows: At weekly intervals, disease incidence was calculated from a sample of 30 plants. The first threshold of 70 percent disease incidence is reached if mildew pustules have been diagnosed on 21 of 30 plants in the sample. The second threshold of 70 percent disease incidence is reached if, on the topmost green leaves, an average of 21 of 30 plants show at least one powdery mildew pustule. The fungicide is sprayed whenever the threshold level is reached (Verreet et al., 2000). The use of fungicide should be avoided when the threshold value is not exceeded. This fungicide application strategy will be highly economical and useful in the control of diseases. However, this method is highly knowledge intensive, and many farmers may not be able to follow it. In fact, according to a published report, many farmers in the United Kingdom were not able to correctly diagnose disease symptoms in wheat. Only 13 percent of farmers could correctly recognize early yellow rust symptoms, and only 24 percent of farmers could recognize glume blotch (Hewitt, 1998).

Routine Prophylactic Fungicide Application

Prophylactic application involves routine preventative fungicide applications at set intervals. There is no monitoring of disease, and fungicides may be applied irrespective of disease levels. This method is very simple, and there is no need for planning. It would be ideal in crops in which a particular disease occurs regularly during the particular season. For example, late blight occurs regularly in potato, and five or more routine rounds of prophylactic sprays will provide effective control of the disease. When a greater number of diseases occur in a crop, ready-formulated mixtures of fungicides can be used. Fungicides with long-lasting protection will be ideal for this method of application. However, these fungicides are normally more expensive.The major disadvantage of this method is that fungicides may be used even when there is little or no risk of a disease outbreak, so money may be wasted. For example, weekly sprays of captan or thiram starting from early in the season did not reduce the incidence of fruit rot caused by *Botrytis cinerea* in strawberry until at least the fourth week of harvest, nine to ten weeks after application began. Bloom applications of iprodione applied twice during the second peak flowering period controlled the disease effectively. However, bloom application of iprodione during the first flowering period did not control the disease effectively. This suggests application of fungicides at the correct stage of the crop is important in disease control (Legard et al., 2001). Another drawback is that an unexpected disease epidemic may occur between sprays when the crop is effectively left unsprayed.

FUNGICIDE RESISTANCE

What Is Fungicide Resistance?

After introduction of systemic fungicides, fungicide resistance has been observed in various plant pathogens. Fungi that are sensitive to a particular fungicide becomes less sensitive to that fungicide, which is called *fungicide resistance*. Fungicide resistance has been reported in *Phytophthora infestans, Peronospora parasitica,* and *Bremia lactucae* against phenylamides (metalaxyl compounds); *Ustilago nuda* and *Ustilago maydis* against carboxamides (carboxin compounds); *Erysiphe graminis* f. sp. *hordei* against hydroxypyrimidine; *E. graminis* f. sp. *tritici, E. graminis* f. sp. *hordei,* and *Pyrenophora teres* against triazoles (propiconazole); and *Botrytis cinerea, Botrytis fabae, Septoria tritici, Pyrenopeziza brassicae, Cercospora beticola, Fusarium nivale, Fusarium culmorum,* and *Pseudocercosporella herpotrichoides* against benzimidazoles (carbendazim, benomyl, thiophanate-methyl compounds). There are only a few reports of resistance to surface-protectant fungicides such as copper fungicides, organomercurials, organotin, and dithiocarbamates.

Fungi may become resistant to fungicides through several mechanisms. The fungicide may act at a particular site in fungi. A change at the site of inhibitor action may result in a decreased affinity to the fungicide. Decreased uptake or decreased accumulation of the fungicide in the fungus may result in fungicide resistance. Sometimes increased production of an enzyme that is normally inhibited by the fungicide may compensate for the inhibitory effect of the fungicide. The fungicide may also be detoxified before it reaches the site of action, resulting in fungicide resistance in fungi. The fungicide will be normally converted into a toxic principle to inhibit the fungi. However, at certain conditions, the fungicide would not have been converted into the toxic principle, and this will result in fungicide resistance in fungi. Fungicides may block a site in an important metabolic pathway in fungi. When circumvention of the blocked site by the operation of an alternative pathway occurs, fungicide resistance can be observed in fungi.

How Fungicide-Resistant Pathogens Develop in the Field

Fungicide-resistant isolates may arise by mutation of a single gene. Systemic fungicides usually act at a single site, so resistance to systemic fungicides occurs frequently. It is unlikely that two or more mutations will occur simultaneously in fungi so that the fungi become resistant to multisite fungicides (mostly protectants). The fungicide-resistant cells may appear spontaneously in the field, normally at the rate of 1 in 10^{-4} to 10^{-9}, even when no fungicides are applied in the field. When the fungicide is applied to the crop, all but resistant strains of the pathogen may be killed. This will allow the resistant strains to multiply. However, it has been demonstrated that the resistant strains do not survive for a long time. By the time the effects of the fungicide diminish, probably within a fortnight, the sensitive (wild) strains reappear as the dominant partners in the pathogen population. However, resistant strains may continue to predominate if the same fungicide is repeatedly applied at short intervals. During the constant presence of the fungicide, the sensitive isolates may not have time to reestablish. Resistant strains may multiply faster than the sensitive wild isolates which may result in a condition in which control of the disease with the particular fungicide is no longer effective in the field.

Development of Cross-Resistance in Fungal Pathogens

When a fungal pathogen develops resistance to a fungicide with a particular mode of action, it may show resistance to other fungicides with the same mode of action. This phenomenon is called *cross-resistance*. The phenylamide fungicides metalaxyl, benalaxyl, ofurace, and oxadixyl show a similar mode of action against fungi. The isolates of *Phytophthora infestans* resistant to metalaxyl show resistance to other phenlamides, benalaxyl, ofurace, and oxadixyl. Sometimes resistance to one fungicide may result in increased sensitivity to another fungicide. This is called negative cross-resistance. *Blumeria graminis* isolates, which show resistance to triazoles, show increased sensitivity to ethirimol compared to the wild isolates without any resistance to the fungicides.

Management of Fungicide Resistance in Fungal Pathogens

The following strategies are suggested to avoid development of fungicide resistance in fungal pathogens:

1. Consecutive applications of fungicides with the same mode of action should be avoided.
2. Seed treatments with systemic fungicides should not be followed by foliar sprays of fungicides with the same mode of action.
3. The spray schedule should include some multisite fungicides.
4. Fungicides should be sprayed only when necessary. Fungicides should not be sprayed when there is little disease. Very early and late applications should be avoided as much as possible. Prophylactic sprays should be avoided; if this is not possible, prophylactic sprays should be restricted to three sprays.
5. Instead of spraying a fungicide with a single active ingredient, a fungicide with a mixture of active ingredients should be used. The risk of the development of resistant strains is significantly reduced if active ingredients with two modes of action are mixed. Farmers themselves can mix fungicides using the compatibility chart provided by the manufacturers. However, many pesticide-manufacturing companies now produce ready-formulated mixtures. These mixtures consist of site-specific fungicides with different modes of action or multisite fungicides.

REFERENCES

Cu, R. M. and Line, R. F. (1994). An expert advisory system for wheat disease management. *Plant Dis,* 78:209-215.

Ellis, M. C. B. and Tuck, C. R. (2000). The variation in characteristics of air-included sprays with adjuvants. *Aspects of Applied Biology,* 57:155-162.

Frahm, J. and Volk, T. (1994). PRO-PLANT—A computer-based decision support system for cereal disease control. *EPPO Bulletin,* 23:685-694.

Gleason, M. L., MacNab, A. A., Pitblado, R. E., Ricker, M. D., East, M. D., and Latin, R. X. (1995). Disease-warning systems for processing tomatoes in Eastern North America: Are we there yet? *Plant Dis,* 79:113-121.

Gouk, S. C., Spink, M., and Laurenson, M. R. (1999). FireWork—A windows-based computer program for prediction of fire blight on apples. *Acta Horticulturae,* 489:407-412.

Grayson, B. T., Webb, J. D., Batten, D. M., and Edwards, D. (1996). Effects of adjuvants on the therapeutic activity of dimethomorph controlling vine downy mildew. I. A survey of adjuvant types. *Pesticide Science,* 46:199-206.

Hewitt, H. G. (1998). *Fungicides in Crop Protection.* CAB International, University Press, Cambridge, U.K.

Knight, S. C., Anthony, V. M., Brady, A. M., Greenland, A. J., Heaney, S. P., Murray, D. C., Powell, K. A., Schultz, M. A., Spinks, C. A., Worthington, P. A., and Youle, D. (1997). Rationale and perspectives on the development of fungicides. *Annu Rev Phytopathol,* 35:349-372.

Kwon, Y. W., Lee, J. K., and Chung, B. J. (1989). Interaction of adjuvants with rice leaf surface in spraying fungicide. In P. N. B. Chow (Ed.), *Adjuvants and Agrochemicals.* CRC Press, Boca Raton, FL, pp. 63-74.

Legard, D. E., Xiao, C. L., Mertely, J. C., and Chandler, C. K. (2001). Management of Botrytis fruit rot in annual winter strawberry using captan, thiram, and iprodione. *Plant Dis,* 85:31-39.

Lightner, G. W. and Steiner, P. W. (1990). Computerization of blossom blight prediction model. *Acta Horticulturae,* 273:159-162.

McCallan, S. E. A. (1967). History of fungicides. In D. C. Torgeson (Ed.), *Fungicides.* Academic Press, New York, pp. 1-37.

Ragsdale, N. N. and Sisler, H. D. (1991). The nature, modes of action, and toxicity of fungicides. In D. Pimentel (Ed.), *CRC Handbook of Pest Management in Agriculture, Volume II.* CRC Press, Boca Raton, FL, pp. 461-496.

Research Information Ltd. (1998). *International Pesticide Directory,* Sixteenth Edition. Research Information Ltd., Herts, U. K.

Richards, M. D., Holloway, P. J., and Stock, D. (2000). Effects of some polymeric additives on spray deposition, coverage, and deposit structure. *Aspects of Applied Biology,* 57:185-192.

Riehm, E. (1914). Prüfung einiger Mittel Zur Bekämpfung des Steinbrandes, Zentbl. *Bakt ParasitKde (Abt.11)* 40:424.

Robertson, J. (1824). On the mildew and some other diseases incident to fruit trees. *Trans Hort Soc London,* 5:175-185.

Sim, T. IV, Willis, W. G., and Eversmeyer, M. G. (1988). Kansas disease survey. *Plant Dis,* 72:832-836.

Stock, D. (1996). Achieving optimal biological activity from crop protection formulations: Design or chance? In *Proceedings of the Brighton Crop Protection Conference 1996—Pests and Diseases, Volume 3.* British Crop Protection Council, Farnham, U.K., pp. 791-800.

Tashima, S., Kawaguchi, A., Kashino, H., Matsumoto, K., Ando, I., Takeda, R., and Shiraishi, T. (1999). Long-term control of rice blast disease by seedling box application of controlled release granules containing metominostrobin. *J Pesticide Sci,* 24:287-289.

Thieron, M., Pontzen, R., and Kurahashi, Y. (1998). Carpropamid: A rice fungicide with two modes of action. *Pflanzenschutz Nachrichten Bayer,* 51:259-280.

Verreet, J. A., Klink, H., and Hoffmann, G. M. (2000). Regional monitoring for disease prediction and optimization of plant protection measures: The IPM wheat model. *Plant Dis,* 84:816-826.

Waxman, M. F. (1998). *Agrochemical and Pesticide Safety Handbook*. Lewis Publishers, Boca Raton, FL.
Zadoks, J. C. (1981). EPIPRE: A disease and pest management system for winter wheat developed in the Netherlands. *EPPO Bulletin,* 11:365-369.

24

Chemical Control—
Fungicide Application Equipment

Application equipment is needed to get the fungicide to the target crop. If proper application equipment is not used, even the most efficient fungicide will not be able to control the disease. Fungicides are available as dustable powder, wettable powder, liquids, emulsifiable concentrate, oil formulations, ultralow-volume formulations, and gases. Fungicides need to be applied in dwarf crops such as peanut as well as in large-sized tall trees such as apples. Hence, different types of application equipment are needed to apply different formulations of fungicides to different types of crops. This chapter describes all available application equipment.

DUSTERS

Hand Dusters

These dusters are primarily used by home gardeners. They may consist of a squeeze bulb, bellows, tube, sliding tube, shaker, or a fan powered by a hand crank. When the dusters are used it will be difficult to control drifting, causing wastage of fungicides. However, they are easy to operate. Because no mixing with water is needed, the user does not need to make calculations and prepare solutions.

Power Dusters

Power dusters contain a powered fan or blower to propel the fungicide dust to the target. Several types of power dusters are available. Knapsack types, backpack types, and units mounted on or pulled by tractors are available.

SPRAYERS

Push-Pull Hand Sprayer

Hand sprayers operate on compressed air, which is supplied by a manually operated pump. They are primarily used in home gardens. Push-pull hand sprayers contain a hand-operated plunger which forces air out of a cylinder, creating a vacuum at the top of a siphon tube. The suction draws fungicide from the small tank and forces it out with the airflow. The capacity is approximately one quart.

Hose-End Hand Sprayer

In this type of sprayer, a fixed rate of fungicide is mixed with water flowing through the hose to which the sprayer is attached. The mixture is expelled through a high-volume nozzle. The capacity of these sprayers is approximately one quart, but because the concentrate mixes with water, they may deliver about 20 gallons of finished spray solution.

Trigger-Pump Sprayer

The fungicide is mixed with the required quantity of water in the sprayer and forced through the nozzle when the pump is squeezed. Trigger-pump sprayers are useful for applying only small quantities of fungicide and are intended for small jobs.

Backpack Sprayer

This sprayer is similar to a push-pull sprayer. However, it contains a self-contained tank and a pump. A mechanical agitator is attached to the pump plunger. The sprayer is carried on the operator's back.

Compressed Air Sprayer

This is a hand-carried sprayer that contains a self-contained manual pump, which creates pressure and operates the sprayer. The capacity of the sprayer is approximately one to three gallons.

Bucket (or Trombone) Sprayer

This sprayer contains a double-action hydraulic pump, which is operated with a push-pull motion. The fungicide is sucked into the cylinder and pushed out through the hose and nozzle with a stroke. The separate tank is a small bucket.

Estate Sprayer

This sprayer is mounted on a two-wheeled cart with handles for pushing. The tank capacity ranges from 15 to 50 gallons. The pump delivers one to four gallons per minute at pressures ranging from 250 to 400 psi. Power is supplied by an air-cooled engine of up to five horsepower.

Power Wheelbarrow Sprayer

This is similar to estate sprayer, but it is driven by power wheel.

Low-Pressure Boom Sprayer

This type of sprayer is mounted on tractors or trucks. Tank size ranges from 50 gallons to 1,000 gallons. The sprayer delivers ten to 60 gallons per acre at pressures ranging from ten to 80 psi. The sprayer is equipped with sprayer booms ranging from ten to 60 feet in length. The booms contain many nozzles.

Low-Pressure Boomless Sprayer

This sprayer does not contain the boom. It contains a central nozzle cluster that produces a horizontal spray pattern. This sprayer can move through narrow places and avoid trees and other obstacles, unlike the boom sprayer.

High-Pressure Hydraulic Sprayer

This sprayer is useful to spray through dense foliage, to the tops of tall trees. The sprayer is mounted on a tractor or truck or is self-propelled. It delivers about 60 gallons per minute, and the application rate is about 100 gallons per acre. The sprayer contains a large tank with a capacity of 500 to 1,000 gallons to cover large areas. The sprayer is also fitted with a boom.

Air-Jet (Air-Blast) Sprayer

This sprayer uses a combination of air and liquid. It is similar to high-pressure sprayers but also contains a high-speed fan. The air jet shatters the drops of fungicide into fine droplets and transports them to the target. This sprayer is usually trailer mounted. An air-jet sprayer covers a swath up to 90 feet wide and reaches trees up to 70 feet tall. A device that permits precision spraying in orchards with an air-blast sprayer has been developed (Balsari and Tamagnone, 1997). A computer commands the appropriate valves, which control nozzle output according to ground speed, row spacing, and tree shape. Only the necessary number of nozzles is turned on to cover a particular size.

Tunnel Sprayer

A tunnel sprayer is a type of air-jet sprayer. Two types of tunnel sprayers have been developed. The ISK-2 tunnel sprayer contains eight air jets which are directed 40° upward. The OSG-NI tunnel sprayer contains two horizontal jets blown by cross-flow fans (Holownicki et al., 1997). The best distribution of spray deposits on leaf surfaces is obtained with the ISK-2 tunnel sprayer. The best practical method for economical fungicide application in orchards is the tunnel sprayer with sensors. The spray-liquid recirculation system in tunnel sprayers reduces fungicide consumption by 20 to 40 percent (Holownicki et al., 1998). Sensors identify the objects to be sprayed.

Spinning Disc Sprayer

This sprayer contains a special type of nozzle, which spins at a high speed and breaks the liquid into uniformly sized droplets by centrifugal force. A small electric or hydraulic motor provides the power to spin. Fungicide droplets are carried to the target crop canopy by gravity or by an airstream. Different sizes of the sprayers are available. Sizes range from a small hand-held sprayer to a large tractor- or trailer-mounted sprayer. The sprayer produces a narrower range of droplet sizes, thus reducing drift.

Electrostatic Sprayer

The fungicide is charged with a positive electric charge as it leaves the nozzles. Plants have a natural negative charge. Hence, the fungicide is attracted to the plants. The spray is directed through or above the crop. The electrostatic sprayer system reduces drift. Because the fungicide is directed only to the crop plants, a reduced quantity of the fungicide is sufficient to cover an acre.

Ultralow-Volume Sprayer

This sprayer sprays specially prepared concentrated fungicides. It can be handheld or mounted on ground equipment or on aircraft.

GRANULE APPLICATORS

Some fungicides are available in granular form. Several types of granule applicators have been developed. They have been designed to apply granules as broadcast with even distribution over the entire area, to apply them by drilling (soil injection or soil incorporation), or to apply them in furrows, in bands, or as side dress. They distribute the granules by forced air, by spinning or whirling discs, or by soil injectors.

SEED TREATERS

Seed treaters are used to coat seeds with seed-treatment fungicide. Four types of seed treaters are available.

Dust Treater

Seed is treated with a dust formulation of the fungicide in this treater. It mixes fungicide with seeds in a mechanical mixing chamber. Seed flow is controlled by adjusting the gate opening on the feed hopper. The amount of fungicide added is controlled by a vibrating feeder, which can be adjusted to achieve the desired dosage.

Slurry Treater

This treater is used to coat seeds with wettable powder formulations of fungicides in the form of a slurry. A specific amount of fungicide is added to a specific weight of seeds in a mechanical mixing chamber. A small amount of water is added. Agitators keep the material mixed during the treating operation. The capacity of slurry tanks ranges from 15 to 35 gallons.

Panogen Liquid Treater

A small amount of liquid fungicide is added to a large quantity of seeds and mixed well in this treater. Some treaters may have dual tanks so that seeds can be treated with more than one fungicide at a time. The Panogen treater meters one treatment cup of fungicide per dump of seed into a revolving mixing drum. The fungicide flows into the drum from a tube and is distributed over the seeds as the seeds rub against the walls of the drum.

Mist-O-Matic Seed Treater

Fungicide is applied as a mist onto the seed in this seed treater. The treater delivers one treatment cup of fungicide per dump of seed. The fungicide flows onto a rapidly whirling disc, which breaks the liquid fungicide into a fine mist. The seeds fall onto a large cone, which spreads the seeds out so that they are evenly coated with the fungicide spray mist.

SOIL APPLICATION EQUIPMENT

Soil Fumigators

Some fungicides are available in gaseous form. They are applied in soil to control soil-borne pathogens. These fungicides are applied to soil by different methods, such as soil injection, soil incorporation, or drenching, using different types of applicators.

Soil Injectors

Soil injectors use a variety of mechanisms to insert the fungicide into the soil to a depth of about six inches. After injection, the area is covered with soil to seal in the applied fungicide.

Soil Incorporators

The fumigant fungicide is sprayed onto the soil surface by using soil incorporators. The applied area is immediately plowed to a depth of about five inches using power-driven rotary cultivator and then compacted with a drag, or float, or cultipacker.

OTHER TYPES OF EQUIPMENT

Fruit Tree Injector

A portable high-pressure fruit tree injector has been developed. It injects liquid fungicides into the trunk of fruit trees to control fruit-tree diseases. The injector can be used even for large trunks (Liu et al., 1998).

Drenching Equipment

The fungicide is added to water and applied as a drench. It may be applied using a sprinkling can, sprinkler system, or irrigation equipment. The fumigant fungicide can also be sprayed on the soil surface, immediately followed by flooding. Here the water acts as a sealant, and the depth of the water seal ranges from one to four inches of wetted soil.

Pressure-Fed Applicator

This applicator is useful to apply liquid fumigant fungicide. The applicator has a pump and metering device and delivers pressured liquid fumigant to the nozzle opening (orifice).

Gravity-Fed Applicator

In this applicator, the pressure is created by gravity to regulate the output of the liquid fumigant. A constant gravity flow device keeps the pressure at the orifice constant as the tank of fumigant empties. A constant speed is necessary to maintain a uniform delivery rate. Needle valves, orifice discs, and capillary tubes are used to adjust the flow rate.

High-Pressure Fumigator

This applicator is used to apply highly volatile fumigants. The pressure in the tank maintains the pressure at the nozzle orifices. The tank is precharged with sufficient pressure to empty its contents, or an inert pressurized gas is fed into the tank during application to displace the fumigant. A gas-pressure regulator maintains uniform pressure in the system. After application of a highly volatile fumigant, the soil should be sealed with vapor-proof tarps.

PARTS OF APPLICATION EQUIPMENT

Components of application equipment determine the efficacy of fungicide application. The important components of the sprayers are the tank, pump, strainer, hose, pressure gauge, pressure regulator, agitator, control valve, and nozzle. The tank should be made of a corrosion-resistant material. Different types of pumps are used, including roller pumps, gear pumps, diaphragm pumps, piston pumps, and centrifugal pumps. Roller pumps deliver eight to 30 gallons per minute at low to moderate pressure (10 to 300 psi). They are often used on low-pressure sprayers. Gear pumps deliver 5 to 65 gpm at low to moderate pressures (20 to 100 psi). They are often used on special-purpose sprayers. Diaphragm pumps deliver low volume (3 to 10 gpm) at low to moderate pressure (10 to 100 psi). Diaphragm pumps are positive-displacement, self-priming pumps. Centrifugal pumps deliver high volume (about 200 psi) at low pressures (5 to 70 psi) and are not positive-displacement pumps. Piston pumps are used in high-pressure pumps. They are positive-displacement, self-priming pumps, delivering two to 60 gallons per minute at low to high pressures (20 to 800 psi).

Strainers filter the fungicide mixtures and remove dirt, rust flakes, and other foreign materials from the tank mixture. Proper filtering helps to protect the working parts of the sprayer from undue wear and avoids uneven application caused by clogged nozzle tips. Hoses should have a burst strength greater than the peak operating pressures. They should resist oil and solvents present in fungicides. Suction hoses should be larger than pressure hoses. Suction hoses should be reinforced to resist collapse. Pressure gauges monitor the function of the spraying system and must be accurate. The pressure regulator controls the pressure and the quantity of spray material delivered by the nozzles. Three types of pressure regulators are available. Throttling valves are used with centrifugal pumps. Spring-loaded bypass valves are used with diaphragm, roller, gear, and small piston pumps. They open or close in response to changes in pressure. Unloader valves are used on larger piston and diaphragm pumps to avoid damage to the pump when the nozzles are cut off.

Agitators help to keep the spray material uniformly mixed. Several types of agitators are used in the sprayers. Bypass agitators use the returning liquid from the pressure relief valve to agitate the tank. Bypass agitation is used to agitate soluble powders, emulsifiable concentrates, and liquids. They are not useful for wettable powders. Hydraulic agitators are used for wettable powder and flowable formulations and for liquid formulations in 100-gallon or larger tanks with gear, piston, roller, or diaphragm pumps. Mechanical agitators are used on large, high-pressure hydraulic sprayers. They help to mix wettable powder formulations. The mechanical agitator consists of flat propellers mounted on a shaft, which is placed lengthwise along the bottom of the tank. The propellers are rotated by the engine to keep the material well mixed.

Control (cutoff) valves are located between the pressure regulator and the nozzles to provide positive on-off action. Mechanical or electrically operated valves are used. Mechanical valves are within reach of the operator's hand. Electrical operators permit remote control of fungicide flow.

Nozzles are very important for effective spraying of fungicides. They are made up of the nozzle body, cap, strainer, and tip or orifice plate. The nozzle body holds the strainer and tip in proper position. The cap is used to secure the strainer and the tip to the body. Several types of tips that produce a variety of spray patterns can be interchanged. Nozzle tips provide the desired type of spray. Solid-stream nozzles are used to spray in a narrow band or inject fungicides into the soil. Fan-pattern nozzles are used for spraying uniformly to cover surfaces. The regular flat-fan nozzle tip makes a narrow oval pattern. The even net-fan nozzle is used for band spraying. The flooding nozzle delivers wide-angle net-spray droplets. Cluster nozzles are used to extend the effective swath width. Cone-pattern nozzles are used to apply fungicides to foliage. The side-entry hollow-cone (whirl-chamber) nozzle produces a very wide-angle hollow-cone spray pattern at very low pressures. Core-insert cone nozzles produce a solid or hollow-cone spray pattern and operate at moderate pressures. Disc-core nozzles produce a cone-shaped spray pattern, which may be solid or hollow. Adjustable-cone nozzles change their spray angle from a wide cone pattern to a solid stream when the nozzle collar is turned. Nozzle parts may be made of brass, aluminium, stainless steel, tungsten carbide and ceramic, or plastic. Relative wear of the nozzle orifice for different materials varied considerably with usage duration. Nozzles made of hardened stainless steel were most resistant to wear, followed by stainless steel, plastic, and brass. Nozzles with lower flow capacities had a higher percentage of wear than nozzles with higher flow capacities (Duvnjak et al., 1998).

A novel air-delivery nozzle, called Shear-Guard PLUS nozzle, with Dial-A-Drop spray technology allows the applicator to precisely and instantly dial in droplet size while on the go (McCracken et al., 1998).

REFERENCES

Balsari, P. and Tamagnone, M. (1997). An automatic spray control for airblast sprayers: First results. In J. Stafford (Ed.), *Precision Agriculture '97, Volume II.* Bios Scientific Publishers, Oxford, U.K., pp. 619-626.

Duvnjak, V., Banaj, D., Zimmer, R., and Guberac, V. (1998). Influence of nozzle wear on flow rate and stream droplets size. *Bodenkultur,* 49:189-192.

Holownicki, R., Doruchowski, G., and Jaeken, P. (1998). Economical techniques of orchard protection. *Inzynieria Rolnicza,* 5:269-278.

Holownicki, R., Doruchowski, G., and Swiechowski, W. (1997). Uniformity of spray deposit within apple tree canopy as affected by direction of the air-jet in tunnel sprayers. *J Fruit and Ornamental Plant Research,* 5:129-136.

Liu, D.L., Jian, X. C., Zhang, X. L., Zhang, L. X., and Zuo, B. Y. (1998). Application of a portable high pressure fruit tree injector. *AMA, Agricultural Mechanization in Asia, Africa, and Latin America,* 29(4):29-31.

McCracken, T., Bennett, A., and Jonasson, K. (1998). Improved spray efficacy and drift management for spray application using air to simultaneously atomize and propel spray droplets to the crop canopy. *ASAE Annual International Meeting, Orlando, Florida, U.S.A.* American Society of Agricultural Engineers, St. Joseph, MO, p. 12.

25

Chemical Control—Virus Diseases

Chemical control of virus diseases is almost not practical. However, the vectors of viruses can be controlled to a certain extent to reduce the disease spread. The conditions necessary for vector management are discussed in this chapter. Use of plant activators to manage virus diseases is also described.

CONTROL OF VIRUS SPREAD BY VIRICIDES

Chemical control of virus diseases is difficult to achieve. Only ribavirin has been shown to reduce virus diseases. Ribavirin is effective against *Potato S virus (Carlavirus)* and *Odontoglossum ringspot virus (Tobamovirus)* (Research Information Ltd., 1998). Virazole (ribavirin) at 25 mg/liter eliminated *Apple chlorotic leaf spot virus* (ACLSV) from treated apple and pear shoots. However, ribavirin at a concentration of 100 mg was phytotoxic (Cieslinska and Zawadzka, 1999).

CONTROL OF VIRUS SPREAD BY INSECTICIDES

Semipersistently and persistently transmitted viruses are acquired and inoculated only after long feeding probes (generally hours to days). Plants susceptible to infection often are colonized by the vector (Perring et al., 1999). In these cases, spread of virus diseases can be efficiently controlled by insecticides. For a nonpersistently transmitted virus, in which acquisition and inoculation can occur in a matter of seconds, the source and inoculation plants generally are not considered to be hosts by the vector. Hence, it is difficult to expose the vector to a lethal dose of the insecticide prior to virus acquisition (Perring et al., 1999). The efficacy of insecticides on virus disease spread varies depending on whether the virus is brought into a field by insects already infective (primary spread) or whether the spread of virus is from infected to healthy plants within a field (secondary spread). In the case of primary spread, insecticide application to the crop may not be effective, especially if many viruliferous insects enter the field. If the virus involved is a nonpersistently transmitted virus vectored by a noncolonizing, transient vector, insecticide application will be of no use. Virus control may be possible by applying insecticides outside the field to the plants hosting the vector, but this method may be impractical.

When the virus spread is secondary and the vector multiplies in the plant as its host, application of insecticides may be effective in preventing the spread of the virus. Systemic insecticides may be effective in the control of persistently transmitted viruses. Persistently transmitted viruses often are acquired from the phloem, so systemic insecticides in the phloem can kill the vector prior to virus acquisition or inoculation. However, nonpersistently transmitted viruses are acquired from and inoculated to epidermal cells, so the insect does not contact the systemic insecticide (which accumulates in phloem cells). Hence, it may not be possible to control nonpersistent viruses with systemic insecticides. However, some chemicals kill the vector rapidly, repel the vector, or modify vector behavior to prevent probing. Synthetic pyrethroids cause rapid knockdown or mortality of vectors prior to virus inoculation. Some pyrethroids also repel insects. This action can cause increased aphid activity and enhance virus spread (Lowery and Boiteau, 1988). Pyrethroids reduce the probing time of vectors. It may reduce the disease spread because fewer insects acquire the virus when the plants are treated with pyrethroids and viruliferous insects are prevented from inoculating healthy plants.

Mineral oils are known to reduce the spread of nonpersistent viruses through reduction of vector densities. Oils may reduce feeding of the vector and increase time spent by the vector on foliage before initiating probing. Oils may impede the virus infection process, even though the virus is transmitted. Mineral oil does not block transmission, but symptom expression is delayed in sprayed plants (Crane and Calpouzos, 1969).

Several newer insecticides have been reported to reduce virus spread, including spread of nonpersistent viruses. *Tomato yellow leaf curl virus* could be controlled by controlling its vector *Bemisia tabaci* by spraying an imidacloprid insecticide, Confidor (Ahmed et al., 2001). The second generation neonicotinoid insecticide thiamethoxam prevents transmission of *Tomato yellow leaf curl virus* (TYLCV) by the whitefly *B. tabaci* (Mason et al., 2000). The insecticide was applied as drench or foliar treatment. Drench application to tomato plants provided a good level of protection from TYLCV infections (from one to 22 days after treatment application); foliar application resulted in a prompt but short-lasting protection (< eight days). Foliar treatment of plants made infected tomatoes totally ineffective as a virus source for at least eight days. The percentage of viruliferous whiteflies surviving the acquisition on the insecticide-treated plants appeared similar to that of insects fed on untreated plants, suggesting that thiamethoxam activity in preventing TYLCV transmission by *B. tabaci* was simply due to its killing activity, and not by antifeeding or repellent actions. Viruliferous whiteflies exposed to thiamethoxam-treated plants stopped feeding before acquiring enough virus to subsequently inoculate plants (Mason et al., 2000). Deltamethrin + heptenophos (trade name Decisquick, marketed by AgrEvo) effectively controls aphid vectors and prevents virus spread in potatoes, sugar beet, and peas. Aldicarb (Temik), aldicarb + gamma-HCH (Sentry), carbosulfan (Marshal), demeton-*S*-methyl (Metasystox), Dimethoate (Dimethoate), disulfoton (Disulfoton), fenvalerate (Sumicidin), oxamyl (Vydate), phorate (Phorate), pirimicarb (Aphox), deltamethrin (Decis), cypermethrin (Ambush), and thimeton (Ekatin) have been reported to be effective against the virus diseases spread by the aphid vector *Myzus persicae* (Parry, 1990).

CONTROL OF VIRUS DISEASES BY PLANT ACTIVATORS

Another method of controlling virus diseases by chemicals is by using plant activators, which induce systemic resistance against virus infection. Preplant application of the plant activator acibenzolar-S-methyl (Actigard) effectively controls *Tomato spotted wilt virus* (TSWV) in tomato (Csinos et al., 2001). Initial applications made posttransplant had no effect, suggesting that plants must be protected prior to introduction into the field. Therefore, the activator should be applied as preplant application to effectively control the virus infection. For more information on plant activators, see Chapter 28, "Plant Activators."

REFERENCES

Ahmed, N. E., Kanan, H. O., Sugimoto, Y., Ma, Y. Q., and Inanaga, S. (2001). Effect of imidacloprid on incidence of *Tomato yellow leaf curl virus. Plant Dis,* 85:84-87.

Cieslinska, M. and Zawadzka, B. (1999). Preliminary results of investigation on elimination of viruses from apple, pear, and raspberry using thermotherapy and chemotherapy in vitro. *Phytopathologia Polonica,* 17:41-48.

Crane, G. L. and Calpouzos, L. (1969). Suppression of symptoms of sugarbeet virus yellows by mineral oil. *Phytopathology,* 59:697-698.

Csinos, A. S., Pappu, H. R., McPherson, R. M., and Stephenson, M. G. (2001). Management of *Tomato spotted wilt virus* in flue-cured tobacco with acibenzolar-S-methyl and imidacloprid. *Plant Dis,* 85:292-296.

Lowery, D. T. and Boiteau, G. (1988). Effects of five insecticides on the probing, walking, and settling behavior of the green peach aphid and the buckthorn aphid (Homoptera: Aphididae) on potato. *J Econ Entomol,* 81:208-214.

Mason, G., Rancati, M., and Bosco, D. (2000). The effect of thiamethoxam, a second generation neonicotinoid insecticide, in preventing transmission of tomato yellow leaf curl geminivirus (TYLCV) by the whitefly *Bemisia tabaci* (Genadius). *Crop Protection,* 19:473-479.

Parry, D. W. (1990). *Plant Pathology in Agriculture.* Cambridge University Press, Cambridge, U.K.

Perring, T. M., Gruenhagen, N. M., and Farrar, C. A. (1999). Management of plant viral diseases through chemical control of insect vectors. *Annu Rev Entomol,* 44:457-481.

Research Information Ltd. (1998). *International Pesticide Directory, Sixteenth Edition.* Research Information Ltd., Herts, U.K.

26

Cultural Methods

Crop diseases can be managed by manipulating some cultural practices. The important useful cultural practices that help to reduce disease incidence are described in this chapter. Sanitation is important to prevent the introduction of pathogen inoculum into fields, farms, or communities, and to reduce or eliminate inoculum from diseased fields. Soil solarization and mulching are novel approaches, which are now widely practiced in different parts of the United States, Australia, and many other countries. Organic and inorganic soil amendments are increasingly used to manage diseases. Tillage practices may be useful in reducing disease incidence in some crops. Methods of sowing/planting, irrigation practices, and pruning methods also determine disease incidence. Adjustment of the crop sequence providing a fallow period and growing intercrops, living crop covers, and trap crops are other useful approaches in crop-disease management.

PREVENTION OF INTRODUCTION OF INOCULUM

Pathogens may be introduced into a field, farm, or region by means of seeds and vegetative propagating materials. Only seeds free from pathogens should be used for planting. Methods of indexing seeds for the presence of pathogens are described in Chapter 27, "Exclusion and Eradication."

Debris from infected plants may carry the primary inoculum and may serve as source for initial outbreak of diseases. Hence, removal of the debris may be one of the most important cultural practices in management of crop diseases. *Venturia inaequalis* survives the winter in diseased apple leaves on the ground. Pseudothecia and ascospores develop in these fallen leaves. Ascospores produced on diseased leaves in the leaf litter constitute the primary inoculum causing a scab. It has also been reported that viable conidia of *V. inaequalis* can overwinter within apple buds (Becker et al., 1992). Because the primary inoculum for the disease arises from the fallen leaves, crop-debris management may help in reducing disease outbreak. Infested fallen leaves found on the orchard floor should be removed or destroyed to reduce the overwintering inoculum. The degradation of fallen leaves can be enhanced by spraying urea in the fall. The fallen leaves can be deeply incorporated into the orchard soil by tilling and by chopping leaves with flail mowers (Sutton et al., 2000). Shredding the leaf litter in November or April reduces the risk of scab by 80 to 90 percent in the northeastern United States (Sutton et al., 2000). Urea applied to the leaf litter in April (before the bud break) reduced the number of the fungal ascospores trapped by 66 percent (Sutton et al., 2000). It appears that field sanitation, i.e., shredding the leaf litter or treating the leaf litter with urea, is very important in managing apple scab (MacHardy, 2000).

The citrus canker pathogen, *Xanthomonas axonopodis* pv. *citri,* survives in infested leaf litters, and the bacterium can be recovered up to four months after leaf fall. Soil treatments, including burial of leaves, reduce its survival significantly. The bacterium can be recovered from soil beneath diseased trees, but removal of the inoculum source leads to a demise of the bacterial populations within days (Graham et al., 1989). The lettuce bacterial leaf spot pathogen *Xanthomonas campestris* pv. *vitians* was recovered from lettuce plant debris after the one-month summer fallow and the disease developed on all subsequent fall lettuce crops (Barak et al., 2001). It stresses the need for removal of crop debris to manage the disease.

Pruning of infected parts may reduce the disease outbreak in many orchard trees, because the primary inoculum is from infected plant tissues. In fire blight-affected pome fruit trees, blighted blossom clusters and shoots should be pruned. Infected parts should be cut back up to shoot basis and branch ring. Once the disease has spread into the stem, the tree must be eliminated (Richter, 1999). Pruning alone reduced the bacterial infection by 62 percent in pear (Aysan et al., 1999). It may be possible to preserve *Erwinia amylovora*-attacked pear trees by pruning blighted blossom clusters and shoots. Leaf sanitation (removal of senescent and necrotic leaves) reduced *Botrytis* rot incidence in strawberry from 13 to 8 percent (Mertely et al., 2000).

Burning the infected organs may reduce the inoculum in the field. Pear trees in commercial orchards with shoots or blossoms naturally infected with *Erwinia amylovora* were burned for a few seconds with a propane flame torch on the tissue surrounding infected sites (Reuveni et al., 1999). Burning was highly efficient in eliminating *E. amylovora* in infected organs. Burning also prevented the spread of the bacteria from the infected organs to the main limb. No damage to tree viability and vitality was observed as a result of this treatment. Burning the infected organs before internal spread of *E. amylovora* to the main limb occurs can provide a safe and rapid measure to control fire blight in pear (Reuveni et al., 1999).

SOIL SOLARIZATION

Soil solarization, the process of heating soils under transparent plastic tarps to temperatures detrimental to soilborne pathogens, has successfully controlled a variety of plant diseases. Solarization targets mesophyllic organisms, which include most plant pathogens, without destroying the beneficial mycorrhizal fungi and growth-promoting *Bacillus* spp. Increased soil temperatures result in decreased populations of a range of plant pathogens, including fungi, bacteria, and nematodes. If not directly inactivated by heat, soilborne plant pathogens may be weakened and become vulnerable to soil fumigants, to other organisms, or to changes in the soil atmosphere in solarized soil (Pinkerton et al., 2000). The solarized soils are often more suppressive to certain soilborne pathogens than are nonsolarized soils. The effectiveness of solarization against *Verticillium* wilt of safflower and cotton, *Fusarium oxysporum* f. sp. *vasinfectum* wilt of cotton, and several species of *Phytophthora* has been reported (Pinkerton et al., 2000). When the soil was solarized for three weeks after the amendment with farmyard manure, the root rot of melon by *Phomopsis sclerotioides* was controlled to some extent (Itoh et al., 2000). Soil solarization by covering the soil with a 150 μm-thick transparent plastic film for two weeks brought about a high degree of *Pythium* spp. control in greenhouse-grown cucumber (Lopes et al., 2000). Solarization for three weeks using transparent polyethylene sheets for soil mulching provided effective control of wilts caused by *Fusarium oxysporum* f. sp. *lycopersici* and *Verticillium dahliae,* and corky root rot caused by *Pyrenochaeta lycopersici* on tomato plants (Ioannou, 2000). Solarization reduced black dot of potato caused by *Colletotrichum coccodes* by 45 percent when tarping was done for eight weeks and temperatures reached 56°C in the top 5 cm of soil. With a six-week tarping period and lower maximum temperature (50°C), there was no significant reduction in disease incidence (Denner et al., 2000). Soil solarization reduced severity of diseases caused by *Verticillium* spp. on eggplant and *Phytophthora* spp. on snapdragons (Pinkerton et al., 2000). *Agrobacterium* spp. population densities declined within solarized plots, and incidence of crown gall on cherry rootstock planted in solarized plots was reduced significantly (Pinkerton et al., 2000).

ORGANIC AMENDMENTS

Root rots are known to be greater in number and more severe in soil with low organic matter. Incorporation of organic matter reduced root rot in many crops (Tu, 1987). Addition of liquid swine manure to field soils killed

Verticillium dahliae microsclerotia and reduced *Verticillium* wilt of potato (Conn and Lazarovits, 2000). The efficacy of the swine manure was dependent on soil pH. Adjusting the soil pH from 5.0 to 6.5 eliminated the toxicity of the swine manure. Conversely, when soil from a location in which swine manure had no effect was reduced from its initial pH value of 7.5 to below 6, swine manure became effective in that soil (Conn and Lazarovits, 2000). Fresh chicken manure, or chicken manure composted for five weeks before incorporation into the potting mix (25 percent, vol./vol.), significantly reduced *Phytophthora cinnamomi* survival and the development of symptoms on *Lupinus albus* seedlings (Aryantha et al., 2000).

Addition of composts to the field reduces the disease incidence. However, the composting method may determine the efficacy of the composts. Composts were prepared by three different methods (Kannangara et al., 2000). Two of the composts were prepared from separated dairy solids either by windrow (WDS) or vermicomposting (VMC) while the third, obtained from the International Bio Recovery (IBR), was prepared from vegetable refuse using aerobic digestion. Amendment of WDS in the potting mix suppressed *Fusarium oxysporum* f. sp. *radicis-cucumerinum* infection on cucumber, while VMC and IBR had no effect. There was a large increase of fluorescent bacteria near the vicinity of roots particularly in WDS amended potting mixes. Contrasting effectiveness of the WDS and VMC made from the same waste suggests that composting method can influence the disease suppression properties of the finished compost (Kannangara et al., 2000).

Compost-amended potting mixes are used to manage soilborne diseases of floricultural crops caused by *Pythium, Phytophthora, Fusarium,* and *Rhizoctonia*. Several biocontrol agents such as *Trichoderma* spp., *Gliocladium* sp., fluorescent pseudomonads, and *Bacillus* spp. colonize the compost (Hoitink et al., 1991). Pine bark is used for preparing compost on a commercial scale. The raw pine bark is first hammer milled and screened so that all particles are < 1.25 cm in diameter. A small amount of nitrogen (0.6 kg/m³ of bark), mostly as ammonia or urea, is added with water as the first step in the composting process. The wet and nitrified bark is stored in windrows on a concrete pad and turned several times during an eight- to eleven-week period. Temperature, electrical conductivity, pH, and moisture levels are monitored regularly to follow the type of composting. Adjustments are made to maintain optimum conditions for the process (Hoitink et al., 1991). A computer-controlled composting system has also been developed. This type of compost is used at incorporation rates of up to 25 percent. These mixes should be stored in 100 L polyethylene bags or in 2 m³ nylon bags for days to months before their use. At least four days must be allowed for recolonization of high-temperature (40-50°C) compost with a mesophilic microflora to induce the natural suppression of *Pythium* damping-off (Hoitink et al., 1991). Growers in the eastern United States have consistently suppressed Fusarium wilt of cyclamen for over a decade by using composted pine bark mixes (Hoitink et al., 1991).

Incorporation of plant residues into soil suppresses certain soilborne diseases. The efficacy of various organic amendments for controlling soilborne pathogens has been attributed to the formation of toxic volatile compounds or to an increase in antagonistic soil microflora (Pinkerton et al., 2000). Brassicas, which contain glucosinolates that break down to toxic isothiocyanates, and Sudan grasses, which contain dhurrin that breaks down to hydrogen cyanide, have been reported to control fungal pathogens. Soil amendment with brassicas has a beneficial effect on the control of some soilborne pathogens. Residues from brassica crops could be incorporated directly into the soil or brassicas could be used as intercrops (Rosa and Rodrigues, 1999). Cauliflower wilt caused by *Verticillium dahliae* was effectively controlled by incorporating broccoli (*B. oleracea* L. var. *italica*) residues in soil in Salina Valley, California (Koike and Subbarao, 2000). Broccoli residue showed a detrimental effect on the viability of microsclerotia in soil. It also had an inhibitory effect on the root-colonizing potential of surviving microsclerotia (Shetty et al., 2000). Air-dried and crushed mustard *(Brassica juncea)* added to the soil effectively reduced the viability of sclerotia of *Sclerotium cepivorum,* the onion white rot pathogen, and chlamydospores of *Fusarium oxysporum* f. sp. *lycopersici,* the tomato wilt pathogen (Smolinska, 2000). Consequently, the reduction of white rot of onion and tomato wilt was observed. The addition of rapeseed *(Brassica napus)* residues to soil also resulted in a decrease of number of sclerotia of *S. cepivorum* (Smolinska, 2000). Amendment of soil with dried leaves of savoy cabbage (*Brassica oleracea* var. *sabauda*), red cabbage (*B. oleracea* var. *capitata* f. *rubra*), and fringed cabbage (*B. oleracea* var. *acephala*) significantly reduced the appearance of cucumber damping-off caused by *Pythium ultimum* (Burgiel and Schwartz, 2000). The dried leaves possessed fungistatic activity against *P. ultimum* (Burgiel and Schwartz, 2000).

The effect of cruciferous plant residues in reducing the disease incidence may be due to the glucosinolate content of these plants. *Aphanomyces euteiches* root rot of pea was significantly reduced (77 percent) in soil amended with rapeseed meal from *Brassica napus* cv. Dwarf Essex (high glucosinolate concentrations). Amendment with cv. Stonewall (low glucosinolate concentrations) did not control pea root rot (Dandurand et al., 2000), suggesting that a low concentration of glucosinolate may not be sufficient to control the pathogen in soil. The toxicity of glucosinolates may probably be due to isothiocyanates derived from the breakdown of glucosinolates (Rosa and Rodrigues, 1999).

Fresh broccoli or grass was incorporated into soil and covered with plastic sheeting. In this amended soil, anaerobic and strongly reducing soil conditions developed quickly, as indicated by rapid depletion of oxygen and a decrease in redox potential values to as low as –200mV. After 15 weeks, survival of *Fusarium oxysporum* f. sp. *asparagi, Rhizoctonia solani,* and *Verticillium dahliae* in inoculum samples buried 15 cm deep was strongly reduced in amended, covered fields. The pathogens were not or hardly inactivated in amended, noncovered soil or nonamended, covered soil, indicating that inhibitory action of fresh leaves or thermal inactivation due to increased soil temperatures under the plastic cover was not involved in pathogen inactivation (Blok et al., 2000).

MULCHING

Mulches are different from soil amendments in that amendments are incorporated into the soil, while mulches are spread on the soil surface. Mulches may consist of straw or stubble applied to the soil surface, or of plastic foil spread over the soil. Mulches are known to reduce various diseases. Silver-painted mulch suppressed yellow mosaic virus symptoms in squash, while nonreflective (black) mulch did not have any effect on the disease incidence. The reflective mulch was effective only under slight virus pressure (Boyhan et al., 2000). Reflective mulch significantly reduced the vector thrips and *Tomato spotted wilt virus* in tomato (Riley and Pappu, 2000). Black plastic mulch reduced the *Verticillium* wilt symptoms in eggplant (Elmer, 2000).

LIVING GROUND COVERS

Living covers, such as perennial peanuts *(Arachis pintoi),* cinquillo *(Drymaria cordata),* and coriander *(Coriandrum sativum),* grown along with tomato mask the tomato crop from immigrating viruliferous whiteflies *(Bemisia tabaci).* These covers minimize contact between the vector and the tomato plant. The living covers reduce whitefly adult numbers, delay *Tomato yellow mottle virus* dissemination, reduce disease severity, and provide higher yields in tomatoes (Hilje, 2000). The forage groundnut *Arachis pintoi* was grown as a ground cover with bell pepper *(Capsicum annum).* Approximately four times as many whiteflies were observed on bell pepper plants grown without ground cover than on plants with cover. Onset of the gemini virus disease in bell pepper was earlier and frequency was greater in plants without cover than in plants with cover (Rafie et al., 1999).

CATCH CROPS

Control of some soilborne pathogens may be achieved by use of decoy or catch crops. These crops stimulate the germination of resting spores, resulting in limited expression of disease symptoms. Chinese cabbage plants grown in pots inoculated with *Plasmodiophora brassicae* (the clubroot pathogen) that had previously contained leafy daikon (radish, *Raphanus sativus* var. *longipinnatus*) showed less disease incidence compared to control pots in which no plants had been grown before. Numbers of resting spores of *P. brassicae* in soil in pots after cultivation with leafy daikon were reduced by 71 percent compared to control pots (Murakami et al., 2000).

TILLAGE

Some tillage operations may help to reduce disease incidence. Moldboard plowing to a depth of 30 cm reduced black dot of potato caused by *Colletotrichum coccodes* by 34 percent and was twice as effective as plowing to a depth of 60 cm (Denner et al., 2000). Fall chisel plowing plus spring raised seedbed preparation significantly reduced pea root rot (*Fusarium solani* f. sp. *pisi* and *F. oxysporum* f. sp. *pisi*) severity, compared to conventional fall plowing plus spring flat seedbed preparation. Root rot was most severe with fall plowing plus compaction, followed by fall plowing plus fall seedbed preparation (Tu, 1987). Reduced tillage that leaves 30 percent or more of the soil surface covered by crop residue after planting completely or partially controls several wheat root rot and foot rot pathogens, such as *Bipolaris sorokiniana, Fusarium graminearum, F. culmorum, F. avenaceum,* and *Pseudocercosporella herpotrichoides*. Root rot is favored by drought stress. The increased moisture available to the crop under reduced tillage might have led to reduction in disease incidence (Bockus and Shroyer, 1998). The prevalence of Sclerotinia stem rot *(Sclerotinia sclerotiorum)* of soybeans was less in no-till than in minimal-till fields in the north central United States (Workneh and Yang, 2000).

SPACING

Spacing between plants plays an important role in disease incidence. Wider plant spacing reduces Botrytis fruit rot incidence in strawberry (Legard et al., 2000). Good aeration around host tissues often minimizes the incidence and severity of diseases caused by *B. cinerea* (Cooley et al., 1996). Paired rows, as opposed to evenly spaced rows, reduce wheat root rots by allowing the top layer to dry more rapidly because of openness between every other row of plants (Cook and Veseth, 1991). It has been suggested that opening the trees to sunshine and air should be the first measure taken to control sooty blotch and flyspeck of apple caused by four different fungi (Williamson and Sutton, 2000). Pruning creates an environment less favorable to the diseases. The sooty blotch and flyspeck could be reduced by an average of almost 30 percent by severe pruning (Williamson and Sutton, 2000).

MANIPULATION OF SOIL pH

Potato common scab *(Streptomyces scabies)* can be suppressed by lowering the soil pH. When ammonium sulfate was applied into the rows where potato plants were to be planted, the soil pH was lowered and the concentration of water-soluble aluminium was increased. Potato common scab was suppressed in the soil containing water-soluble aluminium in concentrations of 0.2 to 0.3 mg/liter or higher (Mizuno et al., 2000).

INORGANIC AMENDMENTS

Several inorganic compounds alter the severity of disease incidence in the field. Infection of avocado seedlings by *Phytophthora cinnamomi* in infested soil was decreased by 71 percent by the addition of gypsum soil amendments (Messenger et al., 2000b). Sporangial production of *P. cinnamomi* buried in gypsum-amended avocado soil for two days was reduced by as much as 74 percent (Messenger et al., 2000a). Soil extracts from gypsum-amended soil (1, 5, or 10 percent gypsum) reduced in vitro sporangial production. Zoospore production and colony-forming units of *P. cinnamomi* were reduced in soil amended with calcium sulfate, calcium nitrate, or calcium carbonate. It suggests that calcium may be playing a key role in suppressing sporulation and multiplication of the pathogen in soil (Messenger et al., 2000a).

Calcium application effectively controlled bacterial wilt of tomato caused by *Ralstonia solanacearum* (Yamazaki et al., 2000). Adding $CaCO_3$, $Ca(OH)_2$, and $CaSO_4$ to the soil reduced wilt of banana caused by *Fusarium oxysporum* f. sp. *cubense*. Smaller amounts had the greatest effect, and the amounts of calcium compounds used were insufficient to change the pH. Calcium treatment has been shown to inhibit germination of chlamydospores of the pathogen and suggests that calcium may act directly on the pathogen (Peng et al., 1999).

High nitrogen application increases *Rhizoctonia* stem canker in potato; less nitrogen and high potassium and phosphorous application results in decrease in the disease incidence (Crozier et al., 2000). Nitrogen application may have different actions on different pathogens in a host (Hoffland et al., 2000). N application increased susceptibility to the bacterial speck pathogen *Pseudomonas syringae* pv. *tomato* and the powdery mildew pathogen *Oidium lycopersicum* in tomato. It did not have any effect on the wilt pathogen *Fusarium oxysporum* f. sp. *lycopersici*. However, increased application of nitrogen increased resistance to the gray mold pathogen *Botrytis cinerea* (Hoffland et al., 2000).

Different forms of nitrogen may affect the disease incidence differently. The ammoniacal form of nitrogen, as opposed to the nitrate form, reduces wheat take-all disease by producing a more acidic root surface, inhibiting the alkaline-loving pathogen *Gaeumannomyces graminis* (Huber and McCay-Buis, 1993). Fertilization with $(NH_4)_2SO_4$ showed less *Verticillium* symptoms in eggplant than with $Ca(NO_3)_2$ fertilization (Elmer, 2000).

CROP ROTATION AND INTERCROPPING

Diseases can be managed by proper planning of the crop sequence. A rotation including barley, field peas, and wheat for three years following canola *(Brassica napa)* helped to eliminate potential disease sources in canola *(Leptosphaeria maculans)* in Canada (Turkington et al., 2000). Rotation to grain sorghum to produce a one-year break from wheat resulted in effective control of tan spot of wheat. This was true even when the maximum amount of wheat residue was retained on the soil surface (without tillage). The time during the rotation in which there is no susceptible host present may serve to sensitize the soil by utilizing the native microorganisms to weaken and kill residue-borne pathogens (Bockus and Shroyer, 1998). Soils from two orchards in the United States were cultivated with three successive 28-day growth cycles of wheat in the greenhouse and subsequently planted to apple seedlings. Cultivation of wheat in orchard replant soils prior to planting of apple seedlings resulted in reduction in apple root infection by *Rhizoctonia* and *Pythium* spp. In replant soils cultivated with wheat, the antagonistic bacteria *Pseudomonas putida* population dominated (Mazzola and Gu, 2000).

Strawberries are grown as green manure in Italy. Sown in autumn, the strawberry crop is cut and buried when in full flower. It is useful for management of diseases of *Brassica juncea* grown after strawberry. The hydrolyzed glucosides from strawberry are toxic to *Phytophthora, Pythium, Rhizoctonia,* and *Sclerotium* species (Frabboni, 2001). *Fusarium oxysporum*

f. sp. *apii* infection in celery is suppressed with amendments of onion or peppermint crop residues. Celery grown after an onion crop is less affected by the Fusarium yellows fungus (Lacy et al., 1996).

During four consecutive years of experiments in the Netherlands with carrots grown as sole crops or undersown with subterranean clover *(Trifolium subterraneum)*, it was shown that intercropping significantly reduced the cavity spot caused by *Pythium* spp. (Theunissen and Schelling, 2000).

IRRIGATION

Irrigation practices determine the severity of disease incidence in many crops. Overhead sprinkling irrigation enhanced production of oospores of *Phytophthora infestans* in field-grown crops of potato (Cohen, Farkash, et al., 2000). Frequent irrigation increased the *Verticillium* wilt incidence in cauliflower. Deficit irrigation suppressed the wilt incidence; however, it also reduced the yield (Xiao and Subbarao, 2000). Hence, irrigation methods should be properly planned to control diseases without any reduction in crop yield. When compared to pipe irrigation, trickle irrigation (at a depth of 30 cm) reduced damage by plant diseases and increased crop yields of both fennel and cauliflowers in Germany (Eberhard, 2000). Daily irrigation increased *Monosporascus* wilt of melons, while in a less frequently irrigated field, symptom development was delayed and disease incidence was decreased (Cohen, Pivonia, et al., 2000).

HYDROPHOBIC PARTICLE FILMS

Hydrophobic particle film technology has been exploited to manage diseases. Hydrophobic particle film is based on the inert mineral kaolin, which is surface treated with a water-repelling agent. Dust application of hydrophobic particles on plant surfaces results in envelopment of the plant in a hydrophobic particle film barrier that prevents disease inoculum or water from directly contacting the leaf surface. By this method, disease development is suppressed. This technology could offer broad-spectrum protection against several diseases in agricultural crops (Glenn et al., 1999).

REFERENCES

Aryantha, I. P., Cross, R., and Guest, D. I. (2000). Suppression of *Phytophthora cinnamomi* in potting mixes amended with uncomposted and composted animal manures. *Phytopathology*, 90:775-782.

Aysan, Y., Tokgonul, S., Cinar, O., and Kuden, A. (1999). Biological, chemical, cultural control methods and determination of resistant cultivars to fire blight in pear orchards in the Eastern Mediterranean region of Turkey. *Acta Horticulturae,* 489:549-552.

Barak, J. D., Koike, S. T., and Gilbertson, R. L. (2001). Role of crop debris and weeds in the epidemiology of bacterial leaf spot of lettuce in California. *Plant Dis,* 85:169-178.

Becker, C. M., Burr, T. J., and Smith, C. A. (1992). Overwintering of conidia of *Venturia inaequalis* in apple buds in New York orchards. *Plant Dis,* 76:121-126.

Blok, W. J., Lamers, J. G., Termorshuizen, A. J., and Bollen, G. J. (2000). Control of soilborne plant pathogens by incorporating fresh organic amendments followed by tarping. *Phytopathology,* 90:253-259.

Bockus, W. W. and Shroyer, J. P. (1998). The impact of reduced tillage on soilborne plant pathogens. *Annu Rev Phytopathol,* 36:485-500.

Boyhan, G. E., Brown, J. E., Channel-Butcher, C., and Perdue, V. K. (2000). Evaluation of virus resistant squash and interaction with reflective and nonreflective mulches. *Hort-Technology,* 10:574-580.

Burgiel, Z. J. and Schwartz, E. (2000). Research on possibilities of utilization of chosen cruciferous plants in protection of cucumber against damping-off caused by *Pythium ultimum. Phytopathologia Polonica,* 19:59-95.

Cohen, R., Pivonia, S., Burger, Y., Edelstein, M., Gamliel, A., and Katan, J. (2000). Toward integrated management of *Monosporascus* wilt of melons in Israel. *Plant Dis,* 84:496-505.

Cohen, Y., Farkash, S., Baider, A., and Shaw, D. S. (2000). Sprinkling irrigation enhances production of oospores of *Phytophthora infestans* in field-grown crops of potato. *Phytopathology,* 90:1105-1111.

Conn, K. L. and Lazarovits, G. (2000). Soil factors influencing the efficacy of liquid swine manure added to soil to kill *Verticillium dahliae. Can J Plant Pathol,* 22:400-406.

Cook, R. J. and Veseth, R. J. (1991). *Wheat Health Management.* APS Press, St. Paul, MN.

Cooley, D. R., Wilcox, W. F., Kovach, J., and Schloemann, S. G. (1996). Integrated pest management programs for strawberries in the Northern United States. *Plant Dis,* 80:228-235.

Crozier, C. R., Creamer, N. G., and Cubeta, M. A. (2000). Fertilizer management impacts on stand establishment, disease, and yield of Irish potato. *Potato Research,* 43:49-59.

Dandurand, L. M., Mosher, R. D., and Knudsen, G. R. (2000). Combined effects of *Brassica napus* seed meal and *Trichoderma harzianum* on two soilborne plant pathogens. *Can J Microbiol,* 46:1051-1057.

Denner, F. D. N., Millard, C. P., and Wehner, F. C. (2000). Effect of soil solarisation and mouldboard ploughing on black dot of potato, caused by *Colletotrichum coccodes. Potato Research,* 43:195-201.

Eberhard, J. (2000). Trickle irrigation below the soil: An alternative for pipe irrigation? *Gemuse (Munchen),* 36(2):8-10.

Elmer, W. H. (2000). Comparison of plastic mulch and nitrogen form on the incidence of Verticillium wilt of eggplant. *Plant Dis,* 84:1231-1234.

Frabboni, L. (2001). Strawberries, a future with green manuring. *Colture Protette,* 30:39-42.

Glenn, D. M., Puterka, G. J., Vanderzwet, T., Byers, R. E., and Feldhake, C. (1999). Hydrophobic particle films: A new paradigm for suppression of arthropod pests and plant diseases. *J Econ Entomol,* 92:759-771.

Graham, J. H., Gottwald, T. R., Civerolo, E. L., and McGuire, R. G. (1989). Population dynamics and survival of *Xanthomonas campestris* in soil in citrus nurseries in Maryland and Argentina. *Plant Dis,* 73:423-427.

Hilje, I. (2000). Use of living ground covers for managing the whitefly Bemisia tabaci as a geminivirus vector in tomatoes. In *The BCPC Conference: Pests and Diseases. Volume I. Proceedings of an International Conference held at the Brighton Hilton Metropole Hotel, Brighton, U. K., 13-16 November 2000.* British Crop Protection Council, Farnham, U.K., pp. 167-170.

Hoffland, E., Jeger, M. J., and van Beusichem, M. L. (2000). Effect of nitrogen supply rate on disease resistance in tomato depends on the pathogen. *Pl Soil,* 218:239-247.

Hoitink, H. A. J., Inbar, Y., and Boehm, M. J. (1991). Status of compost-amended potting mixes naturally suppressive to soilborne diseases of natural crops. *Plant Dis,* 75:869-873.

Huber, D. M. and McCay-Buis, T. S. (1993). A multiple component analysis of the take-all disease of cereals. *Plant Dis,* 77:437-447.

Ioannou, N. (2000). Soil solarization as a substitute for methyl bromide fumigation in greenhouse tomato production in Cyprus. *Phytoparasitica,* 28:248-256.

Itoh, K., Toyota, K., and Kimura, M. (2000). Effects of soil solarization and fumigation on root rot of melon caused by *Phomopsis sclerotioides* and on soil microbial community. *Japanese J Soil Sci Plant Nutrition,* 71:154-164.

Kannangara, T., Utkhede, R. S., Paul, J. W., and Punja, Z. K. (2000). Effects of mesophilic and thermophilic composts on suppression of *Fusarium* root and stem rot of greenhouse cucumber. *Can J Microbiol,* 46:1021-1028.

Koike, S. T. and Subbarao, K. V. (2000). Broccoli residues can control *Verticillium* wilt of cauliflower. *California Agric,* 54:30-33.

Lacy, M. L., Berger, R. D., Gilbertson, R. L., and Little, E. L. (1996). Current challenges in controlling diseases of celery. *Plant Dis,* 80:1084-1091.

Legard, D. E., Xiao, C. L., Mertely, J. C., and Chandler, C. K. (2000). Effects of plant spacing and cultivar on incidence of Botrytis fruit rot in annual strawberry. *Plant Dis,* 84:531-538.

Lopes, M. E. B. M., Ghini, R., Tessarioli, J., and Patricio, F. R. A. (2000). Control of *Pythium* in greenhouse-grown cucumber using soil solarization. *Summa Phytopathologica,* 26:224-227.

MacHardy, W. E. (2000). Action thresholds for managing apple scab with fungicides and sanitation. *Acta Horticulturae,* 525:123-131.

Mazzola, M. and Gu, Y. (2000). Impact of wheat cultivation on microbial communities from replant soils and apple growth in greenhouse trials. *Phytopathology,* 90:114-119.

Mertely, J. C., Chandler, C. K., Xiao, C. L., and Legard, D. E. (2000). Comparison of sanitation and fungicides for management of Botrytis fruit rot of strawberry. *Plant Dis,* 84:1197-1202.

Messenger, B. J., Menge, J. A., and Pond, E. (2000a). Effects of gypsum on zoospores and sporangia of *Phytophthora cinnamomi* in field soil. *Plant Dis,* 84:617-621.

Messenger, B. J., Menge, J. A., and Pond, E. (2000b). Effects of gypsum soil amendments on avocado growth, soil drainage, and resistance to *Phytophthora cinnamomi. Plant Dis,* 84:612-616.

Mizuno, N., Yoshida, H., and Tadano, T. (2000). Efficacy of single application ammonium sulfate in suppressing potato common scab. *Soil Sci Plant Nutrition,* 46:611-616.

Murakami, H., Tsushima, S., Akimoto, T., Murakami, K., Goto, I., and Shishido, Y. (2000). Effects of growing leafy daikon *(Raphanus sativus)* on populations of *Plasmodiophora brassicae* (clubroot). *Plant Pathol,* 49:584-589.

Peng, H. X., Sivasithamparam, K., and Turner, D. W. (1999). Chlamydospore germination and *Fusarium* wilt of banana plantlets in suppressive and conducive soils are affected by physical and chemical factors. *Soil Biol Biochem,* 31:1363-1374.

Pinkerton, J. N., Ivors, K. L., Miller, M. L., and Moore, L. W. (2000). Effect of soil solarization and cover crops on population of selected soilborne plant pathogens in western Oregon. *Plant Dis,* 84:952-960.

Rafie, A., Diaz, J., and McLeod, P. (1999). Effects of forage groundnut in reducing the sweetpotato whitefly and associated gemini virus disease in bell pepper in Honduras. *Trop Agric,* 76:208-211.

Reuveni, M., Elbaz, S., and Manulis, S. (1999). Control of fire blight in pears by in-situ flamation of blighted shoots and blossoms. *Acta Horticulturae,* 489:573-576.

Richter, K. (1999). Fire blight of pome fruits *(Erwinia amylovora)* and its control. *Erwerbsobstbau,* 41:202-212.

Riley, D. G. and Pappu, H. R. (2000). Evaluation of tactics for management of thrips-vectored *Tomato spotted wilt virus* in tomato. *Plant Dis,* 84:847-852.

Rosa, E. A. S. and Rodrigues, P. M. F. (1999). Towards a more sustainable agriculture system: The effect of glucosinolates on the control of soil-borne diseases. *J Hort Sci Biotechnol,* 74:667-674.

Shetty, K. G., Subbarao, K. V., Huisman, O. C., and Hubbard, J. C. (2000). Mechanism of broccoli-mediated *Verticillium* wilt reduction in cauliflower. *Phytopathology,* 90:305-310.

Smolinska, U. (2000). Survival of *Sclerotium cepivorum* sclerotia and *Fusarium oxysporum* chlamydospores in soil amended with cruciferous residues. *J Phytopathol,* 148:343-349.

Sutton, D. K., MacHardy, W. E., and Lord, W. G. (2000). Effects of shredding or treating apple leaf litter with urea on ascospore dose of *Venturia inaequalis* and disease buildup. *Plant Dis,* 84:1319-1326.

Theunissen, J. and Schelling, G. (2000). Undersowing carrots with clover: Suppression of carrot rust fly *(Psila rosae)* and cavity spot *(Pythium* spp.) infestation. *Biol Agric & Hort,* 18:67-76.

Tu, J. C. (1987). Integrated control of the pea root rot disease complex in Ontario. *Plant Dis,* 71:9-13.

Turkington, T. K., Clayton, G. W., Klein-Gebbinck, H., and Woods, D. L. (2000). Residue decomposition and blackleg of canola: Influence of tillage practices. *Can J Plant Pathol,* 22:150-154.

Williamson, S. M. and Sutton, T. B. (2000). Sooty blotch and flyspeck of apple: Etiology, biology, and control. *Plant Dis,* 84:714-724.

Workneh, F. and Yang, X. B. (2000). Prevalence of Sclerotinia stem rot of soybeans in the North-Central United States in relation to tillage, climate, and latitudinal positions. *Phytopathology,* 90:1375-1382.

Xiao, C. L. and Subbarao, K. V. (2000). Effects of irrigation and *Verticillium dahliae* on cauliflower root and shoot growth dynamics. *Phytopathology,* 90: 995-1004.

Yamazaki, H., Kikuchi, S., Hoshina, T., and Kimura, T. (2000). Calcium uptake and resistance to bacterial wilt of mutually grafted tomato seedlings. *Soil Sci Plant Nutrition,* 46:529-534.

27

Exclusion and Eradication

The most effective method to manage diseases is to exclude them from the field. Quarantine is important in preventing the spread of pathogens across countries, and legislative control is useful to prevent the entry of a pathogen into a new region. If a pathogen has entered into a region, eradication of the established infection is practiced. Because most diseases are spread internationally through seed, seed health testing and seed certification help to prevent the spread of diseases. Various seed health testing methods are described in this chapter.

EXCLUSION

Quarantine services exist in most countries. Plant quarantine is legislative or regulatory control that aims to exclude pathogens from areas where they do not already exist. Legislative control may operate on a national or international level. The legislation prohibits or restricts the introduction of seeds, vegetative propagating materials, plants, or plant parts into a country or a region in a country to exclude pathogens, which may be inadvertently introduced along with those materials. Generally, scientists, the traveling public, and some importers of agricultural products are responsible for the introduction of new pathogens into a region.

Not all pathogens are of quarantine significance. A pathogen species that does not occur in a given country or an exotic strain of a domestic species is of quarantine significance to that country if the pathogen is known to cause economic damage elsewhere or has a life cycle or host/pathogen interaction that shows a potential to cause economic damage under favorable host, inoculum, and environmental conditions (Kahn, 1991). Importation of a pathogen that already occurs in a given country is also of quarantine significance if an ongoing regional or national containment, suppression, or eradication program is directed against that pathogen species.

Actions to be taken to exclude pathogens are authorized by government regulations. In the United States, the Plant Quarantine Act was enacted in 1912. The act provides authority for domestic and foreign quarantines. The

Organic Act of 1944 authorizes the secretary of agriculture to cooperate with states, organizations, and individuals to detect, eradicate, suppress, control, and prevent or retard the spread of plant pests (including pathogens). The Federal Plant Pest Act was enacted in 1957 and regulates the movement (by persons) of plant pests into United States or between states and authorizes emergency actions to prevent the introduction and domestic movement of plant pests not covered by the Plant Quarantine Act of 1912 (Kahn, 1991).

Transport of plant material across international boundaries is regularized by the International Plant Protection Convention of 1951. The convention was organized by the United Nations Food and Agriculture Organization (FAO) with the aim of securing common and effective action to prevent the introduction and spread of pests and diseases of plants and plant products. The convention was signed by 94 nations and conformed to by most other countries. The phytosanitary certificate is an instrument of that treaty. The convention is now regionally organized. In Europe and the Mediterranean, the organizational body is the European and Mediterranean Plant Protection Organization (EPPO), and in North America, it is the North American Plant Protection Organization (NAPPO). These organizations regularly issue bulletins including information on newly identified pests and pathogens and phytosanitary regulations (Parry, 1990).

Most member countries have their own disease legislation with regard to imported plant material. Some countries have formed unions, such as the European Economic Community, and promulgate binding regulations on member countries. The legislation prescribes the form of health certificate to accompany any imported material. It lays down rules for inspection and disposal of material if it contains pathogens. It also provides a list of prohibited imports and a list of restrictions of imports of material from specified areas. In the United Kingdom, Plant Health Order 1987 lists import restrictions into the United Kingdom.

The legislation is normally implemented by customs and excise officers. They check the documentation, and specifically phytosanitary certificates, at ports. Spot checks are carried out by officers of the Plant Health and Seeds Inspectorate on material both entering and leaving the United Kingdom. Suspect materials are put into quarantine for a period of time to detect pathogens, which are present in seed and planting materials.

In spite of quarantine methods, a few pathogens have entered into countries that had not reported the occurrence of such pathogens earlier, shattering the economy of those countries. Rust disease of coffee (*Coffea arabica* and *C. canephora*) wiped out coffee plantations in Sri Lanka in 1880. The disease spread to Central and East Africa by the 1920s and to West Africa in the 1950s. It became severe in Brazil in 1970, and the disease is now prevalent in Mexico, Honduras, Paraguay, Argentina, Peru, Bolivia, Guatemala, Colombia, Costa Rica, India, and almost all countries where coffee is grown. Fire blight of apple is widespread in North America and it is not a quarantine object in the United States. However, in Europe, fire blight of apple is not prevalent in many countries. The disease was first detected in Spain in 1995 and in Hungary in 1996.

Karnal bunt of wheat was first discovered in 1930 at Karnal, a small town in the Haryana state in India. It was subsequently reported in countries around India, Pakistan, Nepal, Afghanistan, and Iraq. It was first reported in Mexico in 1972. In the United States, the disease was first discovered in Arizona in 1996. Subsequently, the disease has been reported in Texas, New Mexico, and California. The Mexican government placed an internal quarantine on Karnal bunt to prevent disease spread within the country in 1984. In 1996, a federal quarantine for Karnal bunt was placed on the states of Arizona, Texas, New Mexico, and California. The U.S. Department of Agriculture (USDA) Animal and Plant Health Inspection Service (APHIS) prohibited importation of seed, grain, straw, and dried plants of wheat, durum, and triticale from Mexico to prevent the entry of Karnal bunt into the United States. Mexico was permanently added to the list of the wheat disease subpart of Quarantine 319 (Babadoost, 2000). A zero-tolerance level of Karnal bunt has been enforced in the United States, Canada, and many other countries. The certification standard in India is zero incidence. APHIS tests the presence of teliospores in wheat seed samples by the centrifuge wash test. Test tubes containing seeds submerged in water are shaken for ten minutes to obtain teliospore suspension, then centrifuged for 20 minutes at 3,000 rpm and the sediment examined under a microscope for the presence of teliospores (Warham, 1992). The size-selective sieving test is also recommended (Peterson et al., 2000). A seed wash of a 50 g seed sample is washed through 50 μM and 20 μM pore size nylon screens to remove unwanted debris and to concentrate and isolate teliospores. The material remaining in the 20 μM pore size is suspended for direct microscopic examination and identified by polymerase chain reaction (PCR) utilizing two pairs of *Tilletia indica*-specific primers.

ERADICATION

National legislation also enforces eradication of exotic pathogens recently introduced along natural or man-made pathways. Canker is the most serious disease in *Citrus* spp. It was first reported in 1913 in Florida, and an extensive citrus canker eradication program was implemented there in 1915. After $6 million had been spent for eradication, Florida was declared free from the disease in 1933. The eradication program was also taken up in Georgia, Alabama, Louisiana, South Carolina, Texas, and Mississippi. In 1947, citrus canker was declared to be eradicated from these states. However, a new form of the disease appeared in 1984 in Florida. Another eradication program was implemented, and by 1991, over 20 million trees had been destroyed at a cost of about $94 million dollars. The Asiatic citrus canker is still prevalent in different parts of the United States (Gottwald et al., 2001). The pathogen of fire blight of apple *(Erwinia amylovora)* entered Hungary in 1996. In 1997 and 1998, further spread of the disease was registered and an eradication program was launched. More than 60,000 trees were uprooted and destroyed across the country. Eradication was performed partly by special brigades and partly with participation of growers. *Erwinia amylovora* was first detected in 1995 in Spain, and several measures were taken to eradicate the bacteria there as well.

In the United Kingdom, national plant disease legislation has been introduced to eradicate specific pathogens. It makes farmers responsible to inform officials about outbreaks of indigenous but geographically localized diseases, known as *notifiable diseases*. Fire blight of apples and pears *(Erwinia amylovora)*, wart disease of potatoes *(Synchytrium endobioticum)*, brown rot *(Ralstonia solanacearum)* and ring rot (*Clavibacter michiganensis* ssp. *sepedonicus*) of potato, plum pox disease of plums *(Plum pox virus)*, red stele disease of strawberries, rhizomania disease of beet *(Beet necrotic yellow vein virus)*, and progressive wilt of hops *(Verticillium alboatrum)* are the important notifiable diseases in the United Kingdom. Occurrence of such diseases must be reported. The diseased material should not be transported or sold and must be destroyed.

LIST OF IMPORTANT SEEDBORNE PATHOGENS

Several fungal, bacterial, viral, and phytoplasmal diseases are transmitted through seeds, including vegetative propagules (Mink, 1993; Johansen et al., 1994; Langerak et al., 1996). The important seedborne pathogens are listed in this section.

Alfalfa—*Alfalfa mosaic virus*
Barley—*Barley stripe mosaic virus, Xanthomonas campestris* pv. *translucens, Xanthomonas campestris* pv. *undulosa, Rhynchosporium secalis, Ustilago segetum* var. *nuda, Pyrenophora teres*
Bean—*Bean common mosaic virus, Bean pod mottle virus, Bean southern mosaic virus, Bean yellow mosaic virus, Pseudomonas savastanoi* pv. *phaseolicola, Curtobacterium flaccumfaciens* pv. *flaccumfaciens,*

Xanthomonas axonopodis pv. *phaseoli, Pseudomonas syringae* pv. *syringae, Colletotrichum lindemuthianum*

Beet—*Phoma betae*

Blackgram—*Urd bean leaf crinkle virus, Blackgram mottle virus*

Broad bean—*Broad bean mottle virus, Broad bean true mosaic virus, Broad bean stain mosaic virus, Broad bean wilt virus*

Carrot—*Xanthomonas campestris* pv. *carotae, Alternaria dauci, A. radicina*

Celery—*Septoria apiicola*

Cherry—*Cherry leaf roll virus, Cherry rasp leaf virus, Cherry X-disease*

Corn—*Maize chlorotic dwarf virus, Maize mosaic virus, Erwinia stewartii, Fusarium moniliforme, Peronosclerospora sacchari, Peronosclerospora sorghi, Sclerospora graminicola*

Cotton—*Xanthomonas axonopodis* pv. *malvacearum, Colletotrichum gossypii*

Cowpea—*Blackeye cowpea mosaic virus, Cowpea aphid borne mosaic virus, Cowpea banding mosaic virus, Cowpea mild mottle virus, Cowpea mosaic virus, Cowpea ringspot virus, Cowpea severe mosaic virus*

Crucifers—*Xanthomonas campestris* pv. *campestris, Phoma lingam, Alternaria brassicicola, Leptosphaeria maculans*

Cucumber—*Cucumber mosaic virus, Cucumber green mottle mosaic virus*

Eggplant—*Eggplant mosaic virus*

Flax—*Alternaria linicola*

Grapevine—*Grapevine fan leaf virus, Grapevine Bulgarian latent virus*

Lettuce—*Lettuce mosaic virus*

Melon—*Melon necrotic spot virus, Muskmelon necrotic ringspot virus*

Mung bean—*Mungbean mosaic virus*

Oats—*Oat mosaic virus, Pyrenophora avenae*

Onion—*Onion yellow dwarf virus*

Pea—*Pea early-browning virus, Pea enation mosaic virus, Pea seedborne mosaic virus, Pseudomonas syringae* pv. *pisi, Ascochyta pisi*

Peach—*Peach rosette mosaic virus, Prunus necrotic ringspot virus, Prune dwarf virus, Peach X-disease*

Peanut—*Peanut clump virus, Peanut mottle virus, Peanut stripe virus, Peanut stunt virus*

Pearl millet—*Sclerospora graminicola*

Plum—*Plum pox virus*

Potato—*Potato virus X, Potato virus Y, Potato virus T, Potato spindle tuber viroid*

Raspberry—*Raspberry ringspot virus, Raspberry bushy dwarf virus*

Red clover—*Red clover mottle virus, Red clover vein mosaic virus*

Rice—*Xanthomonas oryzae* pv. *oryzae*, *X. oryzae* pv. *oryzicola*, *Burkholderia glumae*, *Pseudomonas fuscovaginae*, *Alternaria padwickii*, *Cochliobolus miyabeanus*, *Pyricularia oryzae*, *Tilletia indica*, *Fusarium moniliforme*

Sorghum—*Peronosclerospora sorghi*, *Sclerospora sorghi*, *Sporisorium sorghi*, *Sporisorium cruentum*, *Claviceps sorghi*

Soybean—*Soybean mosaic virus*, *Pseudomonas savastanoi* pv. *glycinea*, *Cercospora kikuchii*, *Diaporthe phaseolorum*

Spinach—*Spinach latent virus*

Squash—*Squash mosaic virus*

Strawberry—*Strawberry latent ringspot virus*

Subterranean clover—*Subterranean clover mottle virus*

Sunflower—*Sunflower mosaic virus*

Tobacco—*Tobacco etch virus*, *Tobacco mosaic virus*, *Tobacco rattle virus*, *Tobacco ringspot virus*, *Tobacco streak virus*

Tomato—*Tomato aspermy virus*, *Tomato black ring virus*, *Tomato bushy stunt virus*, *Tomato ringspot virus*, *Tomato spotted wilt virus*, *Clavibacter michiganensis* ssp. *michiganensis*, *Pseudomonas syringae* pv. *tomato*, *Fusarium oxysporum* f. sp. *lycopersici*

Turnip—*Turnip yellow mosaic virus*

Watermelon—*Watermelon mosaic virus*

Wheat—*Wheat streak mosaic virus*, *Wheat striate mosaic virus*, *Xanthomonas campestris* pv. *translucens*, *Xanthomonas campestris* pv. *undulosa*, *Ustilago nuda*, *Tilletia caries*, *Tilletia controversa*

White clover—*White clover mosaic virus*

SEED HEALTH TESTING AND CERTIFICATION

Several countries have enacted laws for the certification of seeds (including propagating materials) free from pathogens. For example, in the United Kingdom, it is illegal to sell the seed of major agricultural and horticultural crops unless it has been certified as meeting specified minimum standards of quality, including freedom from disease. Crops are inspected by trained inspectors both in the field and after harvest. However, this is not sufficient to identify the pathogen, which persists in symptomless crops and seeds.

Seed health testing has become important in many countries in recent years. For example, in Canada, until the early 1970s, only visual field inspections of growing potato crops and harvested tubers served to identify *Clavibacter michiganensis* ssp. *sepedonicus*-infected lots that needed to be removed from the seed certification program. In 1979, laboratory testing to detect the possible presence of *C. michiganensis* ssp. *sepedonicus* in seed lots that had passed field inspection was initiated in Canada to facilitate international trade, because it was a pathogen of quarantine significance (De Boer and Hall, 2000). By 1985, the advantage of laboratory testing for detecting incipient ring rot infections had become clear, and testing of domestic seed lots was introduced on a voluntary basis in some provinces. By 1992, laboratory indexing of all seed lots for *C. michiganensis* ssp. *sepedonicus* in Canada became mandatory. With privatization of potato testing in Canada, an accreditation program was implemented to ensure that reliable and uniform results were obtained from multiple laboratories. The quality-assurance program of each private laboratory must follow the criteria set by the International Standards Organization (ISO) in their guide. Analysts in private laboratories are required to complete correctly blind "proficiency panel" samples on a semiannual basis to maintain their certified status, which allows them to conduct the tests in an accredited laboratory. These proficiency tests are administered by the Centre of Expertise for Potato Diseases of the Canadian Food Inspection Agency (De Boer and Hall, 2000). Similar seed health testing laboratories are available in the United States, Europe, and many Asian countries. The major purpose of these laboratories is to ensure the supply of seeds (including propagating materials) free from pathogens to the growers in order to exclude pathogens from fields, farms, regions, and countries.

SEED HEALTH TESTING METHODS

Guidelines for the standardization of seed health testing methods were drafted at the first Workshop on Seed Health Testing of the technical Plant Disease Committee (PDC) of the International Seed Testing Association (ISTA), held in Cambridge in 1958 (Langerak et al., 1996). Since then, numerous plant pathologists have worked on the development and standardization of seed health testing methods. These methods were evaluated in comparative testing programs of the PDC. The evaluated methods were compiled and published by ISTA as working sheets. These working sheets describe seed health testing methods for individual pathogens separately for each host and are included in the ISTA *Handbook on Seed Health Testing*.

Standardization of seed health testing methods is important to provide assurances to the seed user that adequate seed health testing was provided. The International Seed Health Initiative (ISHI) was founded in 1993 to address the immediate need for an efficient standardization process to accommodate the international seed trade as well as the level of testing proficiency required in the private sector for the international movement of seed. ISHI is an international consortium of seed industry and seed health testing plant pathologists from the United States, the Netherlands, France, Japan, and Israel (Maddox, 1998). ISHI supports the accreditation of private laboratories to assure the quality assurance of the testing and provide a means for regulatory testing that is both efficient and acceptable for phytosanitary regulation. The members of ISHI are working for worldwide standards in seed health in conjunction with ISTA and other regulatory agencies to provide a database of acceptable testing methods for world phytosanitary goals (Maddox, 1998).

Common Seed Health Testing Methods

Blotter Tests

Seeds are placed on two to three layers of water-soaked blotter papers in petri dishes and incubated under alternating light, provided by fluorescent white or near ultraviolet (NUV) tube lights, and dark periods. Fructifications of fungal pathogens developing on seeds are identified using stereoscopic and/or light microscopes. In a modified blotter test, seeds are placed between several layers of moist paper and incubated either in darkness or exposed to a 12-hour photoperiod. Specific symptoms may develop on germinating seedlings or characteristic fructifications may develop on the seed coat, which can be identified by microscopic inspection. In the 2,4-D blotter test, seed germination is prevented to create conditions for development of mycelium or spores of the pathogen on the seed itself. The herbicide 2,4-D solution is added to the blotter, and seeds are incubated under alternating fluorescent white or near ultra violet light and dark periods. The developing fructifications on the seed coat can be identified by microscopic inspection. In a modified 2,4-D blotter method, instead of adding 2,4-D, the moist blotter with seeds is frozen at −20°C after pre-imbibition at 20°C to prevent seed germination, allowing development of fungal fructifications on the seed coat. The incubation conditions of various blotter tests can be modified depending on the requirements for development of fructifications of individual pathogens (Langerak et al., 1996).

Seed Washing Method

The seed washing method involves placing individual seeds or portions of seeds in water or water plus detergent to promote release of spores or conidia. Staining techniques are employed to distinguish between closely related species of *Tilletia* in wheat. The repetitive-sequence-based polymerase chain reaction (rep-PCR) method is also useful (McDonald et al., 2000).

Embryo Staining Test

This test is used to detect *Ustilago tritici* in wheat and involves visual inspection of internal parts of the seed after separation, clarification, and staining of mycelium fragments in the seed tissue.

Agar Tests

In this method, seeds are plated on agar media containing nutrients. Selective media are also used to identify some specific pathogens. Surface sterilization of seeds with sodium hypochlorite is needed to avoid development of surface contaminants in the agar medium, but it may also inhibit development of pathogens present on the seed coat. Incubation conditions, such as temperature and exposure to light, also determine the development of pathogens on seeds plated on agar media.

Grow-Out (or Growing-On, Seedling Symptom) Tests

Seeds are grown in agar media in test tubes or in sand/soil in pots and incubated under different light and temperature conditions. Development of disease symptoms on seedlings is assessed.

Seed Extract and Dilution Plating

Seedborne bacteria are separated from seeds by soaking, washing, or extraction after crushing or maceration of the seed. The seed extract is then analyzed for the presence of pathogenic bacteria by dilution plating on selective media.

Serological and Nucleic Acid Probe-Based Methods

Recently, serological techniques and DNA-based methods have been developed. These techniques are mostly used to detect viral and bacterial pathogens. The important tests used are the latex agglutination test, the immuno-diffusion test, the microprecipitin test, enzyme-linked immunosorbent assay (ELISA), the immunoblot test, immunofluorescence, dot-immunobinding assay, enzyme-linked fluorescent assay, immunosorbent electron microscopy, radio immunosorbent assay, polymerase chain reaction, and DNA hybridization on DNA extracted from seeds. All these methods have been described in detail in Chapter 13, "Crop Disease Diagnosis."

The commonly used seed health testing methods to detect various pathogens (Langerak et al., 1996; Maddox, 1998) follow:

Wheat

Tilletia caries and *T. controversa*—washing test, repetitive-sequence-based polymerase chain reaction
Ustilago tritici—embryo staining test
Tilletia indica—NaOH soak test, washing test
Stagonospora nodorum—agar test, blotter test, growing-on test, fluorescence test, agar-fluorescence test
Xanthomonas translucens pv. *translucens*—dilution plating, dot immunobinding assay

Barley

Barley stripe mosaic virus—latex agglutination test, immunodiffusion test, immunosorbent electron microscopy, ELISA
Ustilago segetum var. *nuda*—embryo staining test
Xanthomonas translucens pv. *translucens*—dilution plating, dot-immunobinding assay
Pyrenophora teres—blotter test, agar test, growing-on test, deep-freezing test
Rhynchosporium secalis—PCR (Lee et al., 2001)

Rice

Alernaria padwickii—agar test, blotter test
Cochliobolus miyabeanus—blotter test
Pyricularia oryzae—blotter test
Fusarium moniliforme—agar test and blotter test
Tilletia indica—sodium hydroxide soak test
Xanthomonas oryzae pv. *oryzae*—growing-on test, direct immunofluorescence, dilution plating
Xanthomonas oryzae pv. *oryzicola*—growing-on test, direct immunofluorescence, dilution plating

Tomato

> *Clavibacter michiganensis* ssp. *michiganensis*—immunofluorescence with seedling inoculation test, dilution plating, indicator host inoculation, seed wash/liquid plating, PCR
>
> *Pseudomonas syringae* pv. *tomato*—growing-on test, plating enriched seed extract
>
> *Xanthomonas vesicatoria*—dilution plating, immunofluorescence combined with dilution plating, plating enriched seed extract
>
> *Fusarium oxysporum*—agar test
>
> *Tobacco mosaic virus*—indicator plants

Soybean

> *Cercospora kikuchii*—agar test, blotter test
>
> *Diaporthe phaseolorum*—agar test
>
> *Peronospora manshurica*—washing test
>
> *Phomopsis* spp.—blotter test, ELISA, immunoblot test
>
> *Pseudomonas savastanoi* pv. *glycinea*—growing-on test, direct plating, host inoculation, seed wash/liquid plating, immunoassays
>
> *Tobacco ringspot virus*—ELISA, immunosorbent electron microscopy

Bean

> *Colletotrichum lindemuthianum*—blotter test
>
> *Curtobacterium flaccumfaciens*—immunofluorescence, seedling inoculation test, growing-on test
>
> *Pseudomonas savastanoi* pv. *phaseolicola*—dilution plating, immunofluorescence test, immunofluorescence colony staining
>
> *Xanthomonas axonopodis* pv. *phaseoli*—seed wash and host inoculation, seed wash and dilution plating, immunofluorescence test, immunofluorescence colony staining, DNA hybridization, PCR with seed extract
>
> *Bean common mosaic virus*—ELISA, dot-immunobinding assay, immunosorbent electron microscopy, microprecipitin test

Crucifers

> *Xanthomonas campestris* pv. *campestris*—direct plating, immunofluorescence test, seed wash/liquid plating plus pathogenicity test

Phoma lingam—deep freezing blotter test
Alternaria brassicicola—seedling symptom test
Leptosphaeria maculans—2,4-D blotter, freezing blotter, PCR with
    DNA extract from seeds

Cucurbits

Squash mosaic virus—ELISA, grow-out test
Melon necrotic spot virus—ELISA
Cucumber green mottle virus—ELISA
Acidovorax avenae ssp. *citrulli*—grow-out test, PCR, immuno-
    magnetic separation and PCR (Walcott and Gitaitis, 2000)

Lettuce

Lettuce mosaic virus—ELISA, growing-on test, indicator plant test

Sugar beet

Pleospora betae—agar test, blotter test

Peach

Prune dwarf virus—ELISA
Prune necrotic ringspot virus—ELISA

INDEXING PLANT PROPAGATION MATERIALS

Most fruit trees (woody crops) are vegetatively propagated. Use of healthy planting materials will exclude pathogens from the orchard. In recent years, several molecular techniques have been developed to index the plant propagation materials and bud wood materials. Saade and colleagues (2000) have developed techniques for detection of *Prunus necrotic ringspot virus, Prune dwarf virus,* and *Apple mosaic virus* in almond, apricot, cherry, peach, and plum by multiplex reverse-transcription polymerase chain reaction (RT-PCR). This technique will be useful for the analysis of mother plants in certification programs. Merighi and colleagues (2000) developed polymerase chain reaction enzyme-linked immunosorbent assay (PCR-ELISA) for detection of *Erwinia amylovora* in pear. This test will be useful to select nursery stocks and plant propagation materials free from the pathogen. Use of disease-free budwood helps to exclude *Citrus tristeza virus*

(CTV) in citrus cultivation. ELISA and double antibody sandwich-ELISA are useful to detect CTV in propagation materials (Terrada et al., 2000). Lin and colleagues (2000) have described an in situ immunoassay for detection of CTV in citrus bud wood.

REFERENCES

Babadoost, M. (2000). Comments on zero-tolerance quarantine of karnal bunt of wheat. *Plant Dis,* 84:711-712.

De Boer, S. H. and Hall, J. W. (2000). Proficiency testing in a laboratory accreditation program for the bacterial ring rot pathogen of potato. *Plant Dis,* 84:649-653.

Gottwald, T. R., Hughes, G., Graham, J. H., Sun, X., and Riley, T. (2001). The citrus canker epidemic in Florida: The scientific basis of regulatory eradication policy for an invasive species. *Phytopathology,* 91:30-34.

Johansen, E., Edwards, M. C., and Hampton, R. O. (1994). Seed transmission of viruses: Current perspectives. *Annu Rev Phytopathol,* 32:363-386.

Kahn, R. P. (1991). Exclusion as a plant disease control strategy. *Annu Rev Phytopathol,* 29:219-246.

Langerak, C. J., van den Bulk, R. W., and Franken, A. A. J. M. (1996). Indexing seeds for pathogens. *Adv Bot Res,* 23:171-215.

Lee, H. K., Tewari, J. P., and Turkington, T. K. (2001). A PCR-based assay to detect *Rhynchosporium secalis* in barley seed. *Plant Dis,* 85:220-225.

Lin, Y., Rundell, P. A., Xie, L., and Powell, C. A. (2000). In situ immunoassay for detection of *Citrus tristeza virus. Plant Dis,* 84:937-940.

Maddox, D. A. (1998). Implications of new technologies for seed health testing and the worldwide movement of seed. *Seed Sci Res,* 8:277-284.

McDonald, J. G., Wong, E., and White, G. P. (2000). Differentiation of *Tilletia* species by rep-PCR genomic fingerprinting. *Plant Dis,* 84:1121-1125.

Merighi, M., Sandrini, A., Landini, S., Ghini, S., Girotti, S., Malaguti, S., and Bazzi, C. (2000). Chemiluminescent and colorimetric detection of *Erwinia amylovora* by immunoenzymatic determination of PCR amplicons from plasmid pEA29. *Plant Dis,* 84:49-54.

Mink, G. I. (1993). Pollen- and seed-transmitted viruses and viroids. *Annu Rev Phytopathol,* 31:375-402.

Parry, D. W. (1990). *Plant Pathology in Agriculture.* Cambridge University Press, Cambridge, U.K.

Peterson, G. L., Bonde, M. R., and Phillips, J. G. (2000). Size-selective sieving for detecting teliospores of *Tilletia indica* in wheat seed samples. *Plant Dis,* 84:999-1007.

Saade, M., Aparicio, F., Sanchez-Navarro, J. A., Herranz, M. C., Myrta, A., Di Terlizzi, B., and Pallas, V. (2000). Simultaneous detection of the three ilarviruses affecting stone fruits by nonisotopic molecular hybridization and multiplex reverse-transcription polymerase chain reaction. *Phytopathology,* 90:1330-1336.

Terrada, E., Kerschbaumer, R. J., Giunta, G., Galeffi, P., Himmler, G., and Cambra, M. (2000). Fully "recombinant enzyme-linked immunosorbent assays" using genetically engineered single-chain antibody fusion proteins for detection of *Citrus tristeza virus. Phytopathology,* 90:1337-1344.

Walcott, R. R. and Gitaitis, R. D. (2000). Detection of *Acidovorax avenae* subsp. *citrulli* in watermelon seed using immunomagnetic separation and polymerase chain reaction. *Plant Dis,* 84:470-474.

Warham, E. J. (1992). Karnal bunt of wheat. In U.S. Singh, A. N. Mukhopadhyay, J. Kumar, and H. S. Chaube (Eds.), *Plant Diseases of International Importance, Volume I,* Prentice Hall, Englewood Cliffs, NJ, pp. 1-24.

28

Plant Activators

Several plant activators, which activate host defense mechanisms and protect the plants against pathogens, have been reported. These activators appear to activate defense genes in a manner different from the way host resistance genes activate defense genes, possibly using a different signal transduction system. The practical utility of plant activators in crop disease management is discussed in this chapter.

WHAT ARE PLANT ACTIVATORS?

Plants are endowed with several defense genes. However, most of these genes are quiescent in healthy plants. Specific signals are required to activate them. The chemicals that activate the defense genes by providing signals are called *plant defense activators* or *plant activators.* Plant activators induce systemic resistance against diseases. *Induced systemic resistance* (ISR) has been defined as active resistance dependent on the host plant's physical or chemical barriers, activated by biotic or abiotic agents (Kloepper et al., 1992). Induction of ISR has been shown to be due to induction of various defense genes (Vidhyasekaran, 1998, 2001; Vidhyasekaran et al., 2000). The other form of induced resistance in plants, systemic acquired resistance (SAR), refers to a distinct signal transduction pathway that is mediated by salicylic acid and activates defense genes (Dong and Beer, 2000). Salicylic acid and several synthesized analogs of salicylic acid, such as 2,6-dichloroisonicotinic acid (INA), are known to activate SAR. In contrast to SAR, which is mediated by salicylic acid, ISR is mediated by jasmonic acid and ethylene (Dong and Beer, 2000). ISR appears to be induced by biocontrol bacteria, while SAR is induced by infection by pathogens and chemical inducers. There are many exceptions to these definitions. For example, riboflavin, a chemical, activates defense mechanisms and induces resistance; however, the activation is not through the salicylic acid pathway (Dong and Beer, 2000). Several authors prefer to call this type of resistance induced by chemicals "ISR" (Kuc, 2001; Oostendorp et al., 2001). It has been demonstrated that ISR may act through a signaling pathway different from the ones operating in various host-pathogen interactions. Salicylic acid and its synthetic mimics 2,6-dichloroisonicotinic acid, and benzo (1,2,3) thiadiazole-7-carbothioic acid S-methyl ester (BTH) and jasmonic acid are known to induce resistance in barley against the powdery mildew pathogen *Blumeria graminis* f. sp. *hordei*. All the chemically induced genes were not expressed during incompatible barley–*B. graminis* f. sp. *hordei* interactions with gene-for-gene relationships (Besser et al., 2000).

Pretreatment of plants with an inducing compound appears to incite the plant to mount an effective defense response upon subsequent encounters with pathogens, converting what would have been a compatible interaction to an incompatible one (Tuzun, 2001). ISR induces constitutive accumulation of defense gene products prior to pathogen invasion. ISR is known to function against multiple organisms, and there is no specificity observed in the accumulation patterns of defense-related gene products when ISR is induced. Plants in which ISR has been activated appear to move from a latent resistance state to one in which a multigenic, nonspecific form of resistance is active (Tuzun, 2001). ISR is a plant defense state associated with an enhanced ability, or the so-called priming, to resist pathogen attack by stronger activation of cellular responses (Conrath et al., 2001). Plant activators may have a dual role in plant defense-gene activation. First, they may activate some defense genes, which are quiescent. Second, they may not directly activate defense genes, but they may prime them so when the pathogens invade, the defense genes are activated (Conrath et al., 2001). For example, β-aminobutyric acid (BABA) induces systemic resistance, but it did not induce the accumulation of mRNA of the SAR-associated *PR-1* gene. In contrast, BABA potentiated the accumulation of *PR-1* mRNA after attack by virulent pathogenic bacteria (Zimmerli et al., 2000). Carpropamid enhanced the accumulation of the phytoalexins momilactone A and sakuranetin in rice leaves subsequently inoculated with the blast pathogen *Magnaporthe grisea* (Araki and Kurahashi, 1999).

TYPES OF PLANT ACTIVATORS

INA

The plant activator 2,6-dichloroisonicotinic acid has been reported to induce systemic resistance against many diseases. INA was found to protect rice (Metraux et al., 1991), bean (Dann and Deverall, 1995), barley (Kogel et al., 1994; Besser et al., 2000), cucumber (Metraux et al., 1991), sugar beet (Nielsen et al., 1994), and rose (*Rosa* sp.) (Hijwegen et al., 1996) against several pathogens, and was also effective in inducing resistance against

P. syringae pv. *tabaci* in tobacco (Press et al., 1997) and systemic resistance against *Alternaria macrospora* leaf spot in cotton (Colson-Hanks and Deverall, 2000). Two synthetic compounds structurally related to INA, N-phenylsulfony-l2-chlorosonicotinamide and N-cyanomethyl-2-chlorosoisonicotinamide were also reported to induce resistance in rice (Yoshida et al., 1990; Seguchi et al., 1992).

INA has been shown to act via activation of plant defense mechanisms and has a dual mode of action: induction of defense responses prior to infection and potentiation of defense responses postinoculation. In sugar beet, INA does not induce β-1,3-glucanase or chitinase but conditions the plant to induce these pathogenesis-related (PR) proteins faster upon attack by *Cercospora beticola* (Nielsen et al., 1994). Similar results were found in INA-treated cucumber hypocotyls upon challenge by *Colletotrichum lagenarium* (Siegrist et al., 1994). INA may induce salicylic acid, which may signal the induction of defense genes. SAR was characterized by activation of so-called SAR genes (Ward et al., 1991). Application of salicylic acid induces SAR and the accumulation of defense genes (Uknes et al., 1992, 1993; Summermatter et al., 1994). The type of SAR genes induced by salicylic acid were also found to be induced by INA. In tobacco, coordinated induction of SAR genes by INA has been reported. These genes include those encoding PR-1 to PR-9 proteins plus a gene called *SAR 8.2* (Ward et al., 1991). In *Arabidopsis,* the same set of SAR genes and the spectrum of resistance specificity were activated by salicylic acid and INA (Uknes et al., 1992, 1993).

Genetically engineered tobacco and *Arabidopsis* plants unable to accumulate salicylic acid failed to exhibit SAR after induction with pathogens (Vernooij et al., 1994). However, application of INA still induced SAR and SAR gene activity in such plants (Vernooij et al., 1994; Delaney et al., 1995). INA did not induce salicylic acid glucoside accumulation in tobacco (Malamy et al., 1996). However, INA was effective in inducing resistance against *P. syringae* pv. *tabaci* in both wild-type 'Xanthi nc' and transgenic tobacco expressing salicylate hydroxylase (NahG) (Press et al., 1997), suggesting that INA action may not be mediated by salicylic acid.

INA may enter a common pathway downstream of salicylic acid synthesis (Malamy et al., 1996). This model is supported by the isolation and characterization of mutants in *Arabidopsis* that are insensitive to INA as well as salicylic acid (Delaney et al., 1995). INA, as well as salicylic acid, inhibits catalase and ascorbate peroxidase, two key enzymes for H_2O_2 degradation (Durner and Klessig, 1995). INA induces at least some of the SAR genes induced by salicylic acid in different plants (Kogel et al., 1994) and also activates additional regulatory genes in addition to SAR genes (Siegrist et al., 1994). INA was found to be a strong inducer of lipoxygenase (LOX) enzyme activity (Schaffrath et al., 1997; Schweizer et al., 1997). Levels of jasmonic acid were enhanced in leaves of plants treated with INA (Schweizer et al., 1997). Hence, the octadecanoid pathway with (–)-jasmonic acid as a central component may be the INA-activated signaling pathway. However, Schweizer and colleagues (1997) have presented evidences that INA may induce the genes through another, uncharacterized signal transduction pathway. In rice, INA induced massive accumulation of an acid-soluble protein of 31 kDa, ASP 31, which was not induced by salicylic acid, jasmonic acid, ethylene, auxin, kinetin, abscisic acid, gibberellic acid, or by *P. syringae* pv. *syringae* (Schweizer et al., 1997). ASP 31 appears to be a specifically INA-induced protein. Moreover, a number of in vitro translation products (ITPs), ITP24, ITP25, ITP46, and ITP66, accumulated only in INA-treated rice plants (Schweizer et al., 1997) and not in plants inoculated with *P. syringae* pv. *syringae,* an SAR inducer (Smith and Metraux, 1991). Jasmonic acid-induced products did not accumulate in rice plants treated with INA (Schweizer et al., 1997). Exogenously applied jasmonic acid enhanced INA-induced resistance and accumulation of PR-1 and ASP 31 in rice leaves. The fact that ASP 31 accumulation is enhanced by jasmonic acid, although jasmonic acid by itself could not induce accumulation of this protein, suggests the existence of a signaling network between the octadecanoid pathway and another INA-induced pathway (Schweizer et al., 1997).

BTH

Benzo (1,2,3) thiadiazole-7-carbothioic acid S-methyl ester is another plant activator that induces systemic resistance against various pathogens, including fungi, bacteria, and viruses. In monocots, activated resistance by BTH is very long lasting, while the lasting effect is less pronounced in dicots (Oostendorp et al., 2001). BTH is translocated systemically in plants and has been shown to induce resistance in rice (Schaffrath et al., 1997), wheat (Gorlach et al., 1996), barley (Besser et al., 2000), maize (Morris et al., 1998), tobacco (Friedrich et al., 1996), cucumber (Benhamou and Belanger, 1998), *Brassica* (Jensen et al., 1998), and *Arabidopsis* (Lawton et al., 1996). BTH induces systemic resistance in susceptible wheat to *Blumeria graminis* f. sp. *tritici* (Stadnik and Buchenauer, 2000) and resistance against *Beet necrotic yellow vein virus* and *Beet soilborne virus* in beet (Research Information Ltd., 1998). BTH induced synthesis of at least nine acidic and six basic proteins, including three acidic chitinase isozymes, three basic chitinases, three acidic β-1,3-glucanases, and one basic β-1,3-glucanase in sugar beet (Burketova et al., 1999). The signal transduction pathway induced by BTH appears to be different from the pathway involved in SAR induced by incompatible pathogens. Genetically engineered tobacco and *Arabidopsis* plants unable to accumulate salicylic acid failed to exhibit SAR after induction with pathogens. However, application of BTH still induced SAR and SAR gene activity in such plants (Friedrich et al., 1996; Lawton et al., 1996).

Schaffrath and colleagues (1997) inoculated wheat plants with an incompatible pathogen or sprayed them with a 1 mM BTH solution. Inoculation with the nonhost pathogen resulted in accumulation of a set of wheat induced resistance (WIR) transcripts. WIR-1 and its homologue WIR-4 encode a putative cell wall protein (WIR-2), a thaumatin-like protein (PR-5 protein), and WIR-3, a peroxidase. BTH induced another set of transcripts called wheat chemical induction (WCI) transcripts (Schaffrath et al., 1997). WCI-4 encodes a cysteine proteinase, WCI-2 encodes lipoxygenase, and WCI-1, WCI-3, and WCI-5 encode proteins of unknown function (Schaffrath et al., 1997). A nonhost pathogen did not induce WCI transcripts, while BTH did not induce WIR transcripts. In wheat, the biological and chemical (BTH) SAR inducers induce the accumulation of nonidentical sets of transcripts, indicating the existence of different signal transduction pathways.

Similar results have been obtained in rice. SAR induced by *P. syringae* pv. *syringae* in rice results in the accumulation of transcripts homologous to the wheat WIR-1, WIR-2, and WIR-3 clones such as PIR-1, PIR-2, and PIR-3 (PIR denotes pathogen-induced protein from rice). BTH induced SAR in rice. However, BTH treatment does not lead to the accumulation of PIR transcripts in rice (Schaffrath et al., 1997). The types of signal transduction systems activated by BTH appear to be similar to that of INA (Friedrich et al., 1996; Lawton et al., 1996; Schaffrath et al., 1997).

The compound acibenzolar-S-methyl (benzo (1,2,3) thiadiazole-7-carboxylic acid S-methyl ester, [BTH]) is available in Europe as Bion and in the United States as Actigard. It reduces fire blight incidence (Momol, Norelli, Aldwinckle, 1999; Zeller and Zeller, 1999). Control by acibenzolar-S-methyl (100 mg/liter) is usually equal to or better than streptomycin (100 mg/liter) and induces resistance against bacterium. The protection of apple seedlings was constantly associated with the activation of two families of defense-related enzymes, peroxidases and β-1,3-glucanases (Brisset et al., 2000). BTH plus streptomycin provides up to two times better control than either BTH or streptomycin alone (Thomson et al., 1999). Acibenzolar-S-methyl applied as a soil drench or along with the marcot rooting mixture induced resistance in sugarcane against the red rot pathogen, *Colletotrichum falcatum* (Ramesh Sundar et al., 2001). The ISR persisted up to 30 days in the pretreated canes. Increased phenolic content and accumulation of pathogenesis-related proteins, specifically PR-2, PR-3, and PR-5, were observed in sugarcane plants treated with acibenzolar-S-methyl (Ramesh

Sundar et al., 2001). BTH reduced *Potato virus Y* infection in tobacco (Burketova et al., 2000). Tobacco plants were sprayed with 1.2 mM BTH 4 days before inoculation with the virus, and BTH treatment decreased virus multiplication in inoculated leaves. Activity of total ribonucleases and biosynthesis of extracellular acid proteins were increased in both infected and treated leaves. These results suggest that BTH can increase activity of particular ribonuclease enzymes, which are induced by virus infection (Burketova et al., 2000).

Acibenzolar-S-methyl applied singly as pretransplant application or in combination with the insecticide imidacloprid significantly reduced *Tomato spotted wilt virus* (TSWV) incidence in flue-cured tobacco in Georgia (Csinos et al., 2001). Initial applications of acibenzolar-S-methyl made posttransplant did not control TSWV, suggesting plants must be protected prior to introduction into the field. BTH induces resistance against *Alternaria macrospora* leaf spot of cotton (Colson-Hanks and Deverall, 2000). It has also been reported that acibenzolar-S methyl induces resistance in cucumber against *Colletotrichum lagenarium* and *Cladosporium cucumerinum,* and in Japanese pear against *Venturia nashicola* (Ishii et al., 1999).

BABA

DL-3-aminobutyric acid (β-aminobutyric acid, BABA) induced systemic resistance against tomato late blight caused by *Phytophthora infestans* (Jeun et al., 2000), grapevine downy mildew caused by *Plasmopara viticola* (Cohen et al., 1999), pepper blight caused by *Phytophthora capsici* (Lee et al., 2000), and tobacco downy mildew caused by *Peronospora tabacina* (Cohen, 1994). It protected sunflower plants against *Plasmopara helianthi* (= *Plasmopara halstedii*). BABA induced accumulation of free salicylic acid in tomato leaves (Jeun et al., 2000). However, no enhancement of salicylic acid contents was found in the untreated upper leaves expressing systemic acquired resistance after treatment with BABA, suggesting that the locally elevated salicylic acid might be necessary for the synthesis of the still-unknown signal responsible for SAR expression (Jeun et al., 2000). BABA also induces systemic resistance in tobacco (Cohen, 1994). Salicylic acid and INA-treated tobacco plants showed accumulation of PR-1, β-1,3-glucanase, and chitinases. BABA did not induce these PR proteins, but all three of them induced systemic resistance against *Peronospora tabacina* in tobacco (Cohen, 1994). This evidence suggests that BABA induces ISR in a pathway different from that induced by salicylic acid or INA.

BABA protected *Arabidopsis* against *Peronospora parasitica* through activation of natural defense mechanisms of the plant, such as callose deposition, the hypersensitive response, and the formation of trailing necroses (Zimmerli et al., 2000). BABA was still protective against *P. parasitica* in transgenic plants or mutants impaired in the salicylic acid, jasmonic acid, and ethylene signaling pathways. BABA did not induce the accumulation of mRNA of SAR-associated *PR-1* and ethylene- and jasmonic acid-dependent *PDF1.2* genes, suggesting that BABA may act through a pathway other than the salicylic acid, jasmonic acid, and ethylene signaling pathways (Zimmerli et al., 2000). It has been shown that BABA may potentiate the plants to synthesize defense compounds only after pathogen attack (Zimmerli et al., 2000). BABA did not induce PR-1 mRNA but potentiated the accumulation of PR-1 mRNA after attack by virulent pathogenic bacteria in *Arabidopsis* (Zimmerli et al., 2000).

Salicylic Acid

Salicylic acid is an important signal molecule that plays a critical role in plant defense against pathogen invasion. It induces resistance and reduces disease incidence and is effective against fungal, bacterial, and viral diseases. It controls fungal pathogens *Magnaporthe grisea, Fusarium oxysporum* f. sp. *lycopersici,* and *Diplodia pinea,* the bacterial pathogen *Xanthomonas oryzae* pv. *oryzae,* and the virus *Cucumber mosaic virus* (Research Information Ltd., 1998). The addition of 200 μM of salicylic acid to the tomato root system significantly increased the endogenous salicylic acid content of leaves. Concomitant with elevated salicylic acid levels, expression of the tomato pathogenesis-related *PR-1B* gene was strongly induced within 24 hours of the addition of 200 μM of salicylic acid. PR-1B expression was not observed in tomato plants not receiving salicylic acid treatment. Challenge inoculation of salicylic acid-treated tomato plants using conidia of *Alternaria solani* resulted in 83 percent fewer lesions per leaf and a 77 percent reduction in blighted leaf areas compared to control plants not receiving salicylic acid (Spletzer and Enyedi, 1999). Salicylic acid protects barley systemically against powdery mildew caused by *Blumeria graminis* f. sp. *hordei* (Besser et al., 2000). The genes induced by the chemical were different from the genes induced during incompatible barley-*Blumeria graminis* f. sp. *hordei* interactions governing gene-for-gene relationships, which suggests the presence of separate pathways leading to powdery mildew resistance (Besser et al., 2000).

Probenazole

Probenazole (3-allyloxy-1,2-benzisothiazole-1,1-dioxide) induces resistance against blast *(Magnaporthe grisea)* and bacterial blight *(Xantho-*

monas oryzae pv. *oryzae*) in rice (Oostendorp et al., 2001). Its commercial name is Oryzemate. It induces several defense genes (Watanabe et al., 1979). A gene, *RPR1,* was identified as a probenazole-responsive gene (Sakamoto et al., 1999). RPR1 contains a nucleotide binding site and leucine-rich repeats, thus sharing structural similarity with known disease resistance genes. The expression of *RPR1* in rice could be up-regulated by treatment with chemical inducers of SAR and by inoculation with pathogens. *RPR1* was induced during the systemic induced resistance (Sakamoto et al., 1999). Probenazole and its derivative BIT (1,2-benzisothiazol-3-(2H)-one 1,1-dioxide) induced resistance against *Pseudomonas syringae* pv. *tomato* and *Peronospora parasitica* in *A. thaliana* (Yoshioka et al., 2001). They induced expression of several pathogenesis-related genes and accumulation of salicylic acid in *A. thaliana* (Yoshioka et al., 2001). Probenazole or BIT treatment did not induce disease resistance or *PR-1* expression in NahG transgenic plants (salicylic acid is degraded in these plants) or *npr1* mutant *A. thaliana* plants (*npr1* is a key component in the salicylic acid signaling pathway). However, probenazole or BIT treatment induced disease resistance and *PR-1* expression in *Arabidopsis etr1-1* mutant plants, which are insensitive to ethylene, and *coi1-1* mutant plants, which are insensitive to methyl jasmonate (Yoshioka et al., 2001). These results suggest that the salicylic acid pathway is required for probenazole-mediated activation of defense responses, while signal transduction pathways involving ethylene or methyl jasmonate are not required. Probenazole may stimulate the salicylic acid/NPR1-mediated defense signaling pathway upstream of salicylic acid (Yoshioka et al., 2001).

Prohexadione-Ca

The plant growth regulator prohexadione-Ca acts as a plant activator. It has no direct effect on growth of *E. amylovora,* the apple fire blight pathogen. However, it reduced the incidence of fire blight and causes major changes in the flavonoid metabolism of apple. Within four hours after treatment, a new flavonoid (identified as luteoliflavan) occurs in leaf and flower tissue (Roemmelt, Truetter, et al., 1999). Apple trees treated with 250 mg/liter prohexadione-Ca late in the bloom period gave a significantly lower incidence of secondary shoot blight (caused by *E. amylovora*) than unsprayed trees (Fernando and Jones, 1999). Prohexadione-Ca was more effective than Actigard in controlling the apple shoot blight (Momol, Ugine, et al., 1999).

Aluminium Chloride (Synermix)

Hexahydratedaluminium chloride (AlCl$_3$) is the constituent of the new fungicide Synermix (sold by Laboratoires Goemar of France). It acts as a plant activator, activating synthesis of phytoalexins in grapevines. Synermix, when applied in combination with iprodione, increases the efficiency of iprodione in controlling *Botrytis cinerea* in grapevine (Jeandet et al., 2000).

Riboflavin

Riboflavin is a vitamin produced by plants and microbes. Foliar application of riboflavin induced systemic resistance against *Peronospora parasitica* and *Pseudomonas syringae* pv. *tomato* in *Arabidopsis thaliana,* and against *Tobacco mosaic virus* and *Alternaria alternata* in tobacco (Dong and Beer, 2000). It reduces powdery mildew of strawberry plants (Wang and Tzeng, 1998). Riboflavin is involved in antioxidation and peroxidation, processes that affect the production of oxidative burst. The reaction involves complex redox processes that require participation by specific signal molecules, such as H$_2$O$_2$ and nitric oxide, and antioxidants, such as tocopherols and riboflavin (for review, see Dong and Beer, 2000). It acts as plant activator. Riboflavin induced expression of pathogenesis-related genes in tobacco and *Arabidopsis* plants (Dong and Beer, 2000). The response to riboflavin is negated by the protein kinase inhibitor K252a and by mutation of the *NIM1/NPR1* gene (*NIM1/NPR1* encodes a protein that controls expression of various sets of defense genes), but is unaffected in NahG plants (Dong and Beer, 2000). NahG plants expressing the bacterial *nahG* gene, which encodes salicylate hydroxylase that degrades salicylic acid, are incapable of accumulating sufficient salicylic acid to induce systemic acquired resistance when challenged with pathogens. These findings suggest that riboflavin triggers signals that require protein kinases and regulation by *NIM1/NPR1* gene, but does not require accumulation of salicylic acid.

KeyPlex

KeyPlex 350 is a micronutrient preparation. It is a plant activator, activating the plant defense system. Formulations based on KeyPlex 350, applied at 1.90 liters/ 378.53 liters of water per hectare, effectively controlled Sigatoka leaf spot caused by *Mycosphaerella musicola* in Florida (McMillan et al., 2000).

Myco-Sin and Ulmasud

Two products based on mineral powders, Myco-Sin and Ulmasud, reduced apple blossom infections caused by *Erwinia amylovora* in Germany. These products induced synthesis of phenolic acids which resulted in induced resistance (Roemmelt, Plagge, et al., 1999).

Carpropamid

Carpropamid is a new systemic fungicide effective against the rice blast caused by *Pyricularia oryzae*. It induced lignification of infected rice cells and increased accumulation of the phytoalexins momilactone A and sakuranetin in *P. oryzae*-infected leaves (Thieron et al., 1998; Araki and Kurahashi, 1999). It induced resistance against the blast pathogen (Thieron et al., 1998).

Phytogard (K_2HPO_3)

Phytogard is a formulation containing 58 percent potassium phosphonate (Becot et al., 2000). It induces resistance against downy mildew of cauliflower (*Brassica oleracea* var. *botrytis*) caused by *Peronospora parasitica*. It provided complete protection of seedlings, when sprayed with Phytogard at 7 mL/liter of water. The induced resistance persisted for at least 15 days. β-1,3-glucanase activity and PR-2 proteins were induced (Becot et al., 2000).

Phosphate

Phosphorous-containing nutrient solutions of 20 ppm, applied through a hydroponics system, provided ISR against the powdery mildew pathogen *Sphaerotheca fuliginea* in cucumber seedlings (Reuveni et al., 2000). Foliar application of 1 percent solution of monopotassium phosphate effectively protected the foliage against powdery mildew. This treatment was persistent up to 21 days after inoculation. It significantly inhibited powdery mildew development and caused a reduction of 72 percent in sporulation of the fungus compared to control (Reuveni et al., 2000).

Methyl Jasmonate/Jasmonic Acid

Jasmonic acid induces systemic resistance against many pathogens. It controls *Botrytis cinerea* in different crops (Research Information Ltd.,

1998). Jasmonates protect barley plants against *Blumeria graminis* f. sp. *hordei* (Besser et al., 2000) by strengthening plant defense mechanisms that result in effective papillae and host cell death (Besser et al., 2000).

PRACTICAL USES OF PLANT ACTIVATORS

Almost three decades have passed since the effects of salicylic acid and other chemicals inducing disease resistance were first reported. However, large-scale field use of these chemicals has not yet been reported. The major drawback of the chemical induction of defense genes is that their effect is only transient; the induced resistance by chemicals lasts for only a few days. Methods of application such as infiltration by injection or root feeding are impractical under field conditions. When these chemicals are sprayed on foliage, they are easily washed away by rain or dew. They should be applied before infection, as they sensitize the plants for induction of defense genes during the infection process. They are not curative and cannot eliminate already established infection. Most of the chemicals are phytotoxic. However, recently identified chemicals such as INA and BTH persist for a long time, from 30 to 45 days. They can be sprayed or applied through the soil. It appears that these two chemicals have the potential for large-scale field use in management of diseases. The time of application, dosage required, and mode of application must be standardized against each disease (Vidhyasekaran, 2002).

INA and BTH induce resistance by activating genes that are not normally activated during SAR induced by biological agents, which suggests many other defense genes are not activated by INA or BTH. Hence, these chemicals alone cannot contribute to high resistance. In other words, they can only reduce disease intensity and cannot completely control disease (Press et al., 1997). These chemicals should be one of the treatments and not the only treatment to control diseases. For example, efficacy of INA was enhanced with additional application of jasmonic acid in rice (Schweizer et al., 1997). These chemicals may be useful in moderately disease-resistant cultivars rather than in susceptible cultivars and/or when they are combined with other chemicals.

REFERENCES

Araki, Y. and Kurahashi, Y. (1999). Enhancement of phytoalexin synthesis during rice blast infection of leaves by pre-treatment with carpropamid. *J Pesticide Sci,* 24:369-374.

Becot, S., Pajot, E., Corre, D., Monot, C., and Silue, D. (2000). Phytogard (K₂HPO₃) induces localized resistance in cauliflower to downy mildew of crucifers. *Crop Protection*, 19:417-425.

Benhamou, N. and Belanger, R. R. (1998). Induction of systemic resistance to *Pythium* damping-off in cucumber plants by benzothiadiazole—Ultrastructure and cytochemistry of the host response. *Plant J*, 14:13-21.

Besser, K., Jarosch, B., Langen, G., and Kogel, K. H. (2000). Expression analysis of genes induced in barley after chemical activation reveals distinct disease resistance pathways. *Mol Plant Pathol*, 1:277-286.

Brisset, M. N., Cesbron, S., Thomson, S. V., and Paulin, J. P. (2000). Acibenzolar-S-methyl induces the accumulation of defense-related enzymes in apple and protects from fire blight. *Eur J Plant Pathol*, 106:529-536.

Burketova, L., Sindelarova, M., and Sindelar, L. (1999). Benzothiadiazole as an inducer of β-1,3-glucanase and chitinase isozymes in sugar beet. *Biol Plantarum*, 42:279-287.

Burketova, L., Sindelarova, M., and Sindelar, L. (2000). Induced resistance to potato virus Y by benzothiadiazole. In *New Aspects of Resistance Research on Cultivated Plants: Virus Diseases. Proc. Seventh Aschersleben Symp., Aschersleben, Germany, 17-18 November, 1999, Beitrage zur Zuchtungsforschung—Bundesanstalt fur Zuchtungsforschung an Kulturpflanzen*, 6(3):1-3.

Cohen, Y. (1994). 3-Aminobutyric acid induces systemic resistance against *Peronospora tabacina*. *Physiol Mol Plant Pathol*, 44:273-288.

Cohen, Y., Reuveni, M., and Baider, A. (1999). Local and systemic activity of BABA (DL-3-aminobutyric acid) against *Plasmopara viticola* in grapevines. *Eur J Plant Pathol*, 105:351-361.

Colson-Hanks, E. S. and Deverall, B. J. (2000). Effect of 2,6-dichloroisonicotinic acid, its formulation materials and benzothiadiazole on systemic resistance to *Alternaria* leaf spot in cotton. *Plant Pathol*, 49:171-178.

Conrath, U., Thulke, O., Katz, V., Schwindling, S., and Kohler, A. (2001). Priming as a mechanism in induced systemic resistance of plants. *Eur J Plant Pathol*, 107:113-119.

Csinos, A. S., Pappu, H. R., McPherson, R. M., and Stephenson, M. G. (2001). Management of *Tomato spotted wilt virus* in flue-cured tobacco with acibenzolar-S-methyl and imidacloprid. *Plant Dis*, 85:292-296.

Dann, E. K. and Deverall, B. J. (1995). Effectiveness of systemic resistance in bean against foliar and soil-borne pathogens as induced by biological and chemical means. *Plant Pathol*, 44:458-466.

Delaney, T. P., Friedrich, L., and Ryals, J. (1995). *Arabidopsis* signal transduction mutant defective in chemically and biologically induced disease resistance. *Proc Natl Acad Sci USA*, 92:6602-6606.

Dong, H. and Beer, S. V. (2000). Riboflavin induces disease resistance in plants by activating a novel signal transduction pathway. *Phytopathology*, 90:801-811.

Durner, J. and Klessig, D. F. (1995). Inhibition of ascorbate peroxidase by salicylic acid and 2,6-dichloroisonicotinic acid, two inducers of plant defense responses. *Proc Natl Acad Sci*, 92:11312-11316.

Fernando, W. G. D., and Jones, A. L. (1999). Prohexadione calcium—A tool for reducing secondary fire blight infection. *Acta Horticulturae,* 489:597-600.

Friedrich, L., Lawton, K., Reuss, W., Masner, P., Specker, N., Gut Rella, M., Meier, B., Dincher, S., Staub, T., Uknes, et al. (1996). A benzothiadiazole derivative induces systemic acquired resistance in tobacco. *Plant J,* 10:61-70.

Gorlach, J., Volrath, S., Knauf-Beiter, G., Hengry, G., Beckhove, U., Kogel, K.-H., Oostendorp, M., Staub, T., Ward, E., Kessmann, H., and Ryals, J. (1996). Benzothiadiazole, a novel class of inducers of systemic acquired resistance, activates gene expression and disease resistance in wheat. *Plant Cell,* 8:629-643.

Hijwegen, T., Verhaar, M. A., and Zadoks, J. C. (1996). Resistance to *Sphaerotheca pannosa* in roses induced by 2,6-dichloroisonicotinic acid. *Plant Pathol,* 45: 631-635.

Ishii, H., Tomita, Y., Horio, T., Narusaka, Y., Nishimura, K., and Iwamoto, S. (1999). Induced resistance of acibenzolar-S-methyl (CGA-245704) to cucumber and Japanese pear diseases. *Eur J Plant Pathol,* 105:77-85.

Jeandet, P., Adrian, M., Breuil, A. C., Sbaghi, M., Debord, S., Bessis, R., Weston, L. A., and Harmon, R. (2000). Chemical induction of phytoalexin synthesis in grapevines: Application to the control of grey mould in the vineyard. *Acta Horticulturae,* 528:591-596.

Jensen, B. D., Latundedada, A. O., Hudson, D., and Lucas, J. A. (1998). Protection of *Brassica* seedlings against downy mildew and damping-off by seed treatment with CGA-245704, an activator of systemic acquired resistance. *Pestic Sci,* 52:63-69.

Jeun, Y. C., Siegrist, J., and Buchenauer, H. (2000). Biochemical and cytological studies on mechanisms of systemically induced resistance to *Phytophthora infestans* in tomato plants. *J Phytopathol,* 148:129-140.

Kloepper, J. W., Tuzun, S., and Kuc, J. A. (1992). Proposed definitions related to induced disease resistance. *Biocontrol Sci Technol,* 2:349-351.

Kogel, K-H., Beckhove, U., Dreschers, J., Munch, S., and Romme, Y. (1994). Acquired resistance in barley. The resistance mechanism induced by 2,6-dichloroisonicotinic acid is a phenocopy of a genetically based mechanism governing race-specific powdery mildew resistance. *Plant Physiol,* 106:1269-1277.

Kuc, J. (2001). Concepts and direction of induced systemic resistance in plants and its application. *Eur J Plant Pathol,* 107:7-12.

Lawton, K., Friedrich, L., Hunt, M., Weymann, K., Delaney, T., Kessmann, H., Staub, T., and Ryals, J. (1996). Benzothiadiazole induces disease resistance in *Arabidopsis* by activation of the systemic acquired resistance signal transduction pathway. *Plant J,* 10:71-82.

Lee, Y. K., Hong, J. K., Hippe-Sanwald, S., and Hwang, B. K. (2000). Histological and ultrastructural comparisons of compatible, incompatible and DL-β-amino-n-butyric acid-induced resistance responses of pepper stems to *Phytophthora capsici. Physiol Mol Plant Pathol,* 57:269-280.

Malamy, J., Sanchez-Casas, P., Hennig, J., Guo, A., and Klessig, D. F. (1996). Dissection of the salicylic acid signaling pathway in tobacco. *Mol Plant-Microbe Interact,* 9:474-482.

McMillan, R. T. Jr., Graves, W., and McLaughlin, H. J. (2000). KeyPlex effect on Sigatoka of banana. *Proc Florida State Hort Soc,* 112:192-193.

Metraux, J. P., Ahl-Goy, P., Staub, T., Speich, J., Steinemann, A., Ryals, J., and Ward, E. (1991). Induced systemic resistance in cucumber in response to 2,6-dichloro-isonicotinic acid and pathogens. In I. H. Hennecke and D. P. S. Verma (Eds.), *Advances in Molecular Genetics of Plant-Microbe Interactions, Volume I.* Kluwer Academic Publishers, Dordrecht, The Netherlands, pp. 432-439.

Momol, M. T., Norelli, J. L., and Aldwinckle, H. S. (1999). Evaluation of biological control agents, systemic acquired resistance inducers and bactericides for the control of fire blight on apple blossom. *Acta Horticulturae,* 489:553-557.

Momol, M. T., Ugine, J. D., Norelli, J. L., and Aldwinckle, H. S. (1999). The effect of prohexadione calcium, SAR inducers and calcium on the control of shoot blight caused by *Erwinia amylovora* on apple. *Acta Horticulturae,* 489:601-605.

Morris, S. W., Vernooij, B., Titatarn, S., Starrett, M., Thomas, S., Wiltse, C. C., Frederiksen, R. A., Bhandhufalck, A., Hulbert, S., and Uknes, S. (1998). Induced resistance responses in maize. *Mol Plant-Microbe Interact,* 11:543-658.

Nielsen, K. K., Bojsen, K., Collinge, D. B., and Mikkelsen, J.D. (1994). Induced resistance in sugar beet against *Cercospora beticola: Induction by dichloro-isonicotinic acid is independent of chitinase and* β-1,3-glucanase transcript accumulation. *Physiol Mol Plant Pathol,* 45:89-99.

Oostendorp, M., Kunz, W., Dietrich, B., and Staub, T. (2001). Induced disease resistance in plants by chemicals. *Eur J Plant Pathol,* 107:19-28.

Press, C. M., Wilson, M., Tuzun, S., and Kloepper, J. W. (1997). Salicylic acid produced by *Serratia marcescens* 90-166 is not the primary determinant of induced systemic resistance in cucumber or tobacco. *Mol Plant-Microbe Interact,* 10: 761-768.

Ramesh Sundar, A., Velazhahan, R., Viswanathan, R., Padmanabhan, P., and Vidhyasekaran, P. (2001). Induction of systemic resistance to *Colletotrichum falcatum* in sugarcane by a synthetic signal molecule, Acibenzolar-S-methyl (CGA-245704). *Phytoparasitica,* 29:231-242.

Research Information Ltd. (1998). *International Pesticide Directory,* Sixteenth Edition. Research Information Ltd., Herts, U. K.

Reuveni, R., Dor, G., Raviv, M., Reuveni, M. and Tuzun, S. (2000). Systemic resistance against *Sphaerotheca fuliginea* in cucumber plants exposed to phosphate in hydroponics system, and its control by foliar spray of monopotassium phosphate. *Crop Protection,* 19:355-361.

Roemmelt, S., Plagge, J., Truetter, D., and Zeller, W. (1999). Fire blight control in apple using products based on mineral powders. *Acta Horticulturae,* 489:623-624.

Roemmelt, S., Truetter, D., Speakman, J. B., and Rade-Macher, W. (1999). Effects of prohexadione-Ca on the flavonoid metabolism of apple with respect to plant resistance against fire blight. *Acta Horticulturae,* 489:359-363.

Sakamoto, K., Tada, Y., Yokozeki, Y., Akagi, H., Hayashi, N., Fujimura, T., and Ichikawa, N. (1999). Chemical induction of disease resistance in rice is correlated with the expression of a gene encoding a nucleotide binding site and leucine-rich repeats. *Plant Mol Biol,* 40:847-855.

Schaffrath, U., Freydl, E., and Dudler, R. (1997). Evidence for different signaling pathways activated by inducers of acquired resistance in wheat. *Mol Plant-Microbe Interact,* 10:779-783.

Schweizer, P., Buchala, A., and Metraux, J. P. (1997). Gene-expression patterns and levels of jasmonic acid in rice treated with the resistance inducer 2,6-dichloroisonicotinic acid. *Plant Physiol,* 115:61-70.

Seguchi, K., Kurotaki, M., Sekido, S., and Yamaguchi, I. (1992). Action mechanism of N-cyanomethyl-2-chloroisonicotinamide in controlling rice blast disease. *J Pesticide Sci,* 17:107-113.

Siegrist, J., Jeblick, W., and Kauss, H. (1994). Defense responses in infected and elicited cucumber (*Cucumis sativus* L.) hypocotyl segments exhibiting acquired resistance. *Plant Physiol,* 105:1365-1374.

Smith, J. A. and Metraux, J. P. (1991). *Pseudomonas syringae* pv. *syringae* induces systemic resistance to *Pyricularia oryzae* in rice. *Physiol Mol Plant Pathol,* 39:451-461.

Spletzer, M. E. and Enyedi, A. J. (1999). Salicylic acid induces resistance to *Alternaria solani* in hydroponically grown tomato. *Phytopathology,* 89:722-727.

Stadnik, M. J. and Buchenauer, H. (2000). Inhibition of phenylalanine ammonia-lyase suppresses the resistance induced by benzothiadiazole in wheat to *Blumeria graminis* f. sp. *tritici. Physiol Mol Plant Pathol,* 57:25-34.

Summermatter, K., Meuwly, P., Molders, W., and Metraux, J. P. (1994). Salicylic acid levels in *Arabidopsis thaliana* after treatments with *Pseudomonas syringae* or synthetic inducers. *Acta Horticulturae,* 381:367-370.

Thieron, M., Pontzen, R., and Kurahashi, Y. (1998). Carpropamid: A rice fungicide with two modes of action. *Pflanzenschutz-Nachrichten Bayer,* 51:259-280.

Thomson, S. V., Gouk, S. C., and Paulin, J. P. (1999). Efficacy of Bion (Actigard) to control fire blight in pear and apple orchards in U.S.A., New Zealand, and France. *Acta Horticulturae,* 489:589-595.

Tuzun, S. (2001). The relationship between pathogen-induced systemic resistance (ISR) and multigenic (horizontal) resistance in plants. *Eur J Plant Pathol,* 107:85-93.

Uknes, S., Mauch-Mani, B., Moyer, M., Potter, S., Williams, S., Dincher, S., Chandler, D., Slusarenko, A., Ward, E., and Ryals, J. (1992). Acquired resistance in *Arabidopsis. Plant Cell,* 4:645-656.

Uknes, S., Winter, A. M., Delaney, T., Vernooij, B., Morse, A., Friedrich, L., Nye, G., Potter, S., Ward, E., and Ryals, J. (1993). Biological induction of systemic acquired resistance in *Arabidopsis. Mol Plant-Microbe Interact,* 6:692-698.

Vernooij, B., Friedrich, L., Morse, A., Resist, R., Kolditz-Jawhar, R., Ward, E., Uknes, S., Kessmann, H., and Ryals, J. (1994). Salicylic acid is not the translocated signal responsible for inducing systemic acquired resistance but is required in signal transduction. *Plant Cell,* 6:959-965.

Vidhyasekaran, P. (1998). Molecular biology of pathogenesis and induced systemic resistance. *Indian Phytopath,* 51:111-120.

Vidhyasekaran, P. (2002). *Bacterial Disease Resistance in Plants: Molecular Biology and Biotechnological Applications*. The Haworth Press, Binghamton, NY.

Vidhyasekaran, P., Velazhahan, R., and Balasubramanian, P. (2000). Biological control of crop diseases exploiting genes involved in systemic induced resistance. In R. K. Upadhyay, K. G. Mukherji, and B. P. Chamola (Eds.), *Biocontrol Potential and Its Exploitation in Sustainable Agriculture. Volume 1: Crop Diseases, Weeds, and Nematodes*. Kluwer Academic/Plenum Publishers, New York, pp. 1-8.

Wang, S. and Tzeng, D. D. (1998). Methionine-riboflavin mixtures with surfactants and metal ions reduce powdery mildew infection in strawberry plants. *J Am Soc Sci*, 123:987-991.

Ward, E. R., Uknes, S. J., Williams, S. C., Dincher, S. S., Wiederhold, D. L., Alexander, D. C., Ahl-Goy, P., Metraux, J. P., and Ryals, J. (1991). Co-ordinate gene activity in response to agents that induce systemic acquired resistance. *Plant Cell*, 3:1085-1094.

Watanabe, T., Sekizawa, Y., Shimura, H., Suzuki, Y., Matsumoto, M., Iwata, M., and Mase, S. (1979). Effects of probenazole (Oryzemate) on rice plants with reference to controlling rice blast. *J Pestic Sci*, 4:53-58.

Yoshida, H., Konishi, K., Nakagawa, T., Sekido, S., and Yamaguchi, I. (1990). Characteristics of N-phenylsulfonyl-2-chloroisonicotinamide as an anti-rice blast agent. *J Pesticide Sci*, 15:199-203.

Yoshioka, K., Nakashita, H., Klessig, D. F., and Yamaguchi, I. (2001). Probenazole induces systemic acquired resistance in *Arabidopsis* with a novel type of action. *Plant J*, 25:149-157.

Zeller, W. and Zeller, V. (1999). Control of fire blight with the plant activator Bion. *Acta Horticulturae*, 489:639-645.

Zimmerli, L., Jakab, G., Metraux, J. P., and Mauch-Mani, B. (2000). Potentiation of pathogen-specific defense mechanisms in *Arabidopsis* by β-aminobutyric acid. *Proc Natl Acad Sci USA*, 97:12920-12925.

29

Plant Extracts

Several plant extracts have been shown to control diseases. Some of them have been developed as commercial formulations. Plant extracts may have antimicrobial action, or they may induce disease resistance by activating a signal transduction system in plants. The possible uses of plant extracts in crop disease management are described in this chapter.

ANTIMICROBIAL ACTION OF PLANT EXTRACTS

Several plant extracts have been reported to possess antimicrobial action. Proteins from the seeds of 200 Australian native plant accessions were extracted, and most of them were found to show antimicrobial activity. The peptides isolated from *Hardenbergia violacea* and *Macadamia integrifolia* showed strong inhibitory activity in vitro toward plant pathogens (Harrison et al., 1998). A novel antifungal compound, fistulosin (octadecyl 3-hydroxy-indole), isolated from roots of Welsh onion *(Allium fistulosum)* showed high activity against *Fusarium oxysporum,* primarily inhibiting protein synthesis (Phay et al., 1999). Several articles have appeared in recent years describing antimicrobial compounds isolated from various plants. However, only very few of these compounds have been tested for crop-disease management under field conditions. Leaf extracts of *Annona cherimola, Bromelia hemisphaerica,* and *Carica papaya* inhibited sporulation of *Rhizopus stolonifer,* the pathogen of "circula" fruit *(Spondias purpurea)* rot. They also reduced fruit rot infection and development (Bautista-Banos et al., 2000). Garlic juice and grapefruit juice (BioSept) effectively controlled rose powdery mildew *(Sphaerotheca pannosa* var. *rosae)*, and their efficacy was equal to that of the standard fungicide triforine (Wojdyla, 2000). When root extracts of *Mirabilis jalapa,* containing a ribosome inactivating protein called Mirabilis antiviral protein (MAP), were sprayed on potato plants, infection by *Potato virus X* (PVX) and potato spindle tuber viroid was inhibited by almost 100 percent (Vivanco et al., 1999). Similar results were obtained in tomato plants against PVX infection. Antiviral activity of MAP extracts was observed against mechanically transmitted viruses, but the extracts did not

461

show activity against aphid transmitted viruses, such as *Potato virus Y* (PVY) and *Potato leaf roll virus* (PLRV) on potatoes (Vivanco et al., 1999).

Commercial formulations of various plant extracts, such as pepper/mustard (chilli pepper extract and the essential oil of mustard), cassia (extract of cassia tree), clover (70 percent clove oil), and neem (90 percent neem oil) were tested for their efficacy to control muskmelon wilt caused by *Fusarium oxysporum* f. sp. *melonis* (Bowers and Locke, 2000). Soil infested with *F. oxysporum* f. sp. *melonis* was treated with 5 to 10 percent aqueous emulsions of commercial formulations of these plant extracts, and these treatments, excepting neem oil, effectively controlled the wilt disease of muskmelon in the greenhouse. Pepper/mustard, cassia, and clove extracts suppressed disease development (80 to 100 percent healthy plant stand) compared to the untreated infested soil (< 20 percent stand). Neem oil extract was ineffective against *Fusarium* spp. (Bowers and Locke, 2000).

INDUCTION OF DISEASE RESISTANCE BY PLANT EXTRACTS

Several plant extracts induced resistance against diseases by activating signal transduction systems. Milsana, a commercial product of leaf extracts from the giant knot weed *Reynoutria sachalinensis,* significantly reduced the incidence of cucumber *(Cucumis sativus)* powdery mildew *(Sphaerotheca fuliginea)* (Wurms et al., 1999; Daayf et al., 2000). Milsana was more effective than the plant activator benzothiadiazole (BTH) in controlling powdery mildew on long English cucumber cv. Corona (Wurms et al., 1999). Milsana application induced localized resistance. Most haustoria of the pathogen had collapsed in the localized Milsana treatment and were encapsulated by an amorphous material impregnated by electron-opaque substances, probably phenolic substances (Wurms et al., 1999). Milsana treatment increased the levels of *p*-coumaric, caffeic, and ferulic acids and *p*-coumaric acid methyl ester in the cucumber leaves. The amounts of hydroxycinnamic acid increased rapidly in the two test cultivars of cucumber following Milsana treatment (Daayf et al., 2000). All of these phenolic compounds showed antifungal activity against common pathogens of cucumber, such as *Botrytis cinerea, Pythium ultimum,* and *P. aphanidermatum,* which suggests that the plant extract would induce synthesis of antifungal compounds and contribute to disease resistance (Daayf et al., 2000).

Pokeweed (*Phytolacca* spp.) extract has been shown to inhibit *Tobacco mosaic virus, Potato virus X, Potato virus Y, Cucumber mosaic virus, Southern bean mosaic virus,* and *Alfalfa mosaic virus* (Baranwal and Verma,

2000). Several antiviral proteins have been isolated from pokeweed extract. These proteins show ribosome inhibitory action against viruses. A pokeweed antiviral protein (PAP), a 29 kDa protein isolated from *Phytolacca americana,* inhibits translation by catalytically removing a specific adenine residue from the large rRNA of the 60S subunit of eukaryotic ribosomes (Smirnov et al., 1997). The PAP depurinates the sarcin/ricin (S/R) loop of the large rRNA of prokaryotic and eukaryotic ribosomes (Zoubenko et al., 2000). A mutant of pokeweed antiviral protein, PAPn, which has a single amino acid substitution (Gly-75), did not bind ribosomes efficiently, indicating Gly-75 in the N-terminal domain is critical for the binding of PAP to ribosomes. PAPn did not depurinate ribosomes and was nontoxic when expressed in transgenic tobacco plants. The transgenic plants, which expressed nontoxic protein, were also resistant to viral and fungal infection (Zoubenko et al., 2000). These results suggest that PAP may act as inducer of systemic resistance against viruses, rather than by directly inhibiting viruses. PAP induces synthesis of pathogenesis-related proteins and a small increase in salicylic acid levels in tobacco. An intact active site of PAP is necessary for induction of virus resistance in plants. Enzymatic activity of PAP appears to be responsible for generating a signal that renders tobacco plants resistant to virus infection (Smirnov et al., 1997).

Plant extracts have the potential to control various diseases. However, they are less stable on sprayed plants, and frequent application of these products is needed to manage diseases efficiently. Further research is necessary to obtain highly stable compounds.

REFERENCES

Baranwal, V. K. and Verma, H. N. (2000). Antiviral phytoproteins as biocontrol agents for efficient management of plant viruses. In R. K. Upadhyay, K. G. Mukerji, and B. P. Chamola (Eds.), *Biocontrol Potential and its Exploitation in Sustainable Agriculture. Volume 1: Crop Diseases, Weeds, and Nematodes.* Kluwer Academic/Plenum Publishers, New York, pp. 71-79.

Bautista-Banos, S., Hernandez-Polez, M., Diaz-Perez, J. C., and Cano-Ochoa, C. F. (2000). Evaluation of the fungicidal properties of plant extracts to reduce *Rhizopus stolonifer* of "circula" fruit (*Spondias purpurea* L.) during storage. *Postharvest Biol Technol,* 20:99-106.

Bowers, J. H. and Locke, J. C. (2000). Effect of botanical extracts on the population density of *Fusarium oxysporum* in soil and control of *Fusarium* wilt in the greenhouse. *Plant Dis,* 84:300-305.

Daayf, F., Ongena, M., Boulanger, R., El-Hadrami, I., and Belanger, R. R. (2000). Induction of phenolic compounds in two cultivars of cucumber by treatment of healthy and powdery mildew-infected plants with extracts of *Reynoutria sachalinensis. J Chem Ecol,* 26:1579-1593.

Harrison, S. J., Marcus, J. P., Green, J. L., Goulter, K. C., Maclean, D. J., and Manners, J. M. (1998). Isolation and characterization of antimicrobial proteins from Australian native plants. *Proc Royal Soc Queensland,* 107:119-121.

Phay, N., Higashiyama, T., Tsuji, M., Matsuura, H., Fukushi, Y., Yokota, A., and Tomita, F. (1999). An antifungal compound from roots of Welsh onion. *Phytochemistry,* 52:271-274.

Smirnov, S., Shulaev, V., and Tumer, N. E. (1997). Expression of pokeweed antiviral protein in transgenic plants induced virus resistance in grafted wild-type plants independently of salicylic acid accumulation and pathogenesis-related protein synthesis. *Plant Physiol,* 114:1113-1121.

Vivanco, J. M., Querci, M., and Salazar, L. F. (1999). Antiviral and antiviroid activity of MAP-containing extracts from *Mirabilis jalapa* roots. *Plant Dis,* 83:1116-1121.

Wojdyla, A. T. (2000). Influence of some compounds on development of *Sphaerotheca pannosa* var. *rosae. J Plant Protect Res,* 40:106-121.

Wurms, K., Labbe, C., Benhamou, N., and Belanger, R. R. (1999). Effects of Milsana and benzothiadiazole on the ultrastructure of powdery mildew haustoria on cucumber. *Phytopathology,* 89:728-736.

Zoubenko, O., Hudak, K., and Tumer, N. E. (2000). A non-toxic pokeweed antiviral protein mutant inhibits pathogen infection via a novel salicylic acid-independent pathway. *Plant Mol Biol,* 44:219-229.

CROP DISEASES

30

Diseases of Major Cultivated Crops

In the following sections, a comprehensive list of diseases occurring in major crops grown in different parts of the world is presented. The common names of the diseases given in these sections are the American Phytopathological Society approved names (Hansen, 1985, 1988, 1991; APSnet, 1997). Names of fungal pathogens (their accepted names, synonyms, anamorphic and teleomorphic names) with authority are given (Hawksworth et al., 1995; Wrobel and Creber, 1998). Synonyms are given as =. Names of the bacterial pathogens are given based on papers published by Young and colleagues (1996) and Schaad and colleagues (2000). The names include accepted names and all valid synonyms, which are indicated as =. Names of viruses have been given based on the guidelines provided by Van Regenmortel (1999).

APPLE (MALUS × DOMESTICA BORKH.)

Alternaria blotch: *Alternaria mali* Roberts = *A. alternata* (Fr.:Fr) Keissl, apple pathotype
American brown rot: *Monilinia fructicola* (G. Wint.) Honey
American hawthorne rust: *Gymnosporangium globosum* (Farl.) Farl.
Anthracnose canker: *Pezicula malicorticis* (H. Jacks.) Nannf, Anamorph: *Cryptosporiopsis curvispora* (Peck) Gremmen in Boerema and Gremmen
Apple blister bark: Apple fruit crinkle viroid (AFCVd)
Apple chlorotic leafspot: genus *Trichovirus, Apple chlorotic leaf spot virus* (ACLSV)
Apple dimple fruit: Apple scar skin viroid (ASSVd)
Apple flat apple: genus *Nepovirus, Cherry rasp leaf virus* (CRLV) (family *Comoviridae*)
Apple fruit crinkle: Apple fruit crinkle viroid (AFCVd)
Apple mosaic: genus *Ilarvirus, Apple mosaic virus* (ApMV); genus *Ilarvirus, Tulare apple mosaic virus* (TAMV)
Apple proliferation: Phytoplasma

Apple ring rot and canker: *Botryosphaeria berengeriana* De Not. = *Physalospora piricola* Nose

Apple scar skin: Apple scar skin viroid (ASSVd)

Apple stem grooving: genus *Capillovirus, Apple stem grooving virus* (ASGV)

Apple stem pitting: genus *Foveavirus, Apple stem pitting virus* (ASPV)

Apple union necrosis: genus *Nepovirus, Tomato ringspot virus* (ToRSV) (family *Comoviridae*)

Armillaria root rot: *Armillaria mellea* (Vahl:Fr.) P. Kumm.

Bitter rot: *Glomerella cingulata* (Stoneman) Spauld. and H. Schrenk, Anamorph: *Colletotrichum gloeosporioides* (Penz.) Penz. and Sacc. in Penz., *Colletotrichum acutatum* J. H. Simmons

Black pox: *Helminthosporium papulosum* Berg.

Black root rot: *Xylaria mali* Fromme, *X. polymorpha* (Pers.:Fr.) Grev.

Black rot and stem canker: *Botryosphaeria obtusa* (Schwein.) Shoemaker, Anamorph: *Sphaeropsis malorum* Berk. (Mitosporic fungi) = *Physalospora obtusa* (Schwein) Cooke

Blister canker: *Biscogniauxia marginata* (Fr.) Pouzar = *Nummularia discreta* (Schwein.) Tul. and C. Tul.

Blister spot: *Pseudomonas syringae* pv. *papulans* (Rose 1917) Dhanvantari 1977

Blue mold: *Penicillium* spp., *P. expansum* Link

Brooks fruit spot: *Mycosphaerella pomi* (Pass.) Lindau, Anamorph: *Cylindrosporium pomi* C. Brooks

Brown rot: *Monilinia laxa* (Aderhold and Ruhland) Honey

Calyx-end rot: *Sclerotinia sclerotiorum* (Lib.) de Bary

Cedar apple rust: *Gymnosporangium juniperi-virginianae* Schwein

Clitocybe root rot: *Armillaria tabescens* (Scop.) Dennis et al. = *Clitocybe tabescens* (Scop.) Bres.

Crown gall: *Agrobacterium tumefaciens* (Smith and Townsend 1907) Conn 1942

Diaporthe canker: *Diaporthe tanakae* Kobayashi and Sakuma, Anamorph: *Phomopsis tanakae* Kobayashi and Sakuma

Diplodia canker: *Botryosphaeria stevensii* Shoemaker = *Physalospora malorum* Shear et al., Anamorph: *Diplodia mutila* (Fr.:Fr.) Mont.

European brown rot: *Monilinia fructigena* Honey in Whetzel, Anamorph: *Monilia fructigena* Pers.:Fr., *Monilinia laxa* (Aderhold and Ruhland) Honey

Fire blight: *Erwinia amylovora* (Burrill 1882) Winslow, Buchanan, Krumwiede, Rogers, and Smith, 1920

Fisheye rot: *Butlerelfia eustacei* Weresub and Illman = *Corticium centrifugum* (Lév.) Bres.

Fly speck: *Schizothyrium pomi* (Mont.:Fr.) Arx, Anamorph: *Zygophiala jamaicensis* E. Mason

Gray mold rot: *Botrytis cinerea* Pers. Fr., Teleomorph: *Botryotinia fuckeliana* (de Bary) Whetzel

Hairy root: *Agrobacterium rhizogenes* (Riker, Banfield, Wright, Keitt, and Sagen 1930) Conn 1942

Japanese apple rust: *Gymnosporangium yamadae* Miyabe ex Yamada

Leptosphaeria canker: *Diapleella coniothyrium* (Fuckel) Barr = *Leptosphaeria coniothyrium* (Fuckel) Sacc., Anamorph: *Coniothyrium fuckelii* Sacc.

Leucostoma canker: *Leucostoma cincta* (Fr.:Fr.) Hohn., Anamorph: *Cytospora cincta* Sacc., *Valsa auerswaldii* Nitschke = *Leucostoma auerswaldii* (Nitschke) Hohn.

Marssonina blotch: *Diplocarpon mali* Harada and Sawamura, Anamorph: *Marssonina coronaria* (Ellis and J. J. Davis) J. J. Davis

Monilia leaf blight: *Monilinia mali* (Takahashi) Whetzel, Anamorph: *Monilia* sp.

Monochaetia twig canker: *Monochaetia mali* (Ellis and Everh.) Sacc., Teleomorph: *Lepteutypa cupressi* (Nattras et al.) H. J. Swart

Mucor rot: *Mucor* spp., *Mucor piriformis* E. Fischer

Nectria canker: *Nectria galligena* Bres. in Strass., Anamorph: *Cylindrocarpon heteronemum* (Berk. and Broome) Wollenweb

Nectria twig blight: *Nectria cinnabarina* (Tode:Fr.) Fr., Anamorph: *Tubercularia vulgaris* Tode:Fr.

Pacific Coast pear rust: *Gymnosporangium libocedri* (C. Henn.) F. Kern

Peniophora root canker: *Peniophora sacrata* G. H. Cunn.

Perennial canker: *Neofabrae perennans* Kienholz, Anamorph: *Cryptosporiopsis perennans* (Zeller and Childs) Wollenweb.

Phomopsis canker: *Phomopsis mali* Roberts, Teleomorph: *Diaporthe perniciosa* Em. Marchal

Phymatotrichum root rot: *Phymatotrichopsis omnivora* (Duggar) Hennebert = *Phymatotrichum omnivorum* Duggar

Phytophthora crown and root rot: *Phytopthora cactorum* (Lebert and Cohn) J. Schröt., *P. cambivora* (Petr.) Buisman, *P. cryptogea* Pethybr. and Lafferty, *P. megasperma* Drechs., *P. syringae* (Kleb.) Kleb.

Pink mold rot: *Trichothecium roseum* (Pers.:Fr.) Link = *Cephalothecium roseum* Corda

Powdery mildew: *Podosphaera leucotricha* (Ellis and Everh.) E. S. Salmon

Quince rust: *Gymnosporangium clavipes* (Cooke and Peck) Cooke and Peck in Peck

Rosellinia root rot: *Rosellinia necatrix* Prill., Anamorph: *Dematophora necatrix* R. Hartig

Scab: *Venturia inaequalis* (Cooke) G. Wint., Anamorph: *Spilocaea pomi* Fr.:Fr.

Side rot: *Phialophora malorum* (M. N. Kidd and A. Beaumont) McColloch

Silver leaf: *Chondrostereum purpureum* (Pers.:Fr.) Pouzar

Sooty blotch complex: *Peltaster fructicola* Johnson, Sutton, and Hodges, *Geastrumia polystigmatis* Batista and M. L. Farr, *Leptodontium elatius* (G. Mangenot) De Hoog, *Gloeodes pomigena* (Schwein.) Colby

Southern blight: *Sclerotium rolfsii* Sacc., Teleomorph: *Athelia rolfsii* (Curzi) Tu and Kimbrough

Thread blight: *Corticium stevensii* Burt = *Pellicularia koleroga* Cooke = *Hypochnus ochroleucus* Noack

Twig canker: *Phyllosticta solitaria* Ellis and Everh.

Valsa canker: *Valsa ceratosperma* (Tode:Fr.), Anamorph: *Cytospora sacculus* (Schwein.) Gvritischvili

Violet root rot: *Helicobasidium mompa* Tanaka

White root rot: *Scytinostroma galactinum* (Fr.) Donk = *Corticium galactinum* (Fr.) Burt

White rot and stem canker: *Botryosphaeria dothidea* (Moug.) Ces. and De Not., Anamorph: *Fusicoccum aesculi* Corda

X-spot (Nigrospora spot): *Nigrospora oryzae* (Berk. and Broome) Petch

Zonate leaf spot: *Cristulariella moricola* (Hino) Redhead, Teleomorph: *Grovesinia pyramidalis* M. Cline et al.

BANANA (MUSA SPECIES)

Anthracnose: *Colletotrichum musae* (Berk. and M. A. Curtis) Arx

Armillaria rot: *Armillaria mellea* (Vahl:Fr.) P. Kumm., *Armillaria tabescens* (Scop.) Dennis, Orton, and Hora

Black cross: *Phyllachora musicola* C. Booth and D. E. Shaw

Black root rot: *Rosellinia bunodes* (Berk. and Broome) Sacc.

Black Sigatoka: *Mycosphaerella fijiensis* M. Morelet, Anamorph: *Paracercospora fijiensis* (M. Morelet) Deighton

Bract mosaic: genus *Potyvirus, Banana bract mosaic virus* (BBMV) (family *Potyviridae*)

Brown blotch: *Pestalotiopsis leprogena* (Speg.) Steyaert

Brown spot: *Cercospora hayi* Calp.

Bunchy top: genus *Nanavirus, Banana bunchy top virus* (BBTV)

Ceratocystis fruit rot: *Ceratocystis paradoxa* (Dade) C. Moreau, Anamorph: *Chalara paradoxa* (De Seynes) Sacc.

Cigar-end: *Verticillium theobromae* (Turconi) E. W. Mason and S. J. Hughes

Cladosporium speckle: *Cladosporium musae* E. W Mason

Cordana leaf spot: *Cordana johnstonii* M. B. Ellis, *Cordana musae* (Zimm.) Höhn

Corm dry rot: *Junghuhnia vincta* (Berk.) Hood and M. Dick

Deightoniella fruit speckle and leaf spot: *Deightoniella torulosa* (Syd.) M. B. Ellis

Diamond spot: *Cercospora hayi* Calp.

Dwarf Cavendish tip rot: *Nattrassia mangiferae* (Syd. and P. Syd.) B. Sutton and Dyko = *Hendersonula toruloidea* Nattrass

Eyespot: *Dreschslera gigantea* (Heald and F. A. Wolf) S. Ito

Fruit freckle: *Guignardia musae* Racib., Anamorph: *Phyllosticta musarum* (Cooke) Aa

Fruit rot: *Botryosphaeria ribis* Grossenb. and Duggar

Fungal root-rot: *Fusarium solani* (Mart.) Sacc., Teleomorph: *Nectria haematococca* Berk. and Broome, *Fusarium oxysporum* Schltdl.:Fr., and *Rhizoctonia* spp.

Fungal scald: *Colletotrichum musae* (Berk. and M. A. Curtis) Arx.

Fusarium wilt (Panama disease): *Fusarium oxysporum* Schltdl.:Fr. f. sp. *cubense* (E. F. Sm.) W. C. Snyder and H. N. Hansen

Leaf rust: *Uredo musae* Cummins, *Uromyces musae* Henn.

Leaf speckle: *Acrodontium simplex* (G. Mang.) de Hoog

Leaf spots: *Curvularia eragrostidis* (Henn.) J. A. Mey., *Drechslera musae-sapientum* Hansf. and M. B. Ellis, *Leptosphaeria musarum* Sacc. and Berl., *Pestalotiopsis disseminata* (Thüm.) Steyaert

Main stalk rot: *Ceratocystis paradoxa* (Dade) C. Moreau

Malayan leaf spot: *Haplobasidion musae* M. B. Ellis

Marasmiellus rot: *Marasmiellus inoderma* (Berk.) Singer = *Marasmius semiustus* Berk. and M. A. Curt.

Moko, bacterial wilt: *Ralstonia solanacearum* (Smith 1896) Yabauuchi, Kosako, Yano, Hotta, and Nishiuchi 1995 = *Burkholderia solanacearum* (Smith 1896) Yabuuchi, Kosako, Oyaizu, Yano, Hotta, Hashimoto, Ezaki, and Arakawa 1993 = *Pseudomonas solanacearum* (Smith 1896) Smith 1914

Mosaic: genus *Cucumovirus, Cucumber mosaic virus* (CMV) (family *Bromoviridae*)

Pestalotiopsis leaf spot: *Pestalotiopsis palmarum* (Cooke) Steyaert

Phaeoseptoria leaf spot: *Phaeoseptoria musae* Punith.

Pitting: *Pyricularia grisea* Sacc.

Pseudostem heart rot: *Fusarium moniliforme* J. Sheld., Teleomorph:
  Gibberella fujikuroi (Sawada) S. Ito in S. Ito and K. Kimura
Rhizome rot: *Erwinia carotovora* (Jones 1901) Bergey, Harrison, Breed,
  Hammer, and Huntoon 1923, *Erwinia chrysanthemi* Buckholder,
  McFadden, and Dimock 1953
Root and rhizome rot: *Cylindrocarpon musae* C. Booth and R. H. Stover
Sclerotinia fruit rot: *Sclerotinia sclerotiorum* (Lib.) de Bary
Sheath rot: *Nectria foliicola* Berk. and M. A. Curtis
Sooty mold: *Limacinula tenuis* (Earle) Sacc. and Trotter in Sacc.
Speckle: *Mycosphaerella musae* (Speg.) Syd. and P. Syd.
Squirter (black end disease): *Nigrospora sphaerica* (Sacc.) E. W. Mason
Stem-end rot: *Colletotrichum musae* (Berk. and M. A. Curtis) Arx
Streak: genus *Badnavirus, Banana streak virus* (BStV)
Trachysphaera finger rot: *Trachysphaera frutigena* Tabor and Bunting
Tropical speckle: *Ramichloridium musae* (Stahel ex M. B. Ellis) De Hoog
  = *Veronaea musae* Stahel ex M. B. Ellis = *Periconiella musae* Stahel ex
  M. B. Ellis
Verticillium tip rot: *Verticillium theobromae* (Turconi) E. W. Mason and
  S. J. Hughes
Yellow Sigatoka: *Mycosphaerella musicola* J. L. Mulder in J. L. Mulder
  and R. H. Stover, Anamorph: *Pseudocercospora musae* (Zimm.)
  Deighton

BARLEY (HORDEUM VULGARE L. EMEND. BOWDEN)

Anthracnose: *Colletotrichum graminicola* (Ces.) G. W. Wils.,
  Teleomorph: *Glomerella graminicola* Politis
Aster yellows: Aster yellows phytoplasma
Bacterial kernel blight, leaf blight: *Pseudomonas syringae* pv. *syringae*
  van Hall 1902
Bacterial stripe: *Pseudomonas syringae* pv. *striafaciens* (Elliott 1927)
  Young, Dye, and Wilkie 1978
Barley mild mosaic: genus *Bymovirus, Barley mild mosaic virus*
  (BaMMV) (family *Potyviridae*)
Barley stripe: *Drechslera graminea* (Rabenh.) Shoemaker
Barley stripe mosaic: genus *Hordeivirus, Barley stripe mosaic virus*
  (BSMV)
Barley yellow dwarf: genus *Luteovirus, Barley yellow dwarf virus*
  (BYDV)
Barley yellow mosaic: genus *Bymovirus, Barley yellow mosaic virus*
  (BaYMV) (family *Potyviridae*)

Basal glume rot: *Pseudomonas syringae* pv. *atrofaciens* (McCulloch 1920) Young, Dye, and Wilkie 1978

Black chaff and bacterial streak: *Xanthomonas translucens* pv. *translucens* (Jones, Johnson, and Reddy 1917) Vauterin, Hoste, Kersters, and Swings 1995

Brome mosaic: genus *Bromovirus, Brome mosaic virus* (BMV)

Cephalosporium stripe: *Hymenula cerealis* Ellis and Everh. = *Cephalosporium gramineum* Nisikado and Ikata in Nisikado et al.

Cereal northern mosaic: genus *Cytorhabdovirus, Northern cereal mosaic virus* (NCMV) (family *Rhabdoviridae*, Order *Mononegavirales*)

Cereal tillering: genus *Reovirus, Cereal tillering disease virus* (CTDV)

Chloris striate mosaic: genus *Monogeminivirus, Chloris striate mosaic virus* (CSMV)

Covered smut: *Ustilago hordei* (Pers.) Lagerh.

Crown rot: *Fusarium culmorum* (Wm. G. Smith) Sacc., *F. graminearum* Schwabe, Teleomorph: *Gibberella zeae* (Schwein.) Petch, *Bipolaris sorokiniana* (Sacc.) Shoemaker

Crown rust: *Puccinia coronata* Corda

Downy mildew: *Sclerophthora rayssiae* Kenneth et al.

Dwarf bunt: *Tilletia controversa* Kuhn in Rebenh.

Ergot: *Claviceps purpurea* (Fr.:Fr.) Tul., Anamorph: *Sphacelia segetum* Lev.

Eyespot: *Pseudocercosoporella herpotrichoides* (Fron) Deighton, Teleomorph: *Tapesia yallundae* Wallwork and Spooner

False loose smut: *Ustilago avenae* (Pers.) Rostr

Gray snow mold (= Typhula blight): *Typhula incarnata* Fr., *T. ishikariensis* Imai

Halo spot: *Pseudoseptoria donacis* (Pass.) Sutton = *Selenophoma donacis* (Pass.) R. Sprague and A. G. Johnson

Hordeum mosaic: genus *Rymovirus, Hordeum mosaic virus* (HoMV) (family *Potyviridae*)

Leaf rust (= brown rust): *Puccinia hordei* Otth

Leaf spot: *Ascochyta hordei* K. Hara, *Ascochyta graminea* (Sacc.) R. Sprague and A. G. Johnson, *Ascochyta sorghi* Sacc., *Ascochyta tritici* S. Hori and Enjoji

Loose smut: *Ustilago segetum* (Pers.) Roussel var. *nuda* = *Ustilago nuda* (Jensen) Rostr.

Net blotch: *Pyrenophora teres* Drechs., Anamorph: *Drechslera teres* (Sacc.) Shoemaker = *Helminthosporium teres* Sacc., *Drechslera teres* f. *maculata* Smedeg.

Oat blue dwarf: genus *Marafivirus, Oat blue dwarf virus* (OBDV)

Oat pseudorosette: genus *Tenuivirus, Oat pseudorosette virus*

Oat sterile dwarf: genus *Fijivirus, Oat sterile dwarf virus* (OSDV)

Pink snow mold: *Microdochium nivalis* (Fr.) Samuel and I. C. Hallett = *Fusarium nivale* (Fr.) Sorauer, Teleomorph: *Monographella nivalis* (Schaffnit) E. Müller

Powdery mildew: *Blumeria graminis* (DC.) E. O. Speer f. sp. *hordei* Em. Marchel, Anamorph: *Oidium monilioides* (Nees) Link = *Erysiphe graminis* DC, ex M'erat f. sp. *hordei* Em. Marchal

Pythium root rot: *Pythium* spp., *Pythium arrhenomanes* Drechs., *Pythium graminicola* Subramanian, *Pythium tardicrescens* Vanderpool

Rhizoctonia root rot: *Rhizoctonia solani* Kuhn, Teleomorph: *Thanatephorus cucumeris* (A. B. Frank) Donk

Rice black-streaked dwarf: genus *Fijivirus, Rice black-streaked dwarf virus* (RBSDV)

Rice stripe: genus *Tenuivirus, Rice stripe virus* (RSV)

Scab (= head blight): *F. graminearum* Schwabe

Scald: *Rhynchosporium secalis* (Oud.) J. J. Davis

Septoria speckled leaf blotch: *Septoria passerinni* Sacc., *Stagonospora avenae* f. sp. *triticea* T. Johnson

Sharp eyespot: *Rhizoctonia cerealis* Van der Hoeven, Teleomorph: *Ceratobasidium cereale* D. Murray and L. L. Burpee

Snow rot: *Pythium iwayamai* Ito, *P. okanoganense* Lipps, *P. paddicum* Harane

Snow scald: *Myriosclerotinia borealis* (Bubak and Vleugel) L. M. Kohn = *Sclerotinia borealis* Bubak and Vleugel

Southern blight: *Sclerotium rolfsii* Sacc., Teleomorph: *Athelia rolfsii* (Curzi) Tu and Kimbrough

Speckled snow mold: *Typhula idahoensis* Remsberg

Spot blotch: *Cochliobolus sativus* (Ito and Kuribayashi) Drechs. ex Dastur, Anamorph: *Drechslera teres* (Sacc.) Shoemaker

Stagonospora blotch: *Stagonospora avenae* f. sp. *triticea* T. Johnson, Teleomorph: *Phaeosphaeria avenaria* f. sp. *triticea* T. Johnson, *Stagonospora nodorum* (Berk.) Castellani and E. G. Germano = *Septoria nodorum* (Berk.) Berk. in Berk. and Broome, Teleomorph: *Phaeosphaeria nodorum* (E. Muller) Hedjaroude

Stem rust (= black stem rust): *Puccinia graminis* Pers.:Pers. = *Puccinia graminis* f.sp. *tritici* (Pers.) Erikss. et Henn

Take-all: *Gaeumannomyces graminis* (Sacc.) von Arx and Olivier var. *tritici* Walker = *Ophiobolus graminis* Sacc.

Tan spot: *Pyrenophora tritici-repentis* (Died.) Drechs. = *P. trichostoma* (Fr.) Fuckel, Anamorph: *Drechslera tritici-repentis* (Died.) Shoemaker = *Helminthosporium tritici-repentis* Died.

Verticillium wilt: *Verticillium dahliae* Kleb.

Wheat dwarf: genus *Monogeminivirus, Wheat dwarf virus* (WDV)
Wheat soil-borne mosaic: genus *Furovirus, Wheat soil-borne mosaic virus* (SBWMV)
Wheat streak mosaic: genus *Rymovirus, Wheat streak mosaic virus* (WSMV) (family *Potyviridae*)
Wheat yellow leaf: genus *Closterovirus, Wheat yellow leaf virus* (WYLV)
Wirrega blotch: *Drechslera wirreganensis* Wallwork et al.
Yellow rust (= stripe rust): *Puccinia striiformis* Westend

BEAN (PHASEOLUS VULGARIS L.)

Alternaria leaf spot: *Alternaria alternata* (Fr.:Fr.) Keissl.
Angular leaf spot: *Phaeoisariopsis griseola* (Sacc.) Ferraris
Anthracnose: *Colletotrichum lindemuthianum* (Sacc. and Magnus) Lams.-Scribner, Teleomorph: *Glomerella lindemuthiana* Shear
Ashy stem blight (charcoal rot): *Macrophomina phaseolina* (Tassi) Goidanich
Bacterial brown spot: *Pseudomonas syringae* pv. *syringae* van Hall 1902
Bacterial wilt: *Curtobacterium flaccumfaciens* pv. *flaccumfaciens* (Hedges 1922) Collins and Jones 1983 = *Corynebacterium flaccumfaciens* (Hedges 1922) Dowson 1942
Bean common mosaic: genus *Potyvirus, Bean common mosaic virus* (BCMV)
Bean curly dwarf mosaic: genus *Comovirus, Bean curly dwarf mosaic virus* (BCDMV)
Bean dwarf mosaic: *Bean dwarf mosaic virus* (BDMV)
Bean golden mosaic: genus *Potyvirus, Bean common mosaic virus* (BCMV)
Bean mild mosaic: *Bean mild mosaic virus* (BMMV)
Bean pod mottle: genus *Comovirus, Bean pod mottle virus* (BPMV)
Bean rugose mosaic: genus *Comovirus, Bean rugose mosaic virus* (BRMV)
Bean southern mosaic: genus *Sobemovirus, Bean southern mosaic virus* (BSMV)
Bean yellow mosaic: genus *Potyvirus, Bean yellow mosaic virus* (BYMV)
Bean yellow stipple: *Bean yellow stipple virus* (BYSV)
Black node disease: *Phoma exigua* Desmaz. var. *diversispora* (Bubak) Boerema = *P. diversispora* Bubak
Black root rot: *Thielaviopsis basicola* (Berk. and Broome) Ferraris
Cercospora leaf blotch: *Pseudocercospora cruenta* (Sacc.) Deighton = *Cercospora cruenta* Sacc., Teleomorph: *Mycosphaerella cruenta* D.H. Latham, *Cercospora canescens* Ellis and G. Martin

Clover yellow vein: genus *Potexvirus, Clover yellow vein virus* (CYVV)

Common blight, fuscous blight: *Xanthomonas axonopodis* pv. *phaseoli* (Smith 1897) Vauterin, Hoste, Kersters, and Swings 1995 = *Xanthomonas campestris* pv. *phaseoli* (Smith 1897) Dye 1978 = *Xanthomonas phaseoli* (ex Smith 1997) Gabriel, Kingsley, Hunter, and Gottwald 1989

Cucumber mosaic: genus *Cucumovirus, Cucumber mosaic virus* (CMV)

Curly top: genus *Hybrigeminivirus, Beet curly top virus* (BCTV)

Damping-off and stem rot: *Rhizoctonia solani* Kühn, Teleomorph: *Thanatephorus cucumeris* (A. B. Frank) Donk

Fusarium root rot: *Fusarium solani* (Mart.) Sacc. f. sp. *phaseoli* (Burkholder) W. C. Snyder and H. N. Hans.

Fusarium yellows: *Fusarium oxysporum* Schlechtend.:Fr. f. sp. *phaseoli* J. B. Kendrick and W. C. Snyder

Gray mold: *Botrytis cinerea* Pers.:Fr, Teleomorph: *Botryotinia fuckeliana* (de Bary) Whetzel

Halo blight: *Pseudomonas savastanoi* pv. *phaseolicola* (Burkholder 1926) Gardan, Bollet, Abu Ghorrah, Grimont, and Grimont 1992 = *Pseudomonas syringae* pv. *phaseolicola* (Burkholder 1926) Young, Dye, and Wilkie 1978

Leaf spot: *Phoma exigua* Desmaz. var. *exigua* = *Ascochyta phaseolorum* Sacc.

Machismo: Phytoplasma

Peanut mottle: genus *Potyvirus, Peanut mottle virus* (PMV)

Peanut stunt: genus *Cucumovirus, Peanut stunt virus* (PSV)

Phyllosticta leaf spot: *Phyllosticta phaseolina* Sacc.

Phymatotrichum root rot (cotton root rot): *Phymatotrichopsis omnivora* (Duggar) Hennebert = *Phymatotrichum omnivorum* Duggar

Phytopthora blight: *Phytophthora nicotianae* Breda de Haan var. *parasitica* (Dastur) G. M. Waterhouse, *Phytophthora phaseoli* Thaxt.

Powdery mildew: *Erysiphe polygoni* DC.

Pythium blight: *Pythium debaryanum* Auct. non R. Hesse, *P. ultimum* Trow

Red node: genus *Ilarvirus, Tobacco streak virus* (TSV)

Rust: *Uromyces appendiculatus* (Pers.:Pers.) Unger

Southern blight: *Sclerotium rolfsii* Sacc., Teleomorph: *Athelia rolfsii* (Curzi) Tu and Kimbrough

Speckle disease: *Stagonosporopsis hortensis* (Sacc. and Malbr.) Petr. = *Ascochyta boltshauseri* Sacc. in Boltshauser

Stipple streak: genus *Necrovirus, Tobacco necrosis virus* (TNV)

Tobacco ringspot: genus *Tobravirus, Tobacco ringspot virus* (TRS)

Web blight: *Rhizoctonia solani* Kühn

White mold (Sclerotinia rot): *Sclerotinia sclerotiorum* (Lib.) de Bary

CITRUS (CITRUS SPECIES)

Algal spot: *Cephaleuros virescens* Kunze = *Cephaleuros mycoides* Karst. (Alga)

Alternaria leaf and fruit spot: *Alternaria citri* Ell. and Pierce = *Alternaria alternata* (Fr.:Fr) Keissl. pv. *citri*

Black pit: *Pseudomonas syringae* van Hall 1902

Black root rot: *Thielaviopsis basicola* (Berk. and Br.) Ferr., Anamorph: *Chalara elegans* Nag Raj and Kendrick

Black spot: *Guignardia citricarpa* Kiely

Blue mold: *Penicillium italicum* Wehmer

Canker: *Xanthomonas axonopodis* pv. *citri* (Hasse 1915) Vauterin, Hoste, Kersters, and Swings 1995 = *Xanthomonas citri* (ex Hasse 1915) Gabriel, Kingsley, Hunter, and Gottwald 1989 = *Xanthomonas campestris* pv. *citri* (Hasse 1915) Dye 1978

Cercospora leaf and fruit spot: *Cercospora angolensis* Carv. and Mendes

Citrus leaf rugose: genus *Ilarvirus, Citrus rugose virus* (CRV)

Citrus variegated chlorosis: *Xylella fastidiosa* Wells, Raju, Weisburg, Mandelo-Paul, and Brenner 1987

Clitocybe root rot: *Armillaria tabescens* (Scop.:Fr.) Dennis et al. = *Clitocybe tabescens* (Scop.:Fr.) Bres.

Crinkly leaf: genus *Ilarvirus, Citrus crinkly leaf virus* (CCLV)

Diplodia gummosis: *Physalospora rhodina* (Berk. and Curt.) Cke., Anamorph: *Diplodia natalensis* Pole-Evans

Dodder: *Cuscuta* spp.

Dothiorella gummosis: *Botryosphaeria dothidea* (Moug.:Fr.) Ces. and de Not., Anamorph: *Dothiorella gregaria* Sacc.

Exocortis: Citrus exocortis viroid

Fusarium wilt: *Fusarium oxysporum* Schlecht.

Gray mold: *Botrytis cinerea* Pers.:Fr., Teleomorph: *Botryotinia fuckeliana* (de Bary.) Whetzel

Greasy spot: *Mycosphaerella citri* Whiteside, Anamorph: *Cercospora citri grisea* Fisher

Green mold: *Penicillium digitatum* Sacc.

Greening: Citrus greening organism (Nonculturable, phloem-restricted Gram-negative bacteria) "*Candidatus* Liberobacter asiaticum" and "*Candidatus* Liberobacter africanum"

Heart rot: *Ganoderma sessile* Murrill, *G. applanatum* (Pers.) Pat.

Hendersonula branch wilt: *Hendersonula toruloidea* Nattrass

Infectious variegation: genus *Ilarvirus, Citrus infectious variegation virus* (CIVV)

Lime anthracnose: *Gloeosporium limetticola* Clausen

Mal secco: *Phoma tracheiphila* (Petri) Kentschaveli and Gikachvili = *Deuterophoma tracheiphila* Petri

Melanose: *Diaporthe citri* Wolf = *Diaporthe medusaea* Nits., Anamorph: *Phomopsis citri* Fawc.

Phytophthora gummosis, foot rot, root rot, and brown fruit rot: *Phytophthora citrophthora* (R. E. Sm. and E. H. Sm.) Leonian, *P. hibernalis* Carne, *P. parasitica* Dast., *P. syringae* (Kleb.) Kleb.

Pink disease: *Corticium salmonicolor* Berk. and Br.

Postbloom fruit drop disease and anthracnose: *Glomerella cingulata* (Ston.) Spauld. and Schrenk, Anamorph: *Colletotrichum gloeosporioides* (Penz.) Sacc.

Ringspot: *Citrus ringspot virus* (CRSV)

Satsuma dwarf: *Citrus satsuma dwarf virus* (CSDV)

Scab: *Elsinoe fawcettii* Bitanc. and Jenkins

Sclerotinia twig blight, fruit rot, and root rot: *Sclerotinia sclerotiorum* (Lib.) de Bary

Septoria spot: *Septoria citri* Pass.

Shoestring root rot: *Armillaria mellea* (Vahl.:Fr.) Kummer = *Armillariella mellea* (Vahl.:Fr.) Karst.

Sooty blotch: *Gloeodes pomigena* (Schw.) Colby

Sooty mold: *Capnodium citri* Berk. and Desm., *Capnodium citricola* McAlp.

Sour rot: *Geotrichum candidum* Lk. = *Oospora lactis* (Fres.) Sacc.

Stubborn: *Spiroplasma citri* Saglio, I.'hospital, Lafleche, Dupont, Bove, Mouches, Rose, Coan, and Clark 1986

Sweet orange scab: *Elsinoe australis* Bitanc. and Jenkins

Tatter leaf and citrange stunt: genus *Capillovirus, Citrus tatter leaf virus* (CTLV)

Tristeza: genus *Closterovirus, Citrus tristeza virus* (CTV)

Witches'-broom: Phytoplasma

CORN (ZEA MAYS L.)

American wheat striate: genus *Cytorhabdovirus, American wheat striate mosaic virus* (AWSMV)

Anthracnose leaf blight: *Colletotrichum graminicola* (Ces.) G.W. Wils., Teleomorph: *Glomerella graminicola* Politis, *Glomerella tucumanensis* (Speg.) Arx and E. Müller, Anamorph: *Colletotrichum falcatum* Went

Bacterial leaf blight and stalk rot: *Acidovorax avenae* ssp. *avenae* (Manns 1909) Willems, Goor, Thielemans, Gillis, Kersters, and De Ley 1990 = *Pseudomonas avenae* ssp. *avenae* Manns 1909

Bacterial leaf spot: *Xanthomonas vasicola* pv. *holcicola* (Elliott 1930) Vauterin, Hoste, Kersters, and Swings 1995

Bacterial stalk rot: *Enterobacter dissolvens* (Rosen 1922) Brenner, McWhorter, Kai, Steigerwalt, and Farmer 1988 = *Erwinia dissolvens* (Rosen 1922) Burkholder 1948

Bacterial stalk and top rot: *Erwinia carotovora* ssp. *carotovora* (Jones 1901) Bergey, Harrison, Breed, Hammer, and Huntoon 1923, *Erwinia chrysanthemi* pv. *zeae* (Sabet 1954) Victoria, Arboleda, and Munoz 1975

Bacterial stripe: *Burkholderia andropogonis* (Smith 1911) Gillis, Van Van, Bardin, Goor, Hebbar, Willems, Segers, Kersters, Heulin, and Fernandez 1995 = *Pseudomonas andropogonis* (Smith 1911) Stapp 1928

Barley stripe mosaic: genus *Hordeivirus, Barley stripe mosaic virus* (BSMV)

Barley yellow dwarf: genus *Luteovirus, Barley yellow dwarf virus* (BYDV)

Brome mosaic: *Brome mosaic virus* (BMV)

Brown spot: *Physoderma maydis* (Miyabe) Miyabe

Brown stripe downy mildew: *Sclerophthora rayssiae* Kenneth et al. var. *zeae* Payak and Renfro

Cereal chlorotic mottle: *Cereal chlorotic mottle virus* (CCMV)

Charcoal rot: *Macrophomina phaseolina* (Tassi) Goidanich

Chocolate spot: *Pseudomonas syringae* pv. *coronafaciens* (Elliott 1920) Young, Dye, and Wilkie 1978

Common corn rust: *Puccinia sorghi* Schwein.

Common smut: *Ustilago zeae* (Beckm.) Unger = *U. maydis* (DC.) Corda

Corn chlorotic vein banding: *Corn chlorotic vein banding virus* (CCVBV)

Corn stunt: *Spiroplasma kunkelii* Whitcomb, Chen, Williamson, Liao, Tully, Bove, Mouchers, Rose, Coan, and Clark 1986

Crazy top downy mildew: *Sclerophthora macrospora* (Sacc.) Thirumalachar et al. = *Sclerospora macrospora* Sacc.

Cucumber mosaic: genus *Cucumovirus, Cucumber mosaic virus* (CMV)

Cynodon chlorotic streak: *Cynodon chlorotic streak virus* (CCSV)

Didymella leaf spot: *Didymella exitalis* (Morini) E. Muller

Diplodia ear rot: *Diplodia maydis* (Berk.) Sacc.

Diplodia ear rot and stalk rot: *Diplodia frumenti* Ellis and Everh., Teleomorph: *Botryosphaeria festucae* (Lib.) Arx and E. Müller

Diplodia leaf spot: *Stenocarpella macrospora* (Earle) Sutton (Mitosporic fungi) = *Diplodia macrospora* Earle

Ergot: *Claviceps gigantea* Fuentes et al., Anamorph: *Sphacelia* sp.

False smut: *Ustilaginoidea virens* (Cooke) Takah.

Fusarium ear and stalk rot: *Fusarium subglutinans* (Wollenweb. and Reinking) P. E. Nelson et al. = *F. moniliforme* J. Sheld. var. *subglutinans* Wollenweb. and Reinking

Fusarium stalk rot: *Fusarium moniliforme* J. Sheld., Teleomorph: *Gibberella fujikuroi* (Sawada) Ito in Ito and K. Kimura, *Fusarium avenaceum* (Fr.:Fr.) Sacc., Teleomorph: *Gibberella avenacea* R. J. Cooke

Gibberella ear and stalk rot: *Gibberella zeae* (Schwein.) Petch, Anamorph: *Fusarium graminearum* Schwabe

Goss's bacterial wilt: *Clavibacter michiganensis* ssp. *nebraskensis* (Vidaver and Mandel 1974) Davis, Gillaspie, Vidaver, and Harris 1984

Gray leaf spot: *Cercospora sorghi* Ellis and Everh. = *C. sorghi* Ellis and Everh. var. *maydis* Ellis and Everh., *C. zeae-maydis* Tehon and E. Y. Daniels

Green ear downy mildew: *Sclerospora graminicola* (Sacc.) J. Schröt.

Head smut: *Sphacelotheca reiliana* (Kuhn) G. P. Clinton = *Sporisorium holci-sorghi* (Rivolta) K. Vanky

Holcus spot: *Pseudomonas syringae* pv. *syringae* van Hall 1902

Hyalothyridium leaf spot: *Hyalothyridium maydis* Latterell and Rossi

Java downy mildew: *Peronosclerospora maydis* (Racib.) C. G. Shaw = *Sclerospora maydis* (Racib.) Butler

Johnsongrass mosaic: genus *Potyvirus, Johnsongrass mosaic virus* (JGMV)

Late wilt: *Cephalosporium maydis* Samra et al.

Maize bushy stunt: Phytoplasma

Maize chlorotic dwarf: genus *Waikavirus, Maize chlorotic dwarf virus* (MCDV)

Maize chlorotic mottle: genus *Machlomovirus, Maize chlorotic mottle virus* (MCMoV)

Maize dwarf mosaic: genus *Potyvirus, Maize dwarf mosaic virus* (MDMV)

Maize leaf fleck: *Maize leaf fleck virus* (MLFV)

Maize line: *Maize line virus* (MLV)

Maize mosaic: genus *Nucleorhabdovirus, Maize mosaic virus* (MMV)

Maize pellucid ringspot: *Maize pellucid ringspot virus* (MPRV)

Maize rayado fino: *Maize rayado fino virus* (MRFV)

Maize red stripe: *Maize red stripe virus* (MRSV)

Maize ring mottle: *Maize ring mottle virus* (MRMV)

Maize rough dwarf: genus *Fijivirus, Maize rough dwarf virus* (MRDV)

Maize sterile stunt: *Maize sterile stunt virus* (MSSV)

Maize streak: genus *Monogeminivirus, Maize streak virus* (MSV)

Maize stripe: genus *Tenuivirus, Maize stripe virus* (MStV)

Maize tassel abortion: *Maize tassel abortion virus* (MTAV)

Maize vein enation: *Maize vein enation virus* (MVEV)

Maize wallaby ear: *Maize wallaby ear virus* (MWEV)

Maize white leaf: *Maize white leaf virus* (MWLV)

Maize white line mosaic: *Maize white line mosaic virus* (MWLMV)

Millet red leaf: *Millet red leaf virus* (MRLV)

Northern cereal mosaic: *Northern cereal mosaic virus* (NCMV)

Northern corn leaf blight: *Setosphaeria turcica* (Luttrell) K. J. Leonard and E. G. Suggs, Anamorph: *Exserohilum turcicum* (Pass.) K. J. Leonard and E. G. Suggs = *Helminthosporium turcicum* Pass.

Northern corn leaf spot: *Cochliobolus carbonum* R. R. Nelson, Anamorph: *Bipolaris zeicola* (G. L. Stout) Shoemaker = *Helminthosporium carbonum* Ullstrup

Oat pseudorosette: *Oat pseudorosette virus*

Oat sterile dwarf: genus *Fijivirus, Oat sterile dwarf virus* (OSDV)

Phaeocytostroma stalk rot and root rot: *Phaeocytostroma ambiguum* (Mont.) Petr. and Syd. = *Phaeocytosporella zeae* G. L. Stout

Phaeosphaeria leaf spot: *Phaeosphaeria maydis* (P. Henn.) Rane, Payak, and Renfro = *Sphaerulina maydis* P. Henn.

Philippine downy mildew: *Peronosclerospora philippinensis* (W. Weston) C. G. Shaw = *Sclerospora philippinensis* W. Weston

Pyrenochaeta stalk rot and root rot: *Phoma terrestris* E. M. Hans. = *Pyrenochaeta terrestris* (E. M. Hans.) Gorenz et al.

Pythium root rot: *Pythium* spp., *P. arrhenomanes* Drechs., *P. graminicola* Subramanian

Pythium stalk rot: *Pythium aphanidermatum* (Edson) Fitzp. = *P. butleri* L. Subramanian

Red kernel disease: *Epicoccum nigrum* Link

Rhizoctonia root rot and stalk rot: *Rhizoctonia solani* Kühn, *R. zeae* Voorhees

Rice black-streaked dwarf: genus *Fijivirus, Rice black-streaked dwarf virus* (RBSDV)

Rice stripe: genus *Tenuivirus, Rice stripe virus* (RSV)

Rostratum leaf spot: *Setosphaeria rostrata* K. J. Leonard, Anamorph: *Exserohilum rostratum* (Drechs.) K. J. Leonard and E. G. Suggs = *Helminthosporium rostratum* Drechs.

Selenophoma leaf spot: *Selenophoma* sp. (Mitosporic fungi)

Sheath rot: *Gaeumannomyces graminis* (Sacc.) Arx and D. Olivier

Shuck rot: *Myrothecium gramineum* Lib.

Sorghum downy mildew: *Peronosclerospora sorghi* (W. Weston and Uppal) C. G. Shaw = *Sclerospora sorghi* W. Weston and Uppal

Sorghum mosaic: genus *Potyvirus, Sorghum mosaic virus* (SrMV)

Southern corn leaf blight: *Cochliobolus heterostrophus* (Drechs.) Drechs., Anamorph: *Bipolaris maydis* (Nisikado and Miyake) Shoemaker = *Helminthosporium maydis* Nisikado and Miyake

Southern corn rust: *Puccinia polysora* Underw.

Southern leaf spot: *Stenocarpella macrospora* (Earle) Sutton = *Diplodia macrospora* Earle

Spontaneum downy mildew: *Peronosclerospora spontanea* (W. Weston) C. G. Shaw = *Sclerospora spontanea* W. Weston

Stewart's disease (bacterial wilt): *Erwinia stewartii* (Smith 1898) Dye 1963

Sugarcane downy mildew: *Peronosclerospora sacchari* (T. Miyake) Shirai and Hara = *Sclerospora sacchari* T. Miyake

Sugarcane Fiji disease: genus *Fijivirus, Sugarcane Fiji disease virus* (FDV)

Sugarcane mosaic: genus *Potyvirus, Sugarcane mosaic virus* (SCMV)

Tar spot: *Phyllachora maydis* Maubl.

Trichoderma ear rot and root rot: *Trichoderma viride* Pers.:Fr. = *T. lignorum* (Tode), Teleomorph: *Hypocrea* sp.

Tropical corn rust: *Physopella pallescens* (Arth.) Cummins and Ramachar, *P. zeae* (Mains) Cummins and Ramachar = *Angiopsora zeae* Mains

White ear rot, root, and stalk rot: *Stenocarpella maydis* (Berk.) Sutton = *Diplodia zeae* (Schwein.) Lév.

Witchweed disease: *Striga asiatica* (L.) Kuntze = *S. lutea* Lour., *S. hermonthica* (Del.) Benth. (parasitic plants)

Yellow leaf blight: *Phyllosticta maydis* D. C. Arny and R. R. Nelson, Teleomorph: *Mycosphaerella zeae-maydis* Mukunya and Boothroyd, *Ascochyta ischaemi* Sacc.

Zonate leaf spot: *Gloeocercospora sorghi* Bain and Edgerton ex Deighton

COTTON (GOSSYPIUM SPECIES)

Anthocyanosis: *Cotton anthocyanosis virus*

Anthracnose: *Glomerella gossypii* Edgerton, Anamorph: *Colletotrichum gossypii* Southworth

Areolate mildew: *Mycosphaerella areola* J. Ehrlich and F.A. Wolf, Anamorph: *Ramularia gossypii* (Speg.) Cif. = *Cercosporella gossypii* Speg.

Ascochyta blight: *Ascochyta gossypii* Woronichin

Bacterial blight: *Xanthomonas axonopodis* pv. *malvacearum* (Smith 1901) Vauterin, Hoste, Kersters, and Swings 1995 = *X. campestris* pv. *malvacearum* (Smith 1901) Dye 1978

Black root rot: *Thielaviopsis basicola* (Berk. and Broome) Ferraris

Charcoal rot: *Macrophomina phaseolina* (Tassi) Goidanich

Fusarium wilt: *Fusarium oxysporum* Schlechtend.:Fr. f. sp. *vasinfectum* (Atk.) W. C. Snyder and H. N. Hans.

Leaf crumple: genus *Bigeminivirus, Cotton leaf crumple virus* (CLCrV)

Leaf curl: *Cotton leaf curl virus*

Leaf mottle: *Cotton leaf mottle virus*

Leaf spots: *Alternaria macrospora* A. Zimmerm., *A. alternata* (Fr.:Fr.) Keissl., *Mycosphaerella gossypina* (Atk.) Earle

Phyllody: Phytoplasma

Phymatotrichum root rot: *Phymatotrichopsis omnivora* (Duggar) Hennebert = *Phymatotrichum omnivorum* Duggar

Powdery mildews: *Leveillula taurica* (Lev) G. Arnaud, Anamorph: *Oidiopsis sicula* Scalia, *Oidiopsis gossypii* (Wakef.) Raychaudhuri, *Salmonia malachrae* (Seaver) Blumer and E. Muller

Sclerotium stem and root rot: *Sclerotium rolfsii* Sacc., Teleomorph: *Athelia rolfsii* (Curzi) Tu and Kimbrough

Small leaf: Phytoplasma

Southwestern cotton rust: *Puccinia cacabata* Arth. and Holw. in Arth.

Tropical cotton rust: *Phakopsora gossypii* (Lagerh.) Hiratsuka

Verticillium wilt: *Verticillium dahliae* Kleb.

CRUCIFERS

Crops: cabbage (*Brassica oleracea* var. *capitata* L.), cauliflower (*B. oleracea* var. *botrytis* L.), broccoli (*B. oleracea* L. var. *italica*), brussels sprouts (*B. oleracea* var. *gemmifera* Zenk.), knol khol (*B. oleracea* var. *gongylodes* L.), kale (*B. oleracea* var. *sabellica* L.), collards (*B. oleracea* var. *viridis* L.), turnip [*B. rapa* var. *utilis* (D.C.) Metzg.], spinach mustard (*B. rapa* L. var. *perviridis* Bailey), mustard (*B. juncea* L. var. *crispifolia* Bailey), oilseed rape (*B. napus* L. ssp. *oleifera*), black mustard [*B. nigra* (L.) W. Koch], and radish (*Raphanus sativus* L.)

Alternaria black spot: *Alternaria brassicae* (Berk.) Sacc., *A. brassicicola* (Schwein.) Wiltshire, *A. raphani* Groves and Skolko

Anthracnose: *Colletotrichum higginsianum* Sacc. in Higgins

Bacterial leaf spot: *Pseudomonas syringae* pv. *maculicola* (McCulloch 1911) Young, Dye, and Wilkie 1978

Bacterial soft rot Erwinia: *Erwinia carotovora* (Jones 1901) Bergey, Harrison, Breed, Hammer, and Huntoon 1923

Bacterial soft rot Pseudomonas: *Pseudomonas marginalis* pv. *marginalis* (Brown 1918) Stevens 1925

Black leg and Phoma root rot: *Leptosphaeria maculans* (Desmaz.) Ces. and De Not., Anamorph: *Phoma lingam* (Tode:Fr.) Desmaz.

Black root (Aphanomyces): *Aphanomyces raphani* Kendrick

Black rot: *Xanthomonas campestris* pv. *campestris* (Pammel 1985) Dowson 1939

Cauliflower mosaic: genus *Caulimovirus, Cauliflower mosaic virus* (CaMV)

Cercospora leaf spot: *Cercospora brassicicola* Henn.

Clubroot: *Plasmodiophora brassicae* Woronin

Crown gall: *Agrobacterium tumefaciens* (Smith and Townsend 1907) Conn 1942

Damping-off: *Fusarium* spp., *Pythium* spp.

Downy mildew: *Peronospora parasitica* (Pers.:Fr.) Fr.

Fusarium yellows: *Fusarium oxysporum* Schlechtend.:Fr.

Gray mold: *Botrytis cinerea* Pers.:Fr., Teleomorph: *Botryotinia fuckeliana* (de Bary) Whetzel

Light leaf spot: *Pyrenopeziza brassicae* Sutton and Rawl

Phymatotrichum root rot: *Phymatotrichopsis omnivora* (Duggar) Hennebert = *Phymatotrichum omnivorum* Duggar

Phytophthora root rot: *Phytophthora megasperma* Drechs.

Powdery mildew: *Erysiphe polygoni* DC.

Radish mosaic: genus *Comovirus, Radish mosaic virus* (RaMV)

Ring spot: *Mycosphaerella brassicicola* (Duby) Lindau in Engl. and Prantl, Anamorph: *Asteromella brassicae* (Chev.) Boerema and Van Kesteren

Root rot: *Rhizoctonia solani* Kühn, Teleomorph: *Thanatephorus cucumeris* (A. B. Frank) Donk

Scab: *Streptomyces* spp.

Sclerotinia stem rot and watery soft rot: *Sclerotinia sclerotiorum* (Lib.) de Bary

Southern blight: *Sclerotium rolfsii* Sacc., Teleomorph: *Athelia rolfsii* (Curzi) Tu and Kimbrough

Turnip mosaic: genus *Potyvirus, Turnip mosaic virus* (TuMV)

Verticillium wilt: *Verticillium albo-atrum* Reinke and Berthier, *V. dahliae* Kleb.

White leaf spot and gray stem: *Pseudocercosporella capsellae* (Ellis and Everh.) Deighton

White rust: *Albugo candida* (Pers.) Kunze

Xanthomonas leaf spot: *Xanthomonas campestris* pv. *armoraciae* (McCulloch 1929) Dye 1978

Yellows: genus *Luteovirus, Beet western yellows virus* (BWYV)

CUCURBITS

Crops: melon, muskmelon, and cantaloupe (*Cucumis melo* L.), cucumber (*Cucumis sativus* L.), squash, gourd, and pumpkin (*Cucurbita pepo* L.), watermelon [*Citrullus lanatus* (Thunb.) Matsum and Nakai], ash gourd and wax gourd [*Benincasa hispida* (Thumb.) Cogn.], bottle gourd [*Lagenaria siceraria* (Mol.) Standl.], snake gourd (*Trichosanthes anguina* L.), and ribbed gourd [*Luffa acutangula* (L.) Roxb.]

Alternaria leaf blight: *Alternaria cucumerina* (Ellis and Everh.) J. A. Elliott

Alternaria leaf spot: *Alternaria alternata* (Fr.:Fr.) Keissl. f. sp. *cucurbitae*

Angular leaf spot: *Pseudomonas syringae* pv. *lachrymans* (Smith and Bryan 1915) Young, Dye, and Wilkie 1978

Anthracnose: *Colletotrichum orbiculare* (Berk. and Mont.) Arx = *C. lagenarium* (Pass.) Ellis and Halst, Teleomorph: *Glomerella lagenarium* Stevens

Aster yellows: Phytoplasma

Bacterial fruit blotch: *Acidovorax avenae* ssp. *citrulli* (Schaad, Sowell, Goth, Colwell, and Webb 1978) Willems, Goor, Thielemans, Gillis, Kersters, and De Ley 1992 = *Pseudomonas pseudoalcaligenes* ssp. *citrulli* Schaad, Sowell, Goth, Colwell, and Webb 1978

Bacterial leaf spot: *Xanthomonas cucurbitae* (ex Bryan 1926) Vauterin, Hoste, Kersters, and Swings 1995 = *Xanthomonas campestris* pv. *cucurbitae* (Bryan 1926) Dye 1978

Bacterial soft rot: *Erwinia carotovora* ssp. *carotovora* (Jones 1901) Bergey, Harrison, Breed, Hammer, and Huntoon 1923

Bacterial wilt: *Erwinia tracheiphila* (Smith 1895) Bergey, Harrison, Breed, Hammer, and Huntoon 1923

Belly rot: *Rhizoctonia solani* Kühn, Teleomorph: *Thanatephorus cucumeris* (A. B. Frank) Donk

Black root rot: *Thielaviopsis basicola* (Berk. and Broome) Ferraris

Brown spot: *Pantoea ananas* pv. *ananas* (Serrano 1928) Mergaert, Verdonck, and Kersters = *Erwinia ananas* Serrano 1928

Cephalosporium root rot: *Acremonium* spp. = *Cephalosporium* spp.

Cercospora leaf spot: *Cercospora citrullina* Cooke

Charcoal rot: *Macrophomina phaseolina* (Tassi) Goidanich

Chlorotic leaf spot: genus *Potyvirus, Bean yellow mosaic virus* (BYMV)

Choanephora fruit rot: *Choanephora cucurbitarum* (Berk. and Ravenel) Thaxt.

Collapse of melon: *Monosporascus eutypoides* (Petrak) Arx = *Bitrimonospora indica* Sivanesans et al.

Corynespora blight/target spot: *Corynespora cassiicola* (Berk. and M. A. Curtis) C. T. Wei

Crater rot: *Myrothecium roridum* Tode:Fr.

Crown and foot rot: *Fusarium solani* (Mart.) Sacc. f. sp. *cucurbitae* W. C. Snyder and H. N. Hans., Teleomorph: *Nectria haematococca* Berk. and Broome

Cucumber green mottle: genus *Tobamovirus, Cucumber green mottle mosaic virus* (CGMMV)

Cucumber mosaic: genus *Cucumovirus, Cucumber mosaic virus* (CMV)

Cucumber vein yellowing: *Cucumber vein yellowing virus* (CVYV)

Curly top: genus *Hybrigeminivirus, Beet curly top virus* (BCTV)

Downy mildew: *Pseudoperonospora cubensis* (Berk. and M. A. Curtis) Rostovzev

Fusarium wilt: *F. oxysporum* f. sp. *melonis* W. C. Snyder and H. N. Hans. (muskmelon), *Fusarium oxysporum* f. sp. *cucumerinum* J. H. Owen (cucumber), *F. oxysporum* f. sp. *niveum* (E. F. Sm.) W. C. Snyder and H. N. Hans. (watermelon), *F. oxysporum* f. sp. *momordicae* Sun and Huang (bitter melon), *F. oxysporum* Schlechtend.:Fr. f. sp. *benincasae* Gerlagh and Ester (wax gourd), *F. oxysporum* f. sp. *lagenariae* Matuo and Yamamoto (calabash gourd), *F. oxysporum* f. sp. *luffae* Lawai et al. (vegetable sponge)

Gray mold: *Botrytis cinerea* Pers.:Fr., Teleomorph: *Botryotinia fuckeliana* (de Bary) Whetzel

Gummy stem blight: *Didymella bryoniae* (Auersw.) Rehm = *Mycosphaerella melonis* (Pass.) Chiu and J. C. Walker, Anamorph: *Phoma cucurbitacearum* (Fr.:Fr.) Sacc.

Lasiodiplodia fruit rot: *Lasiodiplodia theobromae* (Pat.) Griffon and Maubl. = *Diplodia natalensis* Pole-Evans

Lettuce infectious yellows: genus *Closterovirus, Lettuce infectious yellows virus* (LIYVV)

Melon leaf curl: *Melon leaf curl virus* (MLCV)

Melon necrotic spot: genus *Carmovirus, Melon necrotic spot virus* (MNSV)

Monosporascus root rot: *Monosporascus cannonballus* Pollack and
Uecker
Mosaic: genus *Potyvirus, Papaya ringspot virus W strain* (PRSV-W)
Muskmelon vein necrosis: genus *Carlavirus, Muskmelon vein necrosis
virus* (MkVNV)
Myrothecium canker: *Myrothecium roridum* Tode:Fr.
Net spot: *Leandria momordicae* Rangel
Phoma blight: *Phoma exigua* Desmaz. var. *exigua = Ascochyta
phaseolorum* Sacc.
Phomopsis black stem: *Phomopsis sclerotiodes* Van Kesteren
Phyllosticta leaf spot: *Phyllosticta cucurbitacearum* Sacc.
Phytophthora root rot: *P. capsici* Leonian
Powdery mildew: *Podosphaera* (sect. *Sphaerotheca*) *xanthii* (Castagne)
U. Braun and N. Shishkoff = *Sphaerotheca fuliginea* (Schlechtend.:Fr.)
Pollacci = *Sphaerotheca fusca* (Fr.) S. Blumer, *Erysiphe
cichoracearum* DC.
Purple stem: *Diaporthe melonis* Beraha and O'Brien, Anamorph:
Phomopsis cucurbitae McKeen
Scab: *Cladosporium cucumerinum* Ellis and Arth.
Sclerotinia stem rot: *Sclerotinia sclerotiorum* (Lib.) de Bary
Septoria leaf blight: *Septoria cucurbitacearum* Sacc.
Southern blight: *Sclerotium rolfsii* Sacc., Teleomorph: *Athelia rolfsii*
(Curzi) Tu and Kimbrough
Squash leaf curl: genus *Bigeminivirus, Squash leaf curl virus* (SqLCV)
Squash mosaic: genus *Comovirus, Squash mosaic virus* (SqMV)
Sudden wilt: *Pythium aphanidermatum* (Edson) Fitzp.
Tobacco ringspot: genus *Tobravirus, Tobacco ringspot virus* (TobRSV)
Tomato ringspot: genus *Nepovirus, Tomato ringspot virus* (TRSV)
Tomato spotted wilt: genus *Tospovirus, Tomato spotted wilt virus*
(TSWV)
Ulocladium leaf spot: *Ulocladium consortiale* (Thuem.) E. Simmons
Verticillium wilt: *Verticillium albo-atrum* Reinke and Berthier, *V. dahliae*
Kleb.
Watermelon mosaic: genus *Potyvirus, Watermelon mosaic virus* (WMV)
Web blight: *Rhizoctonia solani* Kühn
Zucchini yellows: genus *Potyvirus, Zucchini yellows mosaic virus*
(ZYMV)

GRAPE (VITIS SPECIES)

Alfalfa mosaic: genus *Alfamovirus, Alfalfa mosaic virus* (AMV)
Alternaria rot: *Alternaria alternata* (Fr.:Fr.) Keissl.

Angular leaf scorch: *Pseudopezicula tetraspora* Korf, R. C. Pearson, and Zhuang, Anamorph: *Phialophora* sp.

Angular leaf spot: *Mycosphaerella angulata* W. A. Jenkins, Anamorph: *Cercospora brachypus* Ellis and Everh.

Anthracnose: *Elsinoe ampelina* Shear, Anamorph: *Sphaceloma ampelinum* de Bary

Arabis mosaic: genus *Nepovirus, Arabis mosaic virus* (ArMV)

Armillaria root rot: *Armillaria mellea* (Vahl:Fr.) P. Kumm., Anamorph: *Rhizomorpha subcorticalis* Pers.

Artichoke Italian latent: genus *Nepovirus, Artichoke Italian latent virus* (AILV)

Bacterial blight: *Xylophilus ampelinus* (Panagopoulos 1969) Willems, Gillis, Kersters, Van den Broecke, and De Ley 1987 = *Xanthomonas ampelina* Panagopoulos 1969

Bitter rot: *Greeneria uvicola* (Berk. and M. A. Curtis) Punithalingam = *Melanconium fuligineum* Lams.-Scrib. and Viala

Black dead arm: *Botryosphaeria stevensii* Shoemaker, Anamorph: *Diplodia mutila* (Fr.:Fr.) Mont.

Black rot: *Guignardia bidwellii* (Ellis) Viala and Ravaz, Anamorph: *Phyllosticta ampelicida* (Engelm.) van der Aa, *G. bidwellii* (Ellis) Viala and Ravaz f. *muscadinii* Luttrell

Bois noir (black wood disease): Phytoplasma

Bratislava mosaic: *Bratislava mosaic virus* (BMV)

Broad bean wilt: genus *Fabavirus, Broad bean wilt virus* (BBWV)

Cercospora leaf spot: *Phaeoramularia dissiliens* (Duby) Deighton = *Cercospora* sp.

Cladosporium leaf spot: *Cladosporium viticola* Cesati

Cladosporium rot: *Cladosporium herbarum* (Pers.:Fr.) Link, Teleomorph: *Mycosphaerella tassiana* (De Not.) Johans

Crown gall: *Agrobacterium tumefaciens* (Smith and Townsend 1907) Conn 1942

Dematophora root rot: *Rosellinia necatrix* Prill., Anamorph: *Dematophora necatrix* R. Hartig

Diplodia cane dieback: *Lasiodiplodia theobromae* (Pat.) Griffon and Maubl. = *Diplodia natalensis* Pole-Evans

Downy mildew: *Plasmopara viticola* (Berk. and M. A. Curtis) Berl and De Toni in Sacc.

Esca (black measles, apoplexy), grapevine decline: Esca is a disease complex or a complex of diseases caused by several pathogens. *Phellinus igniarius* (L.:Fr.) Quél., *Stereum hirsutum* (Willd.:Fr.) S. F. Gray, *Phaeomoniella chlamydospora* Crous et W. Gams = *Phaeoacremonium chlamydosporum, Phaeoacremonium aleophilum, Phaeoacremonium inflatipes,* and *Fomitiporia punctata, Trametes hirsute = Coriolus hirsutus, Trametes versicolor (= Coriolus versicolor), Phaeomoniella chlamydospora,* and several species of *Phaeoacremonium*

Eutypa dieback: *Eutypa lata* (Pers.:Fr.) Tul. and C. Tul. = *E. armeniacae* Hansf. and M. V. Carter, Anamorph: *Libertella blepharis* A. L. Smith = *Cytosporina* sp.

Fanleaf: genus *Nepovirus, Grapevine fanleaf virus* (GFLV)

Flavescence dorée: Phytoplasma

Grapevine Bulgarian latent: genus *Nepovirus, Grapevine Bulgarian latent virus* (GBLV)

Grapevine chrome mosaic: genus *Nepovirus, Grapevine chrome mosaic virus* (GCMV)

Grapevine yellows: Phytoplasma

Gray mold: *Botrytis cinerea* Pers.:Fr., Teleomorph: *Botryotinia fuckeliana* (de Bary) Whetzel

Isariopsis leaf spot: *Mycosphaerella personata* Higgins, Anamorph: *Pseudocercospora vitis* (Lév.) Speg. = *Isariopsis clavispora* (Berk. and Cooke) Sacc.

Leaf blotch: *Briosia ampelophaga* Cavara

Leaf roll: genus *Closterovirus, Grapevine leaf roll-associated virus* (GLRaV)

Leaf spots: *Asperisporium minutulum* (Sacc.) Deighton, *Phaeoramularia heterospora* (Ellis and B. T. Galloway) Deighton

Mélanose: *Septoria ampelina* Berk. and M. A. Curtis

Peach rosette mosaic: genus *Nepovirus, Peach rosette mosaic virus* (PRMV)

Petunia asteroid mosaic: genus *Tombusvirus, Petunia asteroid mosaic virus* (PAMV)

Phomopsis leaf spot (Excoriose): *Phomopsis viticola* (Sacc.) Sacc.

Phymatotrichum root rot: *Phymatotrichopsis omnivora* (Duggar) Hennebert = *Phymatotrichum omnivorum* Duggar

Phytophthora crown and root rot: *P. cinnamomi* Rands, *P. megasperma* Drechs.

Pierce's disease: *Xylella fastidiosa* Wells, Raju, Hung, Weisburg, Mandelco-Paul, and Brenner

Powdery mildew: *Uncinula necator* (Schwein.) Burrill, Anamorph: *Oidium tuckeri* Berk.

Raspberry ringspot: genus *Nepovirus, Raspberry ringspot virus* (RRV)

Ripe rot: *Glomerella cingulata* (Stoneman) Spauld. and H. Schrenk, Anamorph: *Colletotrichum gloeosporioides* (Penz.) Penz. and Sacc. in Penz.

Roesleria root rot: *Roesleria subterranea* (Weinmann) Redhead =
R. *hypogaea* Thuem. and Pass.
Rotbrenner: *Pseudopezicula tracheiphila* (Müll.-Thurg.) Korf and Zhuang
= *Pseudopeziza tracheiphila* Müll.-Thurg., Anamorph: *Phialophora
tracheiphila* (Sacc. and Sacc.) Korf
Rust: *Physopella ampelopsidis* (Dietel and P. Syd.) Cummins and
Ramachar
Sowbane mosaic: genus *Sobemovirus, Sowbane mosaic virus* (SoMV)
Strawberry latent ringspot: genus *Nepovirus, Strawberry latent ringspot
virus* (SLRV)
Tar spot: *Rhytisma vitis* Schwein.
Tobacco mosaic: genus *Tobamovirus, Tobacco mosaic virus* (TMV)
Tobacco necrosis: genus *Necrovirus, Tobacco necrosis virus* (TNV)
Tobacco ringspot virus decline: genus *Tobravirus, Tobacco ringspot virus*
(TRSV)
Tomato black ring: genus *Nepovirus, Tomato black ring virus* (TBRV)
Tomato ringspot virus decline: genus *Nepovirus, Tomato ringspot virus*
(ToRSV)
Verticillium wilt: *Verticillium dahliae* Kleb.
White rot: *Coniella diplodiella* (Speg.) Petr. and Syd.
Yellow speckle: Viroid
Zonate leaf spot: *Cristulariella moricola* (Hino) Redhead =
C. *pyramidalis* A. M.Waterman and R. P. Marshall, Teleomorph:
Grovesinia pyramidalis M. Cline et al.

OATS (AVENA SATIVA L.)

Anthracnose: *Colletotrichum graminicola* (Ces.) G. W. Wils.,
Teleomorph: *Glomerella graminicola* Politis
Bacterial blight (halo blight): *Pseudomonas syringae* pv. *coronafaciens*
(Elliott 1920) Young, Dye, and Wilkie 1978
Bacterial stripe blight: *Pseudomonas syringae* pv. *striafaciens* (Elliott
1920) Young, Dye, and Wilkie 1978
Black chaff and bacterial streak: *Xanthomonas translucens* pv.
translucens (Jones, Johnson, and Reddy 1917) Vauterin, Hoste,
Kersters, and Swings 1995
Covered smut: *Ustilago segetum* (Bull.:Pers.) Roussel = U. *kolleri* Wille
Crown rust: *Puccinia coronata* Corda

Downy mildew: *Sclerophthora macrospora* (Sacc.) Thirum. et al. =
Sclerospora macrospora Sacc.

Ergot: *Claviceps purpurea* (Fr.:Fr.) Tul., Anamorph: *Sphacelia segetum* Lev.

Fusarium foot rot: *Fusarium culmorum* (Wm. G. Sm.) Sacc.

Gray snow mold (= Typhula blight): *Typhula incarnata* Fr.,
T. ishikariensis Imai var. *idahoensis* (Remsb.) Arsvoll and Smith,
T. ishikariensis Imai var. *ishikariensis, T. ishikariensis* Imai var.
canadensis Smith and Arsvoll

Head blight: *Fusarium graminearum* Schwabe, Teleomorph: *Gibberella
zeae* (Schwein.) Petch, *Bipolaris sorokiniana* (Sacc.) Shoemaker,
Teleomorph: *Cochliobolus sativus* (Ito and Kuribayashi) Drechs. ex
Dastur

Leaf blotch: *Drechslera avenacea* (M. A. Curtis ex Cooke) Shoemaker =
Helminthosporium avenaceum M. A. Curtis ex Cooke, *D. avenae*
(Eidam) Scharif = *H. avenae* Eidam, Teleomorph: *Pyrenophora avenae*
Ito and Kuribayashi

Loose smut: *Ustilago avenae* (Pers.) Rostr.

Oat blue dwarf: genus *Marafivirus, Oat blue dwarf virus* (OBDV)

Oat mosaic: genus *Bymovirus, Oat mosaic virus* (OMV)

Oat necrotic mottle: *Oat necrotic mottle virus* (ONMV)

Oat pseudorosette: genus *Tenuivirus, Oat pseudorosette virus* (OPRV)

Oat red leaf: genus *Luteovirus, Barley yellow dwarf virus* (BYDV)

Oat sterile dwarf: genus *Fijivirus, Oat sterile dwarf virus* (OSDV)

Pink snow mold: *Microdochium nivalis* (Fr.) Samuel and I. C. Hallett =
Fusarium nivale Ces. ex Sacc., Teleomorph: *Monographella nivalis*
(Schaffnit) E. Müller

Powdery mildew: *Blumeria graminis* (DC.) E. O. Speer f. sp. = *Erysiphe
graminis* DC. f. sp. *avenae* Em. Marchal, *E. graminis* DC., Anamorph:
Oidium monilioides (Nees) Link

Sharp eyespot: *Rhizoctonia solani* Kuhn, Teleomorph: *Thanatephorus
cucumeris* (A. B. Frank) Donk, *Rhizoctonia cerealis* Van der Hoeven,
Teleomorph: *Ceratobasidium cereale* D. Murray and L. L. Burpee

Speckled blotch (Septoria blight): *Stagonospora avenae* (A. B. Frank)
Bissett = *Septoria avenae* A. B. Frank, Teleomorph: *Phaeosphaeria
avenaria* (G. F. Weber) O. Eriksson f. sp. *avenaria*

Stem rust: *Puccinia graminis* Pers.

Take-all: *Gaeumannomyces graminis* (Sacc.) von Arx and Olivier var.
avenae (E. M. Turner) Dennis

Victoria blight: *Bipolaris victoriae* (F. Meehan and Murphy) Shoemaker,
Teleomorph: *Cochliobolus victoriae* R. R. Nelson

PEA (PISUM SATIVUM L.)

Alternaria blight: *Alternaria alternata* (Fr.:Fr.) Keissl.

Anthracnose: *Colletotrichum gloeosporioides* (Penz.) Penz. and Sacc. in Penz. = *Colletotrichum pisi* Pat.

Aphanomyces root rot: *Aphanomyces euteiches* Drechs. f. sp. *pisi*

Ascochyta blight: *Didymella pinodes* (Berk. and A. Bloxam) Petr., Anamorph: *Ascochyta pinodes* L. K. Jones

Ascochyta leaf and pod spot: *Ascochyta pisi* Lib.

Bacterial blight: *Pseudomonas syringae* pv. *pisi* (Sackett 1916) Young, Dye, and Wilkie 1978

Black leaf: *Fusicladium pisicola* Linford

Brown spot: *Pseudomonas syringae* pv. *syringae* van Hall 1902

Cercospora leaf spot: *Cercospora pisa-sativae* Stevenson

Cladosporium blight: *Cladosporium cladosporioides* (Fresen.) de Vries f. sp. *pisicola* = *Cladosporium pisicola*

Downy mildew: *Peronospora viciae* (Berk.) de Bary

Gray mold: *Botrytis cinerea* Pers.:Fr.

Pea enation mosaic: genus *Enamovirus, Pea enation mosaic virus* (PEMV)

Pea root rot: *Fusarium solani* (Mart.) Sacc. f. sp. *pisi* (F. R. Jones) Snyd. and Hans.

Pea seedborne mosaic: genus *Potyvirus, Pea seedborne mosaic virus* (PSbMV)

Pea streak: genus *Alfamovirus, Alfalfa mosaic virus* (AMV); genus *Carlavirus, Pea streak virus* (PeSV)

Pea stunt: genus *Carlavirus, Red clover vein mosaic virus* (RCVMV)

Powdery mildew: *Erysiphe pisi* DC., Anamorph: *Oidium* sp.

Rhizoctonia seedling blight: *Thanatephorus cucumeris* (A. B. Frank) Donk, Anamorph: *Rhizoctonia solani* Kuhn

Sclerotinia rot: *Sclerotinia sclerotiorum* (Lib.) de Bary

Septoria blotch: *Septoria pisi* Westend.

Thielaviopsis root rot: *Thielaviopsis basicola* (Berk. and Broome) Ferrasis

Wilt: *Fusarium oxysporum* Schlechtend. f. sp. *pisi*

PEACH [PRUNUS PERSICA (L.) BATSCH]

Almond enation: genus *Nepovirus, Tomato black ring virus* (TBRV)

Alternaria rot: *Alternaria alternata* (Fr.:Fr.) Keissl.

Anthracnose: *Colletotrichum gloeosporioides* (Penz.) Penz. and Sacc. in Penz., Teleomorph: *Glomerella cingulata* (Stoneman) Spauld. and H. Schrenk

Apricot chlorotic leaf roll: Phytoplasma

Armillaria crown and root rot: *Armillaria mellea* (Vahl:Fr.) P. Kumm., Anamorph: *Rhizomorpha subcorticalis* Pers.

Bacterial canker: *Pseudomonas syringae* pv. *syringae* van Hall 1902

Bacterial spot: *Xanthomonas arboricola* pv. *pruni* (Smith 1903) Vauterin, Hoste, Kersters, and Swings 1995

Botrytis ripe fruit rot: *Botrytis cinerea* Pers.:Fr., Teleomorph: *Botryotinia fuckeliana* (de Bary) Whetzel

Brown rot: *Monilinia fructicola* (G. Wint.) Honey, *Monilinia laxa* (Aderhold and Ruhland) Honey

Clitocybe root rot: *Armillaria tabescens* (Scop.) Dennis et al. = *Clitocybe tabescens* (Scop.) Bres.

Constriction canker: *Phomopsis amygdali* (Delacr.) Tuset and Portilla = *Fusicoccum amygdali* Delacr.

Crown gall: *Agrobacterium tumefaciens* (Smith and Townsend 1907) Conn. 1942

Cytospora canker: *Cytospora leucostoma* Sacc., Teleomorph: *Leucostoma persoonii* Höhn.

Dark green sunken mottle: genus *Closterovirus, Apple chlorotic leaf spot virus* (ACLSV)

Enation: genus *Enamovirus, Peach enation virus* (PEV)

Gilbertella ripe fruit rot: *Gilbertella persicaria* (E. D. Eddy) Hesseltine

Gummosis: *Botryosphaeria dothidea* (Moug.:Fr.) Ces and De Not. = *Botryosphaeria berengeriana* De Not., Anamorph: *Fusicoccum aesculi* Corda

Leaf curl: *Taphrina deformans* (Berk.) Tul.

Line pattern: genus *Ilarvirus, Prunus necrotic ringspot virus* (PNRSV) + genus *Ilarvirus, Apple mosaic virus* (ApMV)

Little peach (= yellows): Phytoplasma

Necrotic ring spot: genus *Ilarvirus, Prunus necrotic ringspot virus* (PNRSV)

Peach latent mosaic: Peach latent mosaic viroid

Phony peach: *Xylella fastidiosa* Wells, Raju, Hung, Weisburg, Mandelco-Paul, and Brenner 1987

Phymatotrichopsis root rot: *Phymatotrichopsis omnivorum* (Duggar) Hennebert = *Phymatotrichum omnivorum* Duggar

Phytophthora crown and root rot: *Phytophthora cactorum* (Lebert and Cohn) J. Schröt., *P. cambivora* (Petri) Buisman, *P. citricola* Sawada, *P. citrophthora* (R. E. Sm. and E. H. Sm.) Leonian, *P. cryptogea* Pethybr. and Lafferty, *P. drechsleri* Tucker, *P. megasperma* Drechs., *P. syringae* (Kleb.) Kleb.

Plum pox (= Sharka): genus *Potyvirus, Plum pox virus* (PPV)
Powdery mildew: *Podosphaera clandestina* (Wallr.:Fr.) Lév., Anamorph:
  Oidium sp., *Podosphaera leucotricha* (Ellis. and Everh.) E. S. Salmon.,
  Anamorph: *O. leucoconium* Desmaz.
Prunus stem pitting: genus *Nepovirus, Tomato ringspot virus* (TmRSV)
· Red Suture (yellows): Phytoplasma
Rhizoctonia root rot: *Rhizoctonia solani* Kühn, Teleomorph:
  Thanatephorus cucumeris (A. B. Frank) Donk
Rosette: Phytoplasma
Rosette and decline: PNRSV + genus *Ilarvirus, Prune dwarf virus* (PDV)
Rosette mosaic: genus *Nepovirus, Peach rosette mosaic virus* (PRMV)
Rust: *Tranzschelia discolor* (Fuckel) Tranzschel and Litv. f. sp. *persica*
  Bolkan, Ogawa, Michailides, and Kable
Scab: *Cladosporium carpophilum* Thuem. = *Fusicladium carpophilum*
  (Thuem) Oudem, Teleomorph: *Venturia carpophila* E. E. Fisher
Shot hole: *Wilsonomyces carpophilus* (Lév) Adaskaveg, Ogawa, and E. E.
  Butler = *Stigmina carpophila* (Lév.) Ellis
Silver leaf: *Chondrostereum purpureum* (Pers.:Fr.) Pouzar
Verticillium wilt: *Verticillium dahliae* Kleb. (Mitosporic fungi)
Willow leaf rosette: genus *Nepovirus, Strawberry latent ringspot virus*
  (SLRV)
X-Disease: Phytoplasma
Yellow bud mosaic: genus *Nepovirus, Tomato ringspot virus* (TmRSV)
Yellow leaf roll: Phytoplasma
Yellows: Phytoplasma

PEANUT (ARACHIS HYPOGAEA L.)

Alternaria leaf blight: *Alternaria tenuissima* (Kunze:Fr.) Wiltshire
Alternaria leaf spot: *Alternaria arachidis* Kulk.
Alternaria veinal necrosis: *Alternaria alternata* (Fr.:Fr.) Keissl.
Anthracnose: *Colletotrichum arachidis* Sawada, *C. dematium* (Pers.:Fr.)
  Grove, *C. mangenoti* Chevaugeon
Aspergillus crown rot: *Aspergillus niger* Tiegh.
Bacterial wilt: *Ralstonia solanacearum* (Smith 1896) Yabauuchi, Kosako,
  Yano, Hotta, and Nishiuchi 1995
Blackhull: *Thielaviopsis basicola* (Berk. and Broome) Ferraris,
  Synanamorph: *Chalara elegans* Nag Raj and Kendrick
Botrytis blight: *Botrytis cinerea* Pers.:Fr., Teleomorph: *Botryotinia*
  fuckeliana (de Bary) Whetzel

Charcoal rot: *Macrophomina phaseolina* (Tassi) Goidanich = *Rhizoctonia bataticola* (Tassi) E. J. Butler

Choanephora leaf spot: *Choanephora* spp.

Collar rot: *Lasiodiplodia theobromae* (Pat.) Griffon and Maubl. = *Diplodia gossypina* Cooke

Colletotrichum leaf spot: *Colletotrichum gloeosporioides* (Penz.) Penz. and Sacc. in Penz., Teleomorph: *Glomerella cingulata* (Stoneman) Spauld. and H. Schrenk

Cowpea mild mottle: genus *Carlavirus, Cowpea mild mottle virus* (CPMMV)

Cylindrocladium black rot: *Cylindrocladium crotalariae* (C. A. Loos) D. K. Bell and Sobers, Teleomorph: *Calonectria crotalariae* (C. A. Loos) D. K. Bell and Sobers

Cylindrocladium leaf spot: *Cylindrocladium scoparium* Morg., Teleomorph: *Calonectria keyotensis* Terishita

Dodder: *Cuscuta campestris* Yunck. (parasitic plant)

Drechslera leaf spot: *Bipolaris spicifera* (Bainier) Subramanian = *Drechslera spicifera* (Bainier) Arx, Teleomorph: *Cochliobolus spicifer* R. R. Nelson

Early leaf spot: *Cercospora arachidicola* S. Hori, Teleomorph: *Mycosphaerella arachidis* Deighton

Fusarium wilt: *Fusarium oxysporum* Schlechtend.:Fr.

Groundnut crinkle: *Groundnut crinkle virus* (GCV)

Groundnut eyespot: *Groundnut eyespot virus* (GESV)

Groundnut rosette: genus *Umbravirus, Groundnut chlorotic rosette virus* (GCRV); genus *Umbravirus, Groundnut green rosette virus* (GGRV)

Groundnut streak: *Groundnut streak virus* (GSV)

Late leaf spot: *Phaeoisariopsis personata* (Berk. and M. A. Curtis) Arx = *Cercosporidium personatum* (Berk. and M. A. Curtis) Deighton, Teleomorph: *Mycosphaerella berkeleyi* Jenk.

Melanosis: *Stemphylium botryosum* Wallr, Teleomorph: *Pleospora tarda* E. Simmons

Myrothecium leaf blight: *Myrothecium roridum* Tode:Fr.

Olpidium root rot: *Olpidium brassicae* (Woronin) P. A. Dang.

Peanut bud necrosis: genus *Tospovirus, Peanut bud necrosis virus* (PBNV) (family *Bunyaviridae*)

Peanut chlorotic blotch: genus *Potyvirus, Peanut chlorotic blotch virus* (PCBV) (family *Potyviridae*)

Peanut chlorotic fan-spot: *Peanut chlorotic fan-spot virus* (PCFSV)

Peanut chlorotic streak: genus *Caulimovirus, Peanut chlorotic streak virus* (PCSV)

Peanut clump: genus *Furovirus, Peanut clump virus* (PCV)

Peanut green mosaic: *Peanut green mosaic virus* (PGMV)

Peanut mottle: genus *Potyvirus, Peanut mottle virus* (PeMoV) (family *Potyviridae*)

Peanut stripe: genus *Potyvirus, Peanut stripe virus* (PStV) (family *Potyviridae*)

Peanut stunt: genus *Cucumovirus, Peanut stunt virus* (PSV)

Peanut yellow mottle: *Peanut yellow mottle virus* (PYMoV)

Peanut yellow spot: genus *Tospovirus, Peanut yellow spot virus* (PYSV)

Pepper spot and scorch: *Leptosphaerulina crassiasca* (Sechet) C. R. Jackson and D. K. Bell

Pestalotiopsis leaf spot: *Pestalotiopsis arachidis* Satya

Phanerogamic root parasite: *Alectna* sp. (parasitic plant)

Phoma leaf blight: *Phoma microspora* Benk. and M. A. Curtis

Phomopsis foliar blight: *Phomopsis phaseoli* (Desmaz.) Sacc. = *P. sojae* Lehman, Teleomorph: *Diaporthe phaseolorum* (Cooke and Ellis) Sacc.

Phyllosticta leaf spot: *Phyllosticta arachidis-hypogaea* V. G. Rao, *P. sojicola* C. Massal., Teleomorph: *Pleosphaerulina sojicola* Miura

Phymatotrichum root rot: *Phymatotrichopsis omnivora* (Duggar) Hennebert = *Phymatotrichum omnivorum* Duggar

Powdery mildew: *Oidium arachidis* Chorin

Pythium peg and root rot: *Pythium myriotylum* Drechs., *P. aphanidermatum* (Edson) Fitzp., *P. debaryanum* Auct. non R. Hesse, *P. irregulare* Buisman, *P. ultimum* Trow

Rhizoctonia root rot: *Rhizoctonia solani* Kühn

Rust: *Puccinia arachidis* Speg.

Scab: *Sphaceloma arachidis* Bitancourt and Jenk.

Sclerotinia blight: *Sclerotinia minor* Jagger, *S. sclerotiorum* (Lib.) de Bary

Stem rot: *Sclerotium rolfsii* Sacc, Teleomorph: *Athelia rolfsii* (Curzi) Tu and Kimbrough

Tomato spotted wilt: genus *Tospovirus, Tomato spotted wilt virus*

Verticillium wilt: *Verticillium albo-atrum* Reinke and Berthier, *V. dahliae* Kleb.

Web blotch (net blotch): *Phoma arachidicola* Marasas et al. = *Ascochyta adzamethica* Schoschiaschvili, Teleomorph: *Didymosphaeria arachidicola* (Chochrjakov) Alcorn = *Mycosphaerella arachidicola* Chochrjakov

Witches'-broom: Phytoplasma

Zonate leaf spot: *Cristulariella moricola* (Hino) Redhead, Synanamorph: *Sclerotium cinnomomi* Sawada, Teleomorph: *Grovesinia pyramidalis* M. Cline et al.

PEAR (PYRUS COMMUNIS L.)

American hawthorne rust: *Gymnosporangium globosum* (Farl.) Farl.

Anthracnose canker: *Pezicula malicorticus* (H. Jacks) Nannf., Anamorph: *Cryptosporiopsis curvispora* (Peck) Gremmen in Boerema and Gremmen

Armillaria root rot: *Armillaria mellea* (Vahl:Fr.) P. Kumm., Anamorph: *Rhizomorpha subcorticalis* Pers.

Bitter rot: *Glomerella cingulata* (Stoneman) Spauld. and Schrenk, Anamorph: *Colletotrichum gloeosporioides* (Penz.) Penz. and Sacc. in Penz.

Black rot and canker: *Botryosphaeria obtusa* (Schwein.) Shoemaker, Anamorph: *Sphaeropsis malorum* Berk.

Black spot: *Alternaria alternata* (Fr.:Fr.) Keissl.

Blister canker: *Helminthosporium papulosum* Berg.

Blister disease: *Coniothecium chromatosporum* Corda

Botrytis blossom blight: *Botrytis cinerea* Pers.:Fr., Teleomorph: *Botryotinia fuckeliana* (de Bary) Whetzel

Brown rot: *Monilinia fructicola* (Wint.) Honey, *M. laxa* (Aderh. and Ruhl.) Honey

Brown spot: *Stemphylium vesicarium* (Wallr.) E. Simmons, Teleomorph: *Pleospora allii* (Rabenh.) Ces. and De Not.

Cladosporium fruit rot: *Cladosporium herbarum* (Pers.:Fr.) Link, Teleomorph: *Mycosphaerella tassiana* (De Not.) Johans.

Clitocybe root rot: *Armillaria tabescens* (Scop.) Dennis et al. = *Clitocybe tabescens* (Scop.) Bres.

Crown gall: *Agrobacterium tumefaciens* (Smith and Townsend 1907) Conn 1942

Diplodia canker: *Botryosphaeria stevensii* Shoemaker = *Physalospora malorum* Shear et al., Anamorph: *Diplodia mutila* (Fr.:Fr.) Mont.

Elsinoe leaf and fruit spot: *Elsinoe piri* (Woronichin) Jenk., Anamorph: *Sphaceloma pirinum* Jenk.

European canker: *Nectria galligena* Bres. in Strass., Anamorph: *Cylindrocarpon heteronemum* (Berk. and Broome) Wollenweb.

European pear rust: *Gymnosporangium fuscum* R. Hedw. in DC.

Fabraea leaf and fruit spot: *Diplocarpon mespili* (Sorauer) Sutton = *Fabraea maculata* Atk., Anamorph: *Entomosporium mespili* (DC.) Sacc.

Fire blight: *Erwinia amylovora* (Burrill 1882) Winslow, Broadhurst, Buchanan, Krumwiede, Rogers, and Smith 1920

Fly speck: *Schizothyrium pomi* (Mont.:Fr.) Arx, Anamorph: *Zygophiala jamaicensis* E. Mason

Gibberella canker: *Gibberella baccata* (Wallr.) Sacc., Anamorph: *Fusarium lateritium* Nees:Fr.

Kern's pear rust: *Gymnosporangium kernianum* Bethel

Late leaf spot: *Cercospora minima* Tracy and Earle

Mycosphaerella leaf spot: *Mycosphaerella pyri* (Auersw.) Boerema = *M. sentina* (Fr.:Fr.) Schrot., Anamorph: *Septoria pyricola* (Desmaz.) Desmaz.

Nectria twig blight: *Nectria cinnabarina* (Tode) Fr., Anamorph: *Tubercularia vulgaris* Tode

Pacific Coast pear rust: *Gymnosporangium libocedri* (C. Henn.) F. Kern

Pear decline: Phytoplasma

Pear scab: *Venturia pirina* Aderh., Anamorph: *Fusicladium pyrorum* (Lib.) Fuckel

Perennial canker: *Neofabrae perennans* Kienholz, Anamorph: *Cryptosporiopsis perennans* (Zeller and Childs) Wollenweb.

Phyllosticta leaf spot: *Phyllosticta* sp.

Phytophthora crown and root rot: *Phytophthora cactorum* (Lebert and Cohn) J. Schrot.

Powdery mildew: *Podosphaera leucotricha* (Ellis and Everh.) E. S. Salmon

Pseudomonas canker: *Pseudomonas syringae* pv. *syringae* van Hall 1902

Rocky Mountain pear rust: *Gymnosporangium nelsonii* Arth.

Rosellinia root rot: *Rosellinia necatrix* Prill., Anamorph: *Dematophora necatrix* R. Hartig

Side rot: *Phialophora malorum* (M. N. Kidd and A. Beaumont) McColloch

Silver leaf: *Chondrostereum pupureum* (Pers.:Fr.) Pouzar

Sooty blotch: *Gloeodes pomigena* (Schwein.) Colby

Thread blight: *Corticium stevensii* Burt = *Pellicularia koleroga* Cooke = *Hypochnus ochroleucus* Noack

Valsa canker: *Valsa ceratosperma* (Tode:Fr.) Maire, Anamorph: *Cytospora sacculus* (Schwein.) Gvritischvili

Xylaria root rot: *Xylaria* spp.

PLUM (PRUNUS DOMESTICA L.)

Armillaria crown and root rot: *Armillaria mellea* (Vahl:Fr.) P. Kumm., Anamorph: *Rhizomorpha subcorticalis* Pers.

Bacterial canker: *Pseudomonas syringae* pv. *syringae* van Hall 1902

Bacterial spot: *Xanthomonas arboricola* pv. *pruni* (Smith 1903) Vauterin, Hoste, Kersters, and Swings 1995

Black knot: *Apiosporina morbosa* (Schwein.:Fr.) Arx, Anamorph: *Fusicladium* sp.

Brown rot: *Monilinia fructicola* (G. Wint.) Honey, *Monilinia laxa* (Aderhold and Ruhland) Honey

Line pattern: genus *Ilarvirus, Prunus necrotic ringspot virus* (PNRSV) + genus *Ilarvirus, Apple mosaic virus* (ApMV)

Phony disease: *Xylella fastidiosa* Wells, Raju, Hung, Weisburg, Mandelco-Paul, and Brenner 1987

Plum pox (= Sharka): genus *Potyvirus, Plum pox virus* (PPV)

Red Suture (= yellows): Phytoplasma

Rosette mosaic: genus *Nepovirus, Peach rosette mosaic virus* (PRMV)

Rust: *Tranzschelia discolor* (Fuckel) Tranzschel and Litv.

Scab: *Cladosporium carpophilum* Thuem. = *Fusicladium carpophilum* (Thuem) Oudem, Teleomorph: *Venturia carpophila* E. E. Fisher

Silver leaf: *Chondrostereum purpureum* (Pers.:Fr.) Pouzar

Verticillium wilt: *Verticillium dahliae* Kleb.

POTATO (SOLANUM TUBEROSUM L.)

Aster yellows: Phytoplasma

Bacterial wilt (brown rot): *Ralstonia solanacearum* (Smith 1896) Yabauuchi, Kosako, Yano, Hotta, and Nishiuchi 1995

Black dot: *Colletotrichum coccodes* (Wallr.) S. J. Hughes = *C. atramentarium* (Berk. and Broome) Taubenhaus

Blackleg and bacterial soft rot: *Erwinia carotovora* ssp. *atroseptica* (van Hall 1902) Dye 1969, *E. carotovora* ssp. *carotovora* (Jones 1901) Bergey, Harrison, Breed, Hammer and Huntoon 1923, *E. chrysanthemi* Burkholder, McFadden, and Dimock 1953

Cercospora leaf blotch: *Mycovellosiella concors* (Casp.) Deighton = *Cercospora concors* (Casp.) Sacc., *C. solani-tuberosi* Thirumalachar.

Charcoal rot: *Macrophomina phaseolina* (Tassi) Goidanich

Early blight: *Alternaria solani* Sorauer

Fusarium dry rot: *F. sambucinum* Fuckel, Teleomorph: *Gibberella pulicaris* (Fr.:Fr.) Sacc., *F. solani* (Mart.) Sacc. var. *coeruleum* (Lib. ex Sacc.) C. Booth, *F. avenaceum* (Fr.:Fr.) Sacc., Teleomorph: *Gibberella avenacea* R. J. Cooke

Fusarium wilt: *F. avenaceum* (Fr.:Fr.) Sacc., *F. oxysporum* Schlechtend.: Fr., *F. solani* (Mart.) Sacc., Teleomorph: *Nectria haematococca* Berk.

and Broome, *F. solani* (Mart.) Sacc. f. sp. *eumartii* (C. Carpenter)
W. C. Snyder and H. N. Hans = *F. eumartii* C. Carpenter

Gangrene: *Phoma exigua* Desmaz. = *P. solanicola* Prill. and Delacr.,
P. foveata Foister = *P. exigua* var. *foveata* (Foister) Boerema

Gray mold: *Botrytis cinerea* Pers.:Fr., Teleomorph: *Botryotinia fuckeliana*
(de Bary) Whetzel

Late blight: *Phytophthora infestans* (Mont.) de Bary

Leak: *Pythium* spp., *P. debaryanum* Auct. non R. Hesse, *P. ultimum* Trow

Phoma leaf spot: *Phoma andina* Turkensteen

Pink eye: *Pseudomonas fluorescens* Migula

Pink rot: *P. cryptogea* Pethybr. and Lafferty, *P. drechslera* Tucker,
P. erythroseptica Pethybr.

Potato Andean latent: genus *Tymovirus, Potato Andean latent virus*
(APLV)

Potato Andean mottle: genus *Comovirus, Potato Andean mottle virus*
(APMV)

Potato aucuba mosaic: genus *Potexvirus, Potato aucuba mosaic virus*
(PAMV)

Potato early dying, Verticillium wilt: *Verticillium albo-atrum* Reinke and
Berthier, *V. dahliae* Kleb.

Potato leaf roll mosaic: genus *Luteovirus, Potato leaf roll virus* (PLRV)

Potato leaf rolling: genus *Carlavirus, Potato virus M* (PVM)

Potato mild mosaic: genus *Potyvirus, Potato virus A* (PVA)

Potato mop-top: genus *Furovirus, Potato mop-top virus* (PMTV)

Potato rugose mosaic: genus *Potyvirus, Potato virus Y* (PVY, strains O, N,
and C)

Potato spindle tuber: Potato spindle tuber viroid (PSTV)

Potato stem mottle (spraing of tubers): genus *Tobravirus, Tobacco rattle
virus* (TRV)

Potato virus S: genus *Carlavirus, Potato virus S* (PVS)

Potato virus T: genus *Trichovirus, Potato virus T* (PVT)

Potato virus X: genus *Potexvirus, Potato virus X* (PVX)

Potato yellow vein: genus *Bigeminivirus, Potato yellow vein virus* (PYVV)

Powdery mildew: *Erysiphe cichoracearum* DC., Anamorph: *Oidium
asteris-punicei* Peck

Powdery scab: *Spongospora subterranea* (Wallr.) Lagerh

Rhizoctonia canker (black scurf): *Rhizoctonia solani* Kühn, Teleomorph:
Thanatephorus cucumeris (A. B. Frank) Donk

Ring rot: *Clavibacter michiganensis* ssp. *sepedonicus* (Spieckermann and
Kotthoff 1914) Davis, Gillaspie, Vidaver, and Harris 1984

Rosellinia black rot: *Rosellinia* sp.

Septoria leaf spot: *Septoria lycopersici* Speg.

Silver scurf: *Helminthosporium solani* Durieu and Mont. =
Spondylocladium atrovirens (C. Harz) C. Harz ex Sacc.
Skin spot: *Polyscytalum pustulans* (N. M. Owensand Wakef.) M. B. Ellis
Stem rot (southern blight): *Sclerotium rolfsii* Sacc., Teleomorph: *Athelia*
rolfsii (Curzi) Tu and Kimbrough
Tomato spotted wilt: genus *Tospovirus, Tomato spotted wilt virus*
(TSWV) (family *Bunyaviridae*)
Wart: *Synchytrium endobioticum* (Schilberszky) Percival
White mold: *Sclerotinia sclerotiorum* (Lib.) de Bary, *S. minor* Jagger =
S. intermedia Ramsey
Witches'-broom: Phytoplasma

RICE (ORYZA SATIVA L.)

Aggregate sheath spot: *Ceratobasidium oryzae-sativae* Gunnell and
Webster, Anamorph: *Rhizoctonia oryzae-sativae* (Sawada) Mordue
Bacterial blight: *Xanthomonas oryzae* pv. *oryzae* (Ishiyama 1922) Swings,
Van den Mooter, Vauterin, Hoste, Gillis, Mew, and Kersters 1990
Bacterial foot rot: *Erwinia chrysanthemi* Burkholder, McFadden, and
Dimock 1953.
Bacterial leaf streak: *Xanthomonas oryzae* pv. *oryzicola* (Ishiyama 1922)
Swings, Van den Mooter, Vauterin, Hoste, Gillis, Mew, and Kersters 1990
Black kernel: *Curvularia lunata* (Wakk.) Boedijn, Teleomorph:
Cochiobolus lunatus R. R. Nelson and Haasis
Blast: *Magnaporthe grisea* (Hebert) Barr, Anamorph: *Pyricularia grisea*
Sacc. = *Pyricularia oryzae* Cavara
Brown spot: *Cochliobolus miyabeanus* (Ito and Kuribayashi) Drechs. ex
Dastur, Anamorph: *Bipolaris oryzae* (Breda de Haan) Shoemaker =
Helminthosporium oryzae Breda de Haan
Crown sheath rot: *Gaeumannomyces graminis* (Sacc.) Arx and D. Olivier
Downy mildew: *Sclerophthora macrospora* (Sacc.) Thirumalachar et al.
Eyespot: *Drechslera gigantea* (Heald and F. A. Wolf) Ito
False smut: *Ustilaginoidea virens* (Cooke) Takah.
Foot rot (Bakane): *Fusarium moniliforme* Sheldon, Teleomorph:
Gibberella fujikuroi (Sawada) Ito in Ito and K. Kimura
Giallume: genus *Luteovirus, Barley yellow dwarf virus* (BYDV)
Grain rot: *Burkholderia glumae* (Kurita and Tabei 1967) Urakami, Ito-
Yoshida, Araki, Kijima, Suzuki, and Komagata 1994 = *Pseudomonas*
glumae Kurita and Tabei 1967
Kernel smut: *Tilletia barclayana* (Bref.) Sacc. and Syd. in Sacc. =
Neovossia horrida (Takah.) Padwick and A. Khan

Leaf scald: *Microdochium oryzae* (Hashioka and Yokogi) Samuels and
    I. C. Hallett = *Rhynchosporium oryzae* Hashioka and Yokogi
Leaf smut: *Entyloma oryzae* Syd. and P. Syd.
Narrow brown leaf spot: *Cercospora janseana* (Racib.) O. Const. =
    C. oryzae Miyake, Teleomorph: *Sphaerulina oryzina* K. Hara
Rice black streak dwarf: genus *Fijivirus, Rice black streak dwarf virus*
    (RBSDV)
Rice dwarf: genus *Phytoreovirus, Rice dwarf virus* (RDV)
Rice gall dwarf: genus *Phytoreovirus, Rice gall dwarf virus* (RGDV)
Rice grassy stunt: genus *Tenuivirus, Rice grassy stunt virus* (RGSV)
Rice hoja blanca: genus *Tenuivirus, Rice hoja blanca virus* (RHBV)
Rice necrosis mosaic: genus *Bymovirus, Rice necrosis mosaic virus* (RNMV)
Rice ragged stunt: genus *Oryzavirus, Rice ragged stunt virus* (RRSV)
Rice stripe: genus *Tenuivirus, Rice stripe virus* (RStV)
Rice stripe necrosis: genus *Furovirus, Rice stripe necrosis virus* (RStNV)
Rice transitory yellowing: genus *Nucleorhabdovirus, Rice transitory*
    yellowing virus (RTYV)
Rice yellow dwarf: Phytoplasma
Rice yellow mottle: genus *Sobemovirus, Rice yellow mottle virus* (RYMV)
Sheath blight: *Rhizoctonia solani* Kühn– Teleomorph: *Thanatephorus*
    cucumeris (A.B. Frank) Donk
Sheath brown rot: *Pseudomonas fuscovaginae* (ex Tanii, Miyajima, and
    Akita 1976) Miyajima, Tanii, and Akita 1983
Sheath rot: *Sarocladium oryzae* (Sawada) W. Gams and D. Hawksworth =
    Acrocylindrium oryzae Sawada
Sheath spot: *Rhizoctonia oryzae* Ryker and Gooch
Stackburn: *Alternaria padwickii* (Ganguly) M.B. Ellis = *Trichoconis*
    padwickii Ganguly
Stem rot: *Magnaporthe salvinii* (Cattaneo) R. Krause and Webster,
    Synanamorphs: *Sclerotium oryzae* Cattaneo, *Nakataea sigmoidae*
    (Cavara) K. Hara
Tungro: genus *Badnavirus, Rice tungro bacilliform virus* (RTBV); genus
    Waikavirus, Rice tungro spherical virus (RTSV)
Udbatta: *Ephelis oryzae* Syd., Teleomorph: *Balansia oryzae-sativae*
    Hashioka

TOMATO (LYCOPERSICON ESCULENTUM MILL.)

Alternaria stem canker: *Alternaria alternata* (Fr.:Fr.) Keissl. f. sp.
    lycopersici Grogan et al.

Anthracnose: *Colletotrichum coccodes* (Wallr.) S. J. Hughes, *C. dematium* (Pers.) Grove, *C. gloeosporioides* (Penz.) Penz. and Sacc. in Penz., Teleomorph: *Glomerella cingulata* (Stoneman) Spauld. and H. Schrenk

Aster yellows: Phytoplasma

Bacterial canker: *Clavibacter michiganensis* ssp. *michiganensis* (Smith 1910) Davis, Gillaspie, Vidaver, and Harris 1984

Bacterial speck: *Pseudomonas syringae* pv. *tomato* (Okabe 1933) Young, Dye, and Wilkie 1978

Bacterial spot: *Xanthomonas vesicatoria* (ex Doidge 1920) Vauterin, Hoste, Kersters, and Swings 1995 = *Xanthomonas campestris* pv. *vesicatoria* (Doidge 1920) Dye 1978

Bacterial stem rot and fruit rot: *Erwinia carotovora* ssp.*carotovora* (Jones 1902) Bergey, Harrison, Breed, Hammer, and Huntoon 1923

Bacterial wilt: *Ralstonia solanacearum* (Smith 1896) Yabauuchi, Kosako, Yano, Hotta, and Nishiuchi 1995

Black root rot: *Thielaviopsis basicola* (Berk. and Broome) Ferraris, Synanamorph: *Chalara elegans* Nag Raj and Kendrick

Black shoulder: *Alternaria alternata* (Fr.:Fr.) Keissl.

Buckeye fruit and root rot: *Phytophthora capsici* Leonian, *P. dreshsleri* Tucker, *P. nicotianae* Breda de Haan var. *parasitica* (Dastur) G. M. Waterhouse = *P. parasitica* Dastur

Cercospora leaf mold: *Pseudocercospora fuligena* (Roldan) Deighton = *Cercospora fuligena* Roldan

Charcoal rot: *Macrophomina phaseolina* (Tassi) Goidanich

Common mosaic of tomato: genus *Tobamovirus, Tobacco mosaic virus* (TMV)

Corky root rot: *Pyrenochaeta lycopersici* R. Schneider and Gerlach

Curly top: genus *Hybrigeminivirus, Tomato curly top virus* (TCTV)

Didymella stem rot: *Didymella lycopersici* Kleb.

Early blight: *Alternaria solani* Sorauer

Fusarium crown and root rot: *Fusarium oxysporum* Schlechtend.:Fr. f. sp. *radicis-lycopersici* W. R. Jarvis and Shoemaker

Fusarium wilt: *Fusarium oxysporum* Schlechtend.:Fr. f. sp. *lycopersici* (Sacc.) W. C. Snyder and H. N. Hans.

Gray leaf spot: *Stemphylium botryosum* Wallr. f. sp. *lycopersici* Rotem et al., *S. lycopersici* (Enjoji) W. Yamamoto = *S. floridanum* Hannon and G. F. Weber, *S. solani* G. F. Weber

Gray mold: *Botrytis cinerea* Pers.:Fr., Teleomorph: *Botryotinia fuckeliana* (de Bary) Whetzel

Late blight: *Phytophthora infestans* (Mont.) de Bary

Leaf mold: *Fulvia fulva* (Cooke) Cif. = *Cladosporium fulvum* Cooke

Phoma rot: *Phoma destructiva* Plowr. (Anamorphic Pleosporaceae)

Pith necrosis: *Pseudomonas corrugata* (ex Scarlett, Fletcher, Roberts, and Lelliott 1978) Roberts and Scarlett 1981

Potato virus Y: genus *Potyvirus, Potato virus Y* (PVY)

Powdery mildew: *Oidiopsis sicula* Scalia, Teleomorph: *Leveillula taurica* (Lév.) G. Arnaud

Pseudo curly top: genus *Hybrigeminivirus, Tomato pseudo curly top virus* (TPCTV)

Pythium damping-off and fruit rot: *Pythium aphanidermatum* (Edson) Fitzp., *P. arrhenomanes* Drechs., *P. debaryanum* Auct. non R. Hesse, *P. myriotylum* Drechs., *P. ultimum* Trow

Rhizoctonia damping-off and fruit rot: *Rhizoctonia solani* Kühn, Teleomorph: *Thanatephorus cucumeris* (A. B. Frank) Donk

Rhizopus rot: *Rhizopus stolonifer* (Ehrenb.:Fr.) Vuill.

Septoria leaf spot: *Septoria lycopersici* Speg.

Sour rot: *Geotrichum candidum* Link, Teleomorph: *Galactomyces geotrichum* (E. E. Butler and L. J. Petersen) Redhead and Malloch, *G. penicillatum* (do Carmo Sousa) Arx

Southern blight: *Sclerotium rolfsii* Sacc., Teleomorph: *Athelia rolfsii* (Curzi) Tu and Kimbrough

Syringae leaf spot: *Pseudomonas syringae* pv. *syringae* van Hall 1902

Target spot: *Corynespora cassiicola* (Berk. and M. A. Curtis) C. T. Wei

Tomato big bud: Phytoplasma

Tomato bunchy top: Tomato bunchy top viroid

Tomato bushy stunt: genus *Tombusvirus, Tomato bushy stunt virus* (TBSV)

Tomato etch: genus *Potyvirus, Tobacco etch virus* (TEV)

Tomato fern leaf: genus *Cucumovirus, Cucumber mosaic virus* (CMV)

Tomato mosaic: genus *Tobamovirus, Tomato mosaic virus* (ToMV)

Tomato mottle: genus *Bigeminivirus, Tomato mottle virus* (ToMoV)

Tomato necrosis: genus *Alfamovirus, Alfalfa mosaic virus* (AMV)

Tomato planto macho: Tomato planto macho viroid

Tomato spotted wilt: genus *Tospovirus, Tomato spotted wilt virus* (TSWV)

Tomato yellow leaf curl: genus *Bigeminivirus, Tomato yellow leaf curl virus* (TYLCV)

Tomato yellow top: *Tomato yellow top virus* (TYTV)

Verticillium wilt: *Verticillium albo-atrum* Reinke and Berthier, *V. dahliae* Kleb.

White mold: *Sclerotinia sclerotiorum* (Lib.) de Bary, *S. minor* Jagger

WHEAT (TRITICUM SPP.)

Agropyron mosaic: genus *Rymovirus, Agropyron mosaic virus* (AgMV)

Alternaria leaf blight: *Alternaria triticina* Pras. and Prab.

Anthracnose: *Colletotrichum graminicola* (Ces.) G. W. Wils.,
    Teleomorph: *Glomerella graminicola* Politis

Ascochyta leaf spot: *Ascochyta tritici* Hori and Enjoji

Aster yellows: Aster yellows phytoplasma

Bacterial leaf blight: *Pseudomonas syringae* pv. *syringae* van Hall 1902

Barley stripe mosaic: genus *Hordeivirus, Barley stripe mosaic virus*
    (BSMV)

Barley yellow dwarf: genus *Luteovirus, Barley yellow dwarf virus* (BYDV)

Barley yellow striate mosaic: genus *Cytorhabdovirus, Barley yellow
    striate mosaic virus* (BYSMV)

Basal glume rot: *Pseudomonas syringae* pv. *atrofaciens* (McCulloch
    1920) Young, Dye, and Wilkie 1978

Black chaff and bacterial streak: *Xanthomonas translucens* pv.
    translucens (Jones, Johnson, and Reddy 1917) Vauterin, Hoste,
    Kersters, and Swings 1995

Brome mosaic: genus *Bromovirus, Brome mosaic virus* (BMV)

Cephalosporium stripe: *Hymenula cerealis* Ellis and Everh. =
    Cephalosporium gramineum Nisikado and Ikata in Nisikado et al.

Cereal northern mosaic = barley yellow striate mosaic: genus
    Cytorhabdovirus, Northern cereal mosaic virus (NCMV) (family
    Rhabdoviridae, Order *Mononegavirales*)

Cereal tillering: genus *Reovirus, Cereal tillering disease virus* (CTDV)

Cocksfoot mottle: genus *Sobemovirus, Cocksfoot mottle virus* (CoMV)

Common bunts: *Tilletia tritici* (Bjerk.) G. Wint. in Rabenh = *Tilletia
    caries* (DC.) Tul. and C. Tul., *T. laevis* Kühn in Rabenh. = *T. foetida*
    (Wallr.) Liro

Crown rot: *Fusarium culmorum* (Wm. G. Smith) Sacc., *F. graminearum*
    Schwabe, Teleomorph: *Gibberella zeae* (Schwein.) Petch, *Bipolaris
    sorokiniana* (Sacc.) Shoemaker

Dilophospora leaf spot and twist: *Dilophospora alopecuri* (Fr.) Fr.

Downy mildew: *Sclerophthora macrospora* (Sacc.) Thirum. et al =
    Sclerospora macrospora Sacc.

Dwarf bunt: *Tilletia controversa* Kuhn in Rebenh.

Ergot: *Claviceps purpurea* (Fr.:Fr.) Tul., Anamorph: *Sphacelia segetum* Lev.

Eyespot: *Pseudocercosoporella herpotrichoides* (Fron) Deighton,
    Teleomorph: *Tapesia yallundae* Wallwork and Spooner, *T. acuformis*
    (Pers.) Fuckel

Flag smut: *Urocystis agropyri* (Preuss) Schroet.

Gray snow mold (= Typhula blight): *Typhula incarnata* Fr., *Typhula ishikariensis* Imai var. idahoensis (Remsb.) Arsvoll and Smith, *T. ishikariensis* Imai var. *ishikariensis*, *T. ishikariensis* Imai var. *canadensis* Smith and Arsvoll

Halo spot: *Pseudoseptoria donacis* (Pass.) Sutton = *Selenophoma donacis* (Pass.) R. Sprague and A. G. Johnson

Herpotrichoides leaf spot: *Leptosphaeria herpotrichoides* de Not = *Phaeosphaeria herpotrichoides* (de Not.) Holm

Karnal bunt: *Tilletia indica* Mitra = *Neovossia indica* (Mitra) Mundkur

Leaf and glume blotch: *Stagonospora nodorum* (Berk.) Castellani and E. G. Germano = *Septoria nodorum* (Berk.) Berk. in Berk. and Broome, Teleomorph: *Phaeosphaeria nodorum* (E. Muller) Hedjaroude, *Leptosphaeria nodorum* Muller

Leaf (brown) rust: *Puccinia triticina* Eriks. = *Puccinia recondita* Rob. ex Desm. f. sp. *tritici* = *Puccinia recondita* Rob. ex Desm.

Loose smut: *Ustilago tritici* (Pers.) Rostr.

Maize dwarf mosaic: genus *Potyvirus*, *Maize dwarf mosaic virus* (MDMV)

Maize streak: genus *Monogeminivirus*, *Maize streak virus* (MSV)

Microscopica leaf spot: *Leptosphaeria microscopica* Karst., Anamorph: *Phaeoseptoria urvilleana* (Speg.) Sprague

Oat sterile dwarf: genus *Fijivirus*, *Oat sterile dwarf virus* (OSDV)

Phoma spot: *Phoma glomerata* (Cda.) Wr. and Hochapf., *P. insidiosa* Tassi

Pink seed: *Erwinia rhapontici* (Millard 1924) Burkholder 1948 = *Pectobacterium rhapontici* (Millard 1924) Patel and Kulkarni 1951

Pink snow mold: *Microdochium nivalis* (Fr.) Samuel and I. C. Hallett = *Fusarium nivale* Ces. ex Sacc., Teleomorph: *Monographella nivalis* (Schaffnit) E. Müller

Powdery mildew: *Blumeria graminis* (DC.) E. O. Speer f. sp. *tritici* = *Erysiphe graminis* DC. Fr. f. sp. *tritici*

Pythium root rot: *P. aphanidermatum* (Edson) Fitzp., *Pythium graminicola* Subramanian, *Pythium arrhenomanes* Drechs., *Pythium tardicrescens* Vanderpool, *P. volutum* Vant. and Tru, *P. myriotylum* Drechs.

Rice black-streaked dwarf: genus *Fijivirus*, *Rice black-streaked dwarf virus* (RBSDV)

Rice hoja blanca: genus *Tenuivirus*, *Rice hoja blanca virus* (RHBV)

Rice stripe: genus *Tenuivirus*, *Rice stripe virus* (RSV)

Scab (= head blight): *Fusarium graminearum* Schwabe, Teleomorph: *Gibberella zeae* (Schwein.) Petch, *Fusarium culmorum* (Wm. G.

Smith) Sacc., *F. avenaceum* (Fr.) Sacc., Teleomorph: *Gibberella avenaceae* Cook

Septoria speckled leaf blotch: *Septoria tritici* Rob. in Desm., Teleomorph: *Mycosphaerella graminicola* (Feckl.) Sand, *Stagonospora avenae* f. sp. *triticea* T. Johnson

Sharp eyespot: *Rhizoctonia cerealis* Van der Hoeven, Teleomorph: *Ceratobasidium cereale* D. Murray and L. L. Burpe, *Rhizoctonia solani* Kuhn, Teleomorph: *Thanatephorus cucumeris* (A. B. Frank) Donk

Snow rot: *Pythium iwayamai* Ito, *P. okanoganense* Lipps, *P. paddicum* Harane

Snow scald: *Myriosclerotinia borealis* (Bubak and Vleugel) L. M. Kohn = *Sclerotinia borealis* Bubak and Vleugel

Southern blight: *Sclerotium rolfsii* Sacc., Teleomorph: *Athelia rolfsii* (Curzi) Tu and Kimbrough

Speckled snow mold: *Typhula idahoensis* Remsberg

Spike blight: *Rathayibacter tritici* (ex Hutchinson 1917) Zgurskaya, Evtushenko, Akimov, and Kalakoutskii 1993

Spot blotch: *Cochliobolus sativus* (Ito and Kuribayashi) Drechs. ex Dastur, Anamorph: *Drechslera teres* (Sacc.) Shoemaker

Stem rust: *Puccinia graminis* f. sp. *tritici* (Pers.) Erikss. et Henn

Stripe (yellow) rust: *Puccinia striiformis* Westend = *Puccinia striiformis* Westend f. sp. *tritici*

Take-all: *Gaeumannomyces graminis* (Sacc.) von Arx and Oliver var. *tritici* Walker, *Gaeumannomyces graminis* (Sacc.) von Arx and Oliver var. *graminis*, *Gaeumnnomyces* (Sacc.) von Arx and Oliver var. *avenae* (E. M. Turner) Dennis = *Ophiobolus graminis* Sacc.

Tan spot (= yellow leaf spot): *Pyrenophora tritici-repentis* (Died.) Drechs. = *Pyrenophora trichostoma* (Fr.) Fuckel, Anamorph: *Drechslera tritici-repentis* (Died.) Shoemaker = *Helminthosporium tritici-repentis* Died.

Tar spot: *Phyllachora graminis* (Pers.:Fr.) Fekl.

Tobacco mosaic: genus *Tobamovirus*, *Tobacco mosaic virus* (TMV)

Verticillium wilt: *Verticillium dahliae* Kleb.

Wheat chlorotic streak: genus *Cytorhabdovirus*, *Wheat chlorotic streak virus* (WCSV)

Wheat dwarf: genus *Monogeminivirus*, *Wheat dwarf virus* (WDV)

Wheat soil-borne mosaic: genus *Furovirus*, *Wheat soil-borne mosaic virus* (SBWMV)

Wheat spindle streak mosaic: genus *Bymovirus*, *Wheat spindle streak mosaic virus* (WSSMV)

Wheat streak mosaic: genus *Rymovirus*, *Wheat streak mosaic virus* (WSMV)

Wheat yellow leaf: genus *Closterovirus, Wheat yellow leaf virus* (WYLV)
Wheat yellow mosaic: genus *Bymovirus, Wheat yellow mosaic virus*
   (WYMV) (family *Potyviridae*)
White blotch: *Bacillus megaterium* pv. *cerealis* Hosford 1982
Winter crown rot (Coprinus snow mold): *Coprinus psychromorbidus*
   Redhead and Traquair

REFERENCES

APSnet (1997). <www.apsnet.org>.
Hansen, J. D. (1985). Common names for plant diseases. *Plant Dis,* 69:649-676.
Hansen, J. D. (1988). Common names for plant diseases. *Plant Dis,* 72:567-574.
Hansen, J. D. (1991). Common names for plant diseases. *Plant Dis,* 75:225-230.
Hawksworth, D. L., Kirk, P. M., Sutton, B. C., and Pegler, D. N. (1995). *Ainsworth and Bisby's Dictionary of Fungi.* CAB International, U.K.
Schaad, N. W., Vidaver, A. K., Lacy, G. H., Rudolph, K., and Jones, J. B. (2000). Evaluation of proposed amended names of several pseudomonads and xanthomonads and recommendations. *Phytopathology,* 90:208-213.
Van Regenmortel, M. H. V. (1999). How to write the names of virus species. *Arch Virol,* 144:1041-1042.
Wrobel, W. and Creber, G. (1998). *Elsevier's Dictionary of Fungi and Fungal Plant Diseases.* Elsevier, Amsterdam, The Netherlands.
Young, J. M., Saddler, G. S., Takikawa, Y., De Boer, S. H., Vauterin, L., Gardan, L., Gvozdyak, R. I., and Stead, D. E. (1996). Names of plant pathogenic bacteria, 1864-1995. *Rev Plant Pathol,* 75:721-761.

HOST DEFENSE MECHANISMS

Cell Wall Reinforcement

Cell walls of host plants offer resistance to fungal and bacterial pathogens by rapidly modifying their structure when pathogens try to penetrate and disintegrate them. Early induction of cell wall modifications is characteristic of resistant interactions. The possible role of cell walls in host defense mechanisms is described in this chapter.

CELL WALL MODIFICATIONS

A host cell wall offers resistance to pathogen penetration or multiplication due to the interlocking network of macromolecules. The cell wall is in a dynamic state, and when a fungal or bacterial pathogen tries to penetrate or disintegrate the cell wall to facilitate its spread, the cell wall reinforces itself by rapidly modifying its structure. The most common response is formation of cell wall appositions, which are otherwise called *papillae.* Papillae comprise a callose matrix, incorporated with pectic materials, cellulose, suberin, gums, calcium, and silicon. Lignification of papillae occurs. Incorporation of hydroxyproline-rich glycoprotein (HRGP) strengthens the papillae. Several toxic substances, including phenolics, flavonoids, and H_2O_2, accumulate in the papillae.

Upon penetration by some fungal pathogens, such as powdery mildew and rust fungi, another reactive material, the collar, is deposited along the haustorial neck of fungi within the host cell. In some cases, this collar may develop to such an extent that the entire haustorium is encased. The collar develops from papillae and is mostly composed of cellulosic β-1,4-glucans and callose. Callose is a polysaccharide containing a high proportion of 1,3-β-linked glucose and is a polymer of β-1,3-glucans. Uridine diphosphoglucose (UDP-glucose) is converted into a β-1,3-glucan, callose, by β-1,3-glucan synthase (callose synthase). Callose is a minor component of healthy plant tissues. Plants respond to infection by pathogens with the rapid deposition of callose. The presence of collars is usually associated with poor development of the haustoria. The collar may act as a barrier to apoplastic flow in rust fungi.

The extrahaustorial matrix is another component involved in cell wall reinforcement. In the case of powdery mildew and rust fungi, the haustorium is surrounded by an extrahaustorial matrix. The matrix may contain lipids, polysaccharides, glycoproteins, and proteins, but cellulose and pectin are usually absent.

Papillae formation is also associated with bacterial disease resistance. When lettuce leaves were inoculated with the incompatible bacterial pathogen *Pseudomonas savastanoi* pv. *phaseoli,* an apparent convolution of the plasma membrane occurred next to bacterial cells. Fibrillar material accumulates between the convoluted membrane, and progressive thickening and increase in complexity of paramural deposits occur within eight hours after inoculation. Swelling of the endoplasmic reticulum and an increase in numbers of smooth vesicles (papillae) are observed at sites of paramural deposition (Bestwick et al., 1995).

ROLE OF CELL WALL MODIFICATIONS
IN DISEASE RESISTANCE

Papillae may function as an important resistance mechanism. The role of papillae in host defense mechanisms has been demonstrated by the following evidence:

1. Papillae are formed abundantly in only resistant/incompatible host-pathogen interactions. In susceptible interactions, fewer and smaller papillae are formed.
2. When the disease-resistant plants are treated with inhibitors of papilla formation (such as 2-deoxy-D-glucose), papillae formation is inhibited, and the plants become susceptible to pathogens.
3. When papilla formation is induced by treatments with chemicals such as chitosan or some plant extracts, more and larger papillae are formed, and the susceptible plants become resistant.

Earlier papilla initiation is an important factor for papillae-mediated disease resistance. When papilla deposition starts earlier and increases more rapidly, the plants become more resistant. Restriction of growth of *Leptosphaeria maculans* to the infected areas in cotyledons of *Brassica napus-B. nigra* addition line (LA4+) plants was correlated to reinforcement of cell wall barriers, including wall apposition, papillae, and vessel plugging, and the disease resistance was associated with rapidity in plant responses (Roussel et al., 1999). A low penetration efficiency of fungal pathogens is correlated with papillae that are formed in advance of penetration pegs. Frequency of the papilla formation is also important. Treatments such as spraying of water extract of a plant such as *Reynoutria sachaliensis* induce an increased frequency of papillae and confer resistance to pathogens (Schneider and Ullrich, 1994). The size of papillae is also important. Papillae formed in resistant interactions are larger in size than those formed in susceptible interactions. Oversized papillae contribute to disease resistance.

The factors responsible for early induction of papilla formation in resistant interactions have been studied. An aqueous extract from barley seedlings contains the factor that induces papilla formation, called *papilla-regulating extract* (PRE). When coleoptiles of a susceptible variety are floated on PRE solution, papillae are initiated much earlier than in the control when inoculated with the powdery mildew pathogen *(Erysiphe graminis)*. Papillae are initiated before penetration peg initiation in PRE treatments, while they are initiated after penetration peg initiation in controls (Inoue et al., 1994). PRE appears to contain potassium phosphate. Potassium phosphate extracted from uninoculated barley leaves induces papilla mediated resistance against the powdery mildew pathogen in barley. PRE-induced resistance appears to be Ca^{2+} mediated. Phosphate salts (in the presence of Ca^{2+}), such as $Ca(H_2PO_4)_2$, have been shown to induce local and systemic resistance in many plants. Callose synthesis by callose synthase is activated by Ca^{2+}.

REFERENCES

Bestwick, C. S., Bennett, M. H., and Mansfield, J. W. (1995). Hrp mutant of *Pseudomonas syringae* pv. *phaseolicola* induces cell wall alterations but not membrane damage leading to the hypersensitive reaction in lettuce. *Plant Physiol,* 108:503-516.

Inoue, S., Aist, J. R., and Macko, V. (1994). Earlier papilla formation and resistance to barley powdery induced by a papilla-regulating extract (PRE). *Physiol Mol Plant Pathol.,* 44:433-440.

Roussel, S., Nicole, M., Lopez, F., Renard, M., Chevre, A. M., and Brun, H. (1999). Cytological investigation of resistance to *Leptosphaeria maculans* conferred to *Brassica napus* by introgressions originating from *B. juncea* or *B. nigra* B genome. *Phytopathology,* 89:1200-1213.

Schneider, S. and Ullrich, W. R. (1994). Differential induction of resistance and enhanced enzyme activities in cucumber and tobacco caused by treatments with various abiotic and biotic inducers. *Physiol Mol Plant Pathol,* 45:291-304.

32

Hydroxyproline-Rich Glycoproteins

Plant cell walls contain different types of hydroxyproline-rich glyco-proteins. These proteins show extensive modifications when pathogens try to penetrate the plant cell wall, and these modified HRGPs offer resistance to pathogens. The role of HRGPs in disease resistance is described in this chapter.

TYPES OF CELL WALL PROTEINS

A plant cell wall contains up to 10 percent protein. Cell wall protein is exceptionally rich in hydroxyproline (Hyp). There are four types of hydroxy-proline-rich protein classes.

1. Extensins: Extensins are rich in hydroxyproline and serine (Ser) and usually contain the repeating pentapeptide motif Ser-Hyp4. Other commonly found amino acids are valine, tyrosine, lysine, and histidine.
2. Proline (PRO)-rich proteins (PRPs): All PRPs are characterized by the repeating occurrence of PRO-PRO repeats. They contain approximately equimolar quantities of proline and hydroxyproline.
3. Lectins: Hydroxyproline and arabinose are major constituents of lectins. The serine-hydroxyproline-rich glycopeptide domain of lectins bears a striking biochemical resemblance to the extensins.
4. Arabinogalactan proteins (AGPs): They are HRGPs that are highly glycosylated. The protein moiety of AGPs is typically rich in hydroxy-proline, serine, alanine (Ala), threonine, and glycine. They contain Ala-Hyp repeats.

ROLE OF HRGPs IN DISEASE RESISTANCE

HRGPs accumulate in infected tissues, particularly in cell wall apposi-tions (papillae). Accumulation of HRGPs in plant cell walls is one of the earliest changes observed in the cell walls of leaves inoculated with incompatible pathogens (resistant interactions) (Bestwick et al., 1995; Davies et al., 1997). Several new HRGPs have been shown to be synthesized in resistant host-pathogen interactions. For example, Davies and colleagues (1997) reported that when turnip leaves were inoculated with an incompatible strain of *Xanthomonas campestris* pv. *raphani,* two new HRGPs, gp160 and gpS, were induced. In a compatible interaction with an *X. campestris* pv. *campestris* strain, no induction of these glycoproteins was observed. The newly induced gpS comprised three distinct fractions, gpS-1, gpS-2, and gpS-3. The gpS-3 fraction may be an HRGP of the extensin family, while gpS-1 and gpS-2 may be an arabinogalactan protein HRGP of another family. The new HRGPs may prevent penetration of the cell wall by pathogens.

In addition to the synthesis of new HRGPs, HRGPs in plant cell walls become insolubilized in resistant interactions. The insolubilization may be caused by linking HRGPs by isodityrosine residues and by a diphenyl ether linkage, and this process may be due to the action of peroxidase requiring H_2O_2 generated via the oxidative burst (Brisson et al., 1994). H_2O_2 treatment usually induces insolubilization of HRGPs, similar to fungal elicitor treatment. Simultaneous addition of catalase or ascorbate inhibits the effects of elicitor or H_2O_2 (Bradley et al., 1992), which suggests that the elicitor may stimulate H_2O_2-dependent cross-linking of preexisting, soluble forms of HRGPs. The cross-linking of the HRGPs in response to elicitor treatment is initiated within five minutes, and the response is completed within 20 to 30 minutes (Bradley et al., 1992). The burst of H_2O_2 normally occurs within one to two minutes of elicitor treatment in plant cells (Apostol et al., 1989) and provides substrate for peroxidase-mediated oxidative cross-linking of HRGPs. The insolubilized HRGPs may lead to an impenetrable cell wall barrier, impeding pathogen infection.

In many host-pathogen interactions, HRGPs are cross-linked with different cell wall constituents. In the resistant interactions, HRGPs are cross-linked with pectin, lignin, and other phenolics (Bestwick et al., 1995). HRGPs may act as matrices for lignification. When fungal pathogens invade host tissues, HRGPs accumulate and provide a template for the subsequent deposition of lignin. Lignified papillae containing HRGP offer resistance to fungal penetration (Benhamou et al., 1990). The increased cross-linking of HRGPs leads to strengthening of the cell wall to localize pathogens. Cell walls, which undergo ultra-rapid HRGP cross-linking, due to pathogen invasion, are tougher than cell walls of uninfected healthy tissues (Showalter, 1993). All of these changes in HRGPs occur simultaneously and make the cell walls impermeable to pathogens.

Some HRGPs, lectins, are known to agglutinate pathogens, specifically bacterial cells (Leach et al., 1982). Agglutination may result in immobilization of bacterial pathogens in the cell surface. This would contribute to disease resistance.

REFERENCES

Apostol, I., Heinstein, P. F., and Low, P. S. (1989). Rapid stimulation of an oxidative burst during elicitation of cultured plant cells. *Plant Physiol,* 90:109-116.

Benhamou, N., Chamberland, H., and Pauze, F. J. (1990). Implications of pectic components in cell surface interactions between tomato root cells and *Fusarium oxysporum* f. sp. *Radicis-lycopersici:* A cytochemical study by means of a lectin with polygalacturonic acid-binding specificity. *Plant Physiol,* 92:995-1003.

Bestwick, C. S., Bennett, M. H., and Mansfield, J. W. (1995). Hrp mutant of *Pseudomonas syringae* pv. *phaseolicola* induces cell wall alterations but not membrane damage leading to the hypersensitive reaction in lettuce. *Plant Physiol,* 108:503-516.

Bradley, D. J., Kjellbom, P., and Lamb, C. J. (1992). Elicitor- and wound-induced oxidative cross-linking of a proline-rich plant cell wall protein: A rapid defense response. *Cell,* 70:21-30.

Brisson, L. F., Tenhaken, R., and Lamb, C. J. (1994). Functions of oxidative cross-linking of cell wall structural proteins in plant disease resistance. *Plant Cell,* 6:1703-1712.

Davies, H. A., Daniels, M. J., and Dow, J. M. (1997). Induction of extracellular matrix glycoproteins in *Brassica* petioles by wounding and in response to *Xanthomonas campestris. Mol Plant-Microbe Interact,* 10:812-820.

Leach, J. E., Centrell, M. A., and Sequeira, L. (1982). A hydroxyproline-rich bacterial agglutinin from potato: Extraction, purification, and characterization. *Plant Physiol,* 70:1353-1358.

Showalter, A. W. (1993). Structure and function of plant cell wall proteins. *Plant Cell,* 5:9-23.

33

Lignin

Lignin is a complex polymer built from phenolic acids, which are re-duced to the corresponding alcohols. Rapid lignification of plant cell walls appears to be an important host defense mechanism. The role of lignin in bacterial and fungal disease resistance is described in this chapter.

BIOSYNTHESIS OF LIGNIN

Lignin is a complex polymer deposited in secondary cell walls. Hydroxy-proline-rich glycoprotein is known to be a matrix for deposition of lignin. Lignin is a three-dimensional polymer built from three monomers called *monolignols*. The three building units are *p*-coumaryl, coniferyl, and sinapyl alcohols, which are polymerized in lignin via free radical production. Lignin is formed by the random condensation of phenylpropanoid units. Lignin monomers are produced from phenylalanine by a branch of phenyl-propanoid metabolism. The first step of the phenylpropanoid pathway is the deamination of phenylalanine by phenylalanine ammonia-lyase (PAL) to yield *trans*-cinnamic acid. Hydroxylation of cinnamic acid by cinnamic acid-4-hydroxylase (CA4H) produces *p*-coumaric acid. Further hydroxy-lation by *p*-coumaric hydroxylase (CH) results in the synthesis of caffeic acid or 5-hydroxyferulic acid. Methylation of these acids by *O*-methyl transferases (OMT) produces ferulic and sinapic acids, respectively. The three acids (*p*-coumaric, ferulic, and sinapic acids) are coupled to CoA and reduced to the corresponding alcohols. The end products of this pathway are the three monolignols, namely *p*-coumaryl, coniferyl, and sinapyl alcohols (Vidhyasekaran, 1988). Polymerization of these monolignols is catalyzed by peroxidases. Plants possess a number of different peroxidase isozymes. Many of them have been found to be localized to cell walls. Cell wall-bound peroxidases fall into two subgroups, the anionic and cationic. Both the acidic and basic peroxidases have been found to be located in plant cell walls, and both anionic peroxidases and cationic peroxidases are involved in lignification. Cell wall-bound peroxidases are involved not only in the oxi-dative polymerization of hydroxylated cinnamyl alcohols but also in the generation of hydrogen peroxide necessary for lignification. The polymerization of the three cinnamyl alcohols is mediated by the peroxidase-H_2O_2 system.

ROLE OF LIGNIN IN DISEASE RESISTANCE

Deposition of lignin in plant cell walls has been observed in many plant-pathogen interactions. Rapid lignification appears to be an important defense mechanism in plants. Lignin content increased in wheat leaves inoculated with an avirulent strain of *Puccinia recondita* f. sp. *tritici,* but no such increase was observed in leaves inoculated with a virulent strain (Southerton and Deverall, 1990). Lignification was observed in both phloem and xylem of infected cassava plants resistant to *Xanthomonas axonopodis* pv. *manihotis*. This reaction was observed at a higher intensity in resistant plants than in susceptible plants (Kpemoua et al., 1996). When rice leaves were inoculated with incompatible strain of *X. oryzae* pv. *oryzae,* lignin-like compounds accumulated in rice leaves. Similar accumulation of lignin is not seen in a compatible interaction (Reimers and Leach, 1991). In resistant or incompatible interactions between rice and *X. oryzae* pv. *oryzae,* increases in the activities of the three intercellular peroxidases have been correlated with the accumulation of lignin-like compounds and reduction in bacterial multiplication in the leaves (Guo et al., 1993). When primary leaves of 'Red Mexican' bean plants were inoculated with *P. savastanoi* pv. *phaseolicola,* lignin deposition was observed in the inoculated leaf area in the incompatible interaction but not in the compatible interaction (Milosevic and Slusarenko, 1996). Peroxidase activity was also located in the same zone in the incompatible interaction. The increase in peroxidase activity in the incompatible interaction could largely be attributed to the appearance of a novel anionic peroxidase from 18 hours after inoculation onward (Milosevic and Slusarenko, 1996). The novel peroxidase may be involved in the lignification observed in the bacteria-inoculated zone (Milosevic and Slusarenko, 1996).

Lignification in cell walls may serve as a barrier for penetration by fungal pathogens. Cell wall strengthening due to deposition of lignin was an inducible defense mechanism of banana roots against *Fusarium oxysporum* f. sp. *cubense* (De Ascensao and Dubery, 2000). Although lignification of plant cell walls has been observed in many host-bacteria resistance reactions, lignification may not serve as a physical barrier against bacterial pathogens, because bacterial plant pathogens do not penetrate host cells. However, lignified materials may prevent bacterial spread by blocking movement between the epithem and xylem vessels. Reactive compounds associated with the process of lignification may also inhibit pathogen growth (Vidhyasekaran et al., 2001).

REFERENCES

De Ascensao, A. R. D. C. F. and Dubery, L. A. (2000). Panama disease: Cell wall reinforcement in banana roots in response to elicitors from *Fusarium oxysporum* f. sp. *cubense* race four. *Phytopathology,* 90:1173-1180.

Guo, A., Reimers, P. J., and Leach, J. E. (1993). Effect of light on compatible interaction between *Xanthomonas oryzae* pv. *oryzae* and rice. *Physiol Mol Plant Pathol,* 42:413-425.

Kpemoua, K., Boher, B., Nicole, M., Calatayud, P., and Geiger, J. P. (1996). Cytochemistry of defense responses in cassava infected by *Xanthomonas campestris* pv. *manihotis. Canad J Microbiol,* 42:1131-1143.

Milosevic, N. and Slusarenko, A. J. (1996). Active oxygen metabolism and lignification in the hypersensitive response in bean. *Physiol Mol Plant Pathol,* 49:143-158.

Reimers, P. J. and Leach, J. E. (1991). Race-specific resistance to *Xanthomonas oryzae* pv. *oryzae* conferred by bacterial blight resistance gene *Xa-10* in rice *Oryza sativa* involves accumulation of a lignin-like substance in host tissues. *Physiol Mol Plant Pathol,* 38:39-55.

Southerton, S. G. and Deverall, B. J. (1990). Histochemical and chemical evidence for lignin accumulation during the expression of resistance to leaf rust fungi in wheat. *Physiol Plant Pathol,* 36:483-494.

Vidhyasekaran, P. (1988). *Physiology of Disease Resistance in Plants,* Volume I. CRC Press, Boca Raton, FL.

Vidhyasekaran, P., Kamala, N., Ramanathan, A., Rajappan, K., Paranidharan, V., and Velazhahan, R. (2001). Induction of systemic resistance by *Pseudomonas fluorescens* Pf1 against *Xanthomonas oryzae* pv. *oryzae* in rice leaves. *Phytoparasitica,* 29:155-166.

34

Pathogenesis-Related Proteins

Several new proteins appear in plants infected with pathogens, and these are called *pathogenesis-related* (PR) *proteins*. They are induced more in resistant interactions. The molecular mechanism of transcription of PR genes, the signals involved in induction of these genes, and the possible role of PR proteins in disease resistance are discussed in this chapter.

DEFINITION

When plants are infected by pathogens, a number of genes encoding for proteins are transcriptionally activated and new proteins are synthesized. These proteins are called pathogenesis-related proteins. PR proteins have been defined as proteins encoded by the host plants but induced only in pathological situations. These PR proteins are induced by fungal, bacterial, viral, and viroid pathogens. In addition to these pathogens, insect pests, nematodes, and endophytic saprophytic bacteria, such as some strains of *Pseudomonas fluorescens,* induce PR proteins.

Some products from these living organisms also induce PR proteins. Elicitors isolated from fungi and bacteria induce PR proteins in plants. Toxins isolated from fungi, and pectic enzymes, cellulases, and xylanases isolated from fungi and bacteria induce PR proteins. Endogenous elicitors isolated from plant cell walls (oligogalacturonates) have been shown to induce PR proteins in several plants. Several abiotic compounds are also known to induce PR proteins. Heavy metals, polyacrylic acid, mannitol, salicylic acid, methyl salicylate, and methyl jasmonate/jasmonic acid induce synthesis of PR proteins. Several plant activators, such as 2,6-dichloro-isonicotinic acid (INA) and benzothiadiazole (BTH) (Latunde-Dada and Lucas, 2001) are known to induce PR proteins. Several growth regulators, such as ethylene, kinetin, indoleacetic acid, and abscisic acid, induce PR proteins in various plants. Several environmental factors also induce PR proteins.

Temperature, light, ozone, and ultraviolet rays have been shown to influence induction of PR proteins. Osmotic stress induces PR proteins in some plants. Any injury or mechanical wounding induces several PR proteins.

Because PR proteins can be induced by several stresses besides pathogens, these proteins should be redefined as plant proteins that are induced by various types of pathogens as well as stress conditions (Van Loon et al., 1994). However, even this definition needs modifications, because several PR proteins have been detected in healthy tissues without any stress in many plants. In some plants, PR proteins, which appear in leaves only after a stress, constitutively occur in other parts of plants, including seeds, flowers, and roots. Many PR proteins appear constitutively in cultured plant cells. Such proteins, which are induced in one organ of the plant but constitutively present in other parts of the plant, are also considered to be PR proteins. Some PR proteins that are induced proteins in some varieties occur constitutively in other varieties. Even in the same plant, PR proteins appear in lower old leaves without any stress while these proteins could be detected in young leaves only after inoculation with pathogens.

Hence, it was suggested that to be labeled a PR protein, the protein must be expressed upon infection in any one organ of the plant, and at least in any one variety of the plant. However, even PR proteins that could not be detected in unstressed healthy plants by polyacrylamide gel electrophoresis (PAGE) could be detected when Western blot analyses were made using antisera or when cDNA probes were used. This indicates PR proteins may be present in uninfected tissues in amounts too small for detection on gels by general protein stains. Hence, PR proteins are defined as proteins that are readily detected in infected tissues but not in uninfected ones (Van Loon, 1999). Some proteins are induced exclusively during susceptible interaction, and such proteins are not induced in resistant interactions. These proteins are not considered to be PR proteins (Van Loon, 1999). PR proteins should be induced in resistant/incompatible interactions, irrespective of whether they are induced in susceptible/compatible interactions. PR proteins are normally detected in leaves, but in some plants, they are detected in roots, tubers, stems, petioles, and germinating seeds (Vidhyasekaran, 2002).

Some proteins in healthy, unstressed plant tissues show an amino acid (nucleotide) sequence similar to already characterized PR proteins. They may also have enzyme activity similar to some PR proteins. However, they are never induced by pathogens. Such proteins are called PR-like proteins (Van Loon et al., 1994).

CLASSIFICATION OF PR PROTEINS

Several PR proteins occur in plants. PR proteins are classified on the basis of their nucleotide sequence or predicted sequence of amino acids, serological relationship, and/or enzymatic or biological activity (Van Loon et al., 1994). Serological relationship or enzymatic activity or biological activity alone is not taken as the criteria for classification of PR proteins; they are considered to be additional characteristics for classification. The structure of PR proteins is the most important factor for classifying PR proteins into families. PR proteins have been classified into fourteen families based on shared sequence homology. The isoelectric points of PR proteins are considered to classify some PR proteins into subclasses (Koiwa et al., 1994).

PR-1 Proteins

This group of PR proteins has been detected widely in the plant kingdom. Both acidic and basic isoforms (differentiated on the basis of isoelectric points of the PR proteins) of PR-1 proteins have been detected. In tobacco alone, ten PR-1 proteins, PR-1a, PR-1b, PR-1c, PR-1d, PR-1e, PR-1f, PR-1g, PR-1h, PRB-1a, and PRB-1b, have been detected.

PR-2 Proteins

This family of PR proteins shows β-1,3-glucanase activity. β-1,3-Glucanases (glucan endo-1,3-β-glucosidases) catalyze endo-type hydrolytic cleavage of the 1,3-β-D-glucosidic linkages in β-1,3-glucans. β-1,3-Glucan, the substrate for the enzyme β-1,3-glucanase, is widespread in plant tissues and is associated with the formation of callose, leaf and stem hairs, root hairs, pollen grains, ovules, and wound parenchyma cells. As such, endo β-1,3-glucanases are abundant proteins widely distributed in plant species. Constitutive expression of several β-1,3-glucanases in healthy plants has been reported. These enzymes accumulate in normal healthy plants during pollen germination, fertilization, seed maturation, and germination. However, some β-1,3-glucanases have been shown to be induced by pathogens, and such enzymes are considered to be PR proteins. Thus, all β-1,3-glucanases are not PR proteins; only specific β-1,3-glucanases are recognized as PR-2 proteins. PR-2 β-1,3-glucanase exists in multiple forms in a number of plant species. In tobacco, PR-2 proteins are classified into three structural classes based on amino acid sequence identity. The class I β-1,3-glucanases are basic proteins localized in the cell vacuole. These include PR-2e of tobacco *(Nicotiana tabacum)* PR proteins, and Gn1 and Gn2 of *Nicotiana plumbaginifolia*. The class II β-1,3-glucanases are acidic proteins secreted into the extracellular space. These include tobacco PR-2a (PR-2), PR-2b (PR-N), and PR-2c (PR-O). The class III β-1,3-glucanases are also acidic proteins which include PR-2d (= PR-Q') (Leubner-Metzger and Meins, 1999).

PR-3 Proteins

PR-3 proteins show chitinase (β-1,4-glucosaminidase) activity. Chitinases hydrolyze the β-glycosidic bond at the reducing end of glucosaminides, which can be parts of various polymers, such as chitin, chitosan, or peptidoglycan (from bacterial cells) (Neuhaus, 1999). Several chitinases have been reported in almost all plant species studied. Plant chitinases have been classified into seven classes based on sequence homology and the presence or absence of a chitin-binding domain (CBD). Class I contains a CBD, while class II lacks the CBD. Class III chitinases are structurally different from other chitinases. Class IV chitinases differ from class I by several internal deletions within both chitin-binding and catalytic domains (loops 1, 3, and 4). Class V chitinases contain two CBDs. Class VI chitinases are characterized by the presence of a truncated CBD and a long proline-rich spacer. Class VII possesses a catalytic domain homologous to class IV but lacks the CBD (Neuhaus, 1999). In addition to these seven classes, two other classes of plant chitinases have also been identified. The first one is a new type of chitinase with the most related sequences belonging to bacterial chitinases (Melchers et al., 1994), and the other class of chitinases constitutes hevein and Win proteins, which contain a CBD, and PR-4 proteins, which lack the CBD (Linthorst et al., 1991).

PR-3 proteins consist of chitinases belonging to class I through class VII, not including class III (Van Loon et al., 1994). Class III proteins, which are structurally very different from other classes, are considered to be a separate family of PR proteins, called PR-8. The chitinases, which are similar to bacterial chitinases, have been grouped as another family, called PR-11. The PR-11 proteins consist of class I type chitinases. The other group of chitinases similar to hevein and Win proteins are given a separate family number, PR-4, which consists of class I- and class II-type chitinases (Van Loon, 1999). Several chitinases accumulate in plant tissues infected with pathogens. As many as ten chitinases have been shown to accumulate in pepper (Lee and Hwang, 1996) and Norway spruce (Sharma et al., 1993) after infection.

PR-4 Proteins

Two classes of the PR-4 family have been reported. Class I includes hevein of rubber tree and Win protein of potato. The class II PR-4 protein constitutes tobacco PR-4 as the typical member of PR-4 proteins. One of the tobacco PR-4 proteins, CBP20, has been found to have chitinase activity, and the PR-4 family has been included in a chitinase nomenclature.

PR-5 Proteins

The PR-5 group of proteins occurs widely in the plant kingdom. PR-5 proteins have a close resemblance to a sweet-tasting protein, thaumatin, which occurs in the fruit of the West African shrub *Thaumatococcus danielli*. Hence, these proteins are also called thaumatin-like proteins (TLPs). The basic PR-5 proteins are identical to osmotin, the salt stress protein detected in tobacco and hence, they are called osmotins. Three subclasses of PR-5 proteins have been recognized in tobacco based on their isoelectric points: the basic forms (osmotins), neutral forms (osmotin-like proteins, OLPs), and acidic (PR-S) proteins (Koiwa et al., 1994).

Several PR-5 proteins have been detected in many parts of healthy plants. They are induced in some plant tissues only after infection with pathogens (Ruiz-Medrano et al., 1992), and that is why they are called PR proteins. However, some of the PR-5-type proteins detected in floral parts in tobacco could not be detected in any other part of the plants even after infection. A flower-specific gene encoding an osmotin-like protein has been identified in tomato. No expression of the gene was detected in vegetative organs. The protein has 30 to 32 percent amino acid sequence identity to pathogenesis-related osmotins (Chen et al., 1996). However, it cannot be called a PR protein.

The function of PR-5 proteins is not yet known. PR-5 proteins alter permeability of fungal membranes. The PR-5 protein of tobacco (PR-R = PR-S) has identity to a maize-trypsin/α-amylase inhibitor. A PR-5 protein from bean cultivar Pinto, PR-4d, shows β-1,3-glucanase activity. Another PR protein from bean cultivar Saxa, which shows resemblance to tobacco PR-5 protein, Saxa PR-4d, possesses chitinase activity (for more details, see Vidhyasekaran, 2002).

PR-6 Proteins

Some proteinase (protease) inhibitor proteins are considered to be PR-6 proteins. Proteinase inhibitor proteins commonly occur in plants (Vidhyasekaran, 1997), and some of them are induced by pathogens. Proteinase inhibitors are generally categorized according to the class of proteinases they inhibit (Koiwa et al., 1997). Four types of proteinases have been identified as serine, cysteine, aspartic, or metallo proteinases based on the active amino acid in the reaction center. Serine proteinase inhibitors which inhibit trypsin and chymotrypsin contain several families: Kunitz family (Soybean trypsin inhibitor family), Bowman-Birk family, Barley trypsin inhibitor family, Potato inhibitor I family, Potato inhibitor II family, Squash inhibitor family, Ragi I-2/maize trypsin inhibitor family, and Serpin family. Cysteine proteinase inhibitors (phytocystatins) inhibit papain and cathepsin B, H, and L. Aspartic proteinase inhibitors inhibit cathepsin D, and metallo-proteinase inhibitors inhibit papain and cathepsin B, H, and L (Koiwa et al., 1997). Tomato inhibitor I (belonging to the Potato inhibitor I family) is the type member of the PR-6 protein family. Tomato inhibitor II is also commonly induced in tomato due to stresses.

PR-7 Proteins

PR-7 proteins show endoproteinase activity. A PR-7 protein, P69, has been detected in tomato infected by the citrus exocortis viroid. Tornero and colleagues (1997) identified P69B, a second member of the family of plant proteinases induced during the response of tomato plants to pathogen attack. P69B represents a new plant subtilisin-like proteinase based on amino acid sequence conservation and structural organization (Tornero et al., 1997).

PR-8 Proteins

PR-8 proteins show class III chitinase activity possessing lysozyme activity. These chitinases differ from other chitinases in sequence and/or substrate preference (Van Loon et al., 1994). These PR proteins show structural homology to a bifunctional lysozyme/chitinase from *Parthenocissus quinquifolia*. Both acidic and basic forms of PR-8 proteins have been reported.

PR-9 Proteins

PR-9 proteins show peroxidase activity. However all peroxidases are not considered to be PR-9 proteins, since most of them are constitutively expressed and are not induced during pathogenesis. Over 60 peroxidase genes from diverse plant species have been isolated and characterized, including multiple genes from individual species. Only some of the peroxidase genes are induced during pathogen stress. A lignin-forming peroxidase from tobacco is the type member of the PR-9 protein family.

PR-10 Proteins

PR-10 proteins include intracellular defense-related proteins that have a ribonuclease-like structure. The PR-10 family includes parsley "PR1," STH-2 in potato, asparagus AoPR1, pea pI49, and bean pvPR1 and pvPR2. In rice, a PR-10 protein, PBZ1, has been identified (Midoh and Iwata, 1996). This PR protein is induced by *Pyricularia oryzae* and probenazole. All PR-10 proteins are acidic, and in all these proteins a signal peptide was absent, suggesting that this family of proteins is intracellular. The PR-10 gene family may encode ribonucleases.

PR-11 Proteins

PR-11 proteins also show chitinase activity. However, they are distinctly different from other previously described chitinases. The most related sequences of these proteins belong to bacterial chitinases.

PR-12 Proteins (Defensins)

Defensins are a family of small (about 5 kDa), usually basic, peptides, which are rich in disulfide-linked cysteine residues (de Silva, Conceicao, and Broekaert, 1999). Defensins are also considered to be a novel group of thionins, which are called γ-thionins. However, thionins and defensins are structurally unrelated. Terras and colleagues (1995) introduced the term *defensins* for this group of peptides, based on the structural and functional similarities with insect defensins. All defensins (γ-thionins) are less than 50 amino acids in length and contain eight cysteine residues. The defensins have been classified as PR-12 proteins (Van Loon and Van Strien, 1999). Two radish proteins, Rs-AFP1 and Rs-AFP2, are type members of this family and are induced in radish after infection.

PR-13 Proteins (Thionins)

Thionins are small (5 kDa), basic, cysteine-rich proteins commonly found in seeds, roots, and leaves of plants. Although thionins are constitutively expressed in seeds and roots, some leaf thionins are found to be induced during pathogenesis. Those induced thionins are considered to be PR-13 proteins (Van Loon, 1999).

PR-14 Proteins (Lipid Transfer Proteins)

Lipid transfer proteins (LTPs) stimulate the transfer of a broad range of lipids between membranes in vitro. They are generally secreted and externally associated with the cell wall. They may be involved in the secretion or deposition of extracellular lipophilic materials such as cutin or wax. LTPs are distributed at high concentrations in the epidermis of exposed surfaces and in the vascular tissues. Expression of LTP genes has been shown to be induced well above basal levels in some plant-pathogen interactions (Garcia-Olmedo et al., 1995). Some LTPs have been considered to be PR-14 proteins (Van Loon and Van Strien, 1999).

Other PR Proteins

In addition to these characterized PR protein classes, some other unclassified proteins are also considered to be PR proteins. Ribosome-inactivating proteins (RIPs) show specific RNA-*N*-glycosidase activity, which inactivates ribosomes from phylogenetically distant species including animals and microorganisms. Some RIPs are induced in the leaves of some plants in response to pathogens. JIP60, the 60 kDa protein of barley, is induced in leaves due to pathogen stress and is type member of this family. Some chitosanases have been identified as PR proteins. Chitosanases act on chitosan (poly [1-4]-β-D-glucosamine, the deacetylated chitin) without any activity on chitin, and they differ from chitinases by their molecular weight (10 to 24 kDa) and substrate specificity. No chitosanase activity was detected in uninfected roots of Norway spruce *(Picea abies),* but three chitosanases were induced after a fungal infection (Sharma et al., 1993). A new glycine-rich protein (GRP) was induced in tobacco infected with tobacco mosaic virus (Hooft van Huijsduijnen et al., 1986). Newly induced enzymes involved in synthesis of phytoalexins are also considered to be PR proteins (Van Loon et al., 1994).

ROLE OF PR PROTEINS IN DISEASE RESISTANCE

The actual role of PR proteins is still not known. However, many of these PR proteins are always associated with resistant interactions. The possible mechanism of inducing resistance by individual PR proteins is discussed in this section.

PR-1 Proteins

PR-1 proteins have not been shown to inhibit fungal or bacterial growth. However, they are found to be associated with host cell wall outgrowths and papillae. PR-1 proteins may restrict the development of pathogens by host cell wall modifications (Benhamou et al., 1991).

PR-2 Proteins

PR-2 proteins show β-1,3-glucanase activity. Fungal cell walls contain β-1,3-glucans. The PR-2 proteins have been shown to lyse the fungal cell walls. Antiviral action of tobacco β-1,3-glucanase (gp35) has also been reported (Edelbaum et al., 1991). β-1,3-Glucan is an elicitor, and PR-2 proteins release this elicitor from fungal cell walls by its enzymatic action. The elicitor induces synthesis of antimicrobial compounds such as phytoalexins, phenolics, lignins, and other toxic PR proteins.

PR-3 Proteins

PR-3 proteins show chitinase activity. Cell walls of most fungal pathogens contain chitin. PR-3 proteins act on fungal cell walls and lyse them. Chitinase inhibits growth of fungi even at 1 to 2 μg/ml concentrations. A transgenic sorghum expressing high levels of chitinase exhibited less stalk rot development when exposed to conidia of *Fusarium thapsinum* (Waniska et al., 2001). Chitin is another important elicitor of defense mechanisms, and PR-3 proteins release the chitin elicitor molecules from the fungal cell wall. The elicitor, by inducing antimicrobial compounds, may induce resistance.

PR-4 Proteins

Members of the PR-4 group of proteins are also chitinases. A PR-4 class I protein from tobacco has been shown to exhibit antifungal activity toward *Fusarium solani,* causing lysis of germ tubes and inhibiting fungal growth (Ponstein et al., 1994).

PR-5 Proteins

Many PR-5 proteins have been shown to have high antifungal activity (Fagoaga et al., 2001). Some of them caused leakage of cytoplasmic material from the fungi and caused hyphal rupture. AP24 from tobacco and

NP24 from tomato cause lysis of sporangia of *Phytophthora infestans*. PR-S from tobacco inhibits hyphal growth of fungi. Osmotin from tobacco causes fungal spore lysis, inhibition of spore germination, reduced germling viability, and damage to membrane permeability. It induces apoptosis in fungi (Narasimhan et al., 2001).

PR-6 Proteins

PR-6 proteins are proteinase inhibitors. Several bacterial and fungal pathogens secrete proteinases. Proteinase-deficient mutants show considerable loss of virulence. Proteinase activity of pathogens is reduced in plants overexpressing PR-6 proteins, and these plants show resistance to pathogens. It suggests that PR-6 proteins may inhibit proteinases of pathogens and confer resistance. PR-6 proteins inhibit only fungal and bacterial proteinases and not plant proteinases.

PR-7 Proteins

PR-7 proteins show proteinase activity. They inhibit fungal growth, probably degrading proteins of pathogens.

PR-8 Proteins

These proteins show chitinase and lysozyme activity. Bacterial cell walls contain peptidoglycan, which comprises alternate β-(1,4)-linked N-acetyl-muramic acid residues. Peptidoglycan is hydrolyzed by lysozyme. PR-8 proteins possess lysozyme activity. In many plants, bacterial disease resistance has been shown to be due to increased synthesis of chitinases. PR-8 proteins also show antifungal action.

PR-9 Proteins

PR-9 proteins possess peroxidase activity. They are involved in lignification. They are also involved in cross-linking of extensin monomers, polysaccharide cross-linking, and suberization (Venere et al., 1993). PR-9 proteins may enhance resistance by the construction of a cell wall barrier, which may impede bacterial pathogen ingress, spread, and development of fungal pathogens in the apoplast.

PR-10 Proteins

PR-10 proteins contribute for disease resistance; however, their antimicrobial activity has not been established.

PR-11 Proteins

PR-11 proteins show chitinase activity and display antifungal activity in vitro (Melchers et al., 1994; Ohl et al., 1994).

PR-12 Proteins (Defensins)

PR-12 proteins show antibacterial activity. A potato tuber defensin inhibits *Ralstonia solanacearum* and *Clavibacter michiganensis* (Moreno et al., 1994).

PR-13 Proteins

PR-13 proteins (thionins) are highly toxic to both fungal and bacterial pathogens. Thionins act on cell membranes. The toxicity is exerted by a direct, detergent-like interaction with the lipid bilayers of biological membranes (Bohlmann, 1994). Thionins may inhibit certain enzymes (such as cytochrome c) of bacteria and inhibit bacterial growth.

PR-14 Proteins

Lipid transfer proteins may be involved in deposition of extracellular lipophilic materials such as cutin wax in the plant cell wall. The modified cell walls may contribute to disease resistance. LTPs are distributed at high concentrations in the epidermis of exposed surfaces and in the vascular tissues. In addition to their contribution in host defense mechanisms, LTPs have been shown to inhibit growth of bacterial pathogens in vitro. Cell wall extracts from transgenic tobacco plants overexpressing LTP inhibit growth of the bacterial pathogen *Pseudomonas syringae* pv. *tabaci* (Molina and Garcia-Olmedo, 1997).

Synergistic Action of a Combination of PR Proteins

Toxic action of several individual PR proteins could not be demonstrated. However, when they are combined, they show high toxic action against fungal pathogens. Either chitinases Ch1 or β-1,3-glucanase G2 from pea pods do not inhibit growth of *Fusarium solani* f. sp. *phaseoli* even at a 250 µg/mL concentration. However, a combination of Ch1 and G2 effectively inhibits growth of the fungus even at 10 µg/mL (Mauch et al., 1988).

Tobacco class I chitinase acts synergistically with tobacco class I PR-4 protein inhibiting *Fusarium solani* germlings. Tobacco class V chitinases act synergistically with class I β-1,3-glucanase inhibiting growth of *Alternaria radicina* (Ohl et al., 1994). A combination of chitinase and ribosome-inactivating protein from barley inhibits fungal growth more efficiently than does either enzyme alone (Leah et al., 1991). Not one, but several PR proteins are induced in resistant interactions due to infection (Vidhyasekaran, 1997). This suggests that a multitude of PR proteins may act synergistically and contribute to disease resistance.

REFERENCES

Benhamou, N., Grenier, J., and Asselin, A. (1991). Immunogold localization of pathogenesis-related protein P14 in tomato root cells infected by *Fusarium oxysporum* f.sp. *radicis-lycopersici*. *Physiol Mol Plant Pathol*, 38:237-253.

Bohlmann, H. (1994). The role of thionins in plant protection. *Crit Rev Plant Sci*, 13:1-16.

Chen, R., Wang, F., and Smith, A. G. (1996). A flower-specific gene encoding an osmotin-like protein from *Lycopersicon esculentum*. *Gene*, 179:301-302.

de Silva, Conceicao, A., and Broekaert, W. F. (1999). Plant defensins. In S. K. Datta and S. Muthukrishnan (Eds.), *Pathogenesis-Related Proteins in Plants*. CRC Press, Boca Raton, FL, pp. 247-260.

Edelbaum, O., Sher, N., Rubinstein, M., Novick, D., Tal, N., Moyer, M., Ward, E., Ryals, J., and Sela, I. (1991). Two antiviral proteins, gp 35 and gp 22, correspond to β-1,3-glucanase and an isoform of PR-5. *Plant Mol Biol*, 17:171-173.

Fagoaga, C., Rodrigo, I., Conejero, V., Hinarejos, C., Tuset, J. J., Arnau, J., Pina, J. A., Navarro, L., and Pena, L. (2001). Increased tolerance to *Phytophthora citrophthora* in transgenic orange plants constitutively expressing a tomato pathogenesis related protein PR-5. *Mol Breeding*, 7:175-185.

Garcia-Olmedo, F., Molina, A., Segura, A., and Moreno, M. (1995). The defensive role of nonspecific lipid-transfer proteins in plants. *Trends in Microbiol*, 3:72-74.

Hooft van Huijsduijnen, R. A. M., Van Loon, L. C., and Bol, J. F. (1986). Complementary DNA cloning of six messenger RNAs induced by TMV infection of tobacco and characterization of their translation products, *EMBO J*, 5:2057-2061.

Koiwa, H., Bressan, R. A., and Hasegawa, P. M. (1997). Regulation of protease inhibitors and plant defense. *Trends in Plant Sci*, 2:379-384.

Koiwa, H., Sato, F., and Yamada, Y. (1994). Characterization of accumulation of tobacco PR-5 proteins by IEF-immunoblot analysis. *Plant Cell Physiol*, 35:821-827.

Latunde-Dada, A. O. and Lucas, J. A. (2001). The plant defence activator acibenzolar-S-methyl primes cowpea (*Vigna unguiculata* (L.) Walp.) seedlings for rapid induction of resistance. *Physiol Mol Plant Pathol*, 58:199-208.

Leah, R., Tommerup, H., Svendsen, I., and Mundy, J. (1991). Biochemical and molecular characterization of three antifungal proteins from barley seed. *J Biol Chem*, 266:1564-1573.

Lee, Y. K. and Hwang, B. K. (1996). Differential induction and accumulation of β-1,3-glucanase and chitinase isoforms in the intercellular space and leaf tissues of pepper by *Xanthomonas campestris* pv. *vesicatoria* infection. *J Phytopathol*, 144:79-87.

Leubner-Metzger, G. and Meins, F. Jr. (1999). Functions and regulation of plant β-1,3-glucanases (PR2). In S. K. Datta and S. Muthukrishnan (Eds.), *Pathogenesis-Related Proteins in Plants*, CRC Press, Boca Raton, FL, pp. 49-76.

Linthorst, H. J. M., Danhash, N., Brederode, F. T., Van Kan, J. A. L., De Wit, P. J. G. M., and Bol, J. F. (1991). Tobacco and tomato PR proteins homologous to Win and Pro-Hevein lack the "Hevein" domain. *Mol Plant-Microbe Interact*, 4:586-592.

Mauch, F., Mauch-Mani, B., and Boller, T. (1988). Antifungal hydrolases in pea tissue. II. Inhibition of fungal growth by combinations of chitinase and β-1,3-glucanase. *Plant Physiol*, 88:936-942.

Melchers, L. S., Apotheker-deGroot, M., Van der Knaap, J. A., Ponstein, A. S., Sela-Buurlage, M. B., Bol, J. F., Cornelissen, B. J. C., Van den Elzen, P. J. M., and Linthorst, H. J. M. (1994). A new class of tobacco chitinases homologous to bacterial exo-chitinases displays antifungal activity. *Plant J*, 5:469-480.

Midoh, N. and Iwata, M. (1996). Cloning and characterization of a probenazole-inducible gene for an intracellular pathogenesis-related protein in rice. *Plant Cell Physiol*, 37:9-18.

Molina, A. and Garcia-Olmedo, F. (1997). Enhanced tolerance to bacterial pathogens caused by the transgenic expression of barley lipid transfer protein LTP2. *Plant J*, 12:669-675.

Moreno, M., Segura, A., and Garcia-Olmedo, F. (1994). Pseudothionin-Stl, a potato peptide active against potato pathogens. *Eur J Biochem*, 223:135-139.

Narasimhan, M. L., Damsz, B., Coca, M. A., Ibeas, J. L., Yun, D. J., Pardo, J. M., Hasegawa, P. M., and Bressan, R. A. (2001). A plant defense response effector induces microbial apoptosis. *Mol Cell*, 8:921-930.

Neuhaus, J. M. (1999). Plant chitinases. In S. K. Datta and S. Muthukrishnan (Eds.), *Pathogenesis-Related Proteins in Plants*, CRC Press, Boca Raton, FL, pp. 77-105.

Ohl, S., Apotheker-de Groot, M., van der Knaap, J. A., Ponstein, A. S., Sela-Buurlage, M. B., Bol, J. F., Cornelissen, B. J. C., Linthorst, H. J. M., and Melchers, L. S. (1994). A new-class of tobacco chitinases homologous to bacterial exo-chitinases is active against fungi in vitro. *J Cellular Biochem Suppl*, 18A:90.

Ponstein, A. S., Bres-Vloemans, A. A., Sella-Buurlage, M. B., Cornelissen, B. J. C., and Melchers, L. S. (1994). The "Missing" class I PR-4 protein from tobacco exhibits antifungal activity. *J Cellular Biochem Suppl*, 18A:90.

Ruiz-Medrano, R., Jimenez-Moraila, B., Herrera-Estrella, L., and Rivera-Busta-mante, R. F. (1992). Nucleotide sequence of an osmotin-like cDNA induced in tomato during viroid infection. *Plant Mol Biol,* 20:1199-1202.

Sharma, P., Borja, D., Stougaard, P., and Lonneborg, A. (1993). PR-proteins accu-mulating in spruce roots infected with a pathogenic *Pythium* sp. isolate include chitinases, chitosanases, and β-1,3-glucanases. *Physiol Mol Plant Pathol,* 43: 57-67.

Terras, F. R. G., Eggermont, K., Kovaleva, V., Raikhel, N. V., Osborn, R. W., Kester, A., Rees, S. B., Torrekens, S., Leuven, F. V., Vanderleyden, J., Commue, B. P. A., and Broekaert, W. F. (1995). Small cysteine-rich antifungal proteins from radish: Their role in host defense. *Plant Cell,* 7:573-578.

Tornero, P., Conejero, V., and Vera, P. (1997). Identification of a new pathogen-induced member of the subtilisin-like processing protease family from plants. *JBC Online* 272(22):14412-14419.

Van Loon, L. C. (1999). Occurrence and properties of plant pathogenesis-related proteins. In S. K. Datta and S. Muthukrishnan (Eds.), *Pathogenesis-Related Pro-teins in Plants.* CRC Press, Boca Raton, FL, pp. 1-19.

Van Loon, L. C., Pierpoint, W. S., Boller, T., and Conejero, V. (1994). Recommen-dations for naming plant pathogenesis-related proteins. *Plant Mol Biol Reporter* 12:245-264.

Van Loon, L. C. and Van Strien, E. A. (1999). The families of pathogenesis-related proteins, their activities, and comparative analysis of PR-1 type proteins. *Physiol Mol Plant Pathol,* 55:85-97.

Venere, R. J., Wang, X., Dyer, J. H., and Zheng, L. (1993). Role of peroxidase in cotton resistant to bacterial blight: Purification and immunological analysis of phospholipaseD from castor bean endosperm. *Plant Sci Lett,* 306:486-488.

Vidhyasekaran, P. (1997). *Fungal Pathogenesis in Plants and Crops.* Marcel Dekker, New York.

Vidhyasekaran, P. (2002). *Bacterial Disease Resistance in Plants: Molecular Biology and Biotechnological Applications.* The Haworth Press, Binghamton, NY.

Waniska, R. D., Venkatesha, R. T., Chandrashekar, A., Krishnaveni, S., Bejosano, F. P., Jeoung, J., Jayaraj, J., Muthukrishnan, S., and Liang, G. H. (2001). Antifungal proteins and other mechanisms in the control of sorghum stalk rot and grain mold. *J Agric Food Chem,* 49:4732-4742.

35

Phenolics

Phenolics are the important phytoanticipins in plants. They are detected as soluble forms in cytoplasm and are wall-bound in cell walls. They are inhibitory to pathogens at high concentrations. Their content increases rapidly in resistant interactions, while such an increase is not usually observed in susceptible interactions in the early stages of infection. The role of phenolics in disease resistance is described in this chapter.

SOLUBLE PHENOLICS AND DISEASE RESISTANCE

Phenolics are commonly present in cell walls (wall-bound phenolics) and in the cytoplasm (soluble phenolics). Several kinds of phenolics have been reported in plants. Monophenols, dihydroxy phenols, trihydroxy phenols, phenolics acids, pterocarpans, isoflavans, isoflavones, isoflavanones, glucosides of isoflavonoids, furanocoumarins, and anthocyanidins are the common phenolics in plants. Phenolics are synthesized through the phenylpropanoid pathway. They are normal secondary metabolites in healthy plants. However, they are also induced when pathogens invade a host, similar to phytoalexins and pathogenesis-related (PR) proteins. The concentrations of phenolics existing in healthy plants may not be inhibitory to pathogens (Vidhyasekaran, 1988, 1997, 2002). However, phenolic content increases severalfold when plants are infected by pathogens. At increased concentrations, phenolics may be toxic.

Several phenolics are found in every plant, similar to phytoalexins and PR proteins. For example, more than 20 different phenolic compounds have been detected in French bean. Not all of them are equally toxic to pathogens; toxicity varies from compound to compound. For example, the bean isoflavone isoferreirin is inhibitory *to Cladosporium cucumerinum* at a concentration of 10 µg/mL, while another bean isoflavanone, dalbergioidin, is inhibitory to the pathogen only at concentrations above 75 µg/mL (Adesanya et al., 1986). Coumarin, cinnamic acid, *p*-coumaric acid, vanillic acid, ferulic acid, and caffeic acid are highly inhibitory to *Phytophthora cinnamomi,* while protocatechuic acid, catechin, chlorogenic acid, phloroglucinol, and gallic acid are almost noninhibitory to this pathogen (Cahill and McComb, 1992).

Sensitivity of pathogens to different phenolics also varies from pathogen to pathogen. Pyrogallol is inhibitory to *Phytophthora cambivora* and not to *P. cinnamomi* (Casares et al., 1986). Similar differential toxic action has been reported for phytoalexins and PR proteins. Thus, phenolics are similar to PR proteins and phytoalexins in several aspects. However, phenolics are different from the other two in that they are present in plants before infection by pathogens or they are produced after infection solely from preexisting constituents. Phenolics belong to another group of defense chemicals called *phytoanticipins*. VanEtten and colleagues (1995) defined phytoanticipins as low-molecular-weight antimicrobial compounds that are present in plants before challenge by microorganisms or are produced after infection solely from preexisting constituents. In addition to phenolics, terpenoids, glucosinolates, alkaloids, dienes, saponins, and cyanogenic glucosides belong to the group of phytoanticipins.

The soluble phenolics are toxic to fungal and bacterial pathogens. Phenolics may alter the membrane porosity of fungal cells and inhibit certain enzymes of pathogens or DNA transcription (Vidhyasekaran, 1997). Phenolics may also inhibit production of toxins and pectic enzymes by pathogens. Rapid increase in phenolic synthesis due to infection has been correlated with disease resistance in many host-pathogen interactions. In cotton, younger (five- to six-day-old) seedlings are highly susceptible to *Rhizoctonia solani,* while older (12-day-old) seedlings are resistant. The concentration of phenolic compounds, predominantly catechin, increases more in 12-day-old seedlings than in young seedlings due to infection (Hunter, 1978). Some phenolics, such as salicylic acid, act as signal molecules and trigger host defense mechanisms.

Pathogens are capable of degrading phenolics to nontoxic compounds. For example, *Ascochyta rabiei,* the chickpea pathogen, rapidly degrades the chickpea phenolic biochanin A into nontoxic chemicals (Kraft and Barz, 1985). An early increase in concentration in phenolic content may be necessary to confer resistance against pathogens. Phenolic content increased mostly in resistant interactions, and if phenolics accumulated in susceptible interactions, accumulation was always in much later stages than observed in resistant interactions. The delayed accumulation is characteristic of susceptible interactions.

The role of phenolics in inducing resistance against pathogens has been demonstrated by altering the phenolic content of plants. When susceptible plants are fed with phenolics or precursors of phenolics, the plants show enhanced disease resistance (Vidhyasekaran et al., 1986). Preharvest spray of avocado fruits with thidiazuron (TDZ) or benzylaminopurine (benzyladenine) increased levels of epicatechin and resistance to *Colletotrichum gloeosporioides* (Beno-Moualem et al., 2001). When resistant plants are infused with phenylalanine ammonia-lyase (the key enzyme involved in biosynthesis of phenolics) inhibitors, phenolic synthesis is suppressed and the plants become susceptible to pathogens (Carver, Robbins, et al., 1992).

WALL-BOUND PHENOLICS
AND DISEASE RESISTANCE

The plant cell wall is the first barrier, and penetration of the cell wall appears to be the first requirement for pathogenesis of fungal pathogens. Bacterial pathogens multiply in the apoplast. The apoplast constitutes the cell wall continuum that surrounds the symplast (the symplast constitutes the sum total of the living protoplasm of a plant). In the area that separates two cells is a median substance called the *middle lamella.* Bacterial pathogens, which enter through natural openings or wounds, multiply in the apoplast (intercellular spaces), particularly in the middle lamella. The plant cell wall is considered to be the extracellular matrix secreted into intercellular spaces, and hence the cell wall may be the first barrier the bacterial pathogens encounter in the intercellular spaces.

When pathogens try to penetrate, the host cell wall is reinforced by different chemicals in resistant interactions. One of them is deposition of phenolics. Different phenolic compounds have been detected in plant cell walls. Flavonoids are the important cell wall-bound phenolic compounds. Flavonoids are synthesized in the cytoplasm and translocated toward the paramural areas, where they predominantly accumulate in primary cell walls and middle lamellae. Accumulation of these flavonoids occurred only in the resistant cotton variety inoculated with the pathogen *Xanthomonas axonopodis* pv. *malvacearum,* and no such flavonoid accumulation was observed in the susceptible variety (Dai et al., 1996).

Phenolic compounds are fluorescent. Autofluorescence can be observed in association with attempted penetration of the host cell wall by the primary germ tube and the appressorium of fungal pathogens. Autofluorescence is localized in limited regions of the host cell wall surrounding fungal germ tube contact sites. Phenylalanine ammonia-lyase (a key enzyme in the biosynthesis of phenolics) inhibitors suppress accumulation of localized autofluorogens, suggesting accumulation of phenolics (Carver, Zeyen, et al., 1992). Autofluorescence in cell walls has been shown to be associated only with resistant interactions in many host-pathogen interactions (Godwin et al., 1987; Cohen et al., 1989).

The nature of induced cell wall-bound phenolics varies. In tomato, an array of phenolic compounds accumulate in cell walls after inoculation with *Verticillium albo-atrum*. In the uninoculated cells, only traces of esterified phenolics are detected. The major compound accumulated in the infected cell wall is sinapic acid (Bernards and Ellis, 1991). In potato, the induced cell wall-bound phenolics have been identified as 4-hydroxybenzoic acid, 4-coumaric acid, ferulic acid, 4-coumaroyltyramine and feruloyl-tyramine (Clarke, 1982). In cotton, the cell wall-bound phenolics are anthocyanin compounds (Holton and Cornish, 1995).

REFERENCES

Adesanya, S. A., O'Neill, M. J., and Roberts, M. F. (1986). Structure-related fungitoxicity of isoflavonoids. *Physiol Mol Plant Pathol,* 29:95-103.

Beno-Moualem, D., Vinokur, Y., and Prusky, D. (2001). Cytokinins increase epicatechin content and fungal decay resistance in avocado fruits. *J Plant Growth Regulation,* 20:95-100.

Bernards, M. A. and Ellis, B. E. (1991). Phenylalanine ammonia-lyase from tomato cell cultures inoculated with *Verticillium albo-atrum. Plant Physiol,* 97:1494-1500.

Cahill, D. M. and McComb, J. A. (1992). A comparison of changes in phenylalanine ammonia-lyase activity, lignin and phenolic synthesis in the roots of *Eucalyptus calophylla* (field resistant) and *E. marginata* (susceptible) when infected with *Phytophthora cinnamomi. Physiol Mol Plant Pathol,* 40:315-332.

Carver, T. L. W., Robbins, M. P., Zeyen, R. J., and Dearne, G. A. (1992). Effects of PAL-specific inhibition on suppression of activated defence and quantitative susceptibility of oats to *Erysiphe graminis. Physiol Mol Plant Pathol,* 41:149-163.

Carver, T. L. W., Zeyen, R. J., Robbins, M. P., and Dearne, G. A. (1992). Effects of the PAL inhibitor, AOPP, on oat, barley and wheat cell responses to appropriate and inappropriate formae speciales of *Erysiphe graminis* DC. *Physiol Mol Plant Pathol,* 41:397-409.

Casares, A., Melo, E. M. P. F., Ferraz, J. F. P., and Ricardo, C. P. P. (1986). Differences in ability of *Phytophthora cambivora* and *P. cinnamomi* to dephenolize lignin. *Trans Brit Mycol Soc,* 87:229-235.

Clarke, D. D. (1982). The accumulation of cinnamic acid amides in the cell walls of potato tissue as an early response to fungal attack. In R. K. S.Wood (Ed.), *Active Defense Mechanisms in Plants.* Plenum Press, New York, pp. 321-322.

Cohen, Y., Eyal, H., Hanania, J., and Malik, Z. (1989). Ultrastructure of *Pseudoperonospora cubensis* in muskmelon genotypes susceptible and resistant to downy mildew. *Physiol Mol Plant Pathol,* 34:27-40.

Dai, G. H., Nicole, M., Andary, C., Martinez, C., Bresson, E., Boher, B., Daniel, J. F., and Geiger, J. P. (1996). Flavonoids accumulate in cell walls, middle lamellae and callose-rich papillae during an incompatible interaction between

Xanthomonas campestris pv. *malvacearum* and cotton. *Physiol Mol Plant Pathol,* 49:285-306.

Godwin, J. R., Mansfield, J. W., and Darby, P. (1987). Microscopical studies of resistance to powdery mildew disease in the hop cultivar Wye target. *Plant Pathology,* 36:21-32.

Holton, T. A. and Cornish, E. C. (1995). Genetics and biochemistry of anthocyanin biosynthesis. *Plant Cell,* 7:1071-1083.

Hunter, R. E. (1978). Effects of catechin in culture and in cotton seedlings on growth and polygalacturonase activity of *Rhizoctonia solani. Phytopathology,* 68:1032-1036.

Kraft, B. and Barz, W. (1985). Degradation of the isoflavone biochanin A and its glucoside conjugates by *Ascochyta rabiei. Appl Environ Microbiol,* 50:45-48.

VanEtten, H. D., Sandrock, R. W., Wasmann, C. C., Soby, S. D., McCluskey, K., and Wang, P. (1995). Detoxification of phytoanticipins and phytoalexins by phytopathogenic fungi. *Can J Bot,* 73(Suppl. I):S518-S525.

Vidhyasekaran, P. (1988). *Physiology of Disease Resistance in Plants,* Volume I. CRC Press, Boca Raton, FL.

Vidhyasekaran, P. (1997). *Fungal Pathogenesis in Plants and Crops.* Marcel Dekker, New York.

Vidhyasekaran, P. (2002). *Bacterial Disease Resistance in Plants: Molecular Biology and Biotechnological Applications.* The Haworth Press, Binghamton, NY.

Vidhyasekaran, P., Borromeo, E. S., and Mew, T. W. (1986). Host-specific toxin production by *Helminthosporium oryzae. Phytopathology,* 76:261-265.

36

Phytoalexins

Phytoalexins are antimicrobial compounds synthesized in plants due to infection. Several phytoalexins have been isolated from various plants and characterized. Their role in disease resistance is described in this chapter.

DEFINITION

Muller and Borger (1940) defined phytoalexins as plant antibiotics that are synthesized de novo after the plant tissue is exposed to microbial infection. Paxton (1981) defined phytoalexins as low-molecular-weight antimicrobial compounds that are both synthesized by and accumulated in plants after exposure to microorganisms. Two important criteria have been suggested to label a secondary metabolite a phytoalexin: (1) the secondary metabolite should be produced de novo in response to infection, and (2) the compound should accumulate to antimicrobial concentrations in the area of infection (VanEtten et al., 1995).

Phytoalexins differ from pathogenesis-related proteins in that PR proteins are produced by transcription of quiescent host genes, while phytoalexin production in plants requires a biosynthetic pathway involving coordinated action of several host enzymes (Ingham, 1973). Phytoalexins are induced by fungal, bacterial, and viral pathogens, biotic and abiotic elicitors, various toxic chemicals, and environmental factors. Some compounds may be phytoalexins in one organ and constitutive in another organ of the same plant species. Momilactone A, which is induced in rice leaves as a phytoalexin, occurs constitutively in rice seeds. The same compound may be a phytoalexin in one plant species and a constitutive compound in another species. The flavanone sakuranetin is induced in rice leaves but is constitutive in blackcurrant *(Ribes nigrum)* leaves. The same phytoalexin can be detected in more than one plant species. Medicarpin is the phytoalexin detected in chickpea *(Cicer arietinum),* alfalfa *(Medicago sativa),* peanut *(Arachis hypogaea),* and broad bean *(Vicia faba).* Luteolinidin has been detected as a phytoalexin in sugarcane *(Saccharum officinarum)* and sorghum *(Sorghum bicolor).* Spirobrassinin is the phytoalexin detected in *Brassica campestris* and *Raphanus sativus*. Resveratrol has been detected in peanut and *Festuca versuta* (for more details, see Vidhyasekaran, 1988, 1997, 2002). It appears that induction of phytoalexins is universal in the plant kingdom. Several phytoalexins have been detected in each plant species, and as many as 30 phytoalexins in carnation plant have been reported (Niemann, 1993).

ANTIMICROBIAL ACTION OF PHYTOALEXINS

Phytoalexins are inhibitory to microorganisms, and most are highly fungitoxic. Casbene is inhibitory toward the growth of *Aspergillus niger* even at 10 µg/mL. Phytoalexins inhibit spore germination, germ tube elongation, and mycelial growth of fungi. They may suppress toxin production by pathogens. Trichothecene toxin production by the parsnip pathogen *Fusarium sporotrichioides* is completely inhibited by the parsnip furanocoumarin phytoalexins (Desjardins et al., 1989). Several phytoalexins have been shown to inhibit bacterial pathogens in vitro. The lettuce phytoalexin lettucenin A was found to be highly inhibitory to *P. savastanoi* pv. *phaseolicola*. Clear zones of inhibition of bacterial growth were produced by 0.5 µg samples of lettucenin A applied to lawns of *P. savastanoi* pv. *phaseolicola* (Bennett et al., 1994).

Although phytoalexins are inhibitory to fungal pathogens, the pathogens of a particular host are less sensitive or insensitive to the phytoalexins of the host (VanEtten et al., 2001). Actually, pathogens produce enzymes that detoxify the phytoalexins of their host. For example, *Nectria haematococca, Ascochyta pisi, Rhizoctonia solani, Phoma pinodella,* and *Mycosphaerella pinodes* are all pathogens of pea, and all of them are known to detoxify the pea phytoalexin pisatin. The pathogens produce pisatin demethylase which demethylates pisatin, and the resultant product is much less toxic to the pathogens. *Ascochyta rabiei,* a pathogen of chickpea, produces a nicotinamide adenine dinucleotide phosphate hydrogen (NADPH)-dependent reductase and detoxifies the chickpea phytoalexin maackiain. *Fusarium solani* f. sp. *phaseoli,* the pathogen of bean, degrades four phytoalexins of the host, phaseollin, phaseollidin, phaseollinisoflavan, and kievitone, by producing monooxygenases. Kievitone hydratase produced by the pathogen detoxifies kievitone.

In some cases, the produced phytoalexins may be inactivated in susceptible interactions. Rogers and colleagues (1996) showed that while King's medium B containing 500 µg of camalexin was toxic to *P. syringae* pv. *maculicola,* intracellular fluid extracted from *Arabidopsis* leaves with camalexin added to a concentration of 500 µg/mL was not toxic to *P. syringae* pv. *maculicola.* When *P. syringae* pv. *maculicola* cells were grown in intercellular fluid and then transferred to King's medium B containing camalexin, death rate of cells was indistinguishable from that observed when cells were grown in King's medium B, suggesting that intercellular fluid does not induce resistance to camalexin. Camalexin was not degraded in the intracellular fluid, which suggests that a component of intracellular fluid sequesters camalexin in a nontoxic form (Rogers et al., 1996).

ROLE OF PHYTOALEXINS
IN DISEASE RESISTANCE

No conclusive evidence is available to prove the role of phytoalexins in disease resistance. However, much indirect evidence suggests their role in disease resistance. Plants respond to pathogens producing phytoalexins. Phytoalexins are induced rapidly and more abundantly in resistant interactions. The differences between resistance and susceptibility may be quantitative rather than qualitative. Susceptible plants may possess the machinery necessary for induction of phytoalexins, but it may not be activated in sufficient magnitude to restrict the infection. Rapid induction of phytoalexins in resistant interactions may occur even at the stage of transcription of genes encoding biosynthetic enzymes. For example, both phenylalanine ammonia-lyase (PAL) and chalcone synthase (CHS) mRNAs accumulated in the incompatible interaction by three hours postinoculation, while in the compatible interaction, PAL transcripts began to accumulate about 15 hours postinoculation, and CHS transcripts accumulated very weakly around 18 hours postinoculation (Meier et al., 1993). Phytoalexins were detected in the incompatible interaction by 24 hours postinoculation, while no phytoalexins accumulated in the compatible interactions during this period (Meier et al., 1993). In susceptible interactions, suppression of induction of phytoalexins would have been suppressed. When bean plants were inoculated with the incompatible pathogen *P. syringae* pv. *tabaci,* phytoalexins were induced. Prior infiltration with the compatible pathogen *P. savastanoi* pv. *phaseolicola* suppressed the accumulation of phytoalexins induced by the incompatible pathogen (Jakobek et al., 1993). Accumulation of phytoalexins specifically in resistant interactions suggests a role for phytoalexins in disease resistance.

The importance of the phytoalexins in disease resistance has been demonstrated by assessing growth of pathogens in host tissues in which phytoalexins start to accumulate during pathogenesis. When lettuce leaves were inoculated with the incompatible bacterium *P. savastanoi* pv. *phaseolicola,* the phytoalexin lettucenin A accumulated in the inoculated tissues. However, when these leaves were inoculated with an *hrp* mutant *(hrpD⁻)* of this bacterium, only trace amounts of the phytoalexin were noticed (Bestwick et al., 1995). The wild-type bacterium was quickly eliminated (within 48 hours) from the leaf tissues in which phytoalexin accumulated, while the *hrpD* mutant of the bacterium survived in the leaf tissues in which no (or trace amounts of) phytoalexin accumulated (Bestwick et al., 1995). The levels of phytoalexin (glyceollin) accumulation in various cultivars of soybean after infection with *P. syringae* pv. *glycinea* were inversely correlated with bacterial multiplication (Long et al., 1985).

The accumulation of stilbene phytoalexin decreased in grape berries during ripening, and these ripening berries became susceptible to *Botrytis cinerea* infection (Bais et al., 2000). Application of a plant defense activator, acibenzolar-S-methyl induced systemic resistance against *Colletotrichum destructivum* in cowpea, and this treatment resulted in induction of phytoalexins (Latunde-Dada and Lucas, 2001). Mohr and Cahill (2001) provided evidence that phytoalexins may be more important than phenolic and lignin deposition in disease resistance. In compatible interactions of soybean with *Phytophthora sojae,* leaves and hypocotyls were characterized by hypersensitive reaction (HR), phenolic and lignin deposition, and phytoalexin glyceollin accumulation. Norflurazon treatment restricted the spread of *P. sojae* and increased glyceollin accumulation in compatible tissues. Exogenous abscisic acid addition caused spreading lesions in normally incompatible interactions and reduced glyceollin accumulation. Phenolic deposition and HR were unchanged by either treatment in the incompatible and compatible interactions. This evidence suggests that glyceollin is a major factor in conferring resistance against the pathogen (Mohr and Cahill, 2001). All these studies suggest phytoalexins may have a role in disease resistance.

REFERENCES

Bais, A. J., Murphy, P. J., and Dry, I. B. (2000). The molecular regulation of stilbene phytoalexins biosynthesis in *Vitis vinifera* during grape berry development. *Australian J Plant Physiol,* 27:425-433.

Bennett, M. H., Gallagher, M. D. S., Bestwick, C. S., Rossiter, J. T., and Mansfield, J. W. (1994). The phytoalexin response to lettuce to challenge by *Botrytis cinerea, Bremia lactucae* and *Pseudomonas syringae* pv. *phaseolicola. Physiol Mol Plant Pathol,* 44:321-333.

Bestwick, C. S., Bennett, M. H., and Mansfield, J. W. (1995). Hrp mutant of *Pseudomonas syringae* pv. *phaseolicola* induces cell wall alterations but not membrane damage leading to the hypersensitive reaction in lettuce. *Plant Physiol,* 108:503-516.

Desjardins, A. E., Spencer, G. F., Plattner, R. D., and Beremand, M. N. (1989). Furanocoumarin phytoalexins, trichothecene toxins, and infection of *Pastinaca sativa* by *Fusarium sporotrichioides*. *Phytopathology*, 79:170-175.

Ingham, J. L. (1973). Disease resistance in higher plants: The concept of pre-infectional and post-infectional resistance. *Phytopathol Z*, 78:314-335.

Jakobek, J. L., Smith, J. A., and Lindgren, P. B. (1993). Suppression of bean defense responses by *Pseudomonas syringae*. *Plant Cell*, 5:57-63.

Latunde-Dada, A. O. and Lucas, J. A. (2001). The plant defence activator acibenzolar-S-methyl primes cowpea (*Vigna unguiculata* (L.) Walp.) seedlings for rapid induction of resistance. *Physiol Mol Plant Pathol*, 58:199-208.

Long, M., Barton-Willis, P., Staskawicz, B. J., Dahlbeck, D., and Keen, N. T. (1985). Further studies on the relationship between glyceollin accumulation and the resistance of soybean leaves to *Pseudomonas syringae* pv. *glycinea*. *Phytopathology*, 75:235-239.

Meier, B. M., Shaw, N., and Slusarenko, A. J. (1993). Spatial and temporal accumulation of defense gene transcripts in bean *(Phaseolus vulgaris)* leaves in relation to bacteria-induced hypersensitive cell death. *Mol Plant-Microbe Interact*, 6: 453-466.

Mohr, P. G. and Cahill, D. M. (2001). Relative roles of glyceollin, lignin, and the hypersensitive response and the influence of ABA in compatible and incompatible interactions of soybeans with *Phytophthora sojae*. *Physiol Mol Plant Pathol*, 58:31-41.

Muller, K. O. and Borger, H. (1940). Experimentelle untersuchungen uber die *Phytophthora*-resistenz der kartoffel Arb. Biol. Reichsasnstalt. *Landw Forstw Berlin*, 23:189-231.

Niemann, G. J. (1993). The anthranilamide phytoalexins of the caryophyllaceae and related compounds. *Phytochemistry*, 34:319-328.

Paxton, J. D. (1981). Phytoalexins—A working redefinition. *Phytopathol Z*, 101: 106-109.

Rogers, E. E., Glazebrook, J., and Ausubel, F. M. (1996). Mode of action of the *Arabidopsis thaliana* phytoalexin camalexin and its role in *Arabidopsis*-pathogen interactions. *Mol Plant-Microbe Interact*, 9:748-757.

VanEtten, H. D., Sandrock, R. W., Wasmam, C. C., Soby, S. D., Mellusky, K., and Wang, P. (1995). Detoxification of phytoanticipins and phytoalexins by phytopathogenic fungi. *Can J Bot*, 73(Suppl.1):S518-S525.

VanEtten, H. D., Temporini, E., and Wasmann, C. (2001). Phytoalexin (and phytoanticipin) tolerance as a virulence trait: Why is it not required by all pathogens? *Physiol Mol Plant Pathol*, 59:83-93.

Vidhyasekaran, P. (1988). *Physiology of Disease Resistance in Plants*, Volume I. CRC Press, Boca Raton, FL.

Vidhyasekaran, P. (1997). *Fungal Pathogenesis in Plants and Crops*. Marcel Dekker, New York.

Vidhyasekaran, P. (2002). *Bacterial Disease Resistance in Plants: Molecular Biology and Biotechnological Applications*. The Haworth Press, Binghamton, NY.

PATHOGENESIS

37

Physiology of Pathogenesis

Adhesion of fungal and bacterial cells to the host surface triggers the pathogenesis process. Several molecules of both the pathogen and the host are involved in the early recognition process. Various enzymes and toxins are involved in the disease development. This process is described in this chapter.

ADHESION

Pathogenesis appears to start when fungal spores or bacterial cells land on the surface of the plant. Adhesion of fungi and bacteria to the leaf surface seems to be important in pathogenesis. Mutants with adhesion-deficient macroconidia of *Nectria haematococca* were found to be avirulent (Jones and Epstein, 1990). The mutants of *Pseudomonas savastanoi* pv. *phaseolicola* with greatly reduced adherence to the leaf surface of bean showed lower incidence of halo blight disease (Romantschuk et al., 1991). Non-attaching mutants of *Agrobacterium tumefaciens* are known to be avirulent (Matthysse et al., 1996). Preformed and induced adhesive materials have been detected in fungal spores. The preformed adhesive material is a protein, and the induced material is a glycoprotein. The glycoprotein adhesive material is released from ungerminated *Colletotrichum graminicola* conidia at an extremely early time in the infection process (Mercure et al., 1994). Cutinases and other esterases have been shown to be involved in adhesion of some fungi (Pascholati et al., 1992). Hydrophobins are relatively small cysteine-rich proteins, and these hydrophobins are involved in attachment and development of appressoria of fungi (Vidhyasekaran, 1997; Wosten, 2001). Fimbriae and pili are the filamentous nonflagellar appendages of many bacterial pathogens which may enable the bacteria to attach to plant cells. The bulk of the filament of fimbriae is made up of helically arranged subunits of a major protein ranging in size from 9 to 22 kDa in different fimbrial types. These proteins may be responsible for adhesion of bacterial cells to plant cells (Vidhyasekaran, 2002). Various strains of *P. syringae* and *P. savastanoi* adsorb to leaves of the host plants using their pili (Romantschuk et al., 1991). Some extracellular polysaccharides of bacteria have been implicated in adhesion. Mutants of *Agrobacterium tumefaciens* that are defective in neutral glucan production or export are also defective in attachment to plant cells, suggesting that extracellular neutral glucan is required for attachment and pathogenesis (Thomashow et al., 1987). Lipopolysaccharides have been implicated in the initial attachment of *A. tumefaciens* to the plant cell surface. Rhicadhesin produced by *A. tumefaciens* is also involved in adhesion, and some plant molecules may also be involved in adhesion. Vitronectin-like proteins have been reported in plants, and the binding of *A. tumefaciens* by vitronectin has been reported (Wagner and Matthysse, 1992).

NUTRITION

Fungal and bacterial pathogens obtain their nutrition from host cells. They depend on the host for the synthesis of various constituents of their cells. Fungal and bacterial cells consist of chitin, cellulose, glucose, mannans, uronic acids, rhamnose, xylose, proteins, lignin, lipids, nucleic acids, and pigments. The cell constituents are made up of carbon, nitrogen, phosphorous, sulfur, potassium, magnesium, zinc, iron, manganese, copper, molybdenum, calcium, scandium, vanadium, and gallium. In short, the fungal and bacterial cells are made up of carbohydrates, fats, proteins, nucleic acids, and minerals. Host plants supply carbon, nitrogen, and minerals from which the fungi and bacteria themselves synthesize all the complex cell structures for their growth.

Bacterial pathogens multiply in the apoplast and obtain nutrients from the plant cell by causing leakage in cell membranes. A group of bacterial genes are involved in inducing electrolyte leakage. These include *hrp* (hypersensitive *r*esponse and *p*athogenicity) and *avr* (avirulence) genes. *hrp* genes are required for the elicitation of hypersensitive necrosis on nonhost plants and on resistant cultivars of the susceptible host, and are also required for pathogenicity on susceptible cultivars. In both compatible and incompatible interactions, electrolyte leakage is observed. In incompatible interactions, increase in electrolyte leakage may be more during the first few hours (10 to 30 hours), but subsequent increase in electrolyte leakage is almost similar to compatible interaction (Brisset and Paulin, 1991). *hrp* genes are present in all bacterial pathogens, irrespective of whether they are virulent or avirulent. *hrp* genes appear to be necessary for growth of bacteria in planta. Bacterial pathogens are able to grow in both host and nonhost plants, although multiplication may be less in nonhost plants (Vidhyasekaran, 2002). Saprophytic bacteria, which do not possess *hrp* genes, are unable to multiply in planta. When saprophytic bacteria carry *hrp* genes, they are capable of multiplying in the host (Mulya et al., 1996). The major function of *hrp* genes appears to be in inducing electrolyte leakage to provide nutrients to pathogens.

Several *hrp* genes have been detected in a single bacterial pathogen. For example, *hrp* genes encoding 25 Hrp proteins, such as HrpA, HrpB, HrpC, HrpD, HrpE, HrpF, HrpG, HrpH, HrpI, HrpJ, HrpJ-3, HrpJ-4, HrpJ-5, HrpK, HrpL, HrpO, HrpR, HrpS, HrpU, HrpU1, HrpU2, HrpW, HrpX, HrpY, and HrpZ, have been detected in *Pseudomonas syringae* pv. *syringae* strain 61 (Huang et al., 1995). *hrp* genes appear to be highly conserved in several bacterial pathogens. The *Ralstonia solanacearum* strain GMI 1000 *hrp* gene cluster showed structural homology with all *Xanthomonas* isolates tested, namely *X. campestris* pv. *campestris, X. vesicatoria, X. arboricola* pv. *juglandis, X. axonopodis* pv. *begoniae,* and *X. translucens* pv. *graminis* (Boucher et al., 1988). The existence of homology between the *hrp* clusters of *Erwinia amylovora* and *P. syringae* has been reported (Beer et al., 1991). Homologues to Hrp H and Hrp I of *P. syringae* pv. *syringae* have been identified in the *hrp* clusters of *X. vesicatoria* (Fenselau et al., 1992) and *R. solanacearum* (Gough et al., 1992).

avr genes are pathogen genes critical in determining whether a bacterial strain will be virulent or avirulent on a specific host (Wanner et al., 1993). *avr* genes also appear to be conserved among several bacterial pathogens. All *avr* genes of *Xanthomonas* spp. sequenced to date are 95 to 98 percent identical to one another. Conservation of *avr* genes even among different genera of bacterial pathogens has been reported. Some *avr* genes appear to encode a function required for pathogenicity, and some *avr* genes have been shown to be virulence factors involved in promoting parasitism (Gopalan et al., 1996). *hrp* genes appear to be needed for transcription of *avr* genes in bacterial pathogens.

Some *hrp* and *avr* genes encode elicitors. The electrolyte (nutrient) leakage may be induced by both *hrp* and *avr* gene-encoded elicitors. Both types of elicitors are secreted into plant cells by the action of *hrp* genes. A moderate leakage in the host plant may provide the pathogens with various nutrients originating from the plant cell and permit growth of the invading pathogen.

In their life cycle, fungal pathogens enter into two distinct phases, biotrophic and necrotrophic. In the biotrophic phase, they invade the host cells inter- and intracellularly and establish themselves in the living cells. In the second phase, they kill the cells and obtain nutrition from the dying cells. This phase is called the *necrotrophic* phase. In the case of obligate pathogens, the biotrophic phase may be much longer, while in the case of facultative saprophytes (normally pathogenic but can also live as a saprophyte), a long biotrophic phase may be followed by a necrotrophic phase. For facultative parasites (normally saprophytes but can also live as a pathogen), the necrotrophic phase will be much longer with a short biotrophic phase. However, in all pathogens, both phases may exist either as a short or long spell (Vidhyasekaran, 1993).

PLANT BARRIERS AGAINST PATHOGENS

In the biotrophic phase, pathogens establish themselves in host cells, but there are many barriers against their establishment. The first barrier between the host plant and the attacking pathogen is the cuticle, which consists of a structural polymer, cutin, impregnated with wax. Suberin is another protective barrier on the surface of plants, and the host cell wall is another important barrier against the pathogen. Cell walls consist of pectic substances, hemicelluloses, cellulose, and hydroxyproline rich-glycoproteins. Below the cell wall, membranes are found. The membrane provides a semipermeable barrier between the cell and its external environment. The membrane consists of lipids and proteins. During the biotrophic phase, the pathogens invade the cells by degrading the cuticle, cell wall, and cell membranes by producing a series of enzymes. After the biotrophic phase, the pathogens produce toxins, which kill the cell and enter into the necrotrophic phase. Thus, enzymes and toxins are important tools of pathogens for their pathogenesis.

ENZYMES

Pathogens produce various enzymes to breach the plant barriers to obtain nutrients from plant cells. Enzymes are complex proteins that catalyze reactions. The following are the important enzymes involved in pathogenesis.

Cutinases

Several fungal pathogens produce cutinases, which degrade the cuticle layer of the host. Cutinase is considered to be a serine esterase. Variation in cutinases produced by different pathogens has been observed, as well as variation in cutinase produced by a single fungal species. Cutinases produced by various fungal pathogens vary remarkably in immunological properties and primary sequences. Fungal cutinase is not a constitutive enzyme, but it is inducible. Cutinase is released immediately upon contact of the fungal conidia with a host substratum. Much evidence has demonstrated the importance of cutinases in pathogenesis. Koller and colleagues (1991) used a cutinase inhibitor, *O*-methyl-*O*-butyl-*O*-(3,5,6-trichloro-2-pyridyl) phosphate to assess the importance of cutinase in penetration. When the infection droplets containing spores of *Venturia inaequalis* and the inhibitor were incubated on apple leaves, formation of stromata beneath the cuticle was completely prevented. In samples without the inhibitor, subcuticular stromata were clearly visible after 24 hours of germination. Virulence of pathogens has been found to be related to their ability to produce cutinases. Virulence of the isolates of *Fusarium solani* f. sp. *pisi,* the pea pathogen, was related to their ability to produce cutinase (Koller et al., 1982). Cutinase-deficient mutants of various pathogens show reduced pathogenicity. Introduction of the cutinase gene into a cutinase-less mutant of *F. solani* restored pathogenicity (Kolattukudy et al., 1989). When the conidia of avirulent isolate T-30 of *F. solani* f. sp. *pisi* were supplemented with cutinase isolated from a virulent isolate of *F. solani* prior to inoculation of intact pea-stem surfaces, the conidia caused severe infection (Koller et al., 1982). All these studies have shown the importance of cutinases in fungal pathogenesis. Bacterial pathogens do not penetrate the host cuticle layer; they enter through wounds and natural openings to cause infection. Hence, there is no need for bacterial pathogens to produce cutinases.

Pectic Enzymes

After penetration through the cuticle, the invading fungus would encounter the pectinaceous barrier. Even when fungi and bacteria gain entry into plants through stomata and wounds, they would have to degrade the intercellular pectinaceous polymers to initiate infection in the plant. Plant pathogens produce multiple forms of different types of pectic enzymes. The pectic substances are of two types, pectin (pectinic acid) and pectic acid (polygalacturonic acid). The activity of the pectic enzymes varies in their specificity to the substrate (pectin or pectic acid), in their mechanism to split the α-1,4-glycosidic bond (hydrolytic or lyitic), and in the type of cleavage (random or terminal). The pectic enzymes are classified as pectin methylesterase (PME), which converts pectin into polygalacaturonic acid, exopolygalacturonase (exo PG), which cleaves polygalacturonic acids in a terminal manner releasing monomeric products (galacturonic acids), and endopolygalacturonase (endo PG), which cleaves polygalacturonic acids in a random manner releasing oligogalacturonic acid. Exo- and endo-pectin methyl galacturonases (PMG) act similar to polygalacturonases, but the perferred substrate is pectin instead of polygalacturonic acid. Both polygalacturonases and polymethylgalacturonases act by hydrolytic action. In contrast, polygalacturonate *trans*-eliminase (PGTE; pectate lyase [PL]) and pectin *trans*-eliminase (PTE; pectin lyase) induce lyitic degradation of the glycosidic linkage, resulting in an unsaturated bond between carbons 4 and 5 of a uronide moiety in the reaction product. The *trans*-eliminases break the glycoside linkage at carbon number 4 and simultaneously eliminate the H from carbon number 4, resulting in oligouronides that contain an unsaturated galacturonyl unit. Both exo- (terminal) and endo- (random) type reactions are seen in the *trans*-eliminases, and they are called exo-PGTE/exo PL, endo-PGTE /endo PL, exo-PTE/exo pectinlyase, and endo-PTE/endo pectinlyase (Vidhyasekaran, 1993).

Both fungal and bacterial pathogens produce various pectic enzymes in vitro and in vivo, and many of them have been purified and characterized (Vidhyasekaran, 1997, 2002). Histological evidence has been provided to show the importance of pectic enzymes in cell wall penetration by fungal pathogens. Penetration of epidermal cells was observed two to seven hours after inoculation with *Pythium ultimum* in cucumber (Cherif et al., 1991). The transmission electron micrographs revealed the alteration of pectic materials in the middle lamella matrices. The pectin breakdown was not restricted to areas neighboring invading hyphae but also occurred at some distance from the point of fungal attack, suggesting that pectin-degradation enzymes produced by *P. ultimum* freely diffuse extracellularly, thus facilitating pathogen ingress through loosened host walls (Cherif et al., 1991). The importance of pectic enzymes in pathogenesis has been demonstrated using antibodies developed against the enzymes. *Fusarium solani* f. sp. *pisi,* the pea pathogen, produces pectate lyase. Antibodies against the enzyme were developed. When suspension of the conidia of the fungus was prepared in these antibodies and inoculated on pea stem, no infection of the pathogen was observed. Penetration of the fungus into the host was strongly suppressed by the antibodies (Crawford and Kolattukudy, 1987). Fungal isolates deficient in pectic enzymes have been shown to be less virulent.

Bacterial pathogens produce colonies within intercellular spaces or xylem vessels. These pathogens in the intercellular spaces and xylem vessels may face the complex pectic polysaccharides, and unless the pectic polysaccharides are degraded by pectic enzymes, they may not be able to breach the cell well barrier. Bacterial pathogens produce pectic enzymes to breach the pectic barrier. Pectin degradation has been demonstrated during bacterial pathogenesis. *Clavibacter michiganensis,* the tomato pathogen, quickly spreads along the xylem vessels of the host. Swelling, shredding, and partial wall dissolution were typical features in areas adjacent to sites of bacterial accumulation (Benhamou, 1991). Pectin-like molecules and galactose residues were the major components of the fibrillar or amorphous material accumulating at those sites where bacteria were actively growing (Benhamou,

1991). Because the bacterial pathogens must breach the complex pectic barrier, they need to produce different types of pectic enzymes with varying specificity to the different types of pectic substances. Production of various pectic enzymes by bacterial pathogens belonging to various genera, including *Pseudomonas, Xanthomonas, Erwinia, Ralstonia, Burkholderia, Agrobacterium,* and *Clavibacter,* has been reported (Vidhyasekaran, 2002).

Cellulases

Cellulose is one of the major components of the plant cell wall. Although pectin is the main constituent of the middle lamella, cellulose is the main constituent of xylem tissue. Cellulose is composed of glucose units in the chain configuration, connected by β-1,4-glycosidic bonds. Cellulase is a general term for a synergistic system of three major categories of enzymes that hydrolyze the β-1,4-glucosidic bonds: endoglucanase (endo β-1,4-glucanase), exoglucanase (β-1,4-cellobiohydrolase), and β-glucosidase. These are classified according to their mode of action and substrate specificity. Endoglucanase cleaves β-glucosidic bonds in an amorphous region of cellulose, creating sites for exoglucanase to cleave cellobiose from the non-reducing ends. Glucosidase would hydrolyze the resultant cellobiose to glucose, preventing cellobiose buildup and exoglucanase inhibition.

Several cellulolytic enzymes are known to be produced by fungal pathogens. *Venturia inaequalis,* the apple scab pathogen, produces in culture 12 cellulase isozymes with isoelectric points in the range of 3.7 to 5.6 (Kollar, 1994). These enzymes could also be isolated from apple leaves infected with the pathogen (Kollar, 1994). *Gaeumannomyces graminis* var. *tritici,* the causal agent of take-all disease of wheat, secretes two groups of enzymes which degrade cellulosic polymers (Dori et al., 1995). The first group contains an endoglucanase and a β-glucosidase with acidic pIs of 4.0 and 5.6, respectively. The second group contains an endoglucanase and a β-glucosidase with basic pIs of 9.3 and >10, respectively. Acidic and basic groups of endoglucanase and β-glucosidase were also obtained from inoculated wheat roots and appeared to be similar to those isolated from the culture fluid (Dori et al., 1995). Several studies have provided evidence to show the involvement of cellulolytic enzymes in fungal pathogenesis. Histological studies using tobacco roots infected by *Phytophthora parasitica* var. *nicotianae* demonstrated the involvement of cellulases in fungal pathogenesis (Benhamou and Cote, 1992). Incubation of the infected tissues with the gold-complexed exoglucanase resulted in the deposition of gold particles over both fungal and plant cell walls. Labeling decreased in the inoculated tissues, indicating degradation of cellulose (Benhamou and Cote, 1992).

Xylem degradation has been documented in many bacterial pathogen-infected tissues. Several bacterial pathogens are known to produce cellulolytic enzymes. *Clavibacter michiganensis* ssp. *michiganensis* produces an endoglucanase (Meletzus et al., 1993). *Clavibacter michiganensis* ssp. *sepedonicus* produces exoglucanase, endoglucanase, and β-glucosidase (Baer and Gudmestad, 1995).

Hemicellulases

Hemicelluloses are covalently linked to pectic polysaccharides and non-covalently bound to cellulose fibrils of cell walls. The important plant cell wall hemicelluloses are xylan, araban, and galactan. Xylans are the most abundant hemicelluloses, constituting up to 40 percent of the primary cell walls of some plants. Many fungal pathogens, such as *Sclerotium rolfsii, Verticillium albo-atrum, Fusarium oxysporum* f. sp. *lycopersici,* and *Sclerotinia sclerotiorum* produce high levels of arabanases, xylanases, and galactanases in vitro. *Phytophthora infestans,* the pathogen of potato, produces endo-galactanases in vitro and in infected tissues and depletes galactan (Jarvis et al., 1981). Arabinose is a major component of the arabinans. It is also found as a side chain constituent of the xyloglucan, xylan rhamnogalacturonan I, and rhamnogalacturonan II polymers of plant cell walls. The removal of the arabinose-containing side chains from these major plant cell wall polymers by arabinofuranosidase might facilitate the digestion of these polymers. *Sclerotinia trifoliorum,* a pathogen of legumes, produces an arabinofuranosidase in culture (Rehnstrom et al., 1994). Three arabinofuranosidase-deficient mutants were obtained. The virulence of the arabinofuranosidase-deficient mutants on alfalfa stems was less than the wild isolates (Rehnstrom et al., 1994). The bacterial pathogen *Clavibacter michiganensis* ssp. *michiganensis* produces xylanase to degrade xylans (Beimen et al., 1992).

Proteases

Plant cell walls contain up to 10 percent protein which is exceptionally rich in hydroxyproline. Fungal pathogens are known to produce various proteases (Movahedi and Heale, 1990). Several bacterial pathogens produce proteases. *Xanthomonas campestris* pv. *campestris, X. campestris* pv. *armoraciae, P. syringae* pv. *tomato, E. carotovora,* and *Erwinia chrysanthemi* produce proteases. The pathogens may produce multiple proteases. *Erwinia carotovora* ssp. *carotovora* isolate 177 produces a metalloprotease which degrades lectin (Heilbronn et al., 1995).

TOXINS

Toxins play an important role in the necrotrophic phase of pathogenesis. Toxins are defined as microbial metabolites, which are harmful to plants at physiologically low concentrations. These toxins induce characteristic symptoms of diseases. Several types of toxins have been detected in pathogens. *Host-specific toxin* is a metabolic product of a pathogenic microorganism that is toxic to only the host of that pathogen when used at physiological concentrations. At high concentrations, even nonsusceptible plants are affected. Some authors prefer the word "host-selective toxin" for "host-specific toxin," because this toxin cannot be strictly called specific, as at high concentrations it induces symptom development in resistant plants as well. *Pathotoxin* is the toxin that is the major determinant of disease and pathogenicity. Pathotoxin induces all the typical disease symptoms at physiological concentrations and is correlated with pathogenicity. Host-specific toxins are considered to be pathotoxins. The toxin produced in vivo (in the infected tissues), which functions in disease development but not as a primary agent, is called *vivotoxin. Nonspecific toxin* is the toxin that is toxic to both hosts and nonhosts. Toxin produced by pathogens, which is toxic to plants but not considered to be of primary importance during pathogenesis, is called *phytotoxin.* Nonspecific toxins are considered to be phytotoxins.

The toxin required for pathogenicity is called the *primary determinant.* It is essential for colonization by the producing pathogen and for disease development in the host. The primary determinant is otherwise called the *pathogenicity factor.* The toxin required for virulence is called the *secondary determinant.* The secondary determinant accounts for certain symptoms or contributes to virulence but is not essential for colonization. The secondary determinant is otherwise called the *virulence factor.* Pathogenicity refers to the capacity of an organism to induce disease. Virulence is a quantitative term denoting the relative disease-inducing ability of a microorganism. A change in host range indicates a change in pathogenicity, while a change in severity of disease indicates a change in virulence. Toxin is not required for penetration of fungal pathogens into a host. Both toxin-producer and nonproducer pathogens penetrate host tissues (Vidhyasekaran et al., 1986). Initial colonization of both types of isolates is also the same in many host–pathogen interactions. Only subsequent invasion is different. It is most likely that toxins play an important role only as a virulence factor rather than as a pathogenicity factor. In other words, toxin is important in the necrotrophic phase rather than in the biotrophic phase (Vidhyasekaran, 1993).

Several types of toxins have been detected in fungal and bacterial pathogens. Some pathogens produce host-specific toxins, while many others produce nonspecific toxins. Some important host-specific toxins characterized from pathogens are HV-toxin (or victorin) produced by *Helminthosporium victoriae,* the oat blight pathogen; HM-T toxin produced by *H. maydis* race T, the maize leaf blight pathogen; HS-toxin produced by *H. sacchari,* the sugarcane pathogen; HC-toxin produced by *H. carbonum,* the maize leaf spot pathogen; HO-toxin produced by *H. oryzae,* the rice brown spot pathogen; AAL-toxins (AAL toxins TA and TB; Mesbah et al., 2000) produced by *Alternaria alternata* f. sp. *lycopersici,* the tomato pathogen; AF-toxins I, II, and III produced by *A. alternata* strawberry pathotype; AK-toxins (three related toxins) produced by *A. kikuchiana,* the Japanese pear black spot pathogen; AL-toxins I and II produced by *A. alternata* tomato pathotype; AM-toxins I, II, and III produced by *A. alternata* apple pathotype; ACR-toxin produced by *A. alternata* rough lemon pathotype; ACT-toxin produced by *A. alternata* tangerine pathotype; AT-toxin produced by *A. alternata* tobacco pathotype; PC toxin produced by *Periconia circinata,* the sorghum pathogen; and PM-toxin produced by *Phyllosticta maydis,* the maize pathogen. *Stemphylium vesicarium,* the cause of brown spot of European pear plants, produces two host-specific SV-toxins (Singh et al., 2000). *Rhizoctonia solani,* the rice sheath blight pathogen produces a host-specific toxin, RS-toxin (Vidhyasekaran et al., 1997).

Several other pathogens produce toxins that are involved in symptom development, and most pathogens produce several toxins. *Pyricularia oryzae,* the rice blast pathogen, produces piricularin, picolinic acid, pyriculol, and tenuazonic acid. *Helminthosporium oryzae,* the rice pathogen, produces ophiobolin A, ophiobolin B, 6-epiophiobolin A, anhydroophiobolin A, 6-epianhydroophiobolin A, and ophiobolin 1. *Alternaria helianthi,* the sunflower pathogen, produces radicinin, radianthin, deoxyradicinol, and 3-epideoxyradicinol. *Alternaria carthami* produces brefeldin A, zinniol, and dehydrobrefeldin A. As a single pathogen produces several toxins, a single toxin is produced by several pathogens. Ophiobolin A is produced by the rice pathogen *H. oryzae,* the maize pathogen *H. maydis,* and the foxtail millet pathogen *H. setariae.* Tenuazonic acid is produced by several *Alternaria* spp., *Pyricularia oryzae, Phoma sorghina,* and *Aspergillus* spp. Cercosporin is produced by *Cercospora nicotianae, C. canescens, C. kikuchii, C. zinniae, C. citrullina, C. ricinella,* and *C. apii.* Polanrazines C and E, the toxins produced by *Phoma lingam (Leptosphaeria maculans),* produce necrotic and chlorotic lesions on brown mustard leaves (Pedras and Biesenthal, 2001). Colletotrichin, the toxin produced by a mango leaf isolate of

Colletotrichum gloeosporioides, caused anthracnose-like symptoms on young mango leaves (Jayasankar et al., 1999). *Colletotrichum dematium,* the causal agent of mulberry anthracnose, produces phytotoxins in vitro and in planta (Yoshida et al., 2000). The potato common scab pathogen *Streptomyces scabies* produces two phytotoxins, thaxtomin A and thaxtomin B (King et al., 2000).

Toxins produced by bacterial pathogens cause watersoaking, chlorosis, and necrosis (Bender et al., 1999). All bacterial toxins studied so far are nonspecific toxins. All bacterial toxins increase disease severity; in other words, they act as virulence factors—a quantitative trait. As an exception, for *Pseudomonas syringae* pv. *tabaci* isolate BR2 on bean, tabtoxinine-β-lactam production is required for disease production (Durbin, 1991). Coronatine is produced by *P. savastanoi* pv. *glycinea, P. syringae* pv. *maculicola, P. syringae* pv. *tomato, P. syringae* pv. *atropurpurea,* and *P. syringae* pv. *morsprunorum.* Coronatine induces chlorotic symptoms, and at higher concentrations, coronatine can cause a necrotic reaction in the central portion of the spot.

Phaseolotoxin is produced by the bean halo blight pathogen *P. syringae* pv. *phaseolicola* and induces chlorosis. The toxin inhibits ornithine carbamoyltransferase (OCTase), resulting in deficiency of arginine, which leads to reduction in chlorophyll biosynthesis (Durbin, 1991). Tabtoxinine-β-lactam is produced by the tobacco wild fire pathogen *P. syringae* pv. *tabaci,* the oat bacterial blight pathogen *P. coronafaciens,* and some other *P. syringae* pathovars. *Pseudomonas syringae* pv. *tabaci* produces only the dipeptide, protoxin form, trivially named tabtoxin, which then is enzymatically hydrolyzed to tabtoxinine-β-lactam when it enters the host cell (Uchytil and Durbin, 1980). Tabtoxin has no biological activity. Tabtoxinine-β-lactam is the biologically active form of the toxin. Tabtoxinine-β-lactam inhibits glutamine synthetase, and this inhibition leads to a rapid buildup of ammonia. Accumulation of ammonia may cause the initial chlorosis by rapidly disrupting the chloroplast's internal membrane system (Durbin, 1991). Syringomycin is produced by many strains of *P. syringae* pv. *syringae,* which cause holcus spot of maize, bacterial canker of stone fruits, and brown spot of bean. The toxin affects ion transport in the plasma membrane. *Pseudomonas syringae* pv. *syringae* produces another toxin, syringopeptin (Scholz-Schroeder et al., 2001). Tagetitoxin is produced by *P. syringae* pv. *tagetis* and produces chlorotic and necrotic spots. Extracellular polysaccharides (EPS) are produced by several bacterial pathogens (Vidhyasekaran et al., 1989; Sridhar et al., 2001). Da Silva and colleagues (2001) showed the involvement of an EPS in pathogenicity of *Xylella fastidiosa.* This EPS consists of polymerized tetrasaccharide repeating units assembled by the sequential addition of glucose-1-phosphate, glucose, mannose, and glucuronic acid on a polyprenol phosphate carrier. Wilting may be induced by the EPS in several ways. Polysaccharides may block the vascular elements at the basal end of the stem or may pass through large vessels of the stem and petiole and accumulate in tiny veinlets and ultrafilters of the pit membrane. EPS transported to leaves may disrupt water transfer from veins and accumulate in membranes of parenchymatous cells and block movement of solutions through cell membranes. Polysaccharides may also accumulate on primary cell walls in the leaf mesophyll, interfering with water movement from veins to the evaporative surfaces. Increase in cell membrane permeability may be induced by EPS, which would result in water imbalance and wilting (Vidhyasekaran et al., 1989).

The importance of toxins in pathogenesis has been demonstrated by several methods. When the toxin-producing ability of a pathogen is suppressed, the pathogen loses its virulence. *CFP* (cercosporin facilitator protein), a gene from the soybean fungal pathogen *Cercospora kikuchii,* encodes the putative major facilitator transporter of the fungal toxin cercosporin. Gene disruption of *CFP* in *C. kikuchii* resulted in dramatically reduced cercosporin production, and this led to reduced virulence of the pathogen (Upchurch et al., 2001). Disruption of the AM-toxin synthetase gene *(AMT)* by transformation of wild-type *Alternaria alternata* apple pathotype with disruption vectors resulted in toxin-minus mutants, and these mutants were unable to cause disease symptoms on susceptible apple cultivars (Johnson et al., 2000). Disruption of the *sypA* gene (encoding a peptide synthetase) in *P. syringae* pv. *syringae* strain B301D resulted in the loss of syringopeptin production (Scholz-Schroeder et al., 2001). Another gene, *sypB1,* is involved in syringomycin toxin production by *P. syringae* pv. *syringae.* Syringopeptin *(sypA)* and syringomycin *(syrB1)* mutants were reduced in virulence by 59 and 26 percent, respectively, compared to the parental strain in cherry, whereas the syringopeptin-syringomycin double mutant was reduced by 76 percent in virulence. This suggests that syringopeptin and syringomycin are major virulence determinants of *P. syringae* pv. *syringae* (Scholz-Schroeder et al., 2001). Three virulence-deficient mutants of *Xanthomonas oryzae* pv. *oryzae,* the causal agent of bacterial blight of rice, were isolated by Tn5 mutagenesis (Sridhar et al., 2001). These mutants also became deficient in EPS production. A 40 kb cosmid clone that restored EPS production restored virulence to all three mutants (Sridhar et al., 2001).

ROLE OF TOXINS IN PATHOGENESIS

The major role of toxins appears to be suppression of the defense mechanisms of the host. When a highly virulent isolate of *Helminthosporium oryzae (Cochliobolus miyabeanus)*, the rice brown spot pathogen, was inoculated on rice leaves, the spores germinated, formed appressoria, and penetrated the epidermal layers at 12 hours after inoculation (Vidhyasekaran et al., 1986). Extensive inter- and intracellular mycelial growth was observed in mesophyll tissue of leaves at 48 hours after inoculation. When a nonpathogenic isolate of *H. oryzae* was inoculated on rice leaves, the isolate did not grow beyond the infection peg. The pathogenic isolate induced typical brown spot symptoms with yellow halo at 48 hours, while the nonpathogenic isolate did not produce any visible symptoms. When spore suspensions of the nonpathogenic isolate were prepared in *H. oryzae* toxin and inoculated on rice leaves, extensive inter- and intracellular mycelial growth was observed at 36 hours. The growth rate of the mycelium of the nonpathogenic isolate was similar to that of the virulent isolate. Typical brown spot symptoms appeared on leaves exposed to the toxin (Vidhyasekaran et al., 1986).

Alternaria alternata is a nonpathogen of rice. *Helminthosporium oryzae,* the brown spot pathogen of rice, produces ophiobolin A. When spores of *A. alternata* were suspended in ophiobolin A solution, the spores germinated well and formed appressoria and infection pegs on the rice leaf sheath surface. Extensive inter- and intracellular hyphal growth was observed only in the sheath tissues that had been inoculated in the presence of ophiobolin A. The spores without ophiobolin A could not grow beyond the infection peg and did not grow inside the cell (Xiao et al., 1991). Ophiobolin A at 3 µg/mL induced susceptibility of rice leaf tissues to the nonpathogen *A. alternata* (Xiao et al., 1991). The toxin inactivated calmodulin (Au et al., 2000). The role of calmodulin in host defense mechanism is well known (Vidhyasekaran, 1997, 2002). The toxin produced by *Pyricularia oryzae* induced susceptibility of rice leaves to the nonpathogen *A. alternata* (Fujita et al., 1994).

Helminthosporium oryzae produces a host-specific toxin. When an avirulent isolate of *H. oryzae* was inoculated along with the toxin, the avirulent isolate could penetrate the rice leaves and grow profusely. Typical brown spot lesions also developed (Vidhyasekaran et al., 1986). HV-toxin, produced by *Helminthosporium victoriae* reduces the accumulation of avenalumins, the phytoalexins, in oats (Mayama et al., 1986). Papilla formation at the penetration sites of *A. kikuchiana* functions as a resistance mechanism in Japanese pear. When pear leaves were inoculated with avirulent spores plus AK-toxin, the rate of papilla formation reduced and the avirulent spores could invade susceptible leaves (Otani et al., 1995). Thus, the pathogens appear to suppress the defense mechanisms of the host by producing toxins. It appears that detoxification alone may allow the resistance genes to function in plants. The resistance gene cloned from maize has been shown to be involved in detoxification of the toxin produced by *H. maydis* (Meeley and Walton, 1991; Johal and Briggs, 1992).

Toxins may induce leakage of nutrients from the plant cell and make them available to pathogens in the infection court. The plasma membrane in the plant cells governs all permeability processes from simple diffusion to active transport. Modification of the plasma membrane in infected tissues may result in an efflux of nutrients to the pathogen. Changes in the membrane are mostly induced by toxins produced by pathogens. The toxins induce electrolyte leakage from cells (Vidhyasekaran et al., 1986). Cercosporin, the toxin provided by *Cercospora* species, generates active oxygen species and destroys the membranes of host plants, providing nutrients to support the growth of pathogens (Daub and Ehrenshaft, 2000).

REFERENCES

Au, T. K., Chick, W. S. H., and Leung, P. C. (2000). Initial kinetics of the inactivation of calmodulin by the fungal toxin ophiobolin A. *Int J Biochem Cell Biol,* 32:1173-1182.

Baer, D. and Gudmestad, N. C. (1995). *In vitro* cellulolytic activity of the plant pathogen *Clavibacter michiganensis* subsp. *sepedonicus. Can J Microbiol,* 41: 877-888.

Beer, S. V., Bauer, D. W., Jiang, X. H., Laby, R. J., Sneath, B. J., Wei, Z.-M., Wilcox, D. A., and Zumoff, C. H. (1991). The *hrp* gene cluster of *Erwinia amylovora.* In H. Hennecke and D. P. S. Verma (Eds.), *Advances in Molecular Genetics of Plant-Microbe Interactions, Volume 1.* Kluwer Academic Publishers, Dordrecht, The Netherlands, pp. 53-60.

Beimen, A., Bermpohl, A., Meletzus, D., Eichenlaub, R., and Barz, W. (1992). Accumulation of phenolic compounds in leaves of tomato plants after infection with *Clavibacter michiganense* strains differing in virulence. *Z Naturforsch,* 47C:898-909.

Bender, C. L., Alarcon-Chaidez, F., and Gross, D. C. (1999). *Pseudomonas syringae* phytotoxins: Mode of action, regulation, and biosynthesis of peptide and polyketide synthetases. *Microbiol Mol Biol Rev,* 63:266-292.

Benhamou, N. (1991). Cell surface interactions between tomato and *Clavibacter michiganense* subsp. *michiganense:* Localization of some polysaccharides and hydroxyproline-rich glycoproteins in infected host leaf tissues. *Physiol Mol Plant Pathol,* 38:15-38.

Benhamou, N. and Cote, F. (1992). Ultrastructure and cytochemistry of pectin and cellulose degradation in tobacco roots infected by *Phytophthora parasitica* var. *nicotianae. Phytopathology,* 82:468-478.

Boucher, C. A., Barberis, P. A., and Arlat, M. (1988). Acridine orange selects for deletion of *hrp* genes in all races of *Pseudomonas solanacearum. Mol Plant-Microbe Interact,* 1:282-288.

Brisset, M. N. and Paulin, J. P. (1991). Relationships between electrolyte leakage from *Pyrus communis* and virulence of *Erwinia amylovora*. *Physiol Mol Plant Pathol*, 39:443-453.

Cherif, M., Benhamou, N., and Belanger, R. R. (1991). Ultrastructural and cyto-chemical studies of fungal development and host reactions in cucumber plants infected by *Pythium ultimum*. *Physiol Mol Plant Pathol*, 39:353-375.

Crawford, M. S. and Kolattukudy, P. E. (1987). Pectate lyase from *Fusarium solani* f. sp. *pisi:* Purification, characterization, in vitro translocation of the mRNA, and involvement in pathogenicity. *Arch Biochem Biophys*, 258:196-205.

Da Silva, F. R., Vettore, A. L., Kemper, E. L., Leite, A., and Arruda, P. (2001). Fastidian gum: The *Xylella fastidiosa* exopolysaccharide possibly involved in bacterial pathogenicity. *FEMS Microbiology Lett*, 203:165-171.

Daub, M. E. and Ehrenshaft, M. (2000). The photoactivated *Cercospora* toxin Cercosporin: Contributions to plant disease and fundamental biology. *Annu Rev Phytopathol*, 38:461-490.

Dori, S., Solel, Z., and Barash, I. (1995). Cell wall-degrading enzymes produced by *Gaeumannomyces graminis* var. *tritici in vitro* and *in vivo*. *Physiol Mol Plant Pathol*, 46:189-198.

Durbin, R. D. (1991). Bacterial phytotoxins: Mechanisms of action. *Experientia*, 47:776-783.

Fenselau, S., Balbo, I., and Bonas, U. (1992). Determinants of pathogenicity in *Xanthomonas campestris* pv. *vesicatoria* are related to proteins involved in se-cretion in bacterial pathogens of animals. *Mol Plant-Microbe Interact*, 5:390-396.

Fujita, K., Arase, S., Hiratsuka, H., Honda, Y., and Nozu, M. (1994). The role of toxin(s) produced by germinating spores of *Pyricularia oryzae* in pathogenesis. *J Phytopathol*, 142:245-252.

Gopalan, S., Bauer, D. W., Alfano, J. R., Loniello, A. O., He, S. Y., and Collmer, A. (1996). Expression of the *Pseudomonas syringae* avirulence protein AvrB in plant cells alleviates its dependence on the hypersensitive response and pathoge-nicity (Hrp) secretion system in eliciting genotype-specific hypersensitive cell death. *Plant Cell*, 8:1095-1105.

Gough, C. L., Genin, S., Zischek, C., and Boucher, C. A. (1992). *hrp* genes of *Pseu-domonas solanacearum* are homologous to pathogenicity determinants of ani-mal pathogenic bacteria and are conserved among plant pathogenic bacteria. *Mol Plant-Microbe Interact*, 5:384-389.

Heilbronn, J., Johnston, D. J., Dunbar, B., and Lyon, G. D. (1995). Purification of a metalloprotease produced by *Erwinia carotovora* ssp. *carotovora* and the degra-dation of potato lectin *in vitro*. *Physiol Mol Plant Pathol*, 47:285-292.

Huang, H.-C., Lin, R. H., Chang, C. J., Collmer, A., and Deng, W. L. (1995). The complete *hrp* gene cluster of *Pseudomonas syringae* 61 includes two blocks of genes required for harpin$_{Pss}$ secretion that are arranged colinearly with *Yersinia ysc* homologs. *Mol Plant-Microbe Interact*, 5:733-746.

Jarvis, M. C., Threlfall, D. R., and Friend, J. (1981). Potato cell wall polysaccharides: Degradation with enzymes from *Phytophthora infestans. J Expt Bot*, 32: 1309-1319.

Jayasankar, S., Litz, R. E., Gray, D. J., and Moon, P. A. (1999). Responses of embryogenic mango cultures and seedling bioassays to a partially purified phytotoxin produced by a mango leaf isolate of *Colletotrichum gloeosporioides* Penz. *In vitro Cellular & Developmental Biology—Plant*, 35:475-479.

Johal, G. S. and Briggs, S. P. (1992). Reductase activity encoded by the *HM1* disease resistance gene in maize. *Science*, 258:985-987.

Johnson, R. D., Johnson, L., Itoh, Y., Kodama, M., Otani, H., and Kohmoto, K. (2000). Cloning and characterization of a cyclic peptide synthetase gene from *Alternaria alternata* apple pathotype whose product is involved in AM-toxin synthesis and pathogenicity. *Mol Plant-Microbe Interact*, 13:742-753.

Jones, M. J. and Epstein, L. (1990). Adhesion of macroconidia to the plant surface and virulence of *Nectria haematococca. Appl Environ Microbiol*, 56:3772-3778.

King, R. R., Lawrence, C. H., and Calhoun, L. A. (2000). Microbial glucosylation of thaxtomin A, a partial detoxification. *J Agric Food Chem*, 48:512-514.

Kolattukudy, P. E., Podilla, G. K., and Mohan, R. (1989). Molecular basis of the early events in plant-fungus interaction. *Genome*, 31:342-349.

Kollar, A. (1994). Characterization of specific induction, activity, and isozyme polymorphism of extracellular cellulases from *Venturia inaequalis* detected *in vitro* and on the host plant. *Mol Plant-Microbe Interact*, 7:603-611.

Koller, W., Allan, C. R., and Kolattukudy, P. E. (1982). Role of cutinase and cell wall degrading enzymes in infection of *Pisum sativum* by *Fusarium solani* f. sp. *pisi. Physiol Plant Pathol*, 20:47-60.

Koller, W., Parker, D. M., and Becker, C. M. (1991). Role of cutinase in the penetration of apple leaves by *Venturia inaequalis. Phytopathology*, 81:1375-1379.

Matthysse, A. G., Yarnall, H. A., and Young, N. (1996). Requirement for genes with homology to ABC transport systems for attachment and virulence of *Agrobacterium tumefaciens. J Bacteriol*, 178:5302-5308.

Mayama, S., Tani, T., Ueno, T., Midland, S. L., Sims, J. J., and Keen, N. T. (1986). The purification of victorin and its phytoalexin elicitor activity in oat leaves. *Physiol Mol Plant Pathol*, 29:1-18.

Meeley, R. B. and Walton, J. D. (1991). Enzymatic detoxification of HC-toxin, the host selective cyclic peptide from *Cochliobolus carbonum. Plant Physiol*, 97: 1080-1086.

Meletzus, D., Bermpohl, A., Dreier, J., and Eichenlaub, R. (1993). Evidence for plasmid-encoded virulence factors in the phytopathogenic bacterium *Clavibacter michiganense* subsp. *michiganensis* NCPPB382. *J Bacteriol*, 175:2131-2136.

Mercure, E. W., Kunoh, H., and Nicholson, R. L. (1994). Adhesion of *Colletotrichum graminicola* to corn leaves: A requirement for disease development. *Physiol Mol Plant Pathol*, 45:407-420.

Mesbah, L. A., van der Weerden, G. M., Nukamp, H. J. J., and Hille, J. (2000). Sensitivity among species of Solanaceae to AAL toxins produced by *Alternaria alternata* f. sp. *lycopersici. Plant Pathol*, 49:734-741.

Movahedi, S. and Heale, J. B. (1990). The roles of aspartic proteinase and endo-pectin lyase enzymes in the primary stages of infection and pathogenesis of vari-ous host tissues by different isolates of *Botrytis cinerea* Pers ex.Pers. *Physiol Mol Plant Pathol,* 36:303-324.

Mulya, K., Takikawa, Y., and Tsuyumu, S. (1996). The presence of regions homol-ogous to *hrp* cluster in *Pseudomonas fluorescens* PBG32R. *Ann Phytopath Soc Japan,* 62:355-359.

Otani, H., Kohmoto, K., and Kodama, M. (1995). *Alternaria* toxins and their effects on host plants. *Can J Bot,* 73(Suppl. 1):S453-S458.

Pascholati, S. F., Yoshioka, H., Kunoh, H., and Nicholson, R. L. (1992). Prepara-tion of the infection court by *Erysiphe graminis* f. sp. *hordei* cutinase is a compo-nent of the conidial exudates. *Physiol Mol Plant Pathol,* 41:53-59.

Pedras, M. S. C. and Biesenthal, C. J. (2001). Isolation, structure determination, and phytotoxicity of unusual dioxopiperazines from the phytopathogenic fungus *Phoma lingam. Phytochemistry,* 58:905-909.

Rehnstrom, A. L., Free, S. J., and Pratt, R. G. (1994). Isolation, characterization and pathogenicity of *Sclerotinia trifoliorum* arabinofuranosidase-deficient mutants. *Physiol Mol Plant Pathol,* 44:199-206.

Romantschuk, M., Nurmiaho-Lassila, E. L., and Rantala, E. (1991). Pilus-mediated adsorption of *Pseudomonas syringae* to bean leaves. *Phytopathology,* 81:1245.

Scholz-Schroeder, B. K., Hutchison, M. L., Grgurina, L., and Gross, D. C. (2001). The contribution of syringopeptin and syringomycin to virulence of *Pseudomo-nas syringae* pv. *syringae* strain B301D on the basis of *sypA* and *syrB1* bio-synthesis mutant analysis. *Mol Plant-Microbe Interact,* 14:336-348.

Singh, P., Park, P., Bugiani, R., Cavanni, P., Nakajima, H., Kodama, M., Otani, H., and Kohmoto, K. (2000). Effects of host-selective SV-toxin from *Stemphylium vesicarium,* the cause of brown spot of European pear plants, on ultrastructure of leaf cells. *J Phytopathol,* 148:87-93.

Sridhar, D., Yashitola, J., Vishnupriya, M. R., and Sonti, R. V. (2001). Novel genomic locus with atypical G+C content that is required for extracellular polysaccharide production and virulence in *Xanthomonas oryzae* pv. *oryzae. Mol Plant-Microbe Interact,* 14:1335-1339.

Thomashow, M. F., Karlinsey, J. E., Marks, J. R., and Hurlbert, R. E. (1987). Identi-fication of a new virulence locus in *Agrobacterium tumefaciens* that affects polysaccharide composition and cell attachment. *J Bacteriol,* 169:3209-3216.

Uchytil, T. F. and Durbin, R. D. (1980). Hydrolysis of tabtoxins by plant and bacte-rial enzymes. *Experientia,* 36:301-302.

Upchurch, R. G., Rose, M. S., and Eweida, M. (2001). Overexpression of the cercosporin facilitator protein, CFP, in *Cercospora kikuchii* up-regulates pro-duction and secretion of cercosporin. *FEMS Microbiol Lett,* 204:89-93.

Vidhyasekaran, P. (1993). *Principles of Plant Pathology.* CBS Publishers and Dis-tributors, Delhi.

Vidhyasekaran, P. (1997). *Fungal Pathogenesis in Plants and Crops.* Marcel Dekker, New York.

Vidhyasekaran, P. (2002). *Bacterial Disease Resistance in Plants: Molecular Biol-ogy and Biotechnological Applications.* The Haworth Press, Binghamton, NY.

Vidhyasekaran, P., Alvenda, M. E., and Mew, T. W. (1989). Physiological changes in rice seedlings induced by *Xanthomonas campestris* pv. *oryzae. Physiol Mol Plant Pathol,* 35:391-402.

Vidhyasekaran, P., Borromeo, E. S., and Mew, T. W. (1986). Host-specific toxin production by *Helminthosporium oryzae. Phytopathology,* 76:261-266.

Vidhyasekaran, P., Ruby Ponmalar, T., Samiyappan, R., Velazhan, R., Vimala, R., Ramanathan, A., Paranidharan, V., and Muthukrishnan, S. (1997). Host-specific toxin production by *Rhizoctonia solani,* the rice sheath blight pathogen. *Phytopathology,* 87:1258-1263.

Wagner, V. I. and Matthysse, A. G. (1992). Involvement of a vitronectin-like protein in attachment of *Agrobacterium tumefaciens* to carrot suspension culture cells. *J Bacteriol,* 174:5999-6003.

Wanner, L., Mittal, S., and Davis, K. R. (1993). Recognition of the avirulence gene *avrB* from *Pseudomonas syringae* pv. *glycinea* by *Arabidopsis thaliana. Mol Plant-Microbe Interact,* 6:582-591.

Wosten, H. A. B. (2001). Hydrophobins: Multipurpose proteins. *Annu Rev Microbiol,* 55:625-646.

Xiao, J. Z., Tsuda, M., Doke, N., and Nishimura, S. (1991). Phytotoxins produced by germinating spores of *Bipolaris oryzae. Phytopathology,* 81:58-64.

Yoshida, S., Hiradate, S., Fujii, Y., and Shirata, A. (2000). *Colletotrichum dematium* produces phytotoxins in anthracnose lesions of mulberry leaves. *Phytopathology,* 90:285-291.

Glossary

This glossary was prepared after consulting several papers and the following books: The Commonwealth Mycological Institute (1974), *Plant Pathologist's Pocketbook,* Kew, Surrey, Great Britain; Parry, D. W. (1990), *Plant Pathology in Agriculture,* Cambridge University Press, Cambridge, Great Britain; Vidhyasekaran, P. (1993), *Principles of Plant Pathology,* CBS Publishers, Delhi, India; Hawksworth, D. L., Kirk, P. M., Sutton, B. C., and Pegler, D. N. (1995), *Ainsworth and Bisby's Dictionary of the Fungi,* CAB International, Dew, Surrey, Great Britain; Nicholl, D. S. T. (1994), *An Introduction to Genetic Engineering,* Cambridge University Press, Cambridge, Great Britain.

acervulus: Typically a flat, open bed of stromatic mass of hyphae giving rise to generally short conidiophores. Conidia are borne at the tips of the conidiophores.

adjuvants: Materials added to improve some chemical or physical property of a plant protectant.

agglutinin: An antibody that causes a particulate antigen to clump and settle out of suspension.

allele: Chromosomes are usually arranged as homologous pairs, and different forms of the same genes are called alleles.

alternate host: One of two hosts required by a pathogen to complete its life cycle.

alternative host: One of several plant species hosts of a given pathogen.

amphitrichous: Having one flagellum at each pole.

analytic models: Used to analyze epidemics on a theoretical basis without taking into consideration the effects of external variables.

anamorph: Some fungi have two valid names: The perfect state (*see* PERFECT STATE), i.e., the teleomorph, and an imperfect state (*see* IMPERFECT

STATE), i.e., the anamorph. The former takes precedence for the name of a whole fungus or holomorph.

antheridium: The male gametangium of fungi.

antibiotic: A chemical substance produced by a microorganism that is able to inhibit growth of other microorganisms.

antibody: An immunoglobulin that specifically recognizes and binds to an antigenic determinant of an antigen. Antigens (*see* ANTIGEN) stimulate the immune system of the animal and this leads to the production of specific antibodies, each of which recognizes and binds to its complementary antigen. Antibodies can recognize the plant pathogen by recognizing the antigen specific to the pathogen.

anticodon: The three bases on a tRNA molecule that are complementary to the codon on the mRNA.

antigen: A molecule that is bound by an antibody and can induce an immune response. Cells of living animals, particularly mammals, have the ability to recognize binding sites on proteins, glycoproteins, lipopolysaccharides, and carbohydrate molecules that are not present in their bodies (foreign to that animal). Such molecules are known as antigens.

antiserum: Serum that contains antibodies.

apoplast: The total nonliving cell wall continuum that surrounds the symplast (*see* SYMPLAST). Xylem system constitutes a specialized phase of the apoplast. It also constitutes a continuous permeable system through which water and solutes may move freely.

apothecium: An open ascocarp in which asci are produced (plural: apothecia).

apparent infection rate: The rate of infection, which is calculated by taking into account the amount of plant tissue left to be colonized.

appressorium: A swollen fungal hyphal tip usually associated with the mechanism of adherence to the plant surface prior to penetration.

area under disease progress curve (AUDPC): The amount of disease integrated between two times of interest. Calculated without regard to curve shape, AUDPC provides a valid statistical description of disease progress data.

ascocarp: A fruiting body in which asci are produced.

ascospore: A sexually produced spore borne in an ascus.

ascostroma: A stromatic ascocarp bearing asci directly in locules within the stroma.

ascus: A spore sac containing ascospores (plural: asci).

autoecious rust: A rust pathogen that completes its life cycle on one host.

avirulence genes: Genes that are critical in determining whether a pathogenic strain will be virulent or avirulent on a specific host. Avirulence genes restrict a pathogen's host range by specifying the resistance response on host plants carrying complementary disease resistance genes.

avirulent: Lacking virulence.

bacteria: Primitive organisms classified as prokaryotes, with a primitive type of nucleus lacking a clearly defined membrane. Most of the genetic information in a bacterial cell is carried on a single chromosome with double-stranded deoxyribonucleic acid (DNA) in a closed circular form.

bactericide: A substance causing death of bacteria.

bacteriophage: A virus infecting bacteria.

basidiocarp: A fruiting body that bears basidia.

basidiospore: A spore borne on the outside of a basidium.

basidium: A specialized spore-producing body bearing a definite number of basidiospores (plural: basidia).

biotroph: A pathogen that obtains nutrients from the living cells of the host.

cDNA: DNA which is made by copying mRNA using the enzyme reverse transcriptase.

chlamydospores: The hyphal cells of fungi that become enveloped in a thick wall before they separate from one another or from other hyphal cells adjoining them.

chromista: A kingdom which includes organisms that bear flagella with mastigonemes. Cell walls of these organisms do not contain chitin, but contain cellulose and glucan.

chromosome: A DNA molecule carrying a set of genes.

***cis*-acting element:** A DNA sequence that exerts its effect only when on the same DNA molecule as the sequence it acts on.

cleistothecium: A completely closed ascocarp is called a cleistothecium (plural: cleistothecia).

codon: The three bases in mRNA that specify a particular amino acid during translation.

coenocytic: A continuous mass of cytoplasm and nuclei without any septa.

complementary genes: Nonallelic genes that complement one another.

complementation: Appearance of a wild phenotype in an organism or cell containing two different mutations combined in a hybrid diploid or a heterokaryon.

complementation test: The introduction of two mutant chromosomes into the same cell to determine whether the mutations in question occurred in the same gene. The wild phenotype will be expressed, since each chromosome "makes up for" or "complements" the defect in the other.

complete resistance: Very high resistance that suppresses disease development completely.

conidiophore: A specialized hyphal branch bearing conidia.

conidium: An asexually produced fungal spore (plural: conidia).

copy number: The number of copies of a gene in the genome of an organism.

cosmid: A hybrid plasmid/bacteriophage vector that is made up of plasmid sequences joined to the *cos* sites of phage λ.

cytoplasmic inclusion bodies: Some of the water-soluble reserve food materials manufactured by a cell get dissolved in the cell sap. Insoluble constituents precipitate out as cytoplasmic inclusion bodies. Volutin, glycogen, and fat globules are common cytoplasmic inclusions.

differential variety: A variety that gives reactions which distinguish between race-specific isolates of a pathogen.

disease: A harmful deviation from normal functioning of plant physiological processes. The plant is considered diseased when symptoms, by which the disease may be recognized, are present.

disease progress curve: The progress of a disease may be assessed as disease incidence (proportion of diseased plants in a plant population) or as disease severity (proportion of diseased tissue in a plant). When the results are plotted against time, the curve obtained is commonly S-shaped.

disease-specific *(dsp)* genes: Genes involved only in pathogenicity. This is in contrast to *hrp* genes that are involved both in pathogenicity and hypersensitive response (see *hrp* GENES).

dominant gene: The gene that is phenotypically manifested, for example, it may suppress the development of disease symptoms. It produces a functional product such as disease resistance. Its recessive allele cannot phenotypically manifest and its action is marked in the presence of a dominant gene (see RECESSIVE GENE).

dot-blot: A technique in which small spots or dots of nucleic acid are immobilized on a nitrocellulose membrane for hybridization.

dot-blotting: Similar to southern and northern blotting in analysis of nucleic acids, but the samples are not subjected to electrophoresis. The nucleic acid samples are spotted onto the nitrocellulose membranes or nylon filters and hybridized with a radioactive probe.

durable resistance: Resistance that remains effective in a cultivar that is widely grown for a long period of time in an environment favorable to the disease.

elicitors: Signal molecules of pathogens and plants that trigger defense mechanisms of the host.

empirical models: Epidemiological models that are based on actual experiments and observations rather than theory or simulation. They are complex models that attempt to mimic many environmental factors and other influences.

emulsifiers: Chemicals that assist the formation of a suspension of small droplets of a liquid in another liquid in which the first liquid is insoluble. Emulsifiers help in the formation of an emulsion comprising small spheres of organic solvent/fungicide in the sprayer.

endemic: A disease established in moderate or severe form in a defined area. The disease level in an endemic disease remains almost constant.

endopolygalacturonase (endo PG): An enzyme that cleaves polygalacturonic acids in a random manner and releases oligogalacturonic acid.

enhancer: A sequence that enhances transcription from the promoter of a gene; it may be several thousand base pairs away from the promoter.

epidemic: A widespread temporary increase in the incidence of an infectious disease. A plant disease is described as epidemic when the amount of disease present increases rapidly from a low level to a high one.

epidemiology: The science that describes the progress of a disease as it becomes epidemic (*see* EPIDEMIC).

episomes: Autonomous and dispensable genetic elements similar to plasmids (*see* PLASMID). Unlike plasmids, however, episomes can exist even when integrated with the chromosome. Some bacteria contain episomes.

epistasis: Phenomenon in which one gene suppresses the expression of another gene. It is generally observed when two or more resistance genes are present together. Only the gene conditioning the lowest infection type is expressed. For example, expression of recessive genes is suppressed by the presence of dominant genes.

exopolygalacturonase (exo PG or Peh): An enzyme that cleaves polygalacturonic acids in a terminal manner and releases monomeric products (galacturonic acids).

facultative parasite: A parasite that can also live as a saprophyte. Facultative parasites can be cultured on laboratory media.

facultative saprophyte: A saprophyte that can also live as a parasite.

flagellum: A whiplike appendage of a motile cell (plural: flagella).

foam suppressors: Surface-active substances that form a fast draining foam to provide maximum contact of the fungicide spray to the plant surface.

forma specialis (f. sp.): An intraspecific category for taxa characterized from a physiological standpoint (particularly based on pathogenicity to different species or varieties of host plants) and may be indistinguishable morphologically (plural: formae speciales, ff. sp.).

fructifications: Structures containing spores in fungi.

fumigation: Disinfestation by fumes.

fungi: Organisms that are eukaryotic and heterotrophic. Fungi develop branching filaments (these organisms are more rarely single-celled), reproduce by spores, and their cell walls contain chitin and β-glucans. They are mostly nonflagellate. When present, flagella always lack mastigonemes (i.e., the surfaces of flagella are not covered by hairlike processes).

fungicidal: Able to kill fungal spores or mycelium.

fungicide: A chemical substance that kills fungal spores or mycelium.

fungicide resistance: A decrease in sensitivity to a fungicide due to selection or mutation following exposure to the compound.

fungistatic: Able to stop fungal growth without killing the fungus.

gene: The unit of inheritance, located on a chromosome. It is a sequence of DNA that codes for a diffusible protein product, which diffuses away from its site of synthesis to act elsewhere. In molecular terms, the gene means a region of DNA that encodes one function. Broadly, therefore, one gene encodes one protein.

gene cloning: Generation of recombinant DNA molecules in the laboratory.

gene-for-gene hypothesis: An assumption that corresponding genes for resistance and virulence exist in host and pathogen respectively. According to this hypothesis, for each gene that conditions resistance in the host, there is a corresponding gene that conditions pathogenicity (virulence) in the pathogen. No resistance occurs unless a resistance is present in the host along with a corresponding avirulence gene in the pathogen.

general/partial/field resistance: The disease resistance that provides only partial or field resistance to several races of a pathogen. This resistance affects the apparent infection rate and the amount of disease that finally develops on the plants.

genetic engineering: A technology that involves cutting, modifying, and joining DNA molecules using enzymes such as restriction enzymes and DNA ligase. It is also known as gene cloning, recombinant DNA technology, molecular cloning, or gene manipulation.

germ tube: The initial hyphal growth from a germinating fungal spore.

green bridge: Living plant material used by biotrophs to overwinter.

groundkeeper: A self-sown plant.

haustorium: A specially developed fungal hyphal branch within a living cell of the host for absorption of food (plural: haustoria).

helper viruses: Viruses that help multiplication of satellite RNA (*see* SATELLITE VIRUS).

hemibiotroph: A pathogen living at first as a biotroph but later as a necrotroph.

heteroecious: A pathogen requiring two host species to complete its life cycle.

heterothallism: A condition in which sexual reproduction in fungi occurs between oogonium and antheridium from two different sexually compatible mycelia (thalli).

holomorph: *See* ANAMORPH.

homothallism: A condition in which sexual reproduction in some fungi occurs between oogonium and antheridium, even from the same thallus.

horizontal resistance: Resistance against several races of a pathogen.

host-selective toxin: The toxin that is selectively toxic to some hosts. It is similar to host-specific toxin (see HOST-SPECIFIC TOXIN). Some authors prefer the word "host-selective toxin" for "host-specific toxin," because this toxin cannot be strictly called host-specific, as at high concentrations it induces symptom development in resistant plants as well.

host-specific toxin: A metabolic product of a pathogenic microorganism that is toxic only to the host of the pathogen when used at physiological concentrations. At high concentrations, even nonsusceptible plants are affected.

host-specific virulence (*hsv*) genes: Genes that are required for virulence on specific hosts.

***hrp* genes:** Bacterial genes that are required for growth *in planta* and initiation of disease symptoms on host plants and induction of defense mechanisms as expressed by hypersensitive reaction on nonhost plants. *Hy*persensitive *r*esponse and *p*athogenicity *(hrp)* genes are general virulence genes that are required for virulence on all hosts. The *hrp* genes are not host-specific; their inactivation leads to loss of virulence on all hosts and loss of the nonhost hypersensitive response.

hymenium: A fertile layer consisting of asci (whether naked or enclosed in a fruiting body) or basidia. Sterile, elongated hairs, arising between the asci, often form a part of the hymenium (plural: hymeia).

hyperparasitism: Parasitism on another parasite.

hyperplasia: Overdevelopment of some tissues due to abnormal multiplication of cells.

hypersensitivity: A rapid local reaction of plant tissue to an attack by a pathogen, which results in the death of tissue around infection sites and prevents further spread of infection.

hypertrophy: Abnormal enlargement of an organ or part resulting from an increase in the size of the cells.

hypha: A tubular threadlike filament of fungal mycelium (plural: hyphae).

hyphopodium: A short mycelial branch (plural: hyphopodia).

hypoplasia: Subnormal cell multiplication. It often results in the dwarfing of plants.

immune: Exempt from infection.

imperfect state: The asexual period of a fungal life cycle.

incubation period: The period of time between infection and the appearance of symptoms.

infection: The penetration of the host by a pathogen and the earliest stages of development within the host.

infection court: The initial site of contact between infection and the appearance of symptoms.

infection peg: A slender structure formed by the deposition of substances, such as lignin, around a thin hypha penetrating a host cell.

inhibitins: Antimicrobial substances in plants. Their concentration in uninfected (healthy) plants may normally be low, but may increase enormously after infection (probably in order to combat attack by microorganisms).

inoculum: Spores or other pathogen parts that can cause disease.

inoculum potential: Amount of inoculum required to cause disease.

intercellular: Between cells.

intracellular: Within or through cells.

karyogamy: A phase in which the fusion of two nuclei occurs.

Koch's postulates: The conditions stipulated by Koch to describe an organism as the cause of a disease. The postulates include the following: the organism must be consistently associated with the symptoms of the disease; the organism must be isolated and grown in pure culture, free from all other organisms; the organism from the pure culture must be inoculated onto healthy plants of the same species from which they were originally isolated and must reproduce the same disease as was originally observed; the organism must be reisolated and reinoculated and must once again reproduce the original disease. Koch's postulates cannot be fulfilled in the case of an obligate pathogen: this pathogen cannot be isolated and grown in pure culture; instead, the pathogen may be multiplied on a healthy susceptible host.

latent period: The time between infection and sporulation of the pathogen on the host, or the time from the start of a virus vector's feeding period until the vector is able to transmit the virus to healthy plants.

lodging: Breakage of many plant stems, especially cereals, resulting in many tillers falling down.

major gene: A gene with pronounced phenotypic effects in contrast to its modifiers.

major-gene resistance: Resistance is governed by major genes (*see* MAJOR GENE). Major genes are those that have large, distinct phenotypic expressions showing clear Mendelian segregation.

mastigonemes: The hairlike processes that cover the surface of flagella.

meiosis: A reduction division that reduces the number of chromosomes to haploid.

messenger RNA (mRNA): The ribonucleic acid transcribed from DNA that carries the codons specifying the sequence of amino acids in a protein.

microbial pesticides: Introduced microbial agents for control of pests and diseases.

middle lamella: A median substance found in the area that separates two cells.

minor genes: Those genes that have small effects on the expression of the phenotype for resistance and show quantitative segregation.

minor-gene resistance: The resistance is governed by minor genes (*see* MINOR GENES).

monocyclic: Having only one cycle of infection during a growing season.

monogenic resistance: Resistance is governed by single gene.

monotrichous: Having one polar flagellum.

multiline: A combination of almost genetically identical breeding lines (isogenic) that have all agronomic characters in common but differ in a major-gene resistance.

mycelium: A mass of hyphae that form the vegetative body of a fungus.

mycorrhiza: A symbiotic association of a fungus with the roots of a plant.

mycotoxin: Toxins produced by fungi, which may contaminate foodstuffs.

necrotroph: A pathogen that kills the host cells and lives on the dead remains.

nonspecific toxin: A toxin that is toxic to both hosts and nonhosts.

Northern blotting: A method to detect RNA molecules by electrophoresis.

notifiable disease: A disease that must be reported to the appropriate authorities, as per law.

obligate parasite: An organism that is capable of living only as a parasite.

oogonium: The female gametangium of fungi.

operon: A cluster of genes under the control of a single promoter/regulatory region.

parasexual cycle: Through this mechanism, recombination of hereditary properties occurs through mitosis in fungi.

parasite: An organism that lives on another living organism (host), obtaining its nutrient supply from the host.

pathogen: An organism that causes disease.

pathogenesis-related (PR) proteins: Proteins encoded by the host plant but induced in pathological situations. The PR proteins are induced by various types of pathogens as well as stress conditions.

pathotoxin: The toxin that is a major determinant of disease and pathogenicity. A pathotoxin induces all the typical disease symptoms at a physiological concentration and is correlated with pathogenicity. Host-specific toxins are considered to be pathotoxins.

pathotype: A subdivision of species distinguished by common characteristics of pathogenicity, particularly in relation to host range.

pathovars: (*see* FORMA SPECIALIS) Pathovars and formae speciales are synonymous. Formae speciales in bacteria are known as pathovars.

PCR (polymerase chain reaction): A method for the selective amplification of DNA sequences.

pectate lyases (PL or Pel) and pectin lyases (PTE): They induce lytic degradation of the glycosidic linkage, resulting in an unsaturated bond between carbons 4 and 5 of a uronide moiety in the reaction product. Pectate lyases act on polygalacturonic acids, while pectin lyases act on pectins.

pectic acid (or polygalacturonic acid): Refers to the nonesterified forms, while pectinic acid (or pectin) refers to partially esterified forms of pectic substances (*see* PECTINS).

pectin methyl esterase (PME): The enzyme that catalyzes the conversion of pectin by de-esterification of the methyl ester group with production of methanol. PME removes the methoxyl group from rhamnogalacturonan chains. PME is more active on methylated oligogalacturonates than on pectin.

pectinic acids: *See* PECTINS.

pectins: α-1,4-Galacturonans with various degrees of methyl esterification. The unesterified uronic acid molecules containing up to 75 percent of methoxyl groups are known as pectinic acid, and those with more than 75 percent of methyl groups are known as pectins. Generally the pectinic acids are called pectins.

penetrants: Wetting agents, oils, or oil concentrates that enhance the absorption of a systemic fungicide by the plant.

perfect state: Stage in the life cycle of a fungus characterized by sexual spores.

perithecium: The ascocarp is more or less closed, but at maturity it is provided with a pore through which the ascospores escape.

peritrichous: Having flagella distributed over the whole surface.

persistent viruses: The virus-vector relationship varies widely depending on the duration of the virus in the vector (persistence). In case of persistent viruses, the virus may simply circulate through the body of the vector or may propagate as well. Hence, this relationship can be classified as (1) noncirculative nonpersistent, (2) noncirculative semi-persistent, (3) circulative nonpropagative, and (4) circulative propagative transmission.

physiologic race: A taxon of parasites, particularly characterized by specialization to different cultivars of one host species. Physiologic races and formae speciales do not differ morphologically. Within formae speciales races may exist. Bacteriologists use the term pathovar or pathotype, which are both comparable in definition to formae speciales and race in fungi.

phytoalexin: Plant antibiotics that are synthesized de novo after the plant tissue is exposed to microbial infection. Phytoalexins are also defined as low-molecular-weight, antimicrobial compounds that are both synthesized by and accumulated in plants after exposure to microorganisms. Two important criteria have been suggested to label a plant secondary metabolite a phytoalexin: (1) the secondary metabolite should be produced de novo in response to infection, and (2) the compound should accumulate to antimicrobial concentrations in the area of infection.

phytoanticipins: Low-molecular-weight, antimicrobial compounds present in plants before challenge by microorganisms and/or produced after infection solely from preexisting constituents.

phytoncide: A chemical substance produced by plants that can inhibit the growth of microorganisms.

phytosanitary certificate: A certificate of health that accompanies plants or plant products to be exported.

phytotoxic: Toxic to plants.

phytotoxin: Toxin produced by pathogens that is toxic to plants but not considered to be of primary importance during pathogenesis. Nonspecific toxins are considered to be phytotoxins.

plant activators: The chemicals that activate the defense genes by providing signals.

plant pathology: Study of plant disease.

plasmid: A circular extrachromosomal element and a circular DNA molecule. A plasmid is not essential for cell growth. It is used as a vector of the recombinant DNA in genetic engineering studies.

plasmodium: A naked (wall-less) motile mass of protoplasm with many nuclei, bounded by a plasma membrane. It moves and feeds in an amoeboid fashion (it engulfs food and feeds by ingestion).

plasmogamy: In this phase, the union of two protoplasts takes place, bringing nuclei close together within the same cell.

polycyclic: Disease having more than one cycle of infection during a growing season.

polygenic resistance: The resistance is governed by several genes.

polysomes: Ribosomes act in clusters and are called polyribosomes or polysomes. Ribosomes are the sites of protein synthesis.

population: A group of organisms of the same species occupying a particular space at a particular time. A population is a pool of individuals from which the next generation will be drawn. The spatial and temporal limits of a population are defined by uniform allele frequencies.

promoter: DNA sequence lying upstream from a gene to which RNA polymerase binds.

propagule: That part of an organism by which the organism may be dispersed or reproduced.

prophylaxis: Preventative treatment against disease.

prosenchyma: During certain stages of fungal development, the mycelium becomes organized into loosely or compact woven tissues, as against the loose hyphae ordinarily found in the mycelium. The loosely woven tissue in which the component hyphae with elongated cells lie more or less parallel to one another is called prosenchyma.

protectant fungicide: A fungicide that protects against invasion by a pathogen.

protozoa: A kingdom characterized by the presence of organisms that are predominantly unicellular, plasmodial, or colonial. These organisms are phagotrophic, i.e., they feed by ingestion, engulfing food.

pseudoparenchyma: In certain stages of fungal development, fungal tissues are closely packed, in the form of more or less isodiametric or oval cells resembling the parenchyma cells of higher plants, and they are called pseudoparenchyma.

pycnidium: An asexual globose or flask-shaped hollow fruiting body lined with conidiophores (plural: pycnidia).

pyramiding of genes: Combining different genes in a single plant.

qualitative resistance: The disease symptom is almost completely suppressed, mostly by development of hypersensitive pinpoint fleck reaction.

quantitative resistance: The resistance is based on the amount of disease symptom development, quantified by different infection types such as necrotic flecks, necrotic and chlorotic areas with restricted sporulation, sporulation with chlorosis, abundant sporulation without chlorosis, lesion size, lesion area, etc.

quantitative trait loci (QTL): The genetic loci associated with complex traits.

quarantine: Legislative or regulatory control that aims to exclude pathogens from areas where they do not already exist. It includes the holding of imported material in isolation for a period to ensure freedom from diseases and pests.

race: A taxon of pathogens, particularly characterized by specialization to different cultivars of one host species.

race-nonspecific/nonspecific resistance: The resistance is not specific to only a particular race.

race-specific resistance/specific resistance: The resistance is specific to certain races of the pathogen.

recessive gene: A gene that is phenotypically manifest in the homozygous state but is masked in the presence of its dominant allele or dominant gene (*see* DOMINANT GENE).

resistant: Possessing qualities that prevent or retard the development of a given pathogen.

restriction enzyme: An endonuclease that cuts DNA at sites defined by its recognition sequence.

RFLP (restriction fragment-length polymorphism): A variation in the locations of restriction sites in DNA of different individuals; the fragment length defined by the restriction sites may be of different lengths in different individuals.

rhizomorph: The mycelium of some fungi forms thick strands. In such strands, the hyphae lose their individuality and form complex tissues. The strands are called rhizomorph (rootlike structures). The strand has a thick hard cortex and a growing tip. It looks like a root tip.

ribosomes: Globular structures found in the cytoplasm. They are composed of about one-third protein and two-thirds RNA. Ribosomes are designated 30s, 50s, 70s, etc., depending on their size. The size is determined by the rate, measured in svedberg unit(s), at which a particle sediments when it is centrifuged at high speed in an ultracentrifuge.

satellite RNA: Satellite RNA has no coat protein. It completely depends on viruses for multiplication.

satellite virus: A satellite virus has a coat of its own but has only a small genome. It is dependent on a helper virus for its replication. It needs another virus to supply replicase and probably other enzymes. It encapsidates its RNA within its own coat protein. It acts as a parasite of plant-parasitic viruses.

sclerotium: A hard and compact vegetative resting structure of fungi. It is mostly made up of pseudoparenchymatous tissue (*see* PSEUDOPAREN-CHYMA).

seedling resistance: Resistance detectable only at the seedling stage.

sense codon: Any of the 61 triplet codons in mRNA that specify an amino acid.

sense strand: In duplex DNA, the strand that serves as a template for the synthesis of RNA; it is also known as the anticoding strand.

sigma factor: A polypeptide subunit of RNA polymerase. Sigma factors are composed of two functional domains: a core-binding domain and a DNA-binding domain.

simple-interest disease: A disease that goes through only one cycle of infection during a growing season, analogous to a bank account giving simple interest.

simulation models: Simulation is the process of designing a model of a real system and conducting experiments with this model for purposes of either understanding the behavior of the system or evaluating various strategies for the operation of the system. Simulation models are generated by creating conditions for development of epidemics and carrying out experiments on different aspects of disease increase.

somaclonal variation: The stable and heritable variation displayed among somaclones.

somaclones: Plants regenerated from cell, tissue, and organ cultures.

sorus: A mass of spores (plural: sori).

Southern blotting: A method to detect DNA fragments by electrophoresis.

species: A species includes strains of fungi or bacteria, with approximately 70 percent or greater DNA-DNA relatedness or less ΔTm (divergence [unpaired bases] within related nucleotide sequences is 5 percent or less).

sporangiophore: A specialized hyphal branch bearing sporangia.

sporangiospore: A nonmotile asexual spore produced in a sporangium.

sporangium: A sac-like structure in which its entire contents are converted into one or more spores, which are called sporangiospores.

spore: A specialized propagative or reproductive body.

sporodochium: An asexual fruiting body in which the conidiophores are cemented together and the conidiophores arise from the surface of a cushion-shaped stroma.

spreader: A substance added to a spray to assist in its even distribution over the target.

statistical models: Epidemiological models that are mathematical formulas with values of parameters chosen to adequately describe disease progress for specific data sets but lacking a precise biological interpretation.

sticker: A substance added to a spray to assist in its adhesion to the target.

straggling: Breakage of a few plant stems, especially cereals, resulting in a few tillers falling down.

stroma: A fungal structure, usually made up of prosenchyma (*see* PROSENCHYMA). On or in the stroma, fructifications (*see* FRUCTIFICATIONS) are formed (plural: stromata).

subspecies: Based on minor but consistent phenotypic variations within the species or on genetically determined clusters of species. Subspecies designations are used for genetically close organisms that diverge in phenotype.

superrace: A pathogen race that contains virulence factors to match any resistance factors available in the host.

surfactant: A surface-active material, especially a wetter or spreader used with a spray.

susceptible: Subject to infection; nonimmune.

symplast: The sum total of living protoplasm of a plant including the phloem and protoplast. The total mass of living cells of a plant constitute a continuum, the individual protoplasts being ultimately connected throughout the plant by plasmodesmata.

synnema: A group of conidiophores cemented together forming an elongated spore-bearing structure (plural: synnemata).

systemic fungicide: A fungicide that is absorbed and translocated in the plant.

taxa: Taxonomic groups of any rank (singular: taxon).

taxonomy: Systematic classification.

teleomorph: *See* ANAMORPH.

thyriothecium: An inverted ascocarp having the wall more or less radial in structure.

tolerant: Able to endure infection by a pathogen without showing severe symptoms of disease.

trans-**acting element:** A genetic element that can exert its effect without having to be on the same molecule as a target sequence; this element encodes an enzyme or regulatory protein that can diffuse to the site of action.

transposons: Mobile DNA segments that can insert into a few or several sites in a genome. They are transposable genetic elements (the word transpose means "alter the positions of" or "interchange"). They are also called jumping genes.

true resistance: The resistance is complete.

vacuoles: Cavities in the cytoplasm containing a fluid called cell sap.

vector: An organism that transmits a pathogen.

vertical resistance: Resistance that is specific to some races of the pathogen and that will be susceptible to other races.

virion: The complete and infectious nucleoprotein particle of the virus.

viroids: Naked nucleic acids without a coat protein. They consist of only ribonucleic acid (RNA). These "miniviruses" are the smallest known causal organisms of infectious diseases.

viruses: Infectious agents not visible under the microscope (submicroscopic) and small enough to pass through a bacterial filter.

virusoids: Viruses that contain a viroid-like satellite RNA in addition to a linear single-stranded molecule of genomic RNA.

vivotoxin: The toxin produced in vivo (in the infected tissues) that functions in disease development but not as a primary agent.

volunteer plant: A self-sown plant.

volutins: Metachromatic granules found in bacterial cells that localize in the vacuoles of mature forms. They contain inorganic polyphosphate, lipoprotein, RNA, and magnesium and may serve as phosphate storage structures.

Western blotting: A method to detect protein molecules by electrophoresis.

Index

SPECIAL 25%-OFF DISCOUNT!

Order a copy of this book with this form or online at:

http://www.haworthpress.com/store/product.asp?sku=4918

CONCISE ENCYCLOPEDIA OF PLANT PATHOLOGY

_____in hardbound at $97.46 (regularly $129.95) (ISBN:1-56022-942-X)

_____in softbound at $59.96 (regularly $79.95) (ISBN: 1-56022-943-8)

Or order online and use special offer code HEC25 in the shopping cart.

COST OF BOOKS_____

OUTSIDE US/CANADA/
MEXICO: ADD 20%_____

POSTAGE & HANDLING_____
*(US: $5.00 for first book & $2.00
for each additional book)*
*(Outside US: $6.00 for first book
& $2.00 for each additional book)*

SUBTOTAL_____

IN CANADA: ADD 7% GST_____

STATE TAX_____
*(NY, OH, MN, CA, IN, & SD residents,
add appropriate local sales tax)*

FINAL TOTAL_____
*(If paying in Canadian funds,
convert using the current
exchange rate, UNESCO
coupons welcome)*

☐ **BILL ME LATER:** ($5 service charge will be added)
(Bill-me option is good on US/Canada/Mexico orders only;
not good to jobbers, wholesalers, or subscription agencies.)

☐ Check here if billing address is different from
shipping address and attach purchase order and
billing address information.

Signature_____

☐ **PAYMENT ENCLOSED: $**_____

☐ **PLEASE CHARGE TO MY CREDIT CARD.**

☐ Visa ☐ MasterCard ☐ AmEx ☐ Discover
☐ Diner's Club ☐ Eurocard ☐ JCB

Account # _____

Exp. Date_____

Signature_____

Prices in US dollars and subject to change without notice.

NAME_____

INSTITUTION_____

ADDRESS_____

CITY_____

STATE/ZIP_____

COUNTRY_____ COUNTY (NY residents only)_____

TEL_____ FAX_____

E-MAIL_____

May we use your e-mail address for confirmations and other types of information? ☐ Yes ☐ No
We appreciate receiving your e-mail address and fax number. Haworth would like to e-mail or fax special
discount offers to you, as a preferred customer. **We will never share, rent, or exchange your e-mail address
or fax number.** We regard such actions as an invasion of your privacy.

Order From Your Local Bookstore or Directly From

The Haworth Press, Inc.

10 Alice Street, Binghamton, New York 13904-1580 • USA
TELEPHONE: 1-800-HAWORTH (1-800-429-6784) / Outside US/Canada: (607) 722-5857
FAX: 1-800-895-0582 / Outside US/Canada: (607) 771-0012
E-mailto: orders@haworthpress.com

PLEASE PHOTOCOPY THIS FORM FOR YOUR PERSONAL USE.
http://www.HaworthPress.com   BOF03